Alternative Protein Sources in Aquaculture Diets

THE HAWORTH PRESS
Titles of Related Interest

Tilapia: Biology, Culture, and Nutrition by Chhorn Lim and Carl D. Webster

Freshwater Crayfish Aquaculture in North America, Europe, and Australia: Families Astacidae, Cambaridae, and Parastasidae edited by Jay V. Huner

Introduction to the General Principles of Aquaculture by Hans Ackerfors, Jay V. Huner, and Mark Konikoff

Economics of Aquaculture by Curtis Jolly and Howard A. Clonts

Alternative
Protein Sources
in Aquaculture Diets

Chhorn Lim
Carl D. Webster
Cheng-Sheng Lee
Editors

CRC Press
Taylor & Francis Group
Boca Raton London New York

CRC Press is an imprint of the
Taylor & Francis Group, an **informa** business

CRC Press
Taylor & Francis Group
6000 Broken Sound Parkway NW, Suite 300
Boca Raton, FL 33487-2742

First issued in paperback 2019

PUBLISHER'S NOTE
The development, preparation, and publication of this work has been undertaken with great care.
However, the Publisher, employees, editors, and agents of The Haworth Press are not responsible
for any errors contained herein or for consequences that may ensue from use of materials or
information contained in this work. The Haworth Press is committed to the dissemination of ideas
and information according to the highest standards of intellectual freedom and the free exchange of
ideas. Statements made and opinions expressed in this publication do not necessarily reflect the
views of the Publisher, Directors, management, or staff of The Haworth Press, or an endorsement
by them.

Library of Congress Cataloging-in-Publication Data

Alternative protein sources in aquaculture diets / Chhorn Lim, Carl D. Webster, Cheng-Sheng
Lee, editors.
        p. cm.
Includes index.
ISBN: 978-1-56022-148-7 (hard : alk. paper)
    1. Fishes—Feeding and feeds. 2. Proteins in animal nutrition. 3. Fishes—Nutrition—
Requirements. I. Lim, Chhorn. II. Webster, Carl D. III. Lee, Cheng-Sheng.
SH156.A58 2007
639.8—dc22

                                                                                2007015508

Visit the Taylor & Francis Web site at
http://www.taylorandfrancis.com

and the CRC Press Web site at
http://www.crcpress.com

Alternative Protein Sources in Aquaculture Diets

Chhorn Lim
Carl D. Webster
Cheng-Sheng Lee
Editors

CRC Press
Taylor & Francis Group
Boca Raton London New York

CRC Press is an imprint of the
Taylor & Francis Group, an **informa** business

CRC Press
Taylor & Francis Group
6000 Broken Sound Parkway NW, Suite 300
Boca Raton, FL 33487-2742

First issued in paperback 2019

ISBN-13: 978-1-56022-148-7 (hbk)
ISBN-13: 978-0-367-38774-7 (pbk)

PUBLISHER'S NOTE
The development, preparation, and publication of this work has been undertaken with great care.
However, the Publisher, employees, editors, and agents of The Haworth Press are not responsible
for any errors contained herein or for consequences that may ensue from use of materials or
information contained in this work. The Haworth Press is committed to the dissemination of ideas
and information according to the highest standards of intellectual freedom and the free exchange of
ideas. Statements made and opinions expressed in this publication do not necessarily reflect the
views of the Publisher, Directors, management, or staff of The Haworth Press, or an endorsement
by them.

Library of Congress Cataloging-in-Publication Data

Alternative protein sources in aquaculture diets / Chhorn Lim, Carl D. Webster, Cheng-Sheng
Lee, editors.
    p. cm.
Includes index.
ISBN: 978-1-56022-148-7 (hard : alk. paper)
    1. Fishes—Feeding and feeds. 2. Proteins in animal nutrition. 3. Fishes—Nutrition—
Requirements. I. Lim, Chhorn. II. Webster, Carl D. III. Lee, Cheng-Sheng.
SH156.A58 2007
639.8—dc22

          2007015508

Visit the Taylor & Francis Web site at
http://www.taylorandfrancis.com

and the CRC Press Web site at
http://www.crcpress.com

This book is dedicated to my wife, Brenda; our children, Chheang Chhun, Lisa, Chhorn Jr., and Brendan, our grandson, Bryant; our brothers Korn, Daniel, Thong Hor, Muoy Hor, and Trauy; our sisters, Huoy Khim, Chhay Khim, Seang Hay, and Huoy Teang and Seang Hay; and our nephews and nieces, for their patience and unconditional love.

Chhorn Lim

This book is dedicated to my wife, Caroline; my daughters NancyAnn, Catherine, and Emma; my other "children" Lydia, Shyron, KC, Poppins, Michael, Iggy, and Tillie; my parents (for all of their love, support, and encouragement); my brothers Tom and Pete; and to my beloved Samwise, Barley, Darwin, and Misty, who are waiting patiently for me.

Carl D. Webster

CONTENTS

ABOUT THE EDITORS

Chhorn Lim, PhD, has more than 30 years of experience in aquaculture nutrition and feed development research and as a research administrator. He is the Lead Scientist of Nutrition, Immune System Enhancement, and Physiology at the Aquatic Animal Health Research Unit in Auburn, Alabama. His current research includes determining the effect of dietary nutrients and evaluating the nutritional values and antinutritional factors of alternative protein sources on growth performance, immune response, and resistance of channel catfish and tilapia to infectious diseases. He also serves as an Affiliate Researcher of the University of Hawaii Institute of Marine Biology and Affiliate Professor and Graduate Faculty Member of the Auburn University Department of Fisheries and Allied Aquacultures, Auburn, Alabama. Dr. Lim has authored or co-authored over 100 publications in peer-reviewed technical journals, book chapters, and lay publications, and served as co-editor of three books. He is an editorial board member of the *Journal of Applied Aquaculture* and an associate editor of the *Journal of the World Aquaculture Society.*

Carl D. Webster, PhD, has more than 20 years of experience in aquaculture nutrition and diet development research. He is a Professor and Principal Investigator for Aquaculture at the Aquaculture Research Center, Kentucky State University, where he conducts research on nutrient requirements and practical diet formulations for fish and crustacean species that are currently or potentially cultured. He is also Adjunct Professor and Graduate Faculty of the Department of Animal Sciences at the University of Kentucky, and Affiliate Professor of the Department of Fisheries and Allied Aquacultures at Auburn University. Dr. Webster is a past president of the U.S. Chapter of the World Aquaculture Society. He has been the Program Chair of the

Aquaculture America 1999 conference held in Tampa, Florida and the Aquaculture America 2003 conference held in Louisville, Kentucky. He has authored or co-authored more than 100 publications in peer-reviewed technical journals and lay publications, served as co-editor of three books, and was the Editor of the *Journal of Applied Aquaculture* for 14 years, and is the Editor of the *Journal of the World Aquaculture Society.*

Dr. Cheng-Sheng Lee is Executive Director for the Center for Tropical and Subtropical Aquaculture, one of the regional aquaculture centers under CSREES at the USDA since 1997. Dr. Lee received his BS degree in Fishery Biology in 1970 and his MS degree in Marine Biology in 1972 from National Taiwan University. He received his PhD in aquaculture in 1979. Upon his graduation, Dr. Lee immediately joined the Oceanic Institute and has been working in Hawaii since 1979. He has served as program leader for shrimp programs and finfish programs, and is assistant vice president for the Institute. Dr. Lee has a broad interest in various aspects of aquaculture and concentrates on the control of fish reproduction and early life stages of fish. He has authored or co-authored more than 100 publications in refereed journals or books, and has been in charge of several international projects to develop and transfer culture technology to needed developing countries. Dr. Lee has frequently organized international conferences and currently is conducting two international workshops annually. He has also served as a technical editor for 14 books.

CONTRIBUTORS

G. L. Allan, NSW Fisheries, Taylors Beach, NSW, Australia.

Paul B. Brown, Department of Forestry and Natural Resources, West Lafayette, Indiana.

D. P. Bureau, Fish Nutrition Research Laboratory, University of Guelph, Guelph, Ontario, Canada.

C. Burel, Laboratoire d'Etudes et de Recherches Avicoles et Porcines AFSSA, United'Alimentation Animale, 22440 Ploufragan, France.

C. Young Cho, Fish Nutrition Research Laboratory, University of Guelph, Guelph, Ontario, Canada.

D. Allen Davis, Department of Fisheries and Allied Aquacultures, Auburn University, Auburn, Alabama.

Ian Forster, Oceanic Institute, Waimanalo, Hawaii.

Delbert M. Gatlin III, Department of Wildlife and Fisheries Sciences, Texas A&M University, College Station, Texas.

B. D. Glencross, Department of Fisheries, North Beach, Western Australia, Australia.

Ronald W. Hardy, Aquaculture Research Institute, University of Idaho, Hagerman, Idaho.

Sadasivam J. Kaushik, Fish Nutrition Laboratory, INRA-IFREMER, Universite de Bordeaux 1, 64310 Saint Pee-sur-Nivelle, France.

Menghe H. Li, Thad Cochran National Warmwater Aquaculture Center, Mississippi State University, Stoneville, Mississippi.

Alternative Protein Sources in Aquaculture Diets
© 2008 by The Haworth Press, Taylor & Francis Group. All rights reserved.
doi:10.1300/5892_b

xv

Peng Li, Department of Wildlife and Fisheries Sciences, Texas A&M University, College Station, Texas.

Linda S. Metts, Aquaculture Research Center, Kentucky State University, Frankfort, Kentucky.

Laura A. Muzinic, Ambystoma Genetic Stock Center, University of Kentucky, Lexington, Kentucky.

Helena Peres, Centro Interdisciplinar de Investigacao Marinha e Ambiental, 4050-123 Porto, Portugal.

Robert C. Reigh, Aquaculture Research Station, Louisiana State University, Baton Royge, Louisiana.

Mian N. Riaz, Food Protein Research & Development, Texas A&M University, College Station, Texas.

Edwin H. Robinson, Thad Cochran National Warmwater Aquaculture Center, Mississippi State University, Stoneville, Mississippi.

Shi-Yen Shiau, Department of Food and Nutrition, Providence University, Taichung 43301, Taiwan.

D. M. Smith, CSIRO Marine Research, Cleveland, Queensland, Australia.

Kenneth R. Thompson, Aquaculture Research Center, Kentucky State University, Frankfort, Kentucky.

Jesus A. Venero, Department of Fisheries and Allied Aquacultures, Auburn University, Auburn, Alabama.

Mediha Yildirim-Aksoy, USDA-ARS, Aquatic Animal Health Research Unit, Auburn, Alabama.

Yu Yu, National Renderers Association, Inc., Causeway Bay, Hong Kong.

Preface

Aquaculture, the fastest-growing agricultural sector globally, accounts for almost 50 percent of the fish consumed in the world. Much of the expansion has been due to China, whose production has increased 16 percent annually for the past twenty years. However, even if China is excluded from the production data, global aquaculture has increased at an annual rate of 6.3 percent over the past thirty years. If the projections for human population growth are achieved, an additional 40 million metric tons of aquatic food organisms will be needed by 2030 to maintain the current per capita consumption rate. If the consumption rate increases, then even more production will be required. If capture fisheries remain static, or decline in the future, additional aquatic foods will have to be supplied by aquaculture.

Marine fish meal has long been considered the most important protein ingredient used in aquaculture diets due not only to its high quality and quantity of essential amino acids, but also to the presence of essential n-3 highly unsaturated fatty acids (HUFA) and minerals, and its high palatibility. For the past two decades, quantities of harvested wild-caught fish (such as anchovy, menhaden, herring, and capelin) used to produce fish meal have remained constant, indicating that fish meal production is sustainable. However, as global aquaculture production continues to expand, requiring larger quantities of aquaculture diets to be produced, there has been an increased demand for the use of fish meal in fish and crustacean diets. This has led to a dramatic increase in the price of fish meal, which currently is $US850-1,400/metric ton, depending upon the country where purchased. It has been estimated that, in 2002, aquaculture utilized approximately 34 percent of total global fish meal produced. If the current incorporation levels of fish meal are maintained, fish meal

Alternative Protein Sources in Aquaculture Diets
© 2008 by The Haworth Press, Taylor & Francis Group. All rights reserved.
doi:10.1300/5892_c

required in 2010 is projected to be approximately 48 percent of the world production.

This high cost for fish meal comes at a time when prices paid to producers for aquaculture products have decreased or stagnated. Since diet costs represent 30 to 70 percent of the operating expenditures of an aquaculture enterprise, it is vital for the continued growth and expansion of the industry to reduce diet costs so that sustained profitability can be achieved. One means of reducing diet costs is to partially, or totally, replace fish meal, the single-most expensive macro-ingredient in an aquaculture diet, with less expensive and nutritious plant and animal protein sources.

This book, for the first time, presents the reader detailed information on the use of alternative plant and animal protein sources as partial or total replacement of fish meal, in finfish and crustacean diets, and examines the effect of the use of these ingredients on various biological and physiological parameters of cultured organisms, diet manufacturing processes and diet quality, and the health of the organisms. Each chapter has been written by world-renowned nutritionists and feed technologists to provide the most complete, up-to-date information on the most widely used plant and animal protein ingredients for aquaculture diets, and some ingredients that could be of significant potential.

Acknowledgments

The editors gratefully acknowledge the contributions made by the chapter authors. The preparation of this book has involved the cooperative efforts of many people, to whom we are extremely appreciative. Our gratitude is also extended to our families for their enormous encouragement, patience, and support. Through the funding support from the National Oceanic and Atmospheric Administration (NOAA; Grant # NA07RG0579, Amendment 2) to CSL at Oceanic Institute, Hawaii, we were able to assemble the team of experts to contribute to this book. It is greatly appreciated.

Chapter 1

Farmed Fish Diet Requirements for the Next Decade and Implications for Global Availability of Nutrients

Ronald W. Hardy

INTRODUCTION

Aquaculture has become a significant component of global fish production, contributing approximately 30 percent in 2001 and nearly 50 percent in 2005 of fish products for human consumption (New, 1996; Kilpatrick, 2003; FAO, 2006). The proportion is predicted to increase further, but most likely at a slower rate. Growth has resulted not only from more ponds or marine cages in use, but also from increasing productivity of existing ponds and cages associated with a switch from extensive to semi-intensive culture. This has greatly increased global fish diet production, now approximately 13,000,000 mt, and is expected to further increase global fish diet production to more than 37,000,000 mt by the end of the decade (Barlow, 2000), an increase of 24,000,000 mt (Table 1.1). In addition, the fish diet industry is producing more high-quality diets made from high-quality ingredients. Farmers realize that even though high-quality diets are more expensive than low-quality diets, their use leads to higher productivity resulting from improved pond water quality and higher growth rates of fish. Both these developments, that is, more diet production and high-quality diets, present challenges to the fish diet production industries,

Alternative Protein Sources in Aquaculture Diets
© 2008 by The Haworth Press, Taylor & Francis Group. All rights reserved.
doi:10.1300/5892_01

TABLE 1.1. World fish diet production (mt) by species groups in 2000 and predicted production in 2010.

Species group	Diet production (mt)	
	2000	2010
Salmon/trout	1,636,000	2,300,000
Shrimp	1,570,000	2,450,000
Catfish	505,000	700,000
Tilapia	776,000	2,497,000
Marine finfish	1,049,000	2,304,000
Cyprinids (carp)	6,991,000	27,000,000
Total	13,106,000	37,226,000

mainly as regards global supplies of fish meal and oil, which have been the staples of fish diet formulation for most farmed species of fish.

Fish grown in flowing water or cages, such as salmon and trout, have been fed complete diets, that is, ones that supply all of the nutritional needs of the fish, for more than fifty years. Pond-grown fish, in contrast, have traditionally been supplied with less rigorously formulated diets because they can obtain essential dietary nutrients by consuming plants and insects living in ponds. Only when culture densities become higher than the capability of pond biota to supply enough material does the need to provide pond fish with complete diets arise. The trend toward higher stocking densities in pond aquaculture, coupled with the use of mechanical aerators and other devices to circulate pond water, is driving this sector of aquaculture toward the use of higher-quality diets that are close to being nutritionally complete diets. This trend is placing further demands upon global diet ingredient supplies, especially for fish meal, and is mainly responsible for the fact that the aquaculture sector now uses a much higher proportion of annual global fish meal production than in previous decades. In the mid-1980s, less than 10 percent of annual fish meal production was used by the aquaculture diet sector. Today, that proportion is over 34 percent (Hardy and Tacon, 2002). Similarly, aquaculture now uses nearly 56 percent of annual global fish oil production (Pike and Barlow, 2003), up from <10 percent in 1990, but this change is due both to the adoption of high lipid diets by the salmon farming industry and to the growth of fish diet production.

Growth of Aquaculture and Fish Diet Production

Overall, aquaculture production has increased by approximately 9 percent annually over the past three decades, but production increases in some species groups has been much higher than the average. For example, in the salmonid group, which experienced a 10.6 percent per year growth from 1970 to 2000, production of Atlantic salmon, *Salmo salar,* grew much more than that of rainbow trout *Oncorhynchus mykiss* or Pacific salmon. The percentage growth per year of Atlantic salmon production was 33.5 and 45.5 percent per year for the periods from 1970 to 1980 and 1980 to 1990, respectively (Tacon, 2003). Rainbow trout production increased by 8.4 and 6.6 percent per year over the same periods. Production of Pacific salmon, mainly coho salmon, *Oncorhynchus kisutch,* grew by 22.4 percent per year from 1970 to 2000, mainly because of increased production in Chile. In terms of total production, in 2000, more than 1,530,000 mt of salmonids were produced, with approximately 60 percent of this total being Atlantic salmon, 27 percent rainbow trout, and the remaining 13 percent being Pacific salmon. Expressed in terms of diet requirements, the salmonid sector consumed 1,976,000 mt of diet in 2000, up from approximately 1,040,000 mt in 1990, and 320,000 mt in 1980. In other words, diet consumption by the salmon and trout farming sector increased sixfold in twenty years (Table 1.2).

TABLE 1.2. Estimated past, present[a], and predicted future[a] diet use by various species groups of farmed fish (thousand mt).

Year	Salmonids	Shrimp	Catfish	Marine	Cyprinids
1990	650	800	380	200	3,000
2000	1,976	1,783	505	1,212	7,358
2010	2,517	2,607	700	2,465	14,931
2020	2,918	3,227	900	3,289	20,866

[a]From Tacon (2003).

Even more dramatic growth has been experienced by other sectors of aquaculture. Shrimp production grew by an average of 23 to 25 percent per year from 1970 to 1990, with most of the increase occurring between 1985 and 1990 (Tacon, 2003). Production continued to grow through 2000 but at a slower pace. Even so, average growth per year was over 17 percent from 1970 to 2000. Diet production for the shrimp aquaculture industry followed the same pattern, increasing nearly ten times from 1980 to 2000, from 180,000 mt to 1,783,000 mt. Tremendous growth in production of marine fish species, notably gilthead sea bream, *Sparus aurata,* and European sea bass, *Dicentrachus labrax,* was seen, especially in the 1990s, averaging more than 30 percent per year during that period. Total production in 2000 was slightly more than 1,000,000 mt, requiring 1,212,000 mt of feed. Carp production also grew, especially for species of carp that are grown using pelleted diets. In 2000, carp production exceeded 15,700,000 mt, of which 63 percent was for species requiring 9,944,000 mt of pelleted diet. Taken in total, estimated global fish feed production in 2000 was slightly more than 15,000,000 mt, a six- to sevenfold increase over fish diet production levels in 1980.

Effects of Growth in Global Fish Diet Production on Diet Ingredient Use

Growth of global fish diet production over the past two decades has had a profound effect on use patterns of fish meal and oil, but little effect on total fish meal production or on the annual harvest rates of fish captured to produce fish meal and oil (Pike and Barlow, 2003). Fish meal is used in poultry, swine, ruminant, companion animal, and fish diets. Today's poultry diets used in the United States do not contain fish meal, whereas in the early 1970s, poultry diets contained 3 to 5 percent fish meal to supply "unidentified growth factors" needed for poultry to gain weight to their potential (Scott et al., 1982). During the 1970s, the unidentified growth factors were identified as trace elements, and supplemented into diets as inorganic minerals, resulting in a gradual elimination of fish meal from broiler and hen diets. In other parts of the world, however, where alternate protein ingredients are expensive relative to the price of fish meal, poultry diets still contain fish meal, albeit a small percentage. Over the same period, fish meal use in fish diet production has increased dramatically as aquaculture production, and fish diet production, has grown. In 2002, for example, estimated fish meal use in fish diets was 2,217,000 mt (Pike and Barlow, 2003). Average global fish meal production during the period from 1990 to 2000 was 7,047,000 mt, with a high of 7,440,000 mt in 1994, and a low of 5,342,000 mt in 1998 during an El Niño period that reduced catches of anchovies off the coast of Peru (Figure 1.1). Thus, the percentage of the eleven-year average global production of fish meal that was used in fish diets in 2002 was 31.46 percent.

Predictions of future fish meal use in fish diets differ in different sources. Pike and Barlow (2003) predict that by 2010, fish diet production will increase from 2002's level of 15,794,000 mt to 32,378,000 mt. They also report that in 2002, 2,217,000 mt of fish meal were used, meaning that 14.037 percent of the total weight of fish diets produced in 2002 was fish meal (Table 1.3). Using the same percentage of fish meal in diets predicted to be made in 2010 would require 4,601,321 mt of fish meal, or approximately 65 percent of the average annual global production of fish meal between 1990 and 2000. However, they also estimate total fish meal use in 2010 to be 2,854,000 mt, or 1,747,321 mt of fish meal less than one would calculate based upon use-levels in 2002.

Tacon and Forster (2000), in contrast, predicted that fish meal and oil use in fish diets will decrease from 2,190,000 mt and 590,000 mt

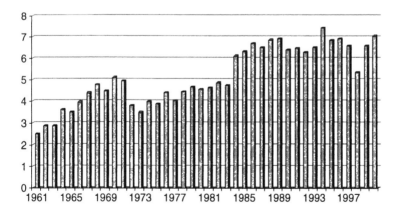

FIGURE 1.1. Annual global fish meal production from 1961 to 2000 (million mt).

TABLE 1.3. Estimated use of fish meal in diets for various species groups in 2000 and 2010 (Pike and Barlow, 2003).

Species group	2000 (%)	2010 (%)	2000 (000 mt)	2010 (000 mt)
Salmon	35	25	455	406
Trout	30	20	180	139
Marine fish	45	40	377	628
Flatfish	55	45	40	145
Shrimp	25	20	487	576
Catfish	2	0	12	0
Carp	4	3	337	602
Other[a]	0	0	629	489
Total			2,117	2,854

[a]Includes eels, milkfish, tilapia, and other carnivorous freshwater species.

in 2000 to 1,550,000 and 520,000 respectively in 2010. These authors contend that fish meal and oil use by the aquaculture feed industry will decrease because prices for meal and oil will increase at the same time that market prices for farmed fish and shrimp decrease, forcing the fish feed industry to replace portions of fish meal and oil in formulations with less expensive ingredients. Tacon and Forster (2000) further contend that consumers will demand that farmed fish be fed diets produced from contaminant-free ingredients and that retailers and consumers will demand that farmed fish be fed environmentally sustainable and environmentally friendly diets. Both of these developments, if they occur, will result in lower percentages of fish meal and fish oil in farmed fish diets than percentages used at present. New (2003) states that aquaculture has the potential to utilize 70 and 100 percent of the total annual production of fish meal and fish oil respectively by 2010. However, New and Wijkström (2002) suggests that criticism of fish meal use by the aquaculture industry and positive research results with diets in which fish meal has been replaced by plant protein concentrates will result in lower use-levels of fish meal and oil than the potential use-levels he calculates for 2010. In the final analysis, economic, rather than social, factors are likely to determine fish meal use levels in aquaculture diets over the coming decade, but food safety issues associated with contaminants in fish meal and oil will become major factors determining how much fish meal and oil is used in fish diets.

Fish Meal and Alternative Protein Sources

Fish meal has been the protein source of choice in diets for farmed fish for several reasons. First, the protein content of fish meals is relatively high, generally 65 to 72 percent, depending upon the fish species used to produce the fish meal. Second, the amino acid profile of fish meal closely matches the dietary requirements of most carnivorous fish species. Third, protein and amino acid apparent digestibility is relatively high in good-quality fish meal for most farmed fish species (Hajen et al., 1993; Sugiura, 2000; Sugiura and Hardy, 2000). Finally, fish meal based diets are highly palatable to most species of farmed fish.

Plant protein sources are similar to fish meal with respect to apparent protein and amino acid digestibility, and protein concentrates are similar to fish meal in protein content. However, amino acid profiles of plant protein sources do not match the dietary requirements of carnivorous fish species as well as do the amino acid profiles of marine proteins. This is no surprise, given the fact that carnivorous fish species eat other fish. The dietary essential amino acid requirements of carnivorous fish are similar to the essential amino acid profiles of prey fish, rather than the essential amino acid profiles of grains and oilseeds. Furthermore, some plant protein sources lower diet intake, presumably by lowering diet palatability when replacement levels are high. In addition, although the apparent digestibility coefficients (ADCs) of plant proteins are generally similar to those of fish meal proteins for carnivorous farmed fish, ADC values for dry matter are nearly always lower, especially in less refined products such as soybean meal. This is presumably due to the presence of indigestible carbohydrate components and fiber in plant proteins.

The list of suitable alternative protein sources to replace fish meal in fish diets is relatively short, and includes products of the poultry and animal rendering industries; marine protein recovered from fish processing and by-catch, protein concentrates made from grains, oilseeds, and pulses; and novel proteins from marine invertebrates and single-cell proteins (Table 1.4). Most of these protein sources have

TABLE 1.4. Prices[a] and price per unit protein (in ascending order) of alternative protein sources compared to menhaden fish meal.

Ingredient	Crude protein (%)	Price ($/mt)	Cost per kg protein ($)
Feather meal	83	225	0.271
Poultry by-product meal	60	225	0.375
Soybean meal	48	185	0.385
Corn gluten meal	60	260	0.433
Blood meal (flash-dried porcine)	89	425	0.478
Meat and bonemeal (porcine)	51	270	0.529
Fish meal (menhaden)	*68*	*630*	*0.926*
Soy protein concentrate	76	1,001	1.317
Wheat gluten	80	1,166	1.458

[a]From Feedstuffs, April 18, 2005, and Nelson and Sons, Murray, UT. For comparison purposes only.

been studied in fish diets, and ranges of suitable replacement for fish meal for major fish species have been estimated. For many of these products ADC values are available.

Changes in Fish Diet Formulations

The amount of protein supplied by fish meal in diet formulations for various species of fish differs significantly, depending on whether or not the species is carnivorous or omnivorous. Salmon and trout, for example, are fed diets that contain 38 to 44 percent crude protein during the grow-out stage, where most is used during a production cycle (Storebakken, 2002). Channel catfish, *Ictalurus punctatus,* in contrast, are fed diets containing 28 to 32 percent crude protein, most of which is supplied by soybean meal (Robinson and Li, 2002). Members of the carp family are fed diets with protein contents varying from 0 to 35 percent, depending on species, where they are farmed, and life-history stage (Shivananda Murthy, 2002; Takeuchi et al., 2002). Fry and fingerling carp are fed diets containing higher protein levels than are post-juvenile fish. Carp diets intended for use in high-input culture systems contain 15 to 25 percent fish meal, and although this is a relatively low fish meal inclusion level, the tremendous increase in high-input carp culture has dramatically increased the amount of fish meal used by this production sector to approximately 17 percent of the total amount of fish meal used in all aquaculture diets in 2000 (Barlow, 2000). Altogether 2,115,000 mt of fish meal was used in diets for fish and shrimp in 2000 (Table 1.3). The percentage of fish meal in the diet of various species groups ranged from 55 percent for marine flatfish (flounder, turbot, halibut) to 3 percent for catfish (channel catfish, African catfish). Carp averaged 5 percent, but this figure includes both high-input and low-input systems. Carp farming is converting to high-input systems, and this will increase the total use of fish meal in this production sector, despite an anticipated reduction in the percentage of fish meal used in diets (Barlow, 2000). Carp diet production is anticipated to increase from approximately 7,000,000 mt in 2000 to 27,000,000 mt by 2010. Soybean meal will likely supply the bulk of protein in carp diets of the future, but fish meal will continue to be used, especially in diets for fry and fingerling carp.

Potential Demand for Various Alternate Protein Ingredients

Potential demand for alternate protein ingredients depends primarily upon their price, protein level, amino acid profile, presence of anti-nutritional factors, or other factors limiting substitution levels. The leading candidates are soy products. If we assume that the fish meal used in fish diets contains 70 percent crude protein, then 1,228,500 mt of protein (1,755,000 mt fish meal × 70 percent protein) from sources other than fish meal will be needed annually in fish diets by 2010 (Table 1.5). If soybean meal (48 percent crude protein) were used to supply this protein, the increase in total use in all aquaculture diets would be 2,559,375 mt. If soy protein concentrate were used, the total would be less, approximately 1,640,000 mt, because of its higher protein content. If other protein concentrates from grains or oilseeds, for example, wheat gluten meal, corn gluten meal, canola protein concentrate, were used, the amounts would be similar.

The amino acid profiles of protein sources derived from oilseeds and grains are less suited to the essential amino acid requirements of farmed fish than is fish meal, as mentioned above (Table 1.6). If we assume that 1,755,000 mt less fish meal will be used in future fish diet

TABLE 1.5. Predicted fish meal use in aquaculture diets in 2000 and 2010.

	Feed production (000 mt)	Fish meal (000 mt)
2000 (est.)	13,098	2,115
2010 (with today's diet formulations)	37,226	4,586
2010 (with lower % fish meal in diets)	0	2,831[a]
Difference	0	1,755[b]

[a]Barlow (2000).

[b]Fish meal equivalent to be supplied by other protein sources.

TABLE 1.6. Amino acid concentration (g/100 g, wet weight basis) of several plant protein ingredients and low-temperature-dried fish meal SPC = soy protein concentrate; CG = corn gluten meal; WG = wheat gluten meal.

Amino acid	SPC	CG	WG	LT fish meal	Dietary requirement of rainbow trout
Arginine	4.04	1.34	2.18	3.35	1.5
Histidine	1.442	0.91	1.35	1.54	0.7
Isoleucine	3.17	2.37	2.78	3.15	0.9
Leucine	5.53	10.26	5.40	5.56	1.4
Lysine	3.84	0.91	1.20	4.69	1.8
Methionine	0.81	1.09	0.98	1.88	1.0[a]
Phenylalanine	2.76	2.79	3.00	2.28	1.8[b]
Threonine	3.03	2.06	2.25	3.42	0.8
Valine	5.59	2.85	3.38	4.09	1.2
Crude protein	64.6	65.9	75.5	73.0	44.0

[a]Plus cystine.

[b]Plus tyrosine.

formulations as predicted by Barlow (2000), as shown in Table 1.5, then we can calculate the amount of the two most limiting essential amino acids, for example, lysine and methionine, that this amount of fish meal contains. These calculations yield values of 87,750 mt of lysine and 35,100 mt of methionine. To supply 87,750 mt of lysine from soybean meal and soy protein concentrate will require 3,274,253 mt and 1,566,964 mt respectively. These numbers exceed the amount of each ingredient needed to supply the protein from 1,755,000 mt of fish meal that will not be used in future fish diets. Synthetic amino acids will likely supply a portion of the potential demand for lysine and methionine in future fish diets.

Approximately 67 percent of the fish meal used by the aquaculture diet industry is used in diets for salmon, trout, shrimp, and marine fish (Table 1.7). These species account for approximately 32 percent of total fish diet production, and 15 percent of total farmed fish production. Barlow (2000) predicts that by 2010 the percentage of annual fish meal production used in diets for these species groups will decrease to 52 percent of the total amount of fish meal used by aquaculture. A portion of this percentage decrease will result from higher total use of fish meal in diets for other species groups, but most of the decrease will be the result of lower percentages of fish meal being used in diet formulations for salmon, trout, shrimp, and marine fish, and concomitant higher use of alternate protein sources. Examining diet formulations used at present, and those likely to be used in the future, sheds light on which of the properties of fish meal are likely to be of highest value in diet formulations of the future.

Until recently, fish meal constituted between 40 and 55 percent of diet formulations for Atlantic salmon (Storebakken, 2002). This number is expected to decrease to approximately 30 percent in the next

TABLE 1.7. World fish meal use in fish diets (2000 estimate).

Species group	Fish meal (mt)	Percentage of total
Salmon	454,000	21.5
Marine finfish	415,000	19.6
Shrimp	372,000	17.6
Cyprinids (carp)	350,000	16.5
Trout	176,000	8.3
Eels	173,000	8.2
Flatfish	69,000	3.3
Other fish	106,000	5.0
Total	2,115,000	

decade, with increasing percentages of plant protein concentrates making up the difference. For rainbow trout, current use-levels range from 25 to 40 percent, depending upon fish meal price relative to alternative protein sources (Hardy, 2002). Use levels are expected to decrease by 5 percent, meaning that percentage fish meal in the diet will likely be 25 percent at most in the near future, assuming that there is sufficient demand for fish meal to keep fish meal prices at the high end of the price range. Using the European sea bass as an example of a farmed marine fish species, current fish meal levels in diets exceed 50 percent (Kaushik, 2002), but future levels are expected to be 40 percent. Extending these expected trends to other farmed species of carnivorous fish, it is clear that diet formulations will shift from high reliance on fish meal to reliance on blends of fish meal and plant protein concentrates, making it more difficult in some cases to balance diets with respect to limiting essential amino acids.

For some fish species, higher inclusion levels of plant protein sources, especially those derived from oilseed meals, will lower the palatability of diets. In some formulations, replacing fish meal with plant protein sources will alter both the mineral balance and bioavailability of minerals in the diet.

The role of fish meal in fish diets is likely to shift over the next decade from that of the primary source of dietary protein to that of a secondary source of dietary protein. However, fish meal will continue to be a primary source of essential amino acids that are limiting in plant protein sources. These essential amino acids are lysine (deficient in corn-derived proteins), methionine (deficient in soybean-derived proteins), and possibly arginine or threonine (deficient in small grains). With a shift in emphasis concerning the role of fish meal in fish diets, high-protein, low-ash fish meals will be increasingly valuable. Value-added processing to lower bone and indigestible protein levels in fish meal will be required to produce such fish meals. They can also be produced from portions of the seafood processing waste stream, such as viscera. Synthetic methionine and lysine will increasingly be used to supplement fish diets containing high amounts of alternative proteins from grains and oilseeds. However, protein concentrates made from grains such as wheat, corn, other small grains are deficient in several amino acids for which there are no inexpensive synthetic replacements. These proteins that are deficient in grain protein concentrates must be blended with other proteins that have a sufficiency of amino acids. Marine proteins can fulfill this role, although this represents a paradigm shift in their use in fish diet formulations.

In conclusion, diet formulations for farmed fish are expected to change in the future, mainly through a reduction in the percentage of fish meal used to produce grow-out diets. The extent of these changes will vary depending on the species of fish, but in general higher percentages of plant proteins will be used in place of fish meal. This will create several problems. Balancing the essential amino acid content of diets will be more difficult, given the fact that soy products are low in methionine, and grain-derived proteins are low in arginine, lysine, and methionine compared with fish meal (Table 1.6). Diet palatability may become an important consideration in diet formulation, especially when oilseed-derived proteins are added to diets.

Another important issue is associated with dietary minerals, both levels in diets and bioavailability. Fish meal is an excellent source of many essential minerals and plant proteins are not. Plant proteins contain phytate, the storage form of phosphorus in seeds, and phytate phosphorus is unavailable to monogastric animals, including fish. Furthermore, phytate is known to interfere with the availability of certain trace elements, especially zinc, making it necessary to over-fortify diets to ensure adequate dietary zinc intake in fish meal diets that contain high levels of phytate, especially in the presence of high dietary calcium levels (Richardson et al., 1985; Gatlin and Phillips, 1989).

We can expect the amount of fish meal used in aquaculture diets to be close to 50 percent of annual global production, but we can also expect an increase in the demand for specialty marine products produced specifically for use in diets for farmed fish. These products will have special characteristics to overcome problems associated with extended use of plant-derived protein concentrates. This will necessitate the expanded recovery and utilization of seafood processing waste and by-catch, with the additional refinement of partitioning of the seafood waste stream into segments that can be further processed to produce specialty products designed to enhance palatability, enrich diets with limiting amino acids, and to increase dietary efficiency, for example, retention of dietary nutrients to support fish growth.

As mentioned, economics will be the principal driver of these changes, although regulations associated with discharges of phosphorus, nitrogen, or fecal solids from farms, or concerns about contaminant levels in some fish meals and oil, for example, those from the North Sea or Baltic Sea, may affect fish meal use patterns in fish diets. Increased emphasis will be placed on dietary nutrient retention, and this will affect future diet formulations and fish meal use. Increasing dietary nutrient retention will require the use of refined diet ingredients in fish diets, in contrast to ingredients simply produced from raw materials. Examples of this include the use of refined starches in place of ground whole wheat, or marine protein concentrate in place of whole fish meal. This will lower levels of indigestible materials in diets, such as fiber from wheat or connective tissue and skin in fish meal. Overall, the amount of fish meal used in fish diets will increase over the next fifteen years, but the rate of increase will be much slower than the rate of increase in fish diet production over the same period. As demand for fish meal increases, the world price of fish meal will also increase, making it profitable to produce specialty diet ingredients from recovered seafood processing waste, or from grains, oilseed, legumes, and other agricultural products for use in diets in the production of specialty (high-value) aquaculture products.

REFERENCES

Barlow, S. (2000). Fishmeal and fish oil. *The Advocate,* 3(2), 85-88.
Food and Agriculture Organization (FAO) (2006). State of World Aquaculture 2006. FAO Fisheries Technical Paper No. 500. Rome, 134 pp.
Gatlin, D.M.I. and Phillips, H.F. (1989). Dietary calcium, phytate and zinc interactions in channel catfish. *Aquaculture,* 79, 259-266.
Hajen, W.E., Higgs, D.A., Beames, R.M., and Dosanjh, B.S. (1993). Digestibility of various dietstuffs by post-juvenile chinook salmon *(Oncorhynchus tshawytscha)* in seawater. 2. Measurement of digestibility. *Aquaculture,* 112, 333-348.
Hardy, R.W. (2002). Rainbow trout, *Oncorhynchus mykiss.* In C.D. Webster and C.E. Lim (Eds.), *Nutrient Requirements and Feeding of Finfish for Aquaculture* (pp. 184-202). New York: CABI Publishing.
Hardy, R.W. and Tacon, A.G.J. (2002). Fish meal historical uses, production trends and future outlook for sustainable supplies. In R.R. Stickney (Ed.), *Responsible Marine Aquaculture* (pp. 311-325). New York: CABI.

Kaushik, S.J. (2002). European sea bass, *Dicentrachus labrax*. In C.D. Webster and C.E. Lim (Eds.), *Nutrient Requirements and Feeding of Finfish for Aquaculture* (pp. 28-39). New York: CABI Publishing.

Kilpatrick, J.S. (2003). Fish processing waste: Opportunity or liability? In. P.J. Bechtel (Ed.) *Advances in Seafood Byproducts 2002 Conference Proceedings* (pp. 1-10). Alaska Sea Grant College Program, University of Alaska Fairbanks.

Kilpatrick, J.S. (2004). Fish processing waste: Opportunity or liability? Proc. 2nd International Seafood Byproduct Conference, November 10-13, 2002, Anchorage, AK. Alaska Sea Grant Report.

New, M.B. (1996). Responsible use of aquaculture diets. *Aquaculture Asia,* 1(1), 3-15.

New, M.B. (2003). Responsible aquaculture: Is this a special challenge for developing countries? *World Aquaculture,* 34(3), 26.

New, M.B. and Wijkström , U.N. (2002). Use of fishmeal and fish oil in aquafeeds: Further thoughts on the fishmeal trap. FAO Circular No. 975, Rome. 61 pp.

Pike, I.H. and Barlow, S.M. (2003). Impact of fish farming on fish stocks. *International Aquafeed Directory 2003,* pp. 24-29.

Richardson, N.L., Higgs, D.A., Beames, R.M., and McBride, J.M. (1985). Influence of dietary calcium, phosphorus, zinc and sodium phytate level on cataract incidence, growth and histopathology in juvenile chinook salmon *(Oncorhynchus tshawytscha). Journal of Nutrition,* 115, 553-567.

Robinson, E.H. and Li, M.H. (2002). Channel catfish, *Ictalurus punctatus*. In C.D. Webster and C.E. Lim (Eds.), *Nutrient Requirements and Feeding of Finfish for Aquaculture* (pp. 293-318). New York: CABI Publishing.

Scott, M.L., Nesheim, M.C., and Young, R.J. (1982). *Nutrition of the Chicken, 3rd edition.* Ithaca, NY: M. L. Scott and Associates.

Shivananda Murthy, H. (2002). Indian major carps. In C.D. Webster and C.E. Lim (Eds.), *Nutrient Requirements and Feeding of Finfish for Aquaculture* (pp. 262-272). New York: CABI Publishing.

Storebakken, T. (2002). Atlantic salmon, *Salmo salar.* In C.D. Webster and C.E. Lim (Eds.), *Nutrient Requirements and Feeding of Finfish for Aquaculture* (pp. 79-102). New York: CABI Publishing.

Sugiura, S.H. (2000). Digestibility. In R.R. Stickney (Ed.), *Encyclopedia of Aquaculture* (pp. 209-218). New York: John Wiley & Sons, Inc.

Sugiura, S.H. and Hardy, R.W. (2000). Environmentally friendly feeds. In R.R. Stickney (Ed.), *Encyclopedia of Aquaculture* (pp. 299-310). New York: John Wiley & Sons, Inc.

Tacon, A.G.J. (2003). Global trends in aquaculture and compound aquafeed production. *International Aquafeed Directory 2003,* pp. 8-23.

Tacon, A.G.J. and Forster, I.P. (2000). Global trends and challenges to aquaculture and aquafeed development in the new millennium. *International Aquafeed Directory and Buyers' Guide 2001,* pp. 4-25.

Takeuchi, T., Satoh, S., and Kiron, V. (2002). Common carp, *Cyprinus carpio.* In C.D. Webster and C.E. Lim (Eds.), *Nutrient Requirements and Feeding of Finfish for Aquaculture* (pp. 245-261). New York: CABI Publishers.

Chapter 2

Development of High Nutrient-Dense Diets and Fish Feeding Systems for Optimum Production and Aquaculture Waste Reduction: A Treatise

C. Young Cho

INTRODUCTION

Over the past decade, the environmental impacts of aquaculture operations have become a matter of concern for the public, various levels of government, and the aquaculture producers themselves. Minimizing environmental impacts is, therefore, a key factor in insuring long-term sustainability of the aquaculture industry.

Collection of wastes, both solid and dissolved, is very difficult and costly in aquaculture. These wastes are rapidly dispersed into the surrounding waters with grave impacts. The release of solid wastes, phosphorus, and nitrogen from aquaculture operations may have significant eutrophication effects on receiving water-bodies and associated ecosystems (Persson, 1991). Phosphorus waste outputs are of great concern in freshwater as phosphorus is generally the most-limiting factor for plant (algae) in that environment. Nitrogen waste outputs are generally a greater concern in the marine environment for the same reason (Persson, 1991). Solid wastes (fecal material and wasted feed) settling to the sediment can have an impact on the benthic ecosystem of inland and marine waters (Gowen et al., 1991). Degradation of organic waste

Alternative Protein Sources in Aquaculture Diets
© 2008 by The Haworth Press, Taylor & Francis Group. All rights reserved.
doi:10.1300/5892_02

by bacteria and other organisms leads to the consumption of oxygen through respiration. Excessive settling of organic matter may result in significant reduction in dissolved oxygen levels and the creation of anoxic conditions that can be damaging to the benthic biota. The hypolimnion of freshwater lakes is most sensitive to this phenomenon as it has a poor capacity of regenerating its oxygen content (Gowen et al., 1991).

Since most aquaculture wastes are ultimately from dietary origin, efforts to reduce waste outputs should focus on nutrition and feeding, mainly through improvements of diet formulation and feeding systems according to the principles of the Nutritional Strategies for Management of Aquaculture Waste (NSMAW; Cho et al., 1991).

Scientific approaches have been used in the feeding of land animals for more than a century. The first feeding standard for farm animals was proposed by Grouven in 1859, and included the total quantities of protein, carbohydrate, and ether extract (fat) found in diets, as determined by chemical analysis. In 1864, E. Wolf published the first feeding standard based on the digestible nutrients in feeds (cited from Lloyd et al., 1978).

Empirical feeding charts for salmonids at different water temperatures were published by Deuel et al. (1952) and were likely intended for use with meat-meal mixture diets widely fed at that time. Since then several methods of estimating daily diet allowance have been reported (Haskell, 1959; Buterbaugh and Willoughby, 1967; Freeman et al., 1967; Stickney, 1979). Unfortunately all methods have been based on the body length increase or live weight gain, and dry weight of feed and feed conversion, rather than on biologically available dietary energy and nutrient contents in relation with protein and energy retention in the body. These methods are no longer suitable for today's energy- and nutrient-dense diets, especially in the light of the large amount of information available on the energy metabolism and partitioning in salmonids.

Feeding standards may be defined as all feeding practices employed to deliver nutritionally balanced and adequate diets to animals, for maintaining normal health and reproduction together with the efficient growth and/or performance of work. Until now, the feeding of fish has been based mostly on instinct and folkloric practices. And the main preoccupation has been looking for "magic" diet formulae. Many "hypes," such as mega fish meal and mega-vitamin C diets, have come and gone, and we are now in the age of the "Norwegian Fish Doughnut" (>36 percent fat diet)! Whichever diet one decides to feed, the amount fed to achieve optimum or maximum gain, while minimizing feed waste, is the ultimate measure of one's productivity in terms of economic benefit and environmental sustainability.

Many problems are encountered when feeding fish, much more than when feeding domestic animals. First, delivery of feed to fish in a water medium requires particular physical properties of feed together with special feeding techniques. It is not possible in the literal sense to feed fish on an "ad libitum" basis, like it is done with most farm animals. The nearest alternative is to feed to "near-satiety" or percent body weight feed per day; however, this can be very subjective. Feeding fish continues to be an "art," and the fish culturist, not the fish, determines "satiety" as well as when and how often fish are fed. The amount of feed not consumed by the fish cannot be recovered and, therefore, all feed dropped in water (feed input, not feed intake) must be assumed eaten for inventory and feed efficiency calculations. This can cause appreciable errors in diet evaluation, as well as in productivity and waste output calculations. Feeding the pre-allocated amounts by hand or mechanical device based on daily energy requirement may be the only logical choice since uneaten feed represents an economical loss, and becomes 100 percent solid and suspended wastes. Feeding a pre-allocated amount of meal may not represent a restricted feeding regime as suggested by Einen et al. (1995) since the amount of feed calculated is based on the amount of energy required by the animal to express its full growth potential.

There are few scientific studies on feeding standards and practices; however, there are many duplications and "desktop" modifications of old feeding charts with little or no experimental basis. Since the mid-1980s, development of high-fat diets has led to most rations being very energy-dense, but feeding charts have changed little to reflect these changes in diet composition, not withstanding the fact that fish, like other animals, eat primarily to meet energy requirements. Most feeding charts available today tend to overestimate ration allowance, and this overfeeding has led to poor feed efficiencies under most husbandry conditions, and this represent a significant, yet avoidable, waste of resources for aquaculture economy. In addition, it will result in considerable self-pollution, which in turn may affect the sustainability of aquaculture operations. Recent governmental regulations imposing feed quota, feed efficiency guidelines, and/or stringent waste output limit may somewhat ease the problem. Sophisticated and expensive systems, such as underwater video camera or feed trapping devices, have been developed to determine the extent of feed wastage and are promoted by many as a solution to overfeeding (Ang et al., 1996). However, regardless of the feeding method used, accurate growth and diet requirement models are needed in order to forecast growth and objectively determine biologically achievable feed efficiency based on feed and carcass composition. These estimates can be used as useful yardsticks to adjust feeding practices or equipment and to compare the results obtained.

The development of scientific feeding systems is one of the most important and urgent subjects of fish nutrition and husbandry because, without this development, nutrient-dense and expensive diets are partially wasted. Sufficient data on nutritional energetics are now available to allow reasonably accurate feeding standards to be computed for different aquaculture conditions (Cho and Bureau, 1998). Presented here is a treatise of a nutritional energetics approach to tabulate ration allowance of high nutrient-dense diets and waste output estimation of fish culture operation as well as the introduction of the *Fish-PrFEQ*[1] computer program. Results obtained from a field station are presented and provide a framework to examine the type of information that can be derived from bioenergetics models and generate a feed requirement for a production scenario.

REDUCTION OF WASTE
THROUGH DIET FORMULATION

Digestibility of the ingredients, and nutrient composition and balance of the diet are the main factors that affect waste outputs by fish. Minimizing waste outputs from aquaculture operations should, therefore, start at the source, that is the diet.

Reducing Solid Waste (SW)

Solid waste outputs by fish fed practical diets consist largely of undigested starch and fiber from grain and various plant products, and minerals from the various ingredients. Undigested protein (nitrogen) and lipids are usually low since protein and lipid ingredients used in fish feeds are, in general, highly digestible. The amount of information on apparent digestibility coefficient (ADC) of dry matter, protein, lipid, carbohydrate, and minerals of common fish diet ingredients is ever expanding. However, estimates of ADC of protein of certain ingredients (e.g., fish meal, soybean meal, feather meal, poultry by-product meal) are quite highly variable in the literature. This variability is probably the result of:

1. Difference in fecal material collection method used (stripping, dissection, Tokyo system, St-Pée system, Guelph system)
2. Experimental errors (suboptimal experimental conditions, leaching, a nalytical errors, erroneous calculations, etc.)
3. Differences in the manufacturing and chemical composition of the ingredients (raw materials, processing technique, heat damage, etc.)
4. Biological differences (fish species, fish size, water temperature)

Sometimes, it may be difficult to determine which published values are most realistic or reliable for the actual ingredients that will be used in the formulation. However, a number of sources (Cho and Kaushik, 1990; Lall, 1991; NRC, 1993; Cho and Bureau, 1997; Guillaume et al., 1999) have summarized information on ADC of dry matter, protein, lipid, phosphorus, and energy of common diet ingredients and this information has proven reliable and helpful.

Reduction of SW outputs from aquaculture operations can be fairly simply done by using highly digestible ingredients with high protein and/or lipid contents, while excluding poorly digested, low-energy, and low-protein ingredients, such as grain by-products rich in starch and fiber. A basic example is shown in Table 2.1 in which 20 percent wheat middlings is deleted from a regular grower formula. Recalculation of this diet without wheat middlings gives a diet with 100/80 times higher energy (20 versus 17 MJ DE/kg) and higher protein (44 percent versus 38 percent), while the DP/DE ratio remains constant at 23 g/MJ. This type of diet was termed "high nutrient-dense (HND) diet" (Cho et al., 1991).

A study conducted at a fish culture station in Ontario showed that the MNR-91H resulted in an output of less than 190 kg total solid

TABLE 2.1. Concept of low and high nutrient-dense (HND) diets formulation.

	Low nutrient-dense	High nutrient-dense	
Ingredients	%	%	% (Recalculated)
Fish meal (68% CP, 10% fat)	30	30	37.50
Corn gluten meal (60% CP)	13	13	16.25
Soybean meal (48% CP)	17	17	21.25
Wheat middlings (17% CP)	26	6	7.50
Vitamin premix	1	1	1.25
Mineral premix	1	1	1.25
Fish oil, marine	12	12	15.00
Total	100	80	100.00
Nutritional specifications			
Digestible energy (DE), MJ/kg	17	19.68	
Digestible protein (DP), %	38	44.00	
Lipid, %	16	19.00	
Ash, %		7	8.00
DP/DE ratio, g/MJ	23	23.00	
Expected gain/Feed ratio	1	1.25	

waste compared with 240 kg with MNR-89G diet per ton of fish produced and less than 3 kg P compared to 4 kg (Table 2.2). The cost of the MNR 91H diet was higher per unit of feed weight, but diet cost per unit of fish produced was similar since feed efficiency was improved.

The first step in the production of diets emitting less SW is, therefore, to eliminate poorly digestible grain or grain by-products used as binders and fillers in the diet formulae and use highly digestible ingredients with good binding properties. Further reduction of SW can then be achieved through careful selection of ingredients, notably the protein sources.

Reducing Dissolved Nitrogen Waste (DNW)

The main factors affecting DNW outputs are those that influence the catabolism and deposition (retention) of amino acids (protein) by the fish. Amino acid composition of the diet is a factor with a determinant

TABLE 2.2. Practical grower (MNR-89G) and HND (MNR-91H) diet formulae for salmonids.

Ingredient	MNR-89G (%)	MNR-91H (%)
Fish meal, (68% CP, 10% fat)	20	35
Blood meal (80% CP)	9	9
Corn gluten meal (60% CP)	17	15
Soybean meal (48% CP)	12	14
Wheat middlings (17% CP)	20	0
Whey (12% CP)	8	10
Vitamin premix (VIT-8905)	0.5	0.5
Mineral premix (MIN-8404)	0.5	0.5
Fish oil, marine	13	16
Total	100	100
Nutritional specifications		
Digestible energy, minimum (MJ/kg)	17	20
Digestible protein/energy ratio (g/MJ)	22	22
Digestible fat (%)	16	20
Total phosphorous (%)	0.9	0.8
Expected feed efficiency, better than	1	1.2
Biologically estimated wastes (kg/t fish produced)		
Total solids		
Maximum	240	190
Nitrogen		
Solid	10	6
Soluble	40	33
Phosphorous		
Solid	4	3
Soluble	2	1.5
Fines (%)	1.5	1

Source: Cho and Bureau, 1998. MNR = Ontario Ministry of Natural Resources.

effect on DNW. Feeding amino acids in excess of requirement will result in the catabolism of the amino acid with associated excretion of ammonia and loss of energy (Kaushik and Cowey, 1991). Diet formulated with protein sources of poorer amino acid profile will result in lower digestible nitrogen retention efficiency and greater DNW (Kaushik, 1994).

Another key factor is the balance between digestible protein (DP) and digestible energy (DE) of the diet (Kaushik, 1994). Numerous studies have shown that decreasing the dietary DP/DE ratio, by increasing dietary nonprotein energy content, resulted in an increase in nitrogen retention efficiency and a decrease in DNW of numerous fish species (Lee and Putnam, 1973; Cho et al., 1976; Watanabe, 1977; Takeuchi et al., 1978; Watanabe et al., 1979; Kaushik and Oliva-Teles, 1985; Cho and Woodward, 1989; Johnsen and Wandsvik, 1991; Einen and Roem, 1997; Arzel et al., 1998; Helland and Grisdale-Helland, 1998b; Hillestad et al., 1998; Santinha et al., 1999; Steffens et al., 1999). The improvement in N retention and decrease in N excretion is due to the utilization of nonprotein energy sources for meeting energy requirements, resulting in a reduction of catabolism of amino acid, in what is commonly referred to "protein sparing." Protein-sparing by dietary lipids has been shown to occur in most fish species (Kaushik, 1994). Protein-sparing by digestible carbohydrate, such as gelatinized starch, has also been demonstrated (Kaushik and Oliva-Teles, 1985) but may be limited, especially when the diet already contains a high level of lipids or a relatively low DP/DE ratio (Lanari et al., 1995; Bureau et al., 1998; Helland and Grisdale-Helland, 1998a).

Overall, experimental data suggest that DP/DE ratio of about 18 g/MJ effectively reduces amino acid catabolism (and consequently DNW) without affecting growth rate and feed efficiency of salmonid fish species. Higher DP/DE are generally required by smaller fish compared to larger ones (Cho and Kaushik, 1990; Einen and Roem, 1997). The formulation of diets that are high in both protein and fat with 18-20 g DP/MJ DE and DE level equal to or exceeding 20 MJ/kg diet are very desirable for effective management of DNW for a large number of fish species.

Total digestible nitrogen retention efficiency rarely exceeds 50 percent in rainbow trout, *Oncorhynchus mykiss* (60 percent in Atlantic salmon, *Salmo salar*) fed diet with low DP/DE ratio (16-18 g DP/MJ DE). It is not clear to what extent this significant catabolism of amino acids, despite ample supply of nonprotein energy (shown by very high lipid deposition), is related to inevitable losses (maintenance requirement, inevitable catabolism) of amino acids or catabolism of amino acids that are in excess of requirement. Most studies on amino acid requirement have focused on the minimum dietary concentration of individual amino acids required to maximize performance. However, the impact of energy content of the diet and overall amino acid composition (or amino acid balance) of the diet on the utilization of amino acids, and consequently DNW, has not been examined in details (Cowey and Cho, 1993; Rodehutscord et al., 2000a).

Reducing Dissolved Phosphorus Waste (DPW)

Phosphorus (P) content of common fish diet ingredients is highly variable. Some practical ingredients contain limited amounts of P (e.g., 0.3 percent P in blood meal) while others contain very significant levels (4 to 5 percent P in meat and bonemeal) (NRC, 1993). P is found under different chemical forms in diet ingredients. Digestibility of the different chemical forms is known to differ widely (Lall, 1991). P contained in organic compound, such as phospholipids and nucleic acids, are apparently highly digestible for fish. P contained in phytate (inositol hexaphosphate), also an organic compound, however, is only slightly digestible to fish since they lack the necessary enzyme (phytase). The digestibility of mineral forms of P, such as dicalcium phosphate, monosodium phosphate, and rock phosphate, varies with degree of solubility of the compound(s) and is, consequently, highly variable. The digestibility of P contained in bone (apatite) is variable between fish species and depends mostly on stomach pH of the animal. For rainbow trout, a fish with a true (acid) stomach, ADC of bone P appears to be between 40 and 60 percent. ADC of bone P appears to be much lower for stomachless fish, such as common carp *(Cyprinus carpio)*. Other factors, such as particle size, feed processing technique, and enzyme treatment, are also known to affect ADC of P (Lall, 1991).

A negative effect of dietary P level on ADC of P has been suggested by some studies (Vielma and Lall, 1998; Rodehutscord et al., 2000b). However, the limited range of data available and the relatively large experimental error often associated with micronutrient digestibility measurements of fish do not allow a valid conclusion to be reached about the existence of such effect at this point in time. ADC for P of common diet ingredients of Lall (1991) measured based on fecal sample collected by stripping and Satoh et al. (1998) with the Tokyo system appear reliable and additive (Bureau and Cho, 1999) and offers a good starting point for the formulation of diet minimizing DPW outputs.

Numerous studies have shown that dietary incorporation of microbial phytase improved the ADC of P of fish fed diets containing phytic acid (Rodehutscord and Pfeffer, 1995; Oliva-Teles et al., 1998; Vielma et al., 1998; Forster et al., 1999). The activity of this enzyme is affected by environmental temperature and its activity may be very limited at low water temperatures (Forster et al., 1999). Moreover, the enzyme is sensitive to heat and may be destroyed during pelleting and extrusion under standard commercial conditions. These factors should be taken into consideration before using phytase in practical diets. The use of phytase makes sense only for diets with digestible P contents below the requirement of the fish and containing significant levels of plant ingredients in which indigestible P is mostly phytate-P.

Both digestibility and quantity will determine the fate of P fed to fish. The undigested fraction of the P of the diet is excreted in the feces by fish. The fraction of P digested by the animal is absorbed where it is deposited in the body of the fish (bones, scales, flesh, etc.) in the growth processes. A number of experimental evidences suggest that there is a requirement to maximize growth, phosphorus deposition, and bone mineralization. Phosphorus requirement of rainbow trout for maximum growth was 0.37 percent digestible P (0.19 g/MJ DE) and 0.53 percent (0.27 g/MJ DE) for maximum phosphorus deposition (Rodehutscord, 1996). The necessity of maximizing mineralization of the skeleton of fish for long-term maintenance of health and performance is still being debated (Rodehutscord, 1996; Asgard and Shearer, 1997; Baeverfjord et al., 1998; Vielma et al., 2000).

Fish receiving only the required amount of digestible P to meet growth requirements excrete only minute amounts of nonfecal P (ca. 5 mg P/kg BW/day) indicating that digestible P intake of the fish is directed almost completely toward deposition (Rodehutscord, 1996; Vielma and Lall, 1998; Bureau and Cho, 1999). There is evidence that efficiency of P utilization tends to decrease as digestible P level increases from the level required for maximum growth to the level required for maximum P deposition (Rodehutscord, 1996; Rodehutscord et al., 2000b). Interpretation of available data suggests that, while feeding a diet with digestible P at the level required to maximize growth results in minimal nonfecal P excretion, feeding a diet with a digestible P level required for maximum P deposition results in significant nonfecal P excretion.

Dissolved phosphorus waste is excreted mostly as phosphate via the urine (Kaune and Hentschel, 1987; Renfro, 1997; Vielma and Lall, 1998; Bureau and Cho, 1999). In mammals, urinary phosphate excretion is determined mostly by plasma phosphate concentration (Bijvoet, 1980). A threshold plasma phosphate concentration exists below which phosphate excretion is minimal and above which phosphate excretion is proportional to the increase in plasma phosphate. As teleost fish and mammals share similar renal physiology (Dantzler, 1989), a similar relationship between plasma phosphate and urinary phosphate excretion should also exist in teleost. This was recently confirmed with rainbow trout (Bureau and Cho, 1999). Given the existence of such a threshold relationship, it might be reasonable to conclude that a digestible P level producing a plasma phosphate concentration near renal P excretion threshold concentration should be acceptable from a biological (the fish) point of view and optimal from a waste management point of view. Recent experimental evidences suggest that this level is approximately 0.4 percent digestible P (0.2 g/MJ DE) for rainbow trout (Rodehutscord et al., 2000b). At that level, growth of rainbow trout is maximized, "acceptable" (not maximal) P deposition is achieved (i.e., acceptable level from the fish point of view since at this level the plasma phosphate pool saturates renal reabsorption systems), and DPW output is minimal. The amount of information available on the ADC of P of common diet ingredients and P utilization and requirement of fish can now allow the formulation of diets resulting in less P waste output for many fish species.

It might be worth noting that not all forms of P excreted by fish may have equal potential to contribute to eutrophication. In order to be utilized by algae and other plants, P must be soluble. Therefore, DPW is highly available to plants and may greatly stimulate eutrophication. In nature, solubilization of solid phosphorus waste (SPW) can either be the result of simple chemical equilibrium or action of enzymes or chemicals (acids) by bacteria and other organisms. To what extent SPW will be mineralized and solubilized in the environment will be dependent on both, on the chemical makeup of SPW and the prevailing conditions. Organic forms of P that are not digested by the fish (e.g., phytin-P) will be excreted as SPW, but can be mineralized by bacteria and other organisms in the aquatic environment (Persson, 1991). Calcium-bound P, as it is found in bones (apatite-P) is in most cases inert under normal environmental conditions (Persson, 1991) and may have no or little potential to stimulate eutrophication.

REDUCING WASTE OUTPUTS THROUGH FEEDING SYSTEM

Prediction of Growth and Energy Retention

Predicting growth performance of a fish culture operation requires first production records of past performance. These records become essential databases for calculating growth coefficients, temperature profiles during growth period, diet intake, and feed efficiency of various seasons. One such production record for rainbow trout from a field station is shown in Table 2.3. A lot of 100,000 fish was grown over a 14-month (410 days) production cycle. Cumulated live weight gain (fish production) was 72 tons with feed consumption of 60 tons which gave an overall feed efficiency (gain/feed) of 1.19 (rang between 1.11 and 1.22). Water temperature ranged from 0.5°C in winter to 21°C in summer, which is typical of most lakes in Ontario. In spite of the wide fluctuation in water temperature, the thermal-unit growth coefficient (TGC) was fairly stable, ranging between 0.177 and 0.204. Total mortality was approximately 9 percent over 410 days. From the production record (Table 2.3), one can extrapolate an overall growth coefficient of 0.191 (0.177-0.204) and this coefficient can be used for the growth prediction of future production plan on assumptions of similar rearing conditions and fish stock. Total diet requirement and setting weekly feeding standards can be computed on the basis of this growth prediction plus the quality of diet being purchased (see Table 2.4).

A more accurate and useful TGC for fish growth prediction in relation to water temperature is based on the exponent 1/3 power of body weight, in contrast to widely known specific growth rate (SGR) based on natural logarithm. Such a cubic coefficient has been applied both to mammals (Kleiber, 1975) and to fish (Iwama and Tautz, 1981).

TABLE 2.3. Rainbow trout production records from a field station.

Month-End	Days	No. fish	Body weight (g/fish)	TGC	Total biomass (kg)	Total feed (kg)	Gain/ Feed	Temp (°C)	Flow rate (L/min)
Initial		100,000	10.0						
May	15	98,900	12.1	0.184	1,191.8	167	1.22	5.0	2,500
Jun	30	95,000	36.5	0.189	3,462.8	2,000	1.18	18.0	6,000
Jul	31	95,000	89.8	0.197	8,534.8	4,300	1.18	19.0	10,000
Aug	31	94,500	177.4	0.175	16,767.1	7,200	1.15	21.0	16,000
Sep	30	94,000	296.3	0.184	27,848.4	9,500	1.18	19.0	20,000
Oct	31	93,500	396.1	0.199	37,031.6	7,800	1.20	11.0	25,000
Nov	30	93,200	451.0	0.197	42,036.0	4,300	1.19	5.5	25,000
Dec	31	93,000	455.9	0.176	42,394.1	400	1.12	0.5	25,000
Jan	31	92,000	460.8	0.178	42,390.8	400	1.14	0.5	25,000
Feb	28	91,500	465.2	0.177	42,568.6	370	1.11	0.5	25,000
Mar	31	91,200	470.4	0.184	42,899.6	420	1.12	0.5	25,000
Apr	30	91,000	475.5	0.188	43,274.1	420	1.12	0.5	25,000
May.	31	91,000	534.7	0.200	48,653.2	4,500	1.20	5.0	30,000
Jun	30	90,800	783.4	0.204	71,130.0	18,500	1.22	18.0	50,000
Total	410 days			0.191		60,277 kg feed	1.19		13.5×10^6 m³ water

Source: Cho and Bureau, 1998.

Note: Fish were grown in 1,200 liter fiberglass tanks with one to two exchanges/hour flow-through water system.

TABLE 2.4. Model prediction of fish body weight and feed requirement based on production records in Table 2.1.

Month-End	No. fish	TGC	Actual production records			Predicted production scenario			Temp (°C)
			Body weight (g/fish)	Total feed (kg)	Gain/ Feed ratio	Body weight (g/fish)[a]	Total feed (kg)[a]	Gain/ Feed ratio	
Initial	100,000		10.0			10.0			
May	98,900	0.184	12.1	167	1.22	12.2	120	1.81	5.0
Jun	95,000	0.189	36.5	2,000	1.18	37.4	1,498	1.68	18.0
Jul	95,000	0.197	89.8	4,300	1.18	87.9	3,446	1.47	19.0
Aug	94,500	0.175	177.4	7,200	1.15	181.9	6,732	1.40	21.0
Sep	94,000	0.184	296.3	9,500	1.18	310.2	9,495	1.35	19.0
Oct	93,500	0.199	396.1	7,800	1.20	406.6	7,775	1.24	11.0
Nov	93,200	0.197	451.0	4,300	1.19	461.5	4,602	1.19	5.5
Dec	93,000	0.176	455.9	400	1.12	466.7	451	1.16	0.5
Jan	92,000	0.178	460.8	400	1.14	471.9	454	1.16	0.5
Feb	91,500	0.177	465.2	370	1.11	477.2	452	1.17	0.5
Mar	91,200	0.184	470.4	420	1.12	482.6	453	1.18	0.5
Apr	91,000	0.188	475.5	420	1.12	488.0	456	1.18	0.5
May	91,000	0.200	534.7	4,500	1.20	544.0	4,627	1.21	5.0
Jun	90,800	0.204	783.4	18,500	1.22	780.8	18,228	1.30	18.0

Source: Cho and Bureau, 1998.

[a]Overall TGC = 0.191 from Table 1 was used to predict body weight and total feed requirement.

The following modified formulae were applied by Cho et al. (1985) and Cho (1990, 1992) for many nutritional experiments:

$$TGC = 100 \times [FBW(g)^{1/3} - IBW(g)^{1/3} / \Sigma[T(°C) \times days];$$

Estimated Final Body Weight (Est. FBW)

$$= [IBW(g)^{1/3} + \Sigma(TGC/100 \times T(°C) \times days]^3;$$

where FBW or IBW is final or initial body weight and T is water temperature in Celsius. It is important to note that the 1/3 exponent must contain at least four decimals (e.g., 0.3333) to maintain good accuracy.

This model equation has been shown by experiments in our laboratory to represent very accurately the actual growth curves of rainbow trout, lake trout, brown trout, Chinook salmon, and Atlantic salmon over a wide range of temperatures. Extensive test data were also presented by Iwama and Tautz (1981). An example of growth, water temperature, and TGC is shown in Figure 2.1. Growth of some salmonid stocks used for our experiments gave the following TGC:

Rainbow trout-A	0.174
Rainbow trout-B	0.153
Rainbow trout-C	0.203
Lake trout	0.139
Brown trout	0.099
Chinook salmon	0.098
Atlantic salmon-A	0.060
Atlantic salmon-B	0.100

As these TGC values and growth rate are dependent on species, stock (genetics), nutrition, environment, husbandry, and other factors, it is essential to calculate the TGC for a given aquaculture condition using past growth records or records obtained from similar stocks and culture conditions (Table 2.4).

Live weight gain is the result of deposition of water, protein, fat, and minerals. The amount of these components deposited per unit of live weight gain is not constant but rather changes with fish species and size, diet used, and other variables. For this reason, knowledge of the composition of the fish grown is another key factor for the accurate determination of diet requirement. Pattern of nutrient deposition

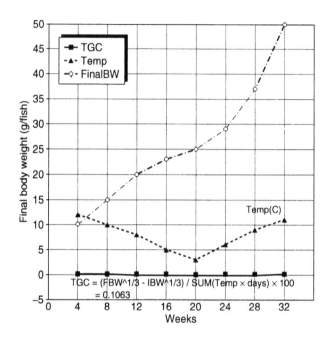

FIGURE 2.1. An example of the relationship among body weight (BW = 10-50 g/fish), water temp (Temp = 3-12°C), and thermal-unit growth coefficient (TGC = 0.17-0.18) of rainbow trout as a function of time. *Source:* Cho and Bureau, 1998.

have received little attention in the past. A limited number of studies on the topic appear to indicate that growth and nutrient deposition follow rational patterns (Shearer, 1994; Azevedo et al., 1998). Growth of rainbow trout and other salmonids over a wide range of body size has been shown to be accurately described by the TGC model (Cho, 1992). Studies with rainbow trout grown at different water temperatures (Azevedo et al., 1998; Rodehutscord and Pfeffer, 1999) have shown that protein and lipid deposition increased linearly with increasing metabolizable energy intake, regardless of body weight or water temperature. These results suggest that nutrient and energy deposition, and consequently carcass composition, follow rational patterns and that simple models can be developed to predict composition of fish of different sizes and the energy cost of nutrient deposition.

As a large proportion of the nutrients (e.g., amino acids, lipids) and, consequently of the dietary energy, consumed by fish is retained as carcass body constituents, carcass energy is a major factor driving dietary energy requirement of the fish. Carcass moisture, protein, and fat contents in various life stages dictate energy level of fish (Bureau et al., 2003). These factors (retention coefficients) are also influenced by species, genetics, age, nutritional status, and husbandry. The water and fat contents of the fish produced are, in general, the most variable factors and have a determinant effect on energy content of the fish. For example, relatively fatty Atlantic salmon and rainbow trout may require more dietary energy per unit of live body weight than leaner salmonids such as brown trout, lake trout, and charr. Fish containing less moisture (more dry matter) and more fat require more energy allocation in feeding standards.

The simplistic assumption of the constant body composition within a growth stage by Einen et al. (1995) is not valid for different species and sizes. Dry matter and energy content of fish can increase dramatically within a growth stage, especially in the case of small fish. Underestimation or overestimation of the diet requirement is likely to occur if constant carcass energy content is assumed in calculations. Reliable measurements of carcass composition of fish at various sizes are essential. Nutrient and energy gains should be calculated at relatively short intervals, at least for small fish. In addition, composition of the diet, notably the digestible protein to digestible energy ratio, and the lipid content of the diet, can have a very significant influence on the composition and energy content of the carcass. Estimation of carcass composition and energy content should rely on data obtained with fish fed diets similar to those one intends to use.

Shearer (1994) concluded that the protein content of growing salmonids is determined solely by fish size, that lipid level is affected by both endogenous (fish size, growth rate) and exogenous (dietary, environmental) factors, and that ash content is homeostatically controlled. An effect of diet composition, notably protein to lipid ratio, on energy gain and carcass composition of rainbow trout has been observed (Rodehutscord and Pfeffer, 1999) but the limited amount of data available does not yet allow the computation of reliable predictive models. Research aiming at the development of models to predict carcass composition at various sizes depending on the composition of the feed is underway.

Estimation of Excretory and Feed Wastes

The maintenance of life processes (integrity of the tissues of the animal, osmoregulation, respiration, circulation, swimming, etc.) and the deposition of body components have costs in terms of nutrient and feed energy. Basic and practical research projects have allowed the development of simple, yet reliable, models (equations) to calculate these costs or wastes. Studies involving the rearing of fish under various conditions (water temperature, feeding level, fish size, etc.) have shown that these biological costs are, surprisingly, fairly constant, and, consequently, fairly easily predicted (Azevedo et al., 1998; Lupatsch et al., 1998; Ohta and Watanabe, 1998; Rodehutscord and Pfeffer, 1999).

Prior to setting goals for reducing waste outputs, one must have access to objective estimates of the amount of different wastes produced. Directly monitoring and quantitatively estimating waste outputs from effluent of aquaculture facilities are an inaccurate and costly process (Cho et al., 1991). It is also nearly impossible for certain types of facilities, such as cage culture operations.

Waste output from aquaculture operations can be estimated using simple principles of nutrition and bioenergetics as applied by Cho et al. (1991, 1994) and it is a "biological" approach, rather than a chemical or limnological one. Ingested feedstuffs must be digested prior to utilization by the fish and the digested protein, lipid, and carbohydrate are the potentially available energy and nutrients for maintenance, growth, and reproduction of the animal. The remainder of the feed (undigested) is excreted in the feces as solid waste (SW), and the by-products of metabolism (ammonia, urea, phosphate, carbon dioxide, etc.) are excreted as dissolved waste (DW), mostly by the kidneys. The aquaculture total wastes (TW) associated with feeding and production is made up of SW and DW, together with apparent feed waste (AFW). Since direct estimation of AFW is very difficult, best estimate can only be obtained by comparison with theoretical feed requirement calculated with bioenergetics models (Cho and Bureau, 1997, 1998).

$$TW = SW + DW + AFW,$$

where SW, DW, and AFW outputs are biologically estimated by:

$$SW = [\text{Feed consumed} \times (1 - ADC)];$$

DW = (Feed consumed × ADC)
    − Fish produced (nutrients retained);
AFW = Actual feed input (AFI)
    − Theoretical feed required (TFR);

in which ADC is the apparent digestibility coefficients of diets. Measurements of ADC and feed intake provide the amount of SW (settled and suspended without AFW) and these values are most critical for accurate quantification of aquaculture waste. For dry matter, nitrogen, and phosphorus, ADC should be determined using reliable methods by research laboratories where special facility, equipment, and expertise are available. More information on the equipment and procedures may be obtained from Cho and Kaushik (1990) and the Web site www.uoguelph.ca/fishnutrition.

Dissolved waste (DW = DNW + DPW) can be calculated by difference between digestible N (DN) or P (DP) intake and retained N (RN) or P (RP) in the carcass if this information is available. These data should also be determined or estimated for each type of diet used by research laboratories. However, controlled feeding and growth trials with particular diets at production sites are also essential to validate ·and fine-tune the coefficients from the laboratory. Dissolved nitrogen waste output depends very much on dietary protein and energy and amino acid balances (Watanabe and Ohta, 1995) and rate of protein deposition by the fish, therefore all coefficients must be determined on a regular basis, particularly when diet formulae are changed.

Accurate estimation of total solid waste (TSW) requires a reliable estimate of AFW. Feeding the fish to appetite or near-satiety is very subjective and, unfortunately, TSW contains a considerable amount of AFW under most fish farming operations. The use of "biomass gain × feed conversion" as an estimate of real feed intake of the fish to calculate waste output as suggested by Einen et al. (1995) can grossly overestimate the real feed intake in many operations where overfeeding is common and can results in an underestimation of the TSW output (Figure 2.2).

It is very difficult scientifically to determine the actual diet intake by fish in spite of many attempts (mechanical, radiological, and biological) that have been made by biologists. Since estimation of AFW is almost impossible, the best estimates can be made based on energy

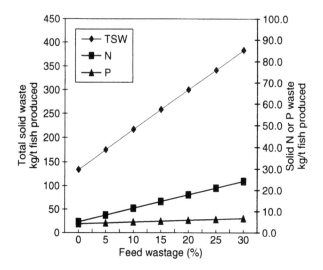

FIGURE 2.2. Effects of feed wastage on total solid, solid nitrogen, and solid phosphorus waste outputs of rainbow trout fed a practical diet and growing from 10 to 100 g live weight at 15°C. *Source:* Bureau et al., 2003.

requirements and expected gain described by Cho (1992) in which the energy efficiency (energy gain/intake) indicates the degree of AFW for a given operation. Theoretical energy and feed requirement (TER and TFR) can be calculated based on nutritional energetics balance as follows:

TER = Retained energy + Excreted energy (including heat loss);

TFR = TER/DE in diet;

and the amount of feed input above the TFR should be assumed to be AFW and all nutrient contents of AFW must be included in solid waste quantification. This approach may yield a relatively conservative estimate.

Biological procedures based on the ADC for SW and comparative carcass analyses for DW provide very reliable estimates. Biological methods are flexible and capable of adaptation to a variety of conditions and rearing environments. It also allows estimation of the TFR and waste output under circumstances where it would be very difficult or impossible to do so with a chemical/limnological method (e.g., cage culture). Properly conducted biological and nutritional approaches to estimate aquaculture waste outputs are not only more accurate, but also much more economical than chemical/limnological methods of effluent analyses (Cho et al., 1991, 1994; Cho and Bureau, 1997).

The waste outputs from the field station are tabulated in Table 2.5 using *Fish-PrFEQ* computer models. SW was estimated at 10,610 kg (fish production 72 t; 60 t feed input over fourteen months). SW represented 90 percent of TSW since AFW (AFI − TFR) was estimated at 1,201 kg or 2.2 percent of feed input (60,277 kg). The TSW outputs were equivalent to 164 kg per ton fish produced. Phosphorus waste was 5.11 kg/t fish produced and nitrogen 30.64 kg. Total water consumption during fourteen months was 13,469 m^3, therefore the average effluent quality can be estimated at: solid 0.877 mg/L, phosphorus 0.027 and nitrogen 0.163 (Table 2.5). The diet used, the detailed procedures to estimate waste production as well as comparative data of chemical and biological estimations from the field experiments at the Ontario Ministry of Natural Resources (OMNR) Fish Culture Stations are described in Cho et al. (1991, 1994).

Diet Selection and Ration Allowance

Selection of diets for aquaculture production is a complex decision by fish culturists and is beyond the scope of this writing. However, all diets selected must contain adequate levels of digestible energy and essential nutrients per kilogram diet and most importantly, also have

TABLE 2.5. Model estimation of waste outputs and effluent quality from the rainbow trout production operation in Table 2.1.

Waste output (Total load estimate)	Solid (kg)	Nitrogen (kg)	Phosphorus (kg)
Apparent feed wastage (2%)[a]	1,201	80.69	12.01
Solid	10,610	356.49	212.19
Dissolved	–	1,764.60	143.23
Total wastes	11,811	2,201.79	367.43
Per ton fish produced	164	30.64	5.11
Av. concentration (mg/L) in effluent (13469 × 10^6 L) during 410 days	0.877	0.163	0.027

[a]Actual feed input − Theoretical feed requirement.

optimally balanced digestible protein and energy ratio for the species being cultured. Without meeting these nutritional conditions of diet, the feeding standard concept in this treatise should not be applied.

Ration allowance (RA or feeding standard) is tabulation of energy and nutrients needs to maintain normal health and reproduction together with the efficient growth and/or performance of work (activity). A considerable portion of dietary energy is expended for maintenance including basal metabolism, which is the minimum energy and nutrients required to maintain basic life processes. The maintenance energy requirement is approximately similar to the heat production of a fasting animal. This amount of dietary energy represents an absolute minimum of "energy-yielding" nutrients and must be covered before any nutrients can be used for growth and reproduction of the animal. Otherwise, body tissues will be catabolized because of a negative energy balance between intake of dietary fuels and energy expenditure. Poikilotherms, such as salmonid fish, require far less maintenance energy (approx. 40 kJ per kg $BW^{0.824}$/day for rainbow trout at 15°C according to Cho and Kaushik, 1990) than do homeotherms (approx. 300 kJ per kg $BW^{0.75}$/day according to Lloyd et al., 1978).

A review of available data suggests that a HE_f of about 36-40 kJ/ $kg^{0.824}$ per day appears accurate for rainbow trout at 15°C, at least for fish between 20 and 150 g live weight with which most of studies have been conducted (Cho et al., 1976; Cho and Slinger, 1980; Kaushik and Gomes, 1988; Cho and Kaushik, 1990; Bureau, 1997).

Cho and Kaushik (1990) estimated the heat increment of feeding (HiE, heat loss to utilize ingested diet) of rainbow trout fed a balanced diet to be approximately 30 kJ/g digestible N or the equivalent of 60 percent HE_f, but the latter relationship does not always hold true. Studies with farm animals suggest that HiE is independent of maintenance and is related to protein and lipid deposition rates separately (Emmans, 1994). Based on experimental results, it was observed that HiE was approximately equivalent of 20 percent of net energy intake, that is, 0.2 (RE + HE_f) and this value is used in the bioenergetics model presented here. Studies are underway to quantify HiE as a function of protein and lipid deposition.

Biological oxygen requirement of feeding fish is equal to the total heat production ($[HE_f + HiE]$/Qox) in which the oxycalorific coefficient (Qox) is 13.64 kJ energy per g oxygen. This represents the absolute minimum quantity of oxygen that must be supplied to the fish by the aquatic system. Oxygen requirement per unit of body weight per hour will vary significantly for different fish sizes and water temperatures.

Tabulation of Total Energy Requirement and Ration Allowance

1. Allocation of approximate maintenance energy requirement (HE_f) at a given body weight (BW), water temperature (T), and period:

   $$HE_f = (-0.0104 + 3.26T - 0.05T^2) \, (kg \, BW^{0.824}) \, kJ \text{ per day} \times \text{days}$$

2. Calculation of expected live weight gain (LWG = FBW − IBW) using TGC and retained energy (RE) based on carcass energy content:

   $$LWG = [IBW(g)^{1/3} + \Sigma \, (TGC/100 \times Temp. \, (°C) \times days)]^3 - IBW(g)$$
   $$RE = (0.004 \, g \, BW^2 + 5.58 \, g \, BW + 7.25) \, kJ \text{ per } g \, BW \times g \, LWG$$

3. Allocation of approximate heat increment of feeding for maintenance and growth:

   $$HiE_{M+G} = (HE_f + RE) \times 0.2$$

4. Allocation of approximate nonfecal energy loss:

   $$ZE + UE = (HE_f + RE + HiE_{M+G}) \times 0.1$$

5. Theoretical (minimum) energy requirement (kJ):

   $$TER = HE_f + RE + HiE_{M+G} + UE + ZE$$

6. Ration Allowance or feeding standard (g):

   $$RA = TER \, / \, kJ \, DE \text{ per } g \text{ diet}$$

The minimum digestible energy requirement that should be fed to the fish is the sum of energy retained (RE) and energy lost as HE_f + HiE + ZE + UE. The *Fish-PrFEQ* software applies this procedure to compute feeding standards. The amount of diet to be fed can be estimated on a weekly or monthly basis, and recalculated if any parameter (growth rate, water temperature, etc.) is changed. The computed quantity of feed should be regarded as a minimum requirement under normal husbandry condition and minor adjustment of the feeding level may be made by fish culturists for local conditions.

Table 2.4 summarizes the monthly fish sizes and ration allowance tabulated by the *Fish-PrFEQ* program for the field station based on the actual production record (see Table 2.3). The diet requirements were calculated using a single TGC (0.191) for whole production cycle (fourteen months) and actual water temperature profile. The nutrient and energy gains used in the calculations were based on carcass composition values for rainbow trout of various sizes obtained in different laboratory trials at the University of Guelph. The main discrepancy is between the actual and predicted feed amount for the first four months with actual feed input being greater than predicted allocation. This may indicate that overfeeding occurred; however, real feed intake by the fish could be somewhere between the predicted amount and the actual input. Using this information, the fish culturists can adjust or fine-tune the feeding strategies in the next production period. In the remaining ten months, RA by the model estimated slightly higher feed requirement (e.g., 7 percent) than the actual feed input. The accuracy of the prediction can be considered acceptable and the largest discrepancies (in terms of predicted and actual) occur at very low temperatures.

Diet requirement is generally governed by how much energy, protein, fat, and minerals the animal deposits in its body and the biological cost of depositing these body components. Fish retain in their body a large proportion of the nutrients/energy fed to them. Their diet and energy requirements are, therefore, very closely related to the rate of body component accretion. Sufficient data on growth process and nutritional energetics are now available to allow reasonably accurate feeding standards to be computed for salmonid fish species as well as a number of marine fish species (Kaushik, 1998; Lupatsch and Kissil, 1998; Lupatsch et al., 1998).

Feeding Strategies

In spite of widespread feeding practice of high-fat (energy) diets for salmonids today, adjustment of old feeding charts has not followed and feed efficiency has not improved accordingly. Many salmonid aquaculture operations still entertain feed conversions (feed/gain) of nearly 1.5 (Costello et al., 1996). These situations lead not only to an increased feed cost, but also create considerable aquaculture pollution problems in effluent receiving rivers, lakes, and coastal waters.

Whatever efforts and techniques are employed to feed to appetite or near-satiety, the actual amount of feed fed under practical conditions can unknowingly be one of the five situations illustrated in Figure 2.3. Aiming at maximum gain and best feed efficiency is desirable, but practising under farming conditions is difficult and almost impossible on a daily basis even with the aid of computer programs and sophisticated feeding equipment. True daily gain and actual feed input are not known until next inventory measurements; therefore maximum gain and minimum feed conversion are mere conceptual figures in daily operations. Real feeding situation will still fall in one of five categories as illustrated in Figure 2.3. The feeding level of category 3 is the theoretical requirement and will give optimum gain and feed efficiency; however, this level in daily situation may be a "moving target." With

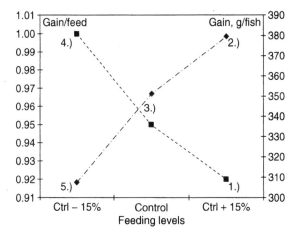

FIGURE 2.3. Effects of feeding level on gain and feed efficiency (gain/feed) of rainbow trout (10 g initial weight) fed a low nutrient-dense diet for thirty-two weeks at 15°C. *Source:* Cho and Bureau, 1998. *Note:* The figure illustrates five feeding categories: (1) Overfeeding-feed waste; (2) Upper range of optimum feeding level-maximum gain; (3) Most optimum feeding level-theoretical requirement; (4) Lower range of optimum feeding level-best feed efficiency; (5) Underfeeding and restricted feeding-lower gain.

the aid of the bioenergetics models, fish culturists can maintain the feeding levels between categories 1 and 5, and aim at near about category 3 on weekly or monthly basis. Since "ad lib" feeding in fish is not possible, the only way to supply requirements of energy and nutrients with minimal waste is a more accurate estimation of ration allowance using the nutritional energetics models and computer program.

Results from carefully conducted feeding trials in our laboratory with rainbow trout and Atlantic salmon (e.g., Azevedo et al., 1998; Bureau, 1997) suggest that feed efficiency reaches its maximum at moderate feed restriction (ca. 50 to 70 percent of near-satiation) and this optimum is maintained up to near-satiation (maximum voluntary feed intake) of the fish. Results obtained elsewhere apparently support this observation (Alanara, 1997). The hypothesis of Einen et al. (1995) that maximum feed efficiency is only attained at maximum feed intake is, therefore, not valid. It might be important to note that as the feed distributed approaches the amount corresponding to near-satiation for the fish, feed wastage may increase because of slower response of the fish to the presentation of feed pellet (Ang et al., 1996). This may result in a reduction of apparent feed efficiency (due to feed wastage), but slightly higher weight gain as observed in Figure 2.3.

Theoretical energy and diet requirement prediction models and computer software cannot replace common sense in feeding fish. The *Fish-PrFEQ* program could represent a convenient and valuable management tool to help improve husbandry practices and may provide considerable benefits if one fine-tunes the model on the basis of his or her own production records and readjustments on the basis of actual performance. Accurate growth and diet requirement prediction models can help in objectively examining one's performance by providing a yardstick with which performance can be compared and results obtained with the feeding system and practice can be validated. With nutritional energetics-based models and programs, production forecast, diet requirement, oxygen requirement, waste output can be estimated a priori. This may prove very useful for aquaculture operations when forecasting production and environmental impacts, negotiating yearly feed and oxygen supply contracts, etc.

Preallocated weekly amounts may be divided into desired number of meals each day, but each meal must be of sufficient quantity for the whole population as long as total ration fed does not exceed the quantity estimated in advance. However, ration allowance may be adjusted according to improvement of fish performance and feed efficiency. Properly sized feed should be dispensed over wide water surface by hand or mechanical devices in such a manner that feed wastage is minimized. With any feeding method, dominant fish will always eat enough feed to express their full growth potential; however, the effort made to ensure adequate feed intake of "weakling" fish may dictate the extent of feed waste. Furthermore, detection of feed waste by underwater camera may already be beyond optimal feeding level. The goal of most of the current feeding systems is fast and maximum body weight gain, and they are less concerned with feed efficiency and wastage. However, this approach is not economical, and will not promote a lasting cohabitation of sustainable aquaculture and a cleaner environment.

Fish-PrFEQ[1] Computer Program

A stand-alone multimedia computer program (*Fish-PrFEQ* = Fish Production, Feeding and Effluent Quality) for the MS Windows platform was written in MS Visual C^{++}.NET language with database functionality. The program has four modules for fish growth prediction, feeding standard/oxygen requirement, production record, and waste output estimation. It is based on the bioenergetics models presented above. Diet composition, body weight, water temperature, flow rate, and mortality are entered by the user, but waste, retention, and other coefficients are parameters that are locked and may only be revised with an authorized program update diskette. These coefficients should be determined by qualified nutritionists from feed manufacturers or research institutions, since specific coefficients are required for each type of diets and species. The use of unrelated coefficients may result in under or overestimation of feed requirement and waste output.

The various outputs are stored and printed using MS Excel so that further manipulation of the output data by users is facilitated. Live weight gain, feeding standard, oxygen need, feed efficiency, growth coefficients, total waste load, and solids, nitrogen, and phosphorus in the effluent are some of the output parameters generated by the *Fish-PrFEQ* program.

Presented previously are relatively simple steps on how to feed fish using scientific principles of the nutritional strategies and management of aquaculture waste (NSMAW). The *Fish-PrFEQ* program will make easier prediction of growth rate, allocation of feed required and estimation of waste outputs, but not necessarily accurate unless the coefficients are fine-tuned. Feeding fish using almost folkloric approaches must become something of the past. The largest portion of fish production costs (more than 40 percent) is expended on feed, and fish feed is among the highest quality and most expensive types of animal feed on the market. Dispensing this expensive commodity through outdated modes is an undeniably wasteful practice. Much more attention and time should be now devoted to feeding systems, quantitatively, rather than qualitatively, to seek better and cheaper feeds! One optimal diet formulation and its nutritional specifications and estimated waste outputs are presented in Table 2.6.

TABLE 2.6. High nutrient-dense (HND) diet formula for the Ontario Ministry of Natural Resource Feed Contract.

Formula id: MNR-95 HG	Grower for salmonids
Ingredient[a]	
Fish meal, 70% CP	18.0 kg
Brewer's dried yeast, 45% CP	6.0 kg
Corn gluten meal, 60% CP	49.0 kg
Whey, dried, 12% CP	12.0 kg
Vitamin premix: VIT-9408[+]	0.5 kg
Mineral premix: MIN-9504[+]	0.5 kg
Fish oil, marine	14.0 kg
NSMAW specifications	
Digestible energy	20 MJ/kg
Digestible protein/energy	20-23 g/MJ
Digestible fat	17-20%
Total phosphorus	<1.0%
Expected feed efficiency (gain/feed)	>1.2
Waste outputs, biologically estimated (kg/ton fish produced)	
Solid	
Total	<150

TABLE 2.6 *(continued)*

Formula id: MNR-95 HG	Grower for salmonids
Nitrogen	
Solid	7 kg
Dissolved	23 kg
Phosphorus	
Solid	3 kg
Dissolved	2 kg
Fines	1%

[+]Do not change concentration.

[a]Diet formula contains 20 MJ DE/kg; 22 g DP/DE; 17% Fat; 0.8% P.

Note: All ingredients must be ground finer than 0.2 mm; Diet must contain more than 45% crude protein and 17% crude fat; less than 6% ash and 0.8% total phosphorus.

NOTE

1. *Fish-PrFEQ* program can be obtained from jinju35@rogers.com

REFERENCES

Alanara, A. (1997). Balancing maximal production and minimum waste in fish farms: A mission impossible? In: *III International Symposium on Nutritional Strategies and Management of Aquaculture Waste,* 2-4 October 1997, Universidade de Trás-os-Montes e Alto Douro, Portugal, 17 pp. (Abstract).

Ang, K.P., R.J. Petrell, and B.E March (1996). Feeding end-points associated with different feeding methods in seacage farming of salmonids. *Bulletin of Aquaculture Association of Canada,* 96, 52-53.

Arzel, J., R. Métailler, P. Le Gall, and J. Guillaume (1998). Relationship between ration size and dietary protein level varying at the expense of carbohydrate and lipid in triploid brown trout fry *(Salmo trutta). Aquaculture,* 162, 259-268.

Asgard, T. and K. Shearer (1997). Dietary phosphorus requirement of juvenile Atlantic salmon *(Salmo salar* L). *Aquaculture Nutrition,* 3, 17-23.

Azevedo, P.A., C.Y. Cho, and D.P. Bureau (1998). Effects of feeding level and water temperature on growth, nutrient and energy utilization and waste outputs of rainbow trout *(Oncorhynchus mykiss). Aquatic Living Resources,* 11, 227-238.

Baeverfjord, G., T. Asgard, and K.D. Shearer (1998). Development and detection of phosphorus deficiency in Atlantic salmon, *Salmo salar* L., parr and post-smolts. *Aquaculture Nutrition,* 4, 1-11.

Bijvoet, O.L.M. (1980). Indices for the measurement of the renal handling of phosphate. In: *Renal Handling of Phosphate,* ed. S.G. Massry and H. Fleisch, pp. 1-37. Plenum Medical Book Company, New York, NY.

Bureau, D.P. (1997). The Partitioning of Energy from Digestible Carbohydrates by Rainbow Trout *(Oncorhynchus mykiss).* Ph.D. Thesis, University of Guelph, Guelph, Ontario, Canada, 170 pp.

Bureau, D.P. and C.Y. Cho (1999). Phosphorus utilization by rainbow trout *(Oncorhynchus mykiss):* Estimation of dissolved phosphorus output. *Aquaculture,* 179, 127-140.

Bureau, D.P., S.J. Gunther, and C.Y. Cho (2003). Chemical composition and preliminary theoretical estimates of waste outputs of rainbow trout reared in commercial cage culture operations in Ontario. *North American Journal of Aquaculture,* 65, 33-38.

Bureau, D.P., J.B. Kirkland, and C.Y. Cho (1998). The partitioning of energy from digestible carbohydrate by rainbow trout *(Oncorhynchus mykiss).* In: *Energy Metabolism of Farm Animals,* ed. K.J. McCracken, E.F. Unsworth and A.R.G. Wylie, pp. 163-166. CAB International Press, Wallingford.

Buterbaugh, G.L. and H. Willoughby (1967). A feeding guide for brook, brown and rainbow trout. *Progressive Fish Culturists,* 29, 210-215.

Cho, C.Y. (1990). Fish nutrition, feeds, and feeding with special emphasis on salmonid aquaculture. *Food Review International,* 6, 333-357.

Cho, C.Y. (1992). Feeding systems for rainbow trout and other salmonids with reference to current estimates of energy and protein requirements. *Aquaculture,* 100, 107-123.

Cho, C.Y., H.S. Bayley, and S.J. Slinger (1976). Energy metabolism in growing rainbow trout: Partition of dietary energy in high protein and high fat diets. In: *Energy Metabolism of Farm Animals,* ed. M. Vermorel, pp. 299-302. Proceedings of the 7th Symposium on Energy Metabolism, Vichy. EAAP Publication No.19, G. de Bussac, Clermont-Ferrand.

Cho, C.Y. and D.P. Bureau (1997). Reduction of waste output from salmonid aquaculture through feeds and feeding. *Progressive Fish Culturists,* 59, 155-160.

Cho, C.Y. and D.P. Bureau (1998). Development of bioenergetics models and the Fish-PrFEQ software to estimate production, feeding ration and waste output in aquaculture. *Aquatic Living Resources,* 11 (4), 199-210.

Cho, C.Y., C.B. Cowey, and T. Watanabe (1985). Finfish nutrition in Asia. In: *Methodological Approaches to Research and Development.* Publication No. IDRC-233e, International Development Research Centre, Ottawa, 154 pp.

Cho, C.Y., J.D. Hynes, K.R. Wood, and H.K. Yoshida (1991). Quantitation of fish culture wastes by biological (nutritional) and chemical (limnological) methods; the development of high nutrient dense (HND) diets. In: *Nutritional Strategies and Aquaculture Waste,* ed. C.B. Cowey and C.Y. Cho, pp. 37-50. Proceedings of the 1st International Symposium on Nutritional Strategies in Management of Aquaculture Waste, Guelph, Ontario, 275 pp.

Cho, C.Y., J.D. Hynes, K.R. Wood, and H.K. Yoshida (1994). Development of high nutrient-dense, low pollution diets and prediction of aquaculture wastes using biological approaches. *Aquaculture,* 124, 293-305.

Cho, C.Y. and S.J. Kaushik (1990). Nutritional energetics in fish: Energy and protein utilization in rainbow trout *(Salmo gairdneri).* In: *Aspects of Food Production, Consumption and Energy Values,* ed. G.H. Bourne, Vol. 61, pp. 132-172. World Review of Nutrition and Dietetics, Basel, Karger.

Cho, C.Y. and S.J. Slinger (1980). Effect of water temperature on energy utilization in rainbow trout *(Salmo gairdneri).* In: *Energy Metabolism,* ed. L.E. Mount, pp. 287-291. EAAP Publication No. 26, Butterworths, London.

Cho, C.Y. and W.D. Woodward (1989). Studies on the protein-to-energy ratio in diets for rainbow trout *(Salmo gairdneri).* In: *Energy Metabolism of Farm Animals,* ed. Y. Van der Honing and W.H. Close, pp. 37-40. Proceedings of the 11th symposium, Lunteren, Netherlands, 18-24 September 1988. EAAP Publication No. 43, 1989, Pudoc Wageningen, The Netherlands.

Costello, M.J., D.T.G. Quigley, and S. Dempsey (1996). Seasonal changes in food conversion ratio as an indicator of fish feeding management. *Bulletin of Aquaculture Society of Canada,* 96, 58-60.

Cowey, C.B. and C.Y. Cho (1993). Nutritional requirements of fish. *Proceedings of the Nutrition Society,* 52, 417-426.

Dantzler, W.H. (1989). Comparative physiology of the vertebrate Kidney. In: *Zoophysiology,* Vol. 22. Springer-Verlag, Berlin.

Deuel, C.R., D.C. Haskell, D.R. Brockway, and O.R. Kingsbury (1952). *New York State Fish Hatchery Feeding Chart,* 3rd Edition. N.Y. Conservation Department, Albany, NY.

Einen, O., I. Holmefjord, T. Asgard, and C. Talbot (1995). Auditing nutrient discharges from fish farms: Theoretical and practical considerations. *Aquaculture Research,* 26, 701-713.

Einen, O. and A.J. Roem (1997). Dietary protein/energy ratios for Atlantic salmon in relation to fish size: Growth, feed utilization and slaughter quality. *Aquaculture Nutrition,* 3, 115-126.

Emmans, G.C. (1994). Effective energy: A concept of energy utilization applied across species. *British Journal of Nutrition,* 71, 801-821.

Forster, I., D.A. Higgs, B.S. Donsanjh, M. Rowshandeli, and J. Parr (1999). Potential for dietary phytase to improve the nutritive value of canola protein concentrate and decrease phosphorus output in rainbow trout *(Oncorhynchus mykiss)* held in 11 degrees C fresh water. *Aquaculture,* 179, 109-125.

Freeman, R.I., D.C. Haskell, D.L. Longacre, and E.W. Stiles (1967). Calculations of amounts to feed in trout hatcheries. *Progressive Fish Culturists,* 29, 194-215.

Gowen, R.J., D.P. Weston, and A. Ervik (1991). Aquaculture and the benthic environment: A review. In: *Nutritional Strategies and Aquaculture Waste,* ed. C.B. Cowey and C.Y. Cho, pp. 187-205. Proceedings of the First International Symposium on Nutritional Strategies in Management of Aquaculture Waste, Guelph, Ontario, 275 pp.

Guillaume, J., S.J. Kaushik, P. Bergot, and R. Métailler (1999). *Nutrition et alimentation des poissons et des crustacés.* Éditions INRA, Paris.

Haskell, D.C. (1959). Trout growth in hatcheries. *New York Fish and Game Journal,* 6, 204-237.

Helland, S.J. and B. Grisdale-Helland (1998a). The influence of dietary carbohydrate and protein levels on energy and nitrogen utilization of Atlantic salmon in seawater. In: *Energy Metabolism of Farm Animals,* ed. K.J. McCracken, E.F. Unsworth, and A.R.G. Wylie, pp. 391-394. CAB International Press, Wallingford.

Helland, S.J. and B. Grisdale-Helland (1998b). The influence of replacing fish meal in the diet with fish oil on growth, feed utilization and body composition of Atlantic salmon *(Salmo salar)* during the smoltification period. *Aquaculture,* 162, 1-10.

Hillestad, M., F. Johnsen, E. Austreng, and T. Asgard (1998). Long-term effects of dietary fat level and feeding rate on growth, feed utilization and carcass quality of Atlantic salmon. *Aquaculture Nutrition,* 4, 89-97.

Iwama, G.K. and A.F. Tautz (1981). A simple growth model for salmonids in hatcheries. *Canadian Journal Fisheries and Aquatic Sciences,* 38, 649-656.

Johnsen, F. and A. Wandsvik (1991). The impact of high energy diets on pollution control in the fish farming industry. In: *Nutritional Strategies and Aquaculture Waste,* ed. C.B. Cowey and C.Y. Cho, pp. 51-64. Proceedings of the 1st International Symposium on Nutritional Strategies in Management of Aquaculture Waste, Guelph, Ontario, 275 pp.

Kaune, R. and H. Hentschel (1987). Stimulation of renal phosphate secretion in the stenohaline freshwater teleost: *Carassius auratus Gibelio* Bloch. *Comparative Biochemistry and Physiology,* 87A, 359-362.

Kaushik, S.J. (1994). Nutritional strategies for the reduction of aquaculture wastes. In: *Proceedings of FOID '94,* ed. K.D. Cho, pp. 115-132. The Third International Conference on Fisheries and Ocean Industrial Development for Productivity Enhancement of the Coastal Waters, 3-4 June 1994, National Fisheries University of Pusan, Pusan.

Kaushik, S.J. (1998). Nutritional bioenergetics and estimation of waste production in non-salmonids. *Aquatic Living Resources,* 11, 211-217.

Kaushik, S.J. and C.B. Cowey (1991). Dietary factors affecting nitrogen excretion by fish. In: *Nutritional Strategies and Aquaculture Waste,* ed. C.B. Cowey and C.Y. Cho, pp. 3-19. Proceedings of the 1st International Symposium on Nutritional Strategies in Management of Aquaculture Waste, Guelph, Ontario, 275 pp.

Kaushik, S.J. and E.F. Gomes (1988). Effect of frequency of feeding on nitrogen and energy balance in rainbow trout under maintenance conditions. *Aquaculture,* 73, 207-216.

Kaushik, S.J. and A. Oliva-Teles (1985). Effects of digestible energy on nitrogen and energy balance in rainbow trout. *Aquaculture,* 50, 89-101.

Kleiber, M. (1975). The fire of life. In: *An Introduction to Animal Energetics.* Robert E. Krieger Publ. Co., Huntington, NY, 453 pp.

Lall, S.P. (1991). Digestibility, metabolism and excretion of dietary phosphorus in fish. In: *Nutritional Strategies and Aquaculture Waste,* ed. C.B. Cowey and C.Y. Cho, pp. 21-36. Proceedings of the 1st International Symposium on Nutritional Strategies in Management of Aquaculture Waste, Guelph, Ontario, 275 pp.

Lanari, D., E. D'Agaro, and R. Ballestrazzi (1995). Effect of dietary DP/DE ratio on apparent digestibility, growth and nitrogen and phosphorus retention in rainbow trout, *Oncorhynchus mykiss* (Walbaum). *Aquaculture Nutrition,* 1, 105-110.

Lee, D.J. and G.B. Putnam (1973). The response of rainbow trout to varying protein/ energy ratios in a test diet. *Journal of Nutrition,* 103, 916-922.

Lloyd, L.E., B.E. McDonald, and E.W. Crampton (1978). *Fundamentals of Nutrition, Second Edition.* W.H. Freeman and Company, San Francisco, CA, 466 pp.

Lupatsch, I. and G.Wm. Kissil (1998). Predicting aquaculture waste from gilthead seabream (*Sparus aurata*) culture using a nutritional approach. *Aquatic Living Resources,* 11, 265-268.

Lupatsch, I., G.Wm. Kissil, D. Sklan, and E. Pfeffer (1998). Energy and protein requirements for maintenance and growth in gilthead seabream (*Sparus aurata* L.). *Aquaculture Nutrition,* 4, 165-173.

National Research Council, NRC (1993). *Nutrient Requirements of Fish.* National Academy Press, Washington, DC.

Ohta, M. and T. Watanabe (1998). Effect of feed preparation methods on dietary energy budgets in carp and rainbow trout. *Fisheries Science,* 64, 99-114.

Oliva-Teles, A., J.P. Pereira, A. Gouveia, and E. Gomes (1998). Utilisation of diets supplemented with microbial phytase by seabass *(Dicentrarchus labrax)* juveniles. *Aquatic Living Resources,* 11, 255-259.

Persson, G. (1991). Eutrophication resulting from salmonid fish culture in fresh and salt waters: Scandinavian experience. In: *Nutritional Strategies and Aquaculture Waste,* ed. C.B. Cowey and C.Y. Cho, pp. 163-185. Proceedings of the 1st International Symposium on Nutritional Strategies in Management of Aquaculture Waste, Guelph, Ontario, 275 pp.

Renfro, J.L. (1997). Hormonal regulation of renal inorganic phosphate transport in the winter flounder, *Pleuronectes americanus. Fish Physiology and Biochemistry,* 17, 377-383.

Rodehutscord, M. (1996). Response of rainbow trout *(Oncorhynchus mykiss)* growing from 50 to 200 g to supplements of dibasic sodium phosphate in a semipurified diet. *Journal of Nutrition,* 126, 324-331.

Rodehutscord, M., F. Borchert, Z. Gregus, M. Pack, and E. Pfeffer (2000a). Availability and utilisation of free lysine in rainbow trout *(Oncorhynchus mykiss).* 1. Effect of dietary crude protein level. *Aquaculture,* 187, 163-176.

Rodehutscord, M., Z. Gregus, and E. Pfeffer (2000b). Availability of phosphorus to rainbow trout Oncorhynchus mykiss. 1. Methodological considerations. In: *IX International Symposium on Nutrition and Feeding of Fish,* 21-25 May 2000, Miyazaki. (Abstract O52).

Rodehutscord, M. and E. Pfeffer (1995). Effects of supplemental microbial phytase on phosphorus digestibility and utilization in rainbow trout *(Oncorhynchus mykiss). Water Science Technology,* 31, 143-147.

Rodehutscord, M. and E. Pfeffer (1999). Maintenance requirement for digestible energy and efficiency of utilisation of digestible energy for retention in rainbow trout, *Oncorhynchus mykiss. Aquaculture,* 179, 95-107.

Santinha, P.J.M., F. Médale, G. Corraze, and E. Gomes (1999). Effects of the dietary protein: Lipid ratio on growth and nutrient utilization in gilthead seabream *(Sparus aurata* L.). *Aquaculture Nutrition,* 5, 147-156.

Satoh, S., M. Takanezawa, and T. Watanabe (1998). Changes of phosphorus absorption from several feed ingredients in rainbow trout during growing stages. *VIII International Symposium on Nutrition and Feeding of Fish,* 1-4 June 1998, Las Palmas, 136 pp. (Abstract).

Shearer, K.D. (1994). Factors affecting the proximate composition of cultured fishes with emphasis on salmonids. *Aquaculture,* 11, 63-88.

Steffens, W., B. Rennert, M. Wirth, and R. Krüger (1999). Effect of two lipid levels on growth, feed utilization, body composition and some biochemical parameters of rainbow trout, *Oncorhynchus mykiss* (Walbaum 1792). *Journal of Applied Ichthyology,* 15, 159-164.

Stickney, R.R. (1979). *Principles of Warmwater Aquaculture.* John Wiley & Sons, New York, NY, 375 pp.

Takeuchi, T., T. Watanabe, and C. Ogino (1978). Optimum ratio of protein to lipid in diets of rainbow trout. *Bulletin of the Japanese Society Science Fisheries,* 46, 683-688.

Vielma, J. and S.P. Lall (1998). Control of phosphorus homeostasis of Atlantic salmon *(Salmo salar)* in fresh water. *Fish Physiology and Biochemistry,* 19, 83-93.

Vielma, J., S.P. Lall, J. Koskela, F.J. Schöner, and P. Mattila (1998). Effects of dietary phytase and cholecalciferol on phosphorus bioavailability in rainbow trout *(Oncorhynchus mykiss). Aquaculture,* 163, 309-323.

Vielma, J., T. Makinen, P. Ekholm, and J. Koskela (2000). Influence of dietary soy and phytase levels on performance and body composition of large rainbow trout *(Oncorhynchus mykiss)* and algal availability of phosphorus load. *Aquaculture,* 183, 349-362.

Watanabe, T. (1977). Sparing action of lipids on dietary protein in fish—Low protein diet with high calorie content. *Technocrat,* 10, 34-39.

Watanabe, T. and M. Ohta (1995). Endogenous nitrogen excretion and nonfecal energy losses in carp and rainbow trout. *Fisheries Sciences,* 61, 53-60.

Watanabe, T., T. Takeuchi, and C. Ogino (1979). Studies on the sparing effect of lipids on dietary protein in rainbow trout *(Salmo gairdneri).* In: *Proceedings of the World Symposium on Finfish Nutrition and Fishfeed Technology,* ed. J.E. Halver and K. Tiews, Vol. I, pp. 113-125. Schriften der Bundesforschungsanstalt für Fischerei, Hamburg.

Chapter 3

Replacement of Fish Meal with Poultry By-Product Meal and Hydrolyzed Feather Meal in Feeds for Finfish

Yu Yu

INTRODUCTION

Global aquaculture has been growing consistently at least 10 percent annually during the past decade, particularly in Asia (Tacon and Hardy, 2000; Hardy, 2003). The key ingredient for carnivorous and omnivorous diets has been fish meal (FM), which has been mainly produced in and exported from Peru and Chili (Hardy, 2003). International experts on sustainable aquaculture have warned of the risk of too heavy a dependence on FM because of: (1) stagnate supply; (2) prohibited price; and (3) low efficiency of fish stock utilization (Hardy, 1996, 2003; Naylor et al., 2003; Pauly et al., 2002). The aquafeed industry and fish nutritionists for the past twenty years have investigated suitable replacements for FM. For carnivorous species, FM replacements of animal origin are generally preferred to those from plant sources due mainly to palatability. Animal-origin protein meals have several other advantages over plant meals, as they are generally free from antinutritional factors; have no phytic acid phosphorus (P); do not contain indigestible complex carbohydrates; usually do not involve genetically modified organisms (GMO); are low in carbohydrates; have a high percentage of crude protein (CP); and are generally good source of energy (crude lipids), some vitamins

Alternative Protein Sources in Aquaculture Diets
© 2008 by The Haworth Press, Taylor & Francis Group. All rights reserved.
doi:10.1300/5892_03

(e.g., B_{12}), and trace minerals such as iron, cobalt, and selenium (Tacon and Hardy, 2000; Francis et al., 2001; NRA, 2003). The present chapter provides a review of the relative value of poultry by-product meal (PBM, IFN 5-03-798) and hydrolyzed feather meal (FeM, IFN 5-03-795) as FM replacement in finfish diets.

USE OF POULTRY BY-PRODUCT MEAL (PBM)

Gross Composition

According to American Feed Control Officials (AFCO, 2003) "PBM consists of the ground, rendered, clean parts of the carcass of slaughtered poultry, such as necks, feet, undeveloped eggs, and intestines, exclusive of feathers, except in such amounts as might occur unavoidably in good processing practices." Typical rendering consists of heat treatment at a temperature of 120°C for forty-five minutes. Practically all commonly known bacteria and virus harmful to poultry and human are destroyed at the completion of rendering process (Firman et al., 2004).

Proximate and amino acid composition of PBM are given in Table 3.1. Owing to the fact that various grades are available for the feed industry (e.g., feed-grade, pet-food grade, refined, low-ash, flash-dry, and defatted), data in Table 3.1 should be used only as references, and not as recipes for diet formulation. Diet formulators must use the actual analyzed value from suppliers of diet formulation to avoid variation in nutrient content and performance of fish (Hardy and Cheng, 2002). Generally, nutritional value of PBM (e.g., digestibility) is inversely related to the ash content (Hardy and Cheng, 2002), but so is the price. Reduction in ash content is mainly through raw material segregation and is unaffected by freshness of raw materials or the processing condition. However, recent advancement in processing technology, such as flash drying and specific enzyme treatment, allows the improvements in feeding value of poultry by-products in addition to that obtained from raw material segregation (Kureshy et al., 2002). These newly developed treatments substantially improve the quality of PBM to a level close to that of FM (Mendoza et al., 2001; Kureshy et al., 2002). Poultry by-product meal produced in North America is generally consistent in freshness, quality, and digestibility. The aqua-feed

TABLE 3.1. Proximate and amino acid composition (% as is) of PBM, FeM, and FM.[a]

	PBM[b]	FeM[c]	FM[d]
Proximate composition			
Crude protein	58-65	80	64.6
Crude fat	12	6	7.9
Calcium	4	0.33	3.93
Phosphorus	2	0.5	2.55
Ash	10-18	2.8	16
Gross energy (kcal/kg)	4,900	5,400	4,500
Amino acid composition			
Arginine	3.94	6.5	3.68
Histidine	1.25	1.7	1.56
Isoleucine	2.01	4.12	3.06
Leucine	3.89	7.12	5
Lysine	3.32	2.1	5.11
Methionine	1.11	0.53	1.95
Phenylalanine	2.26	4.3	2.66
Threonine	2.18	4.15	2.82
Tryptophan	0.48	0.3	0.76
Valine	2.51	5.5	3.51
Cystine	0.66	4.13	0.61
Tyrosine	1.56	2.41	2.15

[a]NRC (1998) and National Renderers Association (2003).

[b]Poultry by-product meal.

[c]Hydrolyzed feather meal.

[d]Fish meal (anchovy meal, mechanically extracted).

industry could use the ash content as the general guide in selecting PBM for different groups of finfishes. For example, low-ash PBM should be used primarily for carnivorous fish, particularly high-value marine fish.

Poultry by-product meal is similar to FM in nutrient composition, but is slightly lower in some essential amino acids (Table 3.1). This would suggest that: (1) efficiency of protein utilization of PBM could be marginally lower than that of FM; (2) it could replace a high proportion of FM in fish diets, but may not be able to completely replace FM without impairing performance; (3) supplementation of limiting essential amino acids with crystalline amino acids may improve the nutritional value of PBM in aqua-feeds; (4) blending of PBM with other high-quality protein meals to meet the amino acid requirement of specific groups of finfish may be a viable option for FM substitution; and (5) the ash content in both PBM and FM may ultimately limit their inclusion rate in low pollution diets.

Apparent Protein Digestibility (APD) and Digestible Energy (DE)

Apparent digestibility of protein and DE measured for various finfishes are given in Table 3.2. For comparison purposes, digestibility values of FM are also listed. Protein digestibility of PBM reported in

TABLE 3.2. Apparent protein and energy digestibility (%) of poultry by-product meal, hydrolyzed feather meal, and fish meal by various species of finfishes.

Species and Reference	PBM[a]		FeM[b]		FM[c]	
	Protein	Energy	Protein	Energy	Protein	Energy
Rainbow trout[d]	65.3	74.3	51.6	58.7	86.4	86.1
Rainbow trout[e]	82.0	80.0	84.0	85.0		
Rainbow trout[f]	89.0	82.0	82.5	78.5		
Rainbow trout[g]	85.0	83.4			87.8	86.1
Rainbow trout[h]	95.9		85.9		92.7	
Rainbow trout[i]	82.5	74.5	76.6	76.9		
Australian silver perch[j]	85.4	93.7	92.8	101	89.0	88.6
Rockfish[k]			71.0	79.0	93.0	94.0
Japanese sea bass[l]	84.5	82.8			91.6	89.2
Red drum[m]	48.7	71.7			82.4	93.6
Hybrid sturgeon[n]	64.7	52.3			81.0	64.5
Carp[o]	47.2	63.9			83.8	93.4
Nile tilapia[p]	73.9	58.8			86.5	79.8

TABLE 3.2 *(continued)*

Species and Reference	PBM[a]		FeM[b]		FM[c]	
	Protein	Energy	Protein	Energy	Protein	Energy
Hybrid tilapia[q]	91.6	87.2			90.3	90.41
Gilthead seabream[r]	80.0	79.0			83.0	80.0
Coho salmon[s]	94.2		79.7		91.4	
Mean	78 (89)[t]	76 (88)	78 (89)	80 (93)	87.6	86
Range	47.2-95.9	58.8-93.7	51.6-92.8	58.7-101	81-93	79.8-84
North American PBM (trout)	88 (98)[t]	82 (95)	84 (93)	82 (95)	90	86

[a]Poultry by-product meal.

[b]Hydrolyzed feather meal.

[c]Fish meal (mostly herring meal).

[d]Smith et al. (1995).

[e]Pfeffer et al. (1995).

[f]Bureau et al. (1999).

[g]Hardy and Cheng (2002).

[h]Suriura et al. (1998).

[i]Cheng et al. (2004).

[j]Allan et al. (2000).

[k]Lee (2002).

[l]Xue et al. (2001).

[m]Gaylord and Gatlin (1996).

[n]Degani (2002).

[o]Degani et al. (1997a).

[p]Hanley (1987).

[q]Degani et al. (1997b).

[r]Lupatsch et al. (1997).

[s]Sugiursa et al. (1998).

[t]Value inside parenthesis is percent of the fish meal value.

several early studies (Cho and Slinger, 1978; Dong et al., 1993; Gaylord and Gatlin, 1996) were low (47.2 to 65.3 percent) as compared to a mean digestibility of 88 percent from recent measurements for rainbow trout *(Oncorhynchus mykiss)* (Bureau et al., 1999; Cheng and Hardy, 2002; Cheng et al., 2004). Possible reasons for the discrepancy are: (1) differences in raw material compositions, quality, and processing conditions; (2) differences in techniques used for collection of fecal samples; (3) differences in test diet formulations; (4) different ages of the experimental fish; and (5) species difference. Protein digestibility varies from 47 percent (carp) to 96 percent (rainbow trout) with a mean of 78 percent. The corresponding value for FM, as obtained from the same trials was 87.6 percent, with a range of 81 to 93 percent (rockfish, *Sebastes schlegeli*). When FM protein digestibility is standardized to 100 percent for each of the trials cited in Table 3.2, the relative digestibility for PBM is 89 percent or approximately 10 percent lower than that of FM. However, if one considers only recent data obtained with rainbow trout, the average APD was 87 or 96 percent of that of FM value.

Apparent energy digestibility of PBM is somewhat lower than that of FM (76 percent versus 86 percent, Table 3.2). Possible reasons are: (1) lower digestibility of poultry fat compared to fish oil due to the difference in degree of unsaturation of the fatty acids and polyunsaturated fatty acid (PUFA) content and (2) difference in the ash content. The average DE of the five recent trials was 82 percent, or 95 percent of that of FM. It appears that high-quality PBM has the closest proximate composition and digestibility of protein and energy to those of FM when compared to most diet ingredients commonly used by the aqua-feed industry; this further supports the feasibility for PBM to replace a high proportion of FM in finfish diets.

Apparent Digestibility of Essential Amino Acids (DEAA)

For PBM and other protein ingredients, amino acid digestibility has not been measured as frequently as that of CP or energy. Data presented in Table 3.3 are measured from four species (rainbow trout, Australian silver perch, gilthead sea bream, and rockfish) and should be used only as reference values for other finfish. The EAAs of PBM are generally well digested by rainbow trout, Australian Silver perch *(Bidyanus bidyanus),* and gilthead sea bream *(Sparus aurata L.)* with

TABLE 3.3. Apparent essential amino acids digestibility of poultry by-product meal, hydrolyzed feather meal, and fish meal by various species of finfishes.

	Apparent digestibility (%)												
	Arg	His	Ile	Leu	Lys	Met	Phe	Thr	Trp	Val	Cys	Tyr	Mean
PBM [a]													
Rainbow trout [b]	86.7	89.4	83.6	86.5	91.6	94.7	85.4	84.6	97.0	84.1	56.0	90.0	85.8
Australian silver perch [c]	88.5	90.7	85.0	86.9	89.3	92.1	85.5	87.8	–	84.5	91.0	86.0	87.9
Gilthead seabream [d]	91.0	80.0	82.0	84.0	86.0	91.0	82.0	89.0	–	83.0	86.0	86.0	85.5
Mean	88.7	86.7	83.5	85.8	89	92.6	84.3	87.1	97.0	83.9	78.0	87.0	87.0
Relative to FM value [e]	(95)	(94)	(92)	(91)	(95)	(98)	(92)	(94)	(130)	(90)	(89)	(93)	(96)
FeM [f]													
Rainbow trout [g]	49.6	74.0	59.3	78.5	69.4	74.6	65.2	48.0	–	44.7	–	69	63.2
Australian silver perch [c]	96.1	92.9	93.5	93.7	89.6	96.3	94.4	92.8	–	92.8	92	91	93.2
Rockfish [h]	74.0	65.0	71.0	68.0	76.0	96.0	71.0	66.0	–	66.0	44	85	71.1
Mean	73.2	77.3	74.6	80.1	78.3	89.0	76.9	68.9		67.8	68	82	76.0
Relative to FM value [e]	(79)	(84)	(82)	(85)	(84)	(94)	(84)	(74)		(73)	(77)	(87)	(84)
FM [e]													
Rainbow trout [b]	93.5	91.7	90.2	92.2	94.7	94.7	88.2	90.2	74.5	91.2	82	95	89.8

TABLE 3.3 *(continued)*

	Apparent digestibility (%)												
	Arg	**His**	**Ile**	**Leu**	**Lys**	**Met**	**Phe**	**Thr**	**Trp**	**Val**	**Cys**	**Tyr**	**Mean**
Australia silver perch[c]	91.3	95.6	93.3	93.8	95.2	91.2	91.8	92.9	—	92.9	88	93	92.6
Rockfish[h]	93.0	88.0	89.0	98.0	90.0	98.0	95.0	96.0	—	96.0	94	93	93.6
Mean	92.6	91.8	90.8	94.7	93.3	94.6	91.7	93.0	74.5	93.4	88	94	91.0

[a]Poultry by-product meal.
[b]Hardy and Cheng (2002).
[c]Allan et al. (2000).
[d]Lupatsch et al. (1997).
[e]Fish meal.
[f]Hydrolyzed feather meal.
[g]Bureau et al. (1998).
[h]Lee (2002).

an average value of 87 percent, which is 96 percent of that of FM. The nutritional differences between PBM and FM, therefore, lie mainly with EAA content rather than digestibility. A good example is the refined PBM used in a recent rainbow trout study (Hardy and Cheng, 2002). It had an ash content of 13 percent (versus 16 percent for FM), lysine of 4.58 percent (versus 5.1 percent for FM), and lysine digestibility of 92.6 percent (versus 94.7 percent for FM). Indeed, low-ash PBM perhaps is the single terrestrial animal-protein ingredient that most closely resembles FM in nutritional composition and digestibility in fish, and has the potential of replacing FM totally in finfish diets. Digestible protein, energy, and EAA of PBM and FM are given in Table 3.4. Poultry by-product meal is about 4 percent higher in DE, but 14 to 40 percent lower in digestible EAA than FM.

TABLE 3.4. Digestible protein, amino acids, and energy content in poultry by-product meal, hydrolyzed feather meal, and fish meal by various species of finfishes.[a]

	Poultry by-product meal PBM	FeM	FM
Digestible crude protein (%)	51-57	71	57
Digestible energy (kcal/kg)	4,018	4,428	3,870
Digestible amino acids			
Arginine	3.5[b] (6.1)[c]	4.8 (6.7)	3.4 (6.0)
Histidine	1.1 (1.9)	1.3 (1.9)	1.4 (2.5)
Isoleucine	1.7 (3.0)	3.1 (4.3)	2.8 (4.9)
Leucine	3.3 (5.9)	5.7 (8.0)	4.7 (8.3)
Lysine	3.0 (5.2)	1.6 (2.3)	4.8 (8.4)
Methionine	1.0 (1.8)	0.5 (0.7)	1.9 (3.2)
Phenylalanine	1.9 (3.4)	3.3 (4.7)	2.4 (4.3)
Threonine	1.9 (3.3)	2.9 (4.0)	3.0 (5.3)
Trytophan	0.5 (0.8)	0.2 (.34)	0.6 (1.0)
Valine	2.1 (3.7)	3.7 (5.3)	3.3 (5.8)
Cystine	0.5 (0.9)	2.8 (4.0)	0.5 (1.0)
Tyrosine	1.4 (2.4)	2.0 (2.8)	2.0 (3.5)

[a]Composition data from Table 3.1 Mean × digestibility coefficient from Table 3.2 (North American rainbow trout data).

[b]Percentage as is.

[c]Percentage of crude protein.

DEAA Profile versus Fish EAA Requirements

Table 3.5 shows the DEAA from PBM in comparison with EAA requirements for major finfishes. The order of limiting EAA in PBM for selected finfish is sulfur-amino acids (methionine plus cystine) followed by threonine, isoleucine, histidine, tryptophan, and lysine. It appears that PBM could replace 100 percent of the FM in diets for salmonid and channel catfish *(lctalurus punctatus),* but with some deficiency (3 to 44 percent) for gilthead sea bream, sea bass *(Dicentrarchus labrax),* Nile tilapia *(Oreochromis niloticus),* and milkfish *(Chanos chanos),* and would require supplementation of several EAA for common carp *(Cyprinus carpio)* diets. Using the ratio of the most limiting EAA in PBM to the EAA requirements, one can calculate the maximum FM replacement rate by PBM in diets for carp (71 percent), Nile tilapia (80 percent), milkfish (56 percent), gilthead sea bream (67 percent), and sea bass (63 percent).

TABLE 3.5. A comparison between digestible essential amino acids of poultry by-product meal (PBM) and the essential amino acid requirements of different species of finfishes (% of protein).[a]

Amino acids	PBM[b]	Salmonids	Catfish	Carp	Tilapia	Milkfish	Seabream	Seabass
Arg	6.1	4.2	4.3	4.4	4.1	5.6	–	–
His	1.9	1.6	1.5	2.4[c]	1.7	2.0	–	–
Ile	3.0	2.0	2.3	3.0	3.1	4.0	–	–
Leu	5.9	3.6	3.5	4.7	3.4	5.1	–	–
Lys	5.2	4.8	5.0	6.0	4.6	4.0	5.0	4.8
Thr	3.3	2	2.1	4.2	3.8	4.9	–	–
Trp	0.8	0.6	0.5	0.8	1.0	0.6	0.6	–
Val	3.7	2.2	3.0	4.1	2.8	3.0	–	–
Met+	2.7	2.4	2.3	3.5	3.2	4.8	4.0	4.4
Phe+	5.8	5.3	4.8	8.2	5.6	5.2	–	–

[a]Bureau (2000).

[b]Data taken from Table 3.4.

[c]Amino acids with italicized values are considered deficient in meeting requirements for that specific species.

Feeding Trial Responses

Results of various feeding trials using common carp, Nile tilapia, channel catfish, fall chinook salmon *(Oncorhynchus tshawytscha)*, rainbow trout, sunshine bass *(Morone chrysops × M. saxatilis)*, palmetto bass *(Morone saxatilis × M. chrysops)*, silver perch, Australian snapper *(Pagrus auratus)*, sea bream, red sea bream *(Pagrus major)*, grouper *(Epinephelus coioides)*, cuneate drum *(Nibea miichthioides)*, and Black Sea turbot *(Scophthalmus mueoticus)*, to measure the performance response to FM substitution with PBM were reviewed (Winfree and Stickney, 1984; Alexis et al., 1985; Quartararo et al., 1998; Fowler, 1991; Gallagher and LaDouceur, 1995; Steffens, 1998; Webster et al., 1999; Bureau, 2000; Chai, 2000; Cui et al., 2002; Takagie et al., 2000; Webster et al., 2000; Abdel-Warith et al., 2001; Hardy and Cheng, 2002; Kureshy et al., 2002; Li et al., 2002; Lin and Yu, 2002; Liu et al., 2002; Wang et al., 2002; Yang et al., 2002; Hao and Yu, 2003; Wang, 2004; Tan et al., 2005; Turker et al., 2005). Information provided includes FM replacement rate, initial weight (IW), weight gain (WG) or specific growth rate, feed intake (FI, g/fish), feed conversion ratio (FCR), trial duration, and body composition (moisture, CP, lipid, and ash) (Table 3.6).

All data in Table 3.6 were transformed to relative values as percent of FM control. Based on the WG response relative to the WG of FM control, major response groups by fish species are arbitrarily established, namely positive (P), equal to FM (E), mildly negative (MN), and negative (N) (Table 3.7). Among all trials evaluated, only a trial conducted in China with gibel carp *(Carassius auratus gibelio)* showed the positive response in WG (+30 percent), and FCR (−15 percent) as dietary FM was replaced by PBM at various rates to the maximum of 100 percent (Yang et al., 2002). On the other hand, reports from Quartararo et al. (1998), Chai (2000), and Turket et al. (2005), suggested that Australian snapper, grouper, and black sea turbot are not suitable for high rates (>50 percent) of FM substitution with PBM. The poor performance of grouper reported by Chai (2000) could be related to the quality and/or the source of PBM. Tan et al. (2005) recently observed similar WG for grouper fed diets with PBM replacing 60 percent of FM compared to the WG of FM control. The poor growth reported for black sea turbot (Turket et al., 2005) could be related to the quality

TABLE 3.6. Response of finfish to fish meal substitution with poultry by-product meal in growth and body composition.

FM replacement rate (%)	Growth			FI (g/fish)	FCR	Body composition (%)[1]				Reference
	IW(g)	SGR	WG			Moisture	CP	Lipid	Ash	
Coho salmon (40% Herring meal)[2]	4.1	1.10	22.5	24	1.1	74.8	17.5	5.5	2.1	Higgs et al. (1979) (168 days)[3]
32	4.4	1.07	22.8	25	1.1	74.4	17.4	5.5	2.2	
66	4.3	1.09	23.2	28	1.2	74.6	18.0	5.0	2.2	
100	4.2	1.06	21.7	28	1.3	75.2	17.6	4.6	2.4	
Fall chinook salmon (37.5% Herring meal)	1.6	2.28[a]	37.6[ab]	41	1.1[a]	73.4	16.7[a]	8.4[b]	2.1[a]	Fowler (1991) (140 days)
25	1.6	2.30[a]	39.0[a]	41	1.1[a]	72.8	16.7[a]	9.0[a]	2.1[a]	
50	1.6	2.25[ab]	36.3[b]	41	1.1[a]	73.1	16.6[a]	9.0[a]	1.9[b]	
75	1.6	2.20[b]	33.7[c]	40	1.2[b]	73.2	16.2[b]	9.3[a]	1.9[b]	
Rainbow trout										
1 (57% Herring meal)	20		227	404	1.8	71.8	16.5	9.4	2.1	Alexis et al. (1985) (197 days)
44	20		259	458	1.8	69.9	16.4	11.3	2.0	
53	20		286	475	1.7	68.7	17.3	12.0	1.8	
61	20		201	364	1.8	70.5	17.2	10.8	1.9	
2 (40% Herring meal)	35	0.26	243	192	0.8					Bureau et al. (2000) (84 days)
40	35	0.24	229	190	0.8					
3 (35% Herring meal)	13.9		37.5[abc]	36	1.0[a]	72.6[ab]	15.1[a]	9.8[a]	2.4[a]	Hardy and Cheng (2002) (42 days)
25	14.0		39.0[bc]	39	1.0[a]	72.2[a]	14.7[b]	10.2[a]	2.6[a]	
50	13.9		37.9[ab]	40	1.1[a]	72.8[ab]	14.9[ab]	9.2[ab]	3.0[b]	
75	13.9		33.3[ac]	40	1.2[a]	73.7[b]	14.3[b]	9.4[ab]	2.3[a]	
100	13.9		31.1[ac]	39	1.3[b]	73.4[b]	14.1[b]	9[b]	2.4[a]	

No.	Treatment									Reference
4	(35% Herring meal)	13.9	37.5[abc]	36	1.0[a]	72.6[ab]	15.1[a]	9.8[a]	2.4[a]	Hardy and Cheng (2002) (42 days)
	25 + Lysine (2.07%)[4]	13.9	43.6[b]	42	1.0[a]	73.0[ab]	14.7[b]	10.3[a]	3.1[b]	
	50 + Lysine (2.07%)[4]	13.9	38.4[bc]	42	1.1[a]	72.9[ab]	14.7[b]	10.8[a]	2.5[a]	
	75 + Lysine (2.07%)[4]	14.0	37.0[ab]	40	1.1[a]	73.5[b]	14.5[b]	10.2[a]	2.2[a]	
	100 + Lysine (2.07%)[4]	13.9	36.8[ab]	40	1.1[a]	73.0[ab]	14.4[b]	10.6[a]	2.3[a]	
5	(35% Herring meal)	13.9	83.3	84	1.0	70.6	16.2[a]	10.5	2.7	Hardy and Cheng (2002) (56 days)
	83 + Lys + Met (.73%)[4]	13.8	85.6	87	1.0	71.1	15.4[b]	10.5	2.4	
	92 + Lys + Met (.73%)[4]	13.9	88.9	95	1.1	70.1	15.4[b]	11.4	2.4	
	100 + Lys + Met (.73%)[4]	14.0	83.5	91	1.1	69.8	15.1[b]	11.7	2.8	
6	(35% Herring meal)	13.9	83.3	84	1.0	70.6	16.2[a]	10.5	2.7	Hardy and Cheng (2002) (56 days)
	83 + Lys + Met (0.83%)[4]	13.9	81.0	91	1.1	70.3	15.2[b]	11.3	2.8	
	92 + Lys + Met (0.83%)[4]	13.8	86.2	92	1.1	69.9	15.2[b]	12.9	2.3	
	100 + Lys + Met(0.83%)[4]	13.9	82.3	87	1.1	70.8	15.1[b]	11.3	2.4	
7	(35% Herring meal)	13.9	83.3	84	1.0	70.6	16.2[a]	10.5	2.7	Hardy and Cheng (2002) (56 days)
	83 + Lys + Met (0.93%)[4]	13.9	87.2	92	1.1	70.6	15.1[b]	11.3	2.6	
	92 + Lys + Met (0.93%)[4]	13.9	83.8	93	1.1	69.9	15.4[b]	11.9	2.8	

TABLE 3.6 (continued)

		Growth					Body composition (%)[1]				
FM replacement rate (%)	IW(g)	SGR	WG	FI (g/fish)	FCR	Moisture	CP	Lipid	Ash	Reference
100 + Lys + Met (0.93%)[4]	14		83.0	96	1.2	70.4	15.1[b]	11.6	2.5	Hardy and Cheng (2002) (42 days)
8 (35% Herring meal)										
100 + Lys + Met (0.93%)[4]	19.2		60.5	59	1.0	71.4	15.6[a]	11.8	2.3	
	·19.2		59.1	62	1.0	70.9	14.6[b]	12.9	2.3	
100 + Lys + His (0.13%)[4]	18.9		54.2	61	1.1	71.1	14.4[b]	12.7	2.5	
100 + Lys + His (0.25%)[4]	19		56.5	62	1.1	71.7	14.6[b]	12.1	2.7	
100 + Lys + His (0.38%)[4]	19.2		55.4	60	1.1	72.3	14.3[b]	11.5	2.3	
100 + Lys + His (0.5%)[4]	19.2		56.8	59	1.0	71.7	14.4[b]	12.0	2.3	
100 + Lys + His (0.63%)[4]	19.2		57.5	60	1.1	71.8	14.3[b]	12.1	2.5	
Nile tilapia										
1										
(35%)[3]	13	2.03	260	484	1.9[a]	70.3[a]	17.0	6.5[a]	5.5[a]	El-Sayed (1998) (150 days)[4]
100	12	2.02	236	529	2.2[b]	67.1[b]	16.5	7.8[b]	6.9[b]	
2										
(7%)	237		312	306	1.0					Lin and Yu (2002) (70 days)
20	348		303	312	1.0					
40	308		316	294	0.9					
60	257		297	306	1.0					

											Reference
3	(44%)	4.64	3.04	31.4	48	1.5					Serwata and Woodgate (2004) (70 days)
	66	4.48	3.31	36.5	56	1.5					
African catfish											
1	(40%)	16.5	3.57[de]	9.6	16	1.6	68.8[a]	15.9[c]	10.3	2.6[a]	Abdel-Warith et al. (2001) (70 days)
	20	16.5	3.68[e]	10.0	16	1.6	70.7[b]	15.8[c]	10.3	3.0[b]	
	40	16.5	3.56[cd]	9.2	15	1.6	70.9[b]	15.4[b]	10.1	3.1[b]	
	60	16.4	3.48[c]	8.6	15	1.7	69.6[a]	15.9[c]	10.6	3.4[c]	
	80	16.5	3.32[b]	7.7	14	1.9	69.4[a]	15.2[b]	11.0	3.5[c]	
	100	16.5	2.83[a]	5.3	12	2.3	69.6[a]	14.7[a]	10.5	3.65[c]	
Channel catfish											
1	(84.5%)	0.03		1.7							Winfree and Stickney (1984) (34 days)
	46	0.03		1.5							
2	(5% Menhaden)	6.4		56.7	76	1.4	75.7	16.6	5.8	1.18	Li et al. (2002) (63 days)
	100	6.4		55.9	77	1.4	75.6	16.3	6.4	1.17	
River catfish											
	0 (14%)	27		122	183	1.5					Hao and Yu 2003 (56 days)
	20	27		128	141	1.1					
	40	27		121	145	1.2					
	60	27		114	194	1.7					
Sunshine bass											
1	(30% Menhaden FM)	15	2.84	396	125	2.1	67.4	15.9	13.3		Webster et al. (1999) (56 days)
	100	15	2.80	384	121	2.1	67.6	16.1	13.5		

TABLE 3.6 (continued)

FM replacement rate (%)	IW(g)	Growth SGR	WG	FI (g/fish)	FCR	Moisture	CP	Lipid	Ash	Reference
2 (30% Menhaden FM)	20	1.79	251[a]	110	2.0[a]	76.1	19.5	2.4		Gallagher and LaDouceur (1995) (77 days)
100	20	1.52	197[b]	108	2.7[b]	76.5	19.4	1.9		
Palmetto bass										
(47% Herring)	5		29.6	39	1.3	6.7	13.1	11.2	5.3	
25	5		27.2	41	1.5	67.8	12.3	10.0	5.2	
50	5		30.5	43	1.4	66.7	12.8	9.6	5.6	
75	5		24.0	34	1.4	68.1	13.3	9.7	5.5	
Silver perch										
(5% Danish FM)	58.5	2.4	461	738	1.6					Allan and Rowland (1999) (143 days)
100	58.5	2.4	443	753	1.7					
Common carp										
1 (51.5%)	40		205	269	1.3	76.9	14.8	4.7	2.2	Steffens (1988) (84 days)
100	40		156	254	1.6	74.8	15.1	7.5	2.2	
2 (15% Peruvian FM)	28.6		159	183	1.2					Liu et al. (2002) (60 days)
33	27.8		128	162	1.3					
66	28.1		139	165	1.2					
100	27.7		136	161	1.2					
100 + Lys + Met	27.7		145	162	1.1					
3 (12% Peruvian FM)	200		233	412	1.8					Wang et al. (2002) (60 days)
33	200		237	417	1.8					
66	200		230	414	1.8					

Body composition (%)[1]

Australian snapper and *Gibel carp* and *Gilthead seabream* feeding-trial data (continued):

No.	Diet										Reference
4	100	200		230	414	1.8					Zhu and Yu (1999) (56 days)
	100 + Lys + Met	200		242	414	1.7					
	(23% Peruvian FM)	75		246	297	1.2					
	100	75		220	296	1.3					
Australian snapper											
1	(64% Danish FM)	77	0.36[a]	41[a]	86	2.1[a]					Quartararo et al. (1998) (115 days)
	53	77	0.33[ab]	36[ab]	83	2.3[a]					
	69	77	0.29[b]	31[b]	81	2.6[a]					
	84	77	0.21[c]	22[c]	86	3.9[b]					
Gibel carp											
1	(53% US white FM)	4.9	2.58	16	27	1.7					Yang, et al. (2002) (56 days)
	40	4.9	3.15	24	36	1.5					
	49	4.9	3.28	26	37	1.4					
	58	4.9	3.24	25	36	1.5					
	66	4.9	3.31	26	38	1.5					
	75	4.9	3.39	28	41	1.5					
	83	4.9	3.22	25	41	1.5					
	92	4.9	3.17	24	39	1.6					
	100	4.9	3.14	23	36	1.6					
2	(54.3% US white FM)	5.3		18	25	1.4					Cui et al. (2000) (56 days)
	15	5.3		18	26	1.4					
	50	5.3		17	27	1.6					
Gilthead seabream											
	(72.9% White FM)	1.61	2.40[d]	10.5[d]	1.6	1.5[c]	69.9[bc]	16.6[bc]	9.3[de]	4.7	Nengas et al. (1999) (84 days)
	75 (PMM)	1.55	2.33[cd]	9.4[cd]	15	1.6[c]	67.7[a]	17.7[d]	10.2[f]	4.7	
	100 (PMM)	1.56	2.28[cd]	9.1[cd]	15	1.6[c]	69.4[ab]	17.1[bcd]	8.8[cde]	4.7	

TABLE 3.6 (continued)

FM replacement rate (%)	Growth			FI (g/fish)	FCR	Body composition (%)[1]				Reference
	IW(g)	SGR	WG			Moisture	CP	Lipid	Ash	
40 (PBMa)	1.63	1.89[b]	6.3[b]	12	1.9[b]	71.9[cd]	16.8[bc]	6.8[a]	4.8	Takag et al. (2000) (232 days)
35 (PBMb)	1.57	2.38[cd]	10.1[cd]	15	1.5[c]	69.3[ab]	17.4[cd]	8.7[cde]	4.7	
50 (PBMb)	1.57	2.32[cd]	9.4[cd]	14	1.5[c]	70.7[bc]	16.8[bcd]	7.8[abc]	4.7	
75 (PBMb)	1.53	1.56[a]	4.2[a]	10	2.3[a]	73[d]	15.3	7.4[ab]	4.9	
Red seabream										
(50% mackerel FM Chilian horse)	280		471	975	2.1	66.7	17.5	11.3	3.7	Takag et al. (2000) (232 days)
30	281		500	950	1.9	66.3	17.6	12.5	3.9	
50	284		495	985	2.0	65.5	17.1	12.5	3.6	
70	277		555	966	1.7	66.4	18.2	11.7	4.1	
90	279		506	966	1.9	65.7	17.4	11.3	3.9	
100	277		443	948	2.1	64.2	16.9	14.2	3.5	
(50% mackerel FM Chilian horse)	53.4		58.8[b]	75	1.3[d]	67.5[ab]	17.7[ab]	11.0[b]	3.7[a]	Takagi et al.(2000) (60 days)
30	53.5		53.0[ab]	70	1.3[d]	67.1[a]	17.5[a]	10.3[ab]	3.9[ab]	
50	53.8		50.2[ab]	71	1.4[c]	68[ab]	18.4[b]	9.9[ab]	4.1[ab]	
70	53.1		51.2[ab]	74	1.5[bc]	68[ab]	18.0[ab]	9.6[ab]	4.0[ab]	
90	53.3		42.2[a]	66	1.6[a]	69[ab]	18.0[ab]	8.1[a]	4.4[b]	
100	55.4		42.7[a]	70	1.6[a]	69.7[b]	18.0[ab]	8.5[a]	4.3[ab]	
Red drum (30% Menhaden FM)	2.3		26.1	27	1.0					Kureshy et al. (2000) (42 days)
17	2.2		23.8	24	1.0					

Cuneate drum										
(35%)										
33	2.3	2.21	27.5	28	1.0	73.7	15.7	6.4	3.8	Wang (2004) (56 days)
50	2.3	1.97	25.5	27	1.1	73.1	16.0	7.0	3.9	
67	2.2	1.98	24.5	26	1.1	73.4	15.7	6.9	3.8	
Grouper										
1										
(50%)										
30	27.1		66.7	70						Chai (2000) (70 days)
50	27.3		55.0	59						
	27.6		55.9	62						
25	27.3		73^a	86	1.2	75	19.8^a	3.9^a	1.5	
50	28.1		66^a	87	1.3	74	19.9^{ab}	5.1^{bc}	1.5	
75	27.8		58^a	79	1.4	74	19.7^c	5.1^{bcd}	1.5	
100	27.5		53^b	75	1.4	74	18.9^{bc}	4.6^{ed}	1.4	
	27.6		46^b	74	1.6	74	19.0^c	4.5^{ed}	1.4	
2										
(40% Anchovy FM)	5.2	2.89^{ab}	24.1^a	26	1.1	74	14.9	4.9	4.0	Tan et al. (2005) (60 days)
20	5.2	2.81^{ab}	23.4^{ab}	24	1.0	74	14.5	4.8	4.1	
30	5.2	2.90^{ab}	24.3^a	26	1.1	75	14.0	4.5	3.9	
40	5.2	2.78^b	22.2^{ab}	23	1.0	74	14.9	4.7	4.0	
60	5.2	2.76^b	22.1^{ab}	24	1.1	74	14.4	4.8	4.1	
80	5.2	2.68^c	18.9^c	21	1.1	73	15.1	5.0	4.1	
100	5.2	2.62^c	18.7^c	21	1.1	75	14.0	4.3	3.9	
Trash fish	5.2	3.01^a	24.9^a	25	1.0	74	14.9	4.6	4.1	
Black sea turbot	18.1	0.85^d	12.0	10	0.9	79.2	16.9	2.6	1.1	Chai (2000) (70 days)
(77.7% Black Sea whiting FM)										
25	18.0	0.81d	11.4	10	0.9	79.3	16.7	2.5	1.1	

TABLE 3.6 (continued)

| FM replacement rate (%) | Growth | | | FI (g/fish) | FCR | Body composition (%)[1] | | | | Reference |
	IW(g)	SGR	WG			Moisture	CP	Lipid	Ash	
50	18.2	0.56[c]	7.3	8	1.2	79.5	16.4	2.3	1.2	
75	18.2	0.39[b]	4.8	7	1.5	80	16.1	2.2	1.5	
100	18.1	0.19[a]	2.2	5	2.5	80	16.0	2.1	1.4	

Note: Values with different letter superscripts are significantly different.

FM = fish meal; IW = initial weight; SGR = specific growth rate or other standardized growth rates; WG = weight gain; FI = feed intake; FCR = feed/conversion ratio

[1]On wet basis.

[2]Percent fish meal in control diet.

[3]Trial length in days.

[4]Level of total amino acid in diet with added crystalline amino acid.

TABLE 3.7. Weight gain responses of finfish to dietary substitution of marine fish meal with poultry by-product meal.

Weight gain responses to increasing substitution rates	Species
Positive[a]	Gibel carp
Equal to fish meal[b]	Coho salmon, fall chinook salmon, rainbow trout, tilapia (Nile and hybrid), sunshine bass, silver perch, red seabream yearlings, red drum, common carp yearlings, channel catfish
Mildly negative[c]	Juvenile common carp, African catfish, river catfish, palmetto bass, gilthead seabream, juvenile red seabream, cuneate drum, grouper
Negative[d]	Australian snapper, Black sea turbot

[a]Positive = Weight gain increases with increasing substitution rates (up to 90%) relative to fish meal control.

[b]Equal to fish meal = Weight gain remains unchanged with increasing substitution rates (up to 90%).

[c]Mildly negative = Weight gain tends to decrease at high substitution rates (>50%).

[d]Negative = Weight gain decreases with increasing substitution rates.

(e.g., palatability and digestibility) of the PBM, rather than the deficiency of EAAs and EFAs even up to 75 percent of FM was substituted. Carp, and several marine finfishes, responded mildly negatively to FM substitution with PBM while salmonid, tilapia, bass, catfish, perch, red drum, and red sea bream had WG generally unaffected by the dietary substitution of FM with PBM at 50 percent or lower and only slightly depressed at higher rates of substitution. The WG response to substitute (Table 3.7) is generally in agreement with the ranking of suitability of FM substitution by PBM based on the calculation of limiting EAA (Table 3.5).

Changes in growth performance (WG, FI, and FCR) and body composition as FM substitution rate increased were calculated by regression analysis for each group of fishes (P, E, MN, and N) (Table 3.8). Except for P group, fish in E group still show a slight increase in FCR (2 percent poorer) when substitution rate is less than 50 percent (Table 3.8), but FCR further increases by 8 percent at higher rates of

TABLE 3.8. Effect of substitution rates of fish meal (FM) with poultry by-product meal on performance and body composition.

Response type	10-50% substitution						51-100% substitution					
	WG[a]	FI[b]	FCR[c]	Body composition			WG[a]	FI[b]	FCR[c]	Body composition		
				CP[d]	Fat	Ash				CP[d]	Fat	Ash
	Relative response (%) to fish fed the control diet (FM)											
Positive	+25	+15	−13	−[e]	–	–	+26	+7	−8	−[e]	–	–
Equal to FM	+1	+4	+2	−1	+7	+6	−2	+5	+8	−4	+7	0
Mildly negative	−5	−7	+2	0	−10	+7	−15	−11	+16	0	−8	+16
Negative	−16	−3	+13	0	+31	+2	−23	−2	+40	–	–	–

[a]Weight gain.

[b]Feed intake.

[c]Feed conversion ratio.

[d]Crude protein.

[e]No effect reported.

substitution. This increase in FCR is largely (60 to 100 percent) related to the increase in FI for PBM diets, although lower digestible EAA content and perhaps EFA supply may also be contributing factors.

In contrast to P and E groups, FI of MN and N groups is modestly reduced. Feed conversion ratio is greatly impaired for MN and N groups, particularly at 51 to 100 percent substitution rates, and this deterioration in FCR can only be partly explained by reduction in FI. Possible reasons for reduced feeding value are (1) inadequate supply of EAA, DE, and EFA and (2) less-than-desired quality and composition of PBM. Mixing PBM with other animal protein meals may provide nutritional complementary benefits in meeting nutrient requirements of the fish as evidenced by several feeding trials (Steffens, 1994; Rodriguez-Serna et al., 1996; Lee et al., 2001, 2002; Yanik et al., 2003).

Body protein content does not appear to change with the dietary FM substitution by PBM, particularly at levels lower than 50 percent (Table 3.8). High rates of substitution may result in slight (<10 percent) reduction in body protein content. Body fat content generally increased with elevated levels of PBM except for salmon, trout (without EAA supplementation), palmetto bass, and juvenile red sea bream. Response of body ash content to PBM substitution varies between species. However, the range of variation of majority species is within ten percentage points from the FM control. Effects on body composition may also be inherently influenced by fish size. Nonetheless, it does not appear that modest inclusion rates of PBM greatly affect fish body composition.

USE OF HYDROLYZED FEATHER MEAL (FeM)

Hydrolyzed feather meal is the product resulting from the treatment under pressure of clean, undecomposed feathers from slaughtered poultry, free of additives, and/or accelerators. Not less than 75 percent of its CP content must be digestible by the pepsin digestibility method (AFCO, 2003). Typical rendering conditions are: 115-140°C and 207 kiloPascals (KPa) for forty-five minutes. The process should be sufficient to destroy most harmful microorganisms.

Proximate and amino acid composition of FeM is given in Table 3.1. The low ash or phosphorus content has merit for low water pollution formulations. However, the high protein content, with an amino acid imbalance (low in lysine and high in cystine) greatly limits its inclusion rate in fish diets.

Apparent digestibility of protein and energy is presented in Table 3.2. Compared to PBM and FM, digestibility of FeM has not been measured widely among fish species. Average protein digestibility is 78 percent (or 89 percent of FM), which is comparable to PBM. Energy digestibility average 80 percent (or 93 percent of FM), and is somewhat (5 percent) better than PBM probably due to its low content of ash. However, the deamination process as the consequence of amino acid imbalance may reduce its energy availability for growth (net energy).

The digestibility values of a FeM produced in North America with rainbow trout were higher than the literature average (84 versus 78 percent for protein, and 82 versus 80 percent for energy), which may reflect the improved processing techniques (Bureau et al., 1998, 1999).

Apparent digestibility of essential amino acids (EAA) is given in Table 3.3. Large differences exist in digestibility measured in different species. Feather meal was well digested by Australian silver perch (Allan et al., 2000), but to much reduced levels by rainbow trout and rockfish (Bureau et al., 1999; Lee, 2002). The average EAA digestibility for FeM was 11 percent lower than that of PBM (76 percent versus 87 percent). Unlike PBM, digestibility of CP and EAA is very similar for FeM. Recent work on enzyme-treated FeM seems to be promising in improving protein digestibility and EAA digestibility for shrimp diets (Mondoza et al., 2001).

Digestible protein, energy, and amino acid (percentage as is and percentage CP) content in FeM are listed in Table 3.4. Opportunity of using crystalline amino acids or other high-quality non-marine protein meals in FeM-based diets is apparent, even though several research works have indicated that supplementing FeM diets with deficient EAA can only partially (up to 50 percent) alleviate reduction in WG compared with FM control group (Pfeffer et al., 1995). More research is needed in amino acid supplementation in order to improve the formulation of low pollution aqua diets.

Comparison of digestible EAA profile in FeM and the EAA requirements of several species is given in Table 3.9. Without supplementation of crystalline amino acids, FeM by itself cannot provide the required amount of lysine and tryptophan for most fishes. Overall, FeM is relatively more suitable for salmonid and catfish than for carp, tilapia, milkfish, and other marine fishes. Calculation on the most limiting digestible EAA indicates that the maximum FM protein substitution rate with FeM is 48, 46, 29, 30, 57, 46, and 48 percent for a salmonid, catfish, carp, tilapia, milkfish, sea bream, and sea bass, respectively. These values are substantially lower than that of PBM, and suggest that the FM protein substitution rate should not exceed 50 percent.

PRACTICAL DIETS

Summary of published data are presented in Table 3.10 (Higgs et al., 1979; Fowler, 1990; Kikuchi et al., 1994; Pfeffer and Henrichfreise, 1994; Wiesmann et al., 1988; Bishop et al., 1995; Hansan et al., 1997; Wiesmann et al., 1988; Aoki et al., 2000; Bureau et al., 2000; Brandson et al., 2001; Jahan et al., 2001; Mendoza et al., 2001; Serwata and

TABLE 3.9. A comparison between digestible essential amino acids of hydro-lyzed feather meal (FeM) and the essential requirements of different species of finfishes (% of protein).[a]

Amino acids	FeM[b]	Salmonid	Catfish	Carp	Tilapia	Milkfish	Seabream	Seabass
Arg	6.7	4.2	4.3	4.4	4.1	5.6	–	–
His	1.9	1.6	1.5	*2.4*[c]	1.7	*2.0*	–	–
Ile	4.3	2.0	2.3	3.0	3.1	4.0	–	–
Leu	8.0	3.6	3.5	4.7	3.4	5.1	–	–
Lys	2.3	*4.8*	*5.0*	*6.0*	*4.6*	*4.0*	*5.0*	*4.8*
Thr	4.0	2.0	2.1	*4.2*	3.8	*4.9*	–	–
Trp	0.3	*0.5*	*0.5*	*0.8*	*1.0*	*0.6*	*0.6*	–
Val	5.3	2.2	3.0	*4.1*	2.8	3.0	–	–
Met + Cys	4.7	2.4	2.3	3.5	3.2	*4.8*	4.0	4.4
Phe + Tyr	7.5	5.3	4.8	*8.2*	5.6	5.2	–	–

[a]Bureau (2000).

[b]Data taken from Table 3.4.

[c]Amino acid with an italicized value is considered deficient in meeting require-ments for that species.

Davies, 2001; Serwata and Woodgate, 2004; Wang, 2004). Based on the ratio of WG response of fish fed FeM as substitutes for FM to that of fish fed FM control, three types of response are identified: (1) equal to FM protein (E; salmon, trout, Indian carps, and Japanese floun-der); (2) mildly negative (MN; Atlantic salmon parr, common carp, tilapia, catfish, red sea bream, and cuneate drum); and (3) negative (N; yellowtail) (Table 3.11). In contrast to WG response to PBM sub-stitution, FeM substitution results in: (1) none of species examined showed a positive WG response; (2) a noticeable negative response trend in WG began at about 40 percent FM protein replacement rate versus 70 percent noted for PBM substitution; and (3) a more severe reduction in WG at 100 percent substitution for FeM (75 percent) versus PBM (40 percent). The slight discrepancy noted for maximum FeM substitution rate without losing WG calculated from limiting di-gestible EAA compared with requirements (50 percent), and observed

TABLE 3.10. Response of finfish to fish meal protein substitution with hydrolyzed feather meal in growth and body composition. Value with different superscripts are significantly different.

	FM replacement rate (%)	IW(g)	SGR	WG (g)	FI (g/fish)	FCR	Moisture	CP	Lipid	Ash	Reference
				Growth			Body composition (%)[1]				
Coho salmon	(52% Herring FM)[2]	0.2	2.07	9.5	13	1.3	73.7	16.3	6.7	2.3	Higgs et al. (1979) (168 days)[3]
	14	0.2	1.98	9.9	13	1.3	73.2	16.2	6.7	2.3	
	31	0.2	1.95	10.1	13	1.3	73.5	16.1	6.5	2.3	
Chinook salmon	(39.3% Herring FM)	1.6	2.16	31.3	34	1.1	73.4	16.8	6.4	2.0	Fowler (1990) (140 days)
	16	1.6	2.12	29.3	33	1.1	73.3	16.7	6.9	1.9	
	48	1.6	2.12	29.3	33	1.1	73.2	16.4	7.8	1.7	
Atlantic salmon parr	(60%)	36		39.4[a]	45	1.1	70.7	15.9[a]	10.4	2.2	Brandsen et al. (2001) (56 days)
	40	36		30.8[b]	33	1.1	70.4	15.3[b]	11.3	2.1	
Rainbow trout											
1	(30% Casein)	41		159	115	0.7					Wiesmann et al. (1988) (84 days)
	100	41		132	119	0.9					
2	0	191		101	74	0.7					
	100	163		136	127	0.9					
3	(30%)	50		148	130	0.9		13.9	15.7		Pfeffer et al. (1994) (98 days)
	33	50		122	115	0.9		13.7	17.6		
	33 + Lys	50		140	130	0.9		14.5	18.2		
	33 + Met	50		136	126	0.9		13.5	19.2		
	33 + Lys + Met	50		139	130	0.9		14.5	17.1		

76

Group	Diet										Reference
4	67	50		108	113	1.1		11.9	17.2		Pfeffer and Henrichereise (1994) (140 days)
	67 + Lys	50		112	111	1.0		13.4	17		
	67 + Met	50		116	121	1.0		11.7	18.2		
	67 + Lys + Met	50		103	101	1.0		13.1	16.3		
	100	50		48	60	1.3		9.3	15.3		
	100 + Lys + Met	50		78	86	1.1		12.8	15.6		
	(59.4%)	60		145	145	1.0[b]		16.8	16.5		
	100 + Lys + Met + Try + His	60		116	142	1.2[a]		15.5	18.2		
5	(50% Herring FM)	17	0.162[ab]	73.5[ab]	52	0.70[ab]	71.5	15.6	9.6	2.1	Bureau et al. (2000) (140 days)
	30–A*	17	0.162[ab]	74.3[ab]	51	0.69[a]	72.3	15.6	9	2.2	
	30–B	17	0.16[bc]	71.1[bc]	52	0.73[bc]	71.7	15.5	9.4	2.2	
	30–C	17	0.161[ab]	73.0[abc]	52	0.70[abc]	72.4	15.7	9	2.2	
6	(50% Herring FM)	17	0.162[ab]	73.5[ab]	52	0.70[ab]	71.5	15.6	9.6	2.1	
	9.6	17	0.164[a]	74.5[a]	52	0.70[a]	72.0	15.5	9.3	2.2	
	14.4	17	0.162[ab]	73.2[abc]	52	0.70[abc]	72.2	15.6	9.4	2.2	
	19.2	17	0.161[ab]	73.3[abc]	52	0.70[abc]	71.5	15.6	9.4	2.2	
	24	17	0.156[c]	70.1[c]	52	0.70[c]	72.3	15.6	9.2	2.1	
Indian Carp	(50.8% Local FM)	1.02	1.77[a]	2.1	5	2.12[a]	76.4[c]	14[ab]	8.9[b]	2.4[b]	Hason et al. (1997) (63 days)
	25	1.02	1.80[a]	2.2	4	2.00[a]	74.1[c]	14.3[a]	9.2[a]	2.3[bc]	
	50	1.02	1.55[a]	1.7	4	2.35[a]	75[c]	13.8[ab]	8.6[b]	2.2[c]	

TABLE 3.10 (continued)

FM replacement rate (%)	IW(g)	Growth				Body composition (%)[1]				Reference
		SGR	WG (g)	FI (g/fish)	FCR	Moisture	CP	Lipid	Ash	
Common Carp										
75	1.02	1.27[b]	1.3	4	2.77[a]	76.1[b]	12.8[c]	8.2[c]	2.4[b]	Jahan et al. (2001) (84 days)
100	1.03	0.68[c]	.6	3	4.99[b]	78.6[a]	13[bc]	5.3[d]	2.9[a]	
(25% Chilean jack mackerel FM)	4.6		92.3[a]	91	0.99[b]	73.5	14.7	8.4	2.8	
20	4.6		82.2[b]	86	1.05[b]	73.8	14.5	8.0	2.6	
40	4.6		75.1[bc]	89	1.19[a]	72.8	14.7	8.8	2.5	
Tilapia										
1 (10% Menhaden FM)	0.013	4.8	0.7							Bishop et al. (1995) (42 days)
33	0.012	4.23	1.1							
66	0.012	5.9	1.3							
100	0.012	4.5	0.6							
2 (44%)	4.64	3.04	31.1	47	1.52					Serwata and Woodgate (2004) (70 days)
66	5.56	1.61	10.8	32	3					
3 (44%)	4.64	3.04	31.1	47	1.52					Serwata and Woodgate (2004) (70 days)
66 (Enzyme treated)	5.46	2.10	16.3	39	2.38					
African catfish (% FM in control diet)	17.4	3.62	114.2	155	1.36					Serwata and Davis (2001) (56 days)

Species	Diet/level										Reference
Channel catfish	25	17.4	3.12	82.5	121	1.46					
	50	17.4	2.44	51.0	85	1.67					
	75	17.4	1.41	20.9	63	3.02					
	100	17.4	0.83	10.3	67	6.51					
	(5% Menhaden)	6.4		56.7	76	1.35	75.7	16.6	5.8	1.18	Li et al. (2002) (63 days)
	100 + Lys	6.4		51.6	74	1.45	75.6	16.4	6.3	1.17	
Yellow tail 1	(30%)	141.4		371.8	721	1.94	72.2	24.5	4.3	1.7	Aoki et al. (2000) (101 days)
	18	142.6		321.4	707	2.20	72.4	24.2	3.8	1.8	
	36	142.0		213	579	2.72	72.7	23.4	3.1	1.7	
Yellow tail 2	(30%)	107.0		483.2	942	1.95	70.3	24.8	5	2.1	Aoki et al. (2000) (97 days)
	18	106.6		421.3	931	2.21	71.9	25.3	3.1	2	
	36	104.9		237.7	737	3.10	73.8	24	2	1.6	
Red seabream	(30%)	29.6		25.7	41	1.59	74.5	21.2	2.5	1.7	Aoki et al. (2000) (70 days)
	18	29.7		21.8	37	1.70	73.4	21.6	2.8	1.8	
	36	29.6		20.7	37	1.79	74.1	21.7	2.2	1.8	
Japanese flounder	(80% white FM)	3.1		22.7	14	0.62	74.6	17.3^{ab}	4.4^{a}	3.6^{ac}	Kikuchi et al. (1994) (56 days)
	19 + AA	2.7		22.8	14	0.63	74.4	17.5^{a}	3.8^{ad}	3.4^{ab}	
	40 + AA	2.9		22.0	16	0.72	74.1	17.2^{ab}	4.0^{a}	3.7^{c}	
	60 + AA	2.8		16.0	14	0.88	75.8	16.7^{bc}	3.3^{ac}	3.6^{ac}	
	80 + AA	2.9		10.6	12	1.11	76.6	16.7^{c}	2.2^{b}	3.7^{c}	
	60	2.9		13.2	13	1.00	75.8	16.6^{ac}	3.3^{cd}	3.9^{d}	

TABLE 3.10 (continued)

FM replacement rate (%)		Growth				Body composition (%)[1]				Reference
	IW(g)	SGR	WG (g)	FI (g/fish)	FCR	Moisture	CP	Lipid	Ash	
Cuneate (35%)										Wang (2004)
Drum 10	27.1	2.2	66.7	68	1.05	73.7	15.7	6.4	3.8	(56 days)
10	27.5	1.9	51.0	65	1.34	74.4	15.4	6.1	3.8	
30	27.1	1.8	46.1	56	1.27	73.8	15.8	6.3	3.9	

Note: Value with different letter superscripts are significantly different.

FM = fish meal; W = initial weight; SGR = specific growth rate or other standardized growth rates; WG = weight gain; FI = feed intake; FCR = feed/conversion ratio.

[1]On wet basis.

[2]Percent fish meal in control diet.

[3]Trial length in days.

TABLE 3.11. Weight gain responses of finfishes to dietary substitution of fish meal protein with hydrolyzed feather meal.

Weight gain responses to substitution rates less than 40%	Species
Equal to fish meal[a]	Coho salmon, chinook salmo, rainbow trout, Indian carp, Japanese flounder
Mildly negative[b]	Atlantic salmon parr, common carp, tilapia, catfish, red seabream, cuneate drum
Negative[c]	Yellowtail

[a]Equal to fish meal = weight gain remains unchanged with substitution rates.

[b]Mildly negative = weight gain tends to decrease at high substitution rates (greater than 50%).

[c]Negative = weight gain decreases with substitution rates.

TABLE 3.12. Effect of substitution rate of fish meal (FM) protein with hydrolyzed feather meal on performance and body composition.

	11-40% substitution						41-100% substitution					
				Body composition						Body composition		
	WG[a]	FI[b]	FCR[c]	CP[d]	Fat	Ash	WG[a]	FI[b]	FCR[c]	CP[d]	Fat	Ash
	Relative response (%) to fish fed the control diet (FM)											
Equal to FM[e]	−2	0	+4	0	+1	+1	−26	−2	+30	−9	−3	−1
Mildly negative	−14	−6	+6	0	+2	−2	−31	−31	−96	−6	−1	+5
Negative	−30	−19	+31	−1	−34	−6						

[a]Weight gain.

[b]Diet intake.

[c]Feed conversion ratio.

[d]Crude protein.

[e]Fish meal.

from feeding trials (40 percent) may be due to reduced FI of diets containing high level of FeM (Table 3.12).

Although substitution of FM protein with FeM up to 40 percent did not adversely affect the protein efficiency ratio (data not shown), higher rates of substitution (>40 percent) resulted in marked reduction in FI and WG. The reduction in feed consumption and WG noted at high substitution rates could be due to relatively low palatability, amino acid imbalance, and deficiency of DE and polyunsaturated fatty acids of the test diets. Supplementation of lysine and methionine at high FM substitution rates had limited effect on the recovery of WG loss in Japanese flounder (Kikuchi et al., 1994), and rainbow trout (Pfeffer et al., 1995). Use in combination with other non-marine animal protein meals as FM protein replacement has generally resulted in no loss in growth performance of several species of finfish (Steffens, 1994; Rodriguez-Serna et al., 1996; Lee et al., 2001, 2002). Further research is needed to evaluate the effectiveness of enzyme treatment, diet attractants, supplementation with deficient amino acids, and fats/oils for high inclusion rates of FeM in finfish diets.

The effect of FM protein substitution with FeM on body composition is presented in Tables 3.10 and 3.12. It appears that high rate of inclusion (>50 percent) of FeM in diets causes a reduction in body protein content. Severe reduction in WG seems to affect body fat content. Body ash content generally is unaffected by the substitution rate.

Analysis of the feeding trial results indicates that, with exception of trout and perhaps tilapia, the FM substitution with FeM should be limited to less than 40 percent of FM protein. The major limitation for greater use of FeM in finfish diets are palatability and amino acid imbalance. Feather meal should be blended with other protein meals in finfish diets to maintain a satisfactory WG and reduction in feed cost.

GENERAL DISCUSSION

Although the fundamental purpose of using PBM and FeM in aquafeeds is to replace FM protein, the inclusion of FM at relatively high levels is no longer considered compulsory as modern formulation is guided by, but not limited to, the following goals for sustainable aquaculture production: (1) minimizing the negative effect on water quality; (2) maximizing the efficiency on feed conversion; (3) optimizing the relative ratio between nutrients; (4) optimizing the level of n-3 fatty acids; (5) minimizing any negative effects on immune function, health, and survival rate; (6) maintaining an acceptable diet palatability; and (7) supporting a positive effect on carcass composition and taste characteristics. In order to meet this criteria, protein meals ideally should be high in digestible protein, amino acids, and energy, and free of antinutritional factors (such as PBM), and low in ash content (such as FeM).

Two protein by-products from the poultry industry, PBM and FeM, not only have reasonably good nutritional value for finfish diets, but also offer a very competitive cost advantage over FM (i.e., cost per unit protein is typically 25 percent and 60 percent lower for PBM and FeM than FM).

In most research trials, FM has been substituted by alternative protein sources on equal CP and energy basis. A few studies have been done on a digestible CP and energy basis, which are scientifically preferred for improved accuracy. However, the prerequisite for the use of digestible nutrient in feed formulation is the availability of digestibility (protein and energy) measured from the relevant species and age of all ingredients available for formulation.

As FM, PBM, and FeM contain different amounts of essential amino acids, particularly lysine and methionine, correct substitution between these meals should include equal levels of limiting amino acids to that of FM, or the requirement levels of the fish. Supplementation of crystalline amino acids is often required, although their bioavailability for finfish could be low compared with terrestrial species. Results from amino acids supplemented trials are combined with amino acids unsupplemented studies in this review for defining the relative nutritional value of PBM and FeM to FM.

Most finfishes have known requirements of essential fatty acids (EFA), such as n-3 series fatty acids (eicosapentaenoic acid, linolenic acid, and docosahexaenoic acid), and n-6 series (linoleic acid and arachinonic acid). However, not all substitution trials have equalized EFA, which may have caused some loss in performance in addition to protein quality differences. In this review, no attempt is made to separate the EFA effect from the overall performance.

In commercial fish diet formulations, a minimum level of FM is usually used along with other protein ingredients (e.g., corn gluten meal, soybean meal, or meat and bonemeal). Thus, interpretation of growth responses from low levels of FM substitution with PBM or FeM could be difficult due to dilution or even confounding effect from other protein ingredients. Studies with high levels of substitution have merit as they allow the expressing of any possible nutrient deficiency or excess in the substituting ingredient, provided that test diets do not pose serious palatability, toxicity, or other health related problems. Regression analysis of stepwise substitution trials could be useful in identifying the trend of response, and also the maximum point of response (either positive or negative).

In order to truly compare the protein quality of FM versus PBM or FeM, semi-purified diets, without other protein ingredients except for FM (in control diet) and PBM or FeM (in test diet), were used in several studies (Chai, 2000; Cui et al., 2000; Yang et al., 2002). Results from these trials are combined with trials of practical diets in this review.

Nutrient requirements, digestive system, and feeding habit are known to vary among fish species (NRC, 1993) and could affect the outcome of response of FM substitution. Therefore, evaluation in this review is species-specific. As juvenile fishes have a higher requirement for most nutrients, and have a greater sensitivity to nutrient deficiency or toxicity from antinutritional factors compared with older fish, most research trials reviewed used juveniles as the experimental animal. It is assumed that the maximum replacement rate observed from juveniles can be safely applied to the grownt fish.

Relative value of PBM and FeM to FM should be established from evaluation of: (1) gross nutrient composition; (2) nutrient digestibility; (3) diet palatability; (4) WG; (5) FCR; (6) protein utilization efficiency (PER); (7) survival; (8) immune response; (9) body composition and sensory measures; and (10) cost-benefit analysis. Among all aforementioned variables, palatability of the diet, sensory evaluation of the fish, and immune response are not commonly reported in the literatures. Weight gain and PER are generally correlated well, thus WG could be used as the indicator of protein utilization for those trials that failed to report PER. Almost all research trials found that only under extreme cases, mortality is not affected by the substitution of PBM or FeM. Based on the availability of data from literature, the evaluation variables in the present review are mainly 1, 2, 4, 5, 9, and 10.

As the design of most substitution trials reported in the literature varies greatly in species, age, culture condition, control diet formulation (i.e., inclusion rate of FM), duration of trial, diet type (semi-purified versus practical), and composition of PBM and FeM, all response measurements from test groups are standardized to a relative (to FM) basis "as percentage of the control (FM) group" (i.e., test/control $\times$ 100). This adjustment within each trial allows a fair comparison across different trials with a wide range of variation in design of the control group. The rate of substitution by PBM or FeM is expressed as "percentage FM replacement" for PBM, and as "percentage FM protein replacement" for FeM, rather than the actual inclusion rate of the two protein meals. The standardization implies that the addition of PBM or FeM is directly at the expense of FM or FM protein, regardless the actual FM use rate. Also, the calculation assumes that other protein ingredients in test diets remain unchanged or with minimum change. These two criteria effectively remove the inherent variation associated with the FM control group.

Narrow range substitution studies (usually at high end of substitution) should be conducted for specific objective only after response trend has been established over a wide range of substitution rates. A good example is the evaluation of amino acids supplementation to trout diets with high rates of FM substitution by PBM (Hardy and Cheng, 2002).

COST COMPARISON

Potential benefits of substitution FM protein with PBM or FeM depends not only on the biological response of finfish, but also the relative cost of PBM and FeM in comparison with FM. Estimation of the relative value could be based on nutrient composition, or WG response from fish. In order to equalize PBM and FM on nutritional value, lysine and methionine need to be added to PBM while some oil is needed for FM. Feather meal requires supplementation of amino acids (e.g., lysine, tryptophan) and phosphorus. Relative WG value for PBM and FeM is estimated by using E group fish at FM protein substitution rate of 80 percent and 40 percent, respectively (Tables 3.6 and 3.10). Approximate relative (or break-even) price of PBM is 86 to 90 percent of FM, and 75 to 80 percent of FM for FeM. These calculations once again demonstrate that PBM and FeM are valuable FM replacements, and should be used in fish diets after thorough consideration of multiple factors, such as species, age, and availability and price of FM, PBM, FeM and dietary supplement (EAA, oil, palatability enhancer, etc.).

RECOMMENDED FM REPLACEMENT RATES

The recommended FM replacement and use rates of PBM and FeM in finfish diets are listed in Tables 3.13 and 3.14. These rates were derived from Tables 3.6, 3.7, 3.8, 3.10, 3.11, and 3.12. It is apparent that PBM can be used as FeM replacement at rates two times that of FeM without noticeable effect on fish performance. With proper supplementation of deficient EAA and balance of DE and EFA, PBM can replace high levels (80 to 100 percent) of FM for a majority of finfish species. The recommended inclusion rates are calculated from practical FM use rate in commercial feeds. Supplementation of lysine is necessary with FeM at moderate to high levels of substitution. Use of feed attractants is also recommended with high levels of FM substitution.

CONCLUSION

Both PBM and FeM are heat-treated (rendered) by-products from poultry slaughtering plants. High-quality (e.g., low-ash) PBM is similar to FM in nutrients composition, digestibility, and also in feeding value to finfish provided the diets are balanced for EAA, EFA, and DE in accordance with fish requirements.

Considering the suitability for high rate of substitution for FM (80 to 100 percent), and the cost savings over FM, PBM perhaps has a greater value for species commonly fed with diet containing high levels of FM (e.g., carnivorous finfishes) than for species on low-FM diet (e.g., omnivorous and herbivores finfishes). Hydrolyzed feather meal can be used in most finfish diet as a partial replacement (<40 percent) of FM protein. Inclusion of FeM is limited by the lack of some EAA and reduced palatability.

TABLE 3.13. Recommended and maximum inclusion rates of poultry by-product meal (PBM) in finfish diets.

Species	Recommended[a]		Maximum[b]	
	% FM[c] replacement	% PBM in diet	% FM[c] replacement	% PBM in diet
Gibel carp	70	11	100	15
Salmon	70	25	80	30
Trout	80	28	100	35
Tilapia (Nile and hybrid)	80	8	100	10
Sunshine bass	70	20	100	30
Silver perch	100	5	100	5
Red seabream	80	40	100	50
Red drum	60	18	80	24
Common carp (Yearling adult)	80	10	100	12
Channel catfish	80	4	100	5
Common carp (Juvenile)	30	5	60	9
African catfish	60	10	80	10
River catfish	50	7	60	9
Palmetto bass	50	15	70	20
Gilthead seabream	60	40	80	50
Cuneate drum	60	24	80	30
Red seabream (Juvenile)	70	35	90	45
Grouper	50	25	70	35
Australian snapper	50	30	60	40

[a]With no or minimal loss in weight gain.

[b]With 5-10% loss in weight gain and feed efficiency.

[c]Fish meal.

This review provides readers with response trend of WG and FCR to FM substitution with PBM and FeM, so that the optimum replacement rate for a specific feeding condition can be calculated from the benefit of cost reduction versus potential loss in performance. Among all species evaluated, salmonid, catfish, and gibel carp accept PBM and FeM diets with minimum adverse effect on growth performance

TABLE 3.14. Recommend and maximum inclusion rates of hydrolyzed feather meal (FeM) in finfish diets.

Species	Recommended[a]		Maximum[b]	
	% FM[c] replacement	% FeM in diet	% FM[c] replacement	% FeM in diet
Salmon	40	18	50	22
Trout	30	10	40	14
Indian carp	40	6	50	7
Japanese flounder	40	20	50	27
Common carp	20	4	30	6
Tilapia (Hybrid)	50	5	80	8
African catfish	75	7	100	10
Cuneate drum	30	12	50	18
Red seabream	8	2	10	3
Yellowtail	5	2	8	3

[a]With no or minimal loss in weight gain.

[b]With 5-10% loss in weight gain and feed efficiency.

[c]Fish meal.

while Australian snapper, yellow tail, and black sea turbot seem to do poorly on PBM- or FeM-containing diets. Fish meal protein substitution with PBM (~60 percent) or FeM (~30 percent) has mild effect on tilapia, bass, perch, sea bream, red drum, Japanese flounder, grouper, and cuneate drum. More research is needed for carcass sensory characteristics and fish immune response to dietary FM substitution with PBM and FeM. Effectiveness of supplementing EAA and feed attractants also warrants further investigations. Use of combination of non-marine animal protein meals and plant protein meals may achieve benefit of cost savings and nutrient balance. Physical (e.g., heat treatment) and chemical (e.g., enzyme) treatments may improve the feeding value of FeM and PBM. Recommended FM protein replacement rates for PBM and FeM are 60 to 80 percent and 30 to 40 percent in finfish diets, respectively. When nutritional value and WG values are compared on an equal basis, the break-even cost relative to FM is 86 to 90 percent for PBM, and 75 percent for FeM.

REFERENCES

Abdel-warith, A.A., P.M. Russell, and S.J. Davies (2001). Inclusion of a commercial poultry by-product meal as a protein replacement of fish meal in practical diets for African catfish *Clarias gariepinus* (Burchell 1822). *Aquaculture Research* 32:296-305.

Alexis, M.N., E. Papaparaskeva-papoutsoglou, and V. Theochari (1985). Formulation of practical diets for rainbow trout *(Salmo gairdneri)* made by partial or complete substitution of fish meal by poultry byproducts and contain plant byproducts. *Aquaculture* 50:61-73.

Allan, G.L., S. Parkinson, M.A. Booth, D.A.J. Stone, S.J. Rowlan, J. Frances, and R. Warner-Smith (2000). Replacement of fish meal in diets for Australian silver perch, *Bidyanus bidyanus:* I. Digestibility of alternative ingredients. *Aquaculture* 186:293-310.

Aoki, H., Y. Sanada, M. Furuichi, R. Kimoto, M. Maita, A. Akimoto, Y. Yamagata et al. (2000). Partial or complete replacement of fish meal by alternate Protein Sources in diets for yellow tail and red seabream. *Suisanzoshoku* 48(1): 53-63.

Association of American Feed Control Officials Inc. (2003). Official Publication. Oxford, IN.

Bishop, C.D., R.A. Angus, and S.A. Watts (1995). The use of feather meal as a replacement for fish meal in the diet of *Oreochromis niloticus* fry. *Bioresource Technology* 54:291-295.

Brandsen, M.P., C.G. Carter, and B.F. Nowak (2001). Alternative protein sources for farmed salmon. *Feed Mix* 9:18-21.

Bureau, D.P. (2000). *Use of Rendered Animal Protein Ingredients in Fish Feed.* Fish Nutrition Research Laboratory Research Report. Department of Animal and poultry Science, University of Guelph, Canada.

Bureau, D.P., C.Y. Cho, H.S. Bayley, and A.M. Harris (1998). Apparent digestibility of amino acids of feather meals, and meat and bone meals for salmonids. *Fats and Protein Research Foundation, Inc.*

Bureau, D.P., A.M. Harris, D.J. Beran, L.A. Simmons, P.A. Azevedo, and C.Y. Cho (2000). Feather meals and meat and bone meal from different origins as protein sources in rainbow trout *(Oncorhynchus mykiss)* diets. *Aquaculture* 181:281-291.

Bureau, D.P., A.M. Harris, and C.Y. Cho (1999). Apparent digestibility of rendered animal protein ingredients for rainbow trout *(Oncorhynchus mykiss). Aquaculture* 180:345-358.

Chai, S.Z. (2000). Studies on improvement of grouper feed-partial substitution of fish meal with different protein sources. MSc Thesis. National Taiwan Ocean University, Keelung, Taiwan.

Cheng, Z.J. and R.W. Hardy (2002). Apparent digestibility coefficients of nutrients and nutritional value of poultry by-product meals for rainbow trout *Oncorhynchus mykiss* measured *in vivo* using settlement. *Journal of World Aquaculture Society* 33(4):458-465.

Cheng, Z.J., R.W. Hardy, and N.J. Huige (2004). Apparent digestibility coefficients of nutrients in brewer's and rendered amiual byproducts for rainbow trout (*Oncorhynchus mykiss* (Walbaum)). *Aquaculture Research* 35:1-9.

Cho, C.Y. and S.J. Slinger (1978). Apparent digestibility measurement in feedstuffs for rainbow trout. *Proceeding of the World Symposium on Finfish Nutrition and Fishfeed Technology* 2:239-245.

Cui, Y.B., S.Q. Xie, X.M. Zhu, Y. Yang, Y.X. Yang, and Y. Yu (2002). *Replacement of fish meal by meat and bone meal and poultry byproduct meal in diets for gibel carp, Carassius auratus gibelio.* Research Report No. 2. National Renderers Association, Inc., Wuhan, China

Dong, F.M., R.W. Hardy, N.F. Haard, F.T. Barrows, B.A. Rasco, W.T. Fairgrieve, and I.P. Forster (1993). Chemical composition and protein digestibility of poultry byproduct meals for salmonid diets. *Aquaculture* 116:149-158

El-Sayed, A-F.M. (1998). Total replacement of fish meal with animal protein sources in Nile tilapia, *Oreochromis niloticus* (L.) feeds. *Aquaculture Research* 29:275-280

Firman, J.D., D. Robbins, and G.G. Pearl (2004). Use of ruminant meat meal and other rendering by-products by the poultry industry. *Director's Digest No. 327.* Fats and Proteins Research Foundation, Inc., Bloomington, IL.

Fowler, L.G. (1990). Feather meal as a dietary protein source during parr-smolt transformation in fall chinook salmon. *Aquaculture* 89:301-314.

Fowler, L.G. (1991). Poultry byproduct meal as a dietary protein source in fall Chinook salmon diets. *Aquaculture* 99:309-321.

Francis, G., H.P.S. Makkar, and K. Becker (2001). Anti-nutritional factors present in plant-derived alternate fish feed ingredients and their effects in fish. *Aquaculture* 199:197-227.

Gallagher, M.L. and M. LaDouceur (1995). The use of blood meal and poultry products as partial replacements for fish meal in diets for juvenile palmetto bass *(Morone saxatilis × M. chrysops). Journal of Applied Aquaculture* 5(3):57-65.

Gaylord, T.G. and D.M. Gatlin III (1996). Determination of digestibility coefficients of various feedstuffs for red drum *(Sciaenops ocellatus). Aquaculture* 139:303-314.

Hao, N.V. and Y. Yu (2003). *Partial replacement of fish meal by meat and bone meal or poultry byproduct meal in diets for river catfish, Pangasianodon hypophthalmus.* Research Report No. 33. National Renderers Association, Inc., HCM City, Vietnam.

Hardy, R.W. (1996). Alternate protein sources for salmon and trout diets. *Animal Feed Science Technology* 59:71-80.

Hardy, R.W. (2003). Fish nutrition and food quality: Future challenges for aquaculture. *Proceedings of education program.* Annual Meeting of the American Feed Industry Association., Hagerman, Idaho.

Hardy, R.W. and Z.J. Cheng (2002). Effect of poultry by-product meal supplemented with L-lysine, DL-methionine and L-histidine as a replacement for fish meal on the performance of rainbow trout. 2002. *Director's Digest No. 317.* Fats and Protein Research Foundation, Inc., Hagerman, Idaho.

Hasan, M.R., M.S. Haq, P.M. Das, and G. Mowlah (1997). Evaluation of poultry-feather meal as a dietary protein source for Indian major carp, *Labeo rohita* fry. *Aquaculture* 151:47-54.

Higgs, D.A., J.R. Markents, D.W. Maequrrie, J.R. McBride, B.S. Dosanjh, C. Nichols, and G. Hoskins (1979). Envelopment of practical dry diets for Coho salmon, *Oncorhynchus kisutch,* using poultry byproduct meal, feather meal, soybean meal and rapeseed meal as major protein sources. Proceedings of the World Symposium *on Finfish Nutrition and Fishfeed Technology* 2:191-215.

Jahan, P., T. Watanabe, S. Satoh, and V. Kiron (2001). Formulation of low phosphorus loading diets for carp (*Cyprinus carpio* L.). *Aquaculture Research* 32:361-368.

Kikuchi, K., T. Furuta, and H. Honda (1994). Utilization of feather meal as a protein source in the diet of juvenile Japanese flounder. *Fisheries Science* 60(2): 203-206.

Kureshy, N., D.A. Davis, and C.R. Arnold (2002). Partial replacement of fish meal with meat and bene meal, flash dried poultry byproduct meal, and enzyme digested poultry byproduct meal in practical diets for juvenile red drum. *North American Journal of Aquaculture* 62:266-272.

Lee, K.J., K. Dabrowski, J.H. Blom, and S.C. Bai (2001). Replacement of fish meal by a mixture of animal by-products in juvenile rainbow trout diets. *North American Journal of Aquaculture* 63:109-117.

Lee, K.J., K. Dabrowski, J.H. Blom, S.C. Bai, and P.C. Stromsburg (2002). A mixture of cottonseed meal, soybean meal and annual byproduct mixture as a fish meal substitute: Growth and tissue gossypol enantiomer in juvenile rainbow trout *(Oncorhynchus mykiss). Journal* of *Animal Nutrition* 86:201-213.

Lee, S.M. (2002). Apparent digestibility coefficients of various feed ingredients for juvenile and grower rockfish *(Sedates schlegeli). Aquaculture* 207:79-95

Li, M.H., B.B. Manning, and E.H. Robinson (2002). Comparison of various animal protein sources for growth, feed efficiency, and body composition of juvenile channel catfish *lctalurus punctatus. Journal of the World Aquaculture Society* 33(4):489-493.

Lin, S. and Y. Yu (2002). *Effect of partial replacement of fish meal by meat and bone meal or poultry by-product meal in commercial diets on growth response of Nile tilapia.* Research Report No. 21. National Renderers Association, Inc., Bangkok, Thailand

Liu, T.Q., X.Q. Qian, and Y. Yu (2002). *Replacement of fish meal with meat and bone meal or poultry by-product meal and the supplementation of amino acids on growth performance of juvenile common carp.* Research Report No. 15. National Renderers Association, Inc., Sichuan, China.

Mendoza, R., A. De Dios, C. Vasquez, E. Cruz, D. Ricque, C. Aguilera, and J. Montemayor (2001). Fishmeal replacement with feather-enzymatic hydrolyzates co-extruded with soya-bean meal in practical diets for the Pacific white shrimp *(Litopeneas vannamei). Aquaculture Nutrition* 7:143-151.

National Renderers Association, Inc. (2003). *Pocket Information Manual. A Buyer's Guide to Rendered Products.* Alexandria, VA.

National Research Council (1993). *Nutrient Requirements of Fish.* National Academy of Science, Washington, DC.

Naylor, R.L., R.J. Goldburg, J.H. Primavera, N. Kautsky, M.C.M. Beveridge, J. Clay, C. Folke, J. Lubcheneo, H. Mooney, and M. Troell (2000). Effect of aquaculture on world fish supplies. *Nature* 405:1017-1023.

Pauly, D., V. Christensen, S. Guenette, T.J. Pitcher, U.R. Sumaila, C.J. Walters, R. Watson et al. (2002). Towards sustainability in world fisheries. *Nature* 418: 689-695.

Pfeffer, E., S. Kinzinger, and M. Rodehutscord (1995). Influence of the proportion of poultry slaughter byproducts and of untreated or hydrothermically treated legume seeds in diets for rainbow trout, *Oncorhynchus mykiss* (Walbaum), on apparent digestibility's of their energy and organic compounds. *Aquaculture Nutrition* 1:111-117.

Pfeffer, E., D. Wiesmann, and B. Henrichfreise (1994). Hydrolyzed feather meal as feed component in diets for rainbow trout *(Oncorhynchus mykiss)* and effects of dietary protein/energy ratio on the efficiency of utilization of digestible energy and protein. *Archies of Animal Nutrition* 46:111-119.

Quartararo, N., G.L. Allan, and J.D. Bell (1998). Replacement of fish meal in diets for Australian snapper, *Pagrus auratus. Aquaculture* 166:279-295.

Rodriguez-Serna, M., M.A. Olvera-Novoa, and C. Carmona-Osalde (1996). Nutritional value of amiual byproduct meal in practical diets for Nile tilapia *Oreochromis niloticus* (L) fry. *Aquaculture Research* 27:67-73.

Serwata, R. and S.J. Davies (2001). Relative value of hydrolyzed feather meal as a fish meal substitute in diets for African catfish *Clarias gariepinus. Book of Abstracts of the World Aquaculture Society,* Baton Rogue, LA, 164 pp.

Serwata, R. and S. Woodgate (2004). *Partial replacement of fish meal with various animal protein meals in juvenile tilapia.* Research Report PDM Group, UK.

Steffens, W. (1988). Utilization of poultry by-product meal of raising carp fingerlings *(Cyprinus carpio) Archies of Animal Nutrition* 38(2):147-152.

Steffens, W. (1994). Replacing fish meal with poultry byproduct meal in diets for rainbow trout, *Oncorhynchus mykiss. Aquaculture* 124:27-34.

Tacon, A.G.J. and R.W. Hardy (2002). Use of rendered products for aquaculture. *Rendered Product Nutrition Symposium,* Mexico City, Mexico. National Renderers Association Publication, Hawaii.

Takagi, S., H. Hosokawa, S. Shimeno, and M. Ukawa (2000). Utilization of poultry byproduct meal in a diet for red seabream *Pagrus major. Nippon Suisan Gakkuishi* 66(3):428-438.

Tan, B.P., S.X. Zheng, and Y. Yu (2005). Replacing fishmeal with poultry by-product meal: Practical diets for grow-out culture of grouper *Epinephelus coioidesn. International Aquafeed* 8(3):26-29.

Turker, A., M. Yigit, S. Ergun, B. Karaali, and A. Erteken (2005). Potential of poultry by-product meal as a substitute for fishmeal in diets for Black Sea turbot *scophthalmus* maeoticus: Growth and nutrient utilization in Winter. *The Israeli Journal of Aquaculture-Bamidgeh* 57(1):49-61.

Wang, C., X.Q. Qian, and Y. Yu (2002). *Replacement of fish meal with meat and bone meal or poultry byproduct meal and the supplementation of amino acids on growth performance of growing comment carp.* Research Report No. 16. National Renderers Association, Inc., Sichuan, China.

Wang, Y. (2004). *Replacement of fish meal with meat and bone meal, poultry by-product meal and feather meal on growth performance of cuneate drum.* Fats and Proteins Research Foundation, Inc., Bloomington, IL.

Webster, C.D., K.R. Thompson, A.M. Morgan, E.J. Grisby, and A.L. Gannam (2000). Use of hempseed meal, poultry byproduct meal, and canola meal in practical diets without fish meal for sunshine bass *(Morone chrysops* × *M. saxatilis). Aquaculture* 188:299-309.

Webster, C.D., L.G. Till, A.M. Morgan, and A. Gannam (1999). Effect of partial and total replacement of fish meal on growth and body composition of sunshine bass *Morone chrysops* × *M. saxatilis* fed practical diets. *Journal of the World Aquaculture Society* 30(4):443-453.

Wiesmann, D., H. Scheid, and E. Pfeffer (1988). Water pollution with phosphorus of dietary origin by intensively fed rainbow trout (*Salmo gairdneru* Rich). *Aquaculture* 69:263-270.

Winfree, R.A. and R.R. Stickney (1984). Formulation and processing of hatchery diets for channel catfish. *Aquaculture* 41:311-323.

Yang, Y., S.Q. Xie, W. Lei, X.M. Zhu, Y.X. Yang, J.K. Liu, and Y. Yu (2002). *Effect of replacement of fish meal by poultry by-product meal in the diets for gibel carp, Carassius auratus gibelio on growth and feed utilization.* Research Report No. 1 National Renderers Association. Inc., Wuhan, China.

Yanik, T., K. Dabrowski, and S.C. Bai (2003). Replacing fish meal in rainbow trout *(Oncorhynchus mykiss)* diets. *The Israeli Journal of Aquaculture* 55(3):179-186.

Chapter 4

Use of Meatpacking By-Products in Fish Diets

Menghe H. Li
Edwin H. Robinson
Chhorn E. Lim

INTRODUCTION

Traditionally fish diets, especially diets for carnivorous species, contain high levels of fish meal because of its high protein quality and palatability, and because of the lack of information on utilization of other protein sources. Since fish meal is of limited supply and more expensive than most other protein sources, reducing its use in fish diets, while maintaining optimum fish performance, will increase profits and improve sustainability of the aquaculture industry. Meat-packing by-products, including meat meal, meat and bonemeal (M&B meal), and blood meal, are less expensive protein sources that can be used to replace all or part of the fish meal in fish diets resulting in less-costly diets.

Global meat production in 2002 reached approximately 245 million tons with pork, beef, and lamb meat accounting for approximately 68 percent of total meat production (FAO, 2003). Annual world animal by-product production is estimated to be approximately 120 million tons (wet weight) based on the estimated proportions of these animals

The manuscript is approved for publication as Journal Article No. BC-10468 of the Mississippi Agricultural and Forestry Experiment Station (MAFES), Mississippi State University. This project is supported under MAFES Project Number MIS-371160.

Alternative Protein Sources in Aquaculture Diets
doi:10.1300/5892_04

that are not suitable for direct human consumption. These materials are transformed into a variety of products, but the majority of these materials are processed into meat meal, M&B meal, and blood meal that may be used in animal diets. These animal by-products are high in protein and generally digestible by most fish.

However, protein quality (as measured by digestible protein and available essential amino acid concentrations) and for some species, palatability, of these protein sources is somewhat inferior to that of fish meal. Protein quality may be improved by using a mixture of meat meal/M&B meal and blood meal because of their complementary essential amino acids. Although M&B meal is a good source of minerals, its high ash content may limit its use in fish diets because of a possible mineral imbalance. Owing to the bovine spongiform encephalopathy or "mad cow" disease, there are government restrictions on the use of certain animal by-products in diets for certain or all farmed animals. This chapter will review existing information concerning the nutritive value and the use of meatpacking by-products in fish diets.

MEATPACKING BY-PRODUCTS

Meat Meal and M&B Meal

Meat meal and M&B meal are the rendered products from mammalian (mainly beef, pork, or lamb) tissues and should not contain added blood, hair, hoof, horn, hide trimmings, manure, and stomach and rumen contents except in amounts as may be unavoidable in good processing practices. They are typically produced by the dry-rendering method, in which the preground raw material is cooked by dry heat at approximately 135-140°C (no steam or hot water is added to the raw material) in a steam-jacketed cooker until the moisture has evaporated, followed by fat removal by draining off and screw press, and grinding. The difference between meat meal and M&B meal is that meat meal should not contain added bone and, therefore, the ash content in meat meal is lower than that in M&B meal.

Blood Meal

Blood meal is prepared from clean, fresh animal blood, excluding hair, stomach contents, and urine, except in trace quantities that are unavoidable. It may be classified into three types based on drying processes. Spray-dried blood meal is prepared by first removing water from fresh, uncoagulated blood by a low temperature (approximately 50°C) evaporator under the vacuum until it contains approximately 30 to 50 percent solids, and then by spray-drying at higher temperatures (250-300°C). Flash-dried (ring-dried) blood meal is processed by first removing a large portion of the water in coagulated blood by centrifugation and then transferring it into rotating rings where high temperatures (400 to 550°C) force off the water, leaving a dry product.

Cooker-dried blood meal is produced by drying the blood in a conventional cooker. The protein quality of blood meal depends largely on drying temperature and time. Overheating during processing of blood meal reduces amino acid availability to animals. High-quality blood meal is of reddish to dark-grayish color. Generally, spray-drying results in a better quality of blood meal than flash-drying and cooker-drying because of lower temperature and shorter retention time used in spray-drying. Cooker-drying is an older process and because of the longer drying time, cooker-dried blood meal usually has a dark black color and is of lowest protein quality.

M&B Meal and Blood Meal Mix

Meat and bonemeal and blood meal are sometimes mixed in certain proportions to mimic the desired nutritional profile (usually fish meal). Mixed products available in the United States typically have a M&B meal:blood meal ratio of approximately 3:2.

CHEMICAL COMPOSITION

Meat meal contains approximately 55 percent crude protein, 7 percent crude fat, and 25 percent ash (Table 4.1). Meat and bonemeal contains approximately 45 to 50 percent crude protein, 8.5 percent crude fat, and 33 to 37 percent ash. Their protein quality is inferior to whole fish meal because they contain less lysine and the consistency of the product may vary considerably. It is a good source of minerals. However, its high ash content may limit its use because of the possibility that a mineral imbalance may occur in the diet.

TABLE 4.1. Composition (% as fed) of selected nutrients for blood meal, meat meal, M&B meal, menhaden fish meal, and dehulled soybean meal. Values in parenthesis are % of crude protein. Data adapted from Dale (1998).

	M&B meal	Meat meal	Blood meal	Fish meal (menhaden)	Soybean meal (dehulled)
Dry matter	93	93	89	92	89
Crude protein	50.0	55.0	80.0	62.0	48.5
Crude fat	8.5	7.2	1.0	9.2	1.0
Crude fiber	2.8	2.5	1.0	1.0	3.0
Ash	33.0	25.0	4.4	19.0	6.0
Total phosphorus	4.7	4.0	0.22	3.0	0.65
Arginine	3.35 (6.70)	3.70 (6.73)	2.35 (2.94)	3.65 (5.89)	3.80 (7.84)
Histidine	0.96 (1.92)	1.10 (2.00)	3.05 (3.81)	1.52 (2.45)	1.30 (2.68)
Isoleucine	1.70 (3.40)	1.90 (3.45)	0.80 (1.00)	2.40 (3.87)	2.60 (5.36)
Leucine	3.20 (6.40)	3.50 (6.36)	10.30 (12.88)	4.40 (7.10)	3.80 (7.84)
Lysine	2.60 (5.20)	3.00 (5.45)	6.90 (8.63)	4.70 (7.58)	3.20 (6.60)
Methionine	0.67 (1.34)	0.75 (1.36)	1.00 (1.25)	1.70 (2.74)	0.75 (1.55)
Cystine	0.33 (0.66)	0.68 (1.24)	1.40 (1.75)	0.50 (0.81)	0.74 (1.53)
Phenylalanine	1.70 (3.40)	1.90 (3.45)	5.10 (6.38)	2.28 (3.68)	2.70 (5.57)
Threonine	1.70 (3.40)	1.81 (3.29)	3.80 (4.75)	2.75 (4.44)	2.00 (4.12)
Tryptophan	0.26 (0.52)	0.35 (0.64)	1.00 (1.25)	0.50 (0.81)	0.70 (1.44)
Valine	2.25 (4.50)	2.60 (4.73)	5.20 (6.50)	2.80 (4.52)	2.70 (5.57)

Blood meal generally contains approximately 80 to 85 percent crude protein, 1 percent crude fat, and 4 percent ash. It is a good source of lysine, but it is deficient in methionine. Meat and bone and blood meal mix typically contains approximately 60 to 65 percent crude protein. Its amino acid composition is more balanced than that of M&B meal or blood meal as a single ingredient. The mixed product contains less ash and fat than M&B meal because blood meal is low in these nutrients.

NUTRIENT AVAILABILITY

Protein Digestibility

Protein digestibility coefficients of protein sources vary among fish species (Table 4.2). Different results also have been reported for the same species in different studies due to differences in experimental conditions, diet compositions, or fecal collection methods. Common carp, *Cyprinus carpio* (Pongmaneerat and Watanabe, 1991) and rainbow trout, *Oncorhynchus mykiss* (Watanabe and Pongmaneerat, 1991; Smith et al., 1995; Sugiura and Hardy, 2000) appear to digest meat meal protein equally well as fish meal. Protein digestibility of meat meal by Australian silver perch, *Bidyanus bidyanus* (Stone et al., 2000) and gilthead seabream, *Sparus aurata* (Nengas et al., 1995; Lupatsch et al., 1997) is lower than that of fish meal. Based on protein digestibility coefficients determined for several species, fish generally do not digest M&B meal protein as well as fish meal, likely due to the higher ash content in the M&B meal (Watanabe and Pongmaneerat, 1991; Nengas et al., 1995).

Protein from blood meal appears to be as easily digested as that from fish meal by Australian silver perch (Allan et al., 2000a), channel catfish, *Ictalurus punctatus* (Brown et al., 1985), palmetto bass, *Morone saxatilis* × *M. chrysops* (Sullivan and Reigh, 1995), rainbow trout (Smith et al., 1995; Bureau et al., 1999; Sugiura and Hardy, 2000), and red drum, *Sciaenops ocellatus* (McGoogan and Reigh, 1996). Protein digestibility of steam-dried blood meal by gilthead seabream was quite low compared to that of fish meal (46 percent versus 96 percent) (Nengas et al., 1995). However, for the same species, protein digestibility of spray-dried blood meal is slightly higher than that of fish meal

TABLE 4.2. Apparent protein digestibility coefficients of selected species. [a]

Feedstuff[b]	Australian silver perch[c]	Channel catfish[d]	Chinook salmon[e]	Common carp[f]	Gilthead seabream[g]	Nile tilapia[h]	Palmetto bass[i]	Sunshine bass[j]	Red drum[k]	Rainbow trout[l]
				Apparent protein digestibility coefficient (%)						
Meat meal	84	–	–	87	72, 79	–	–	–	–	81, 92
M&B meal	72, 73	63, 74, 82	–	54	40	78	73	73	74, 79	54, 78, 87
Blood meal	90	74	29	–	46, 90	–	86	–	100	80, 91, 93
Fish meal	92	78, 81, 89	90	87	83, 96	85	88	81	82, 96	87, 89, 91
Soybean meal	95	81, 85, 95	76	–	87, 91	94	80, 77, 80, 86	82, 88	–	87, 89, 91

[a]Scientific names: Australian silver perch, *Bidyanus bidyanus*; channel catfish, *Ictalurus punctatus*; chinook salmon, *Oncorhynchus tshwaytscha*; common carp, *Cyprinus carpio*; gilthead seabream, *Sparus aurata*; Nile tilapia, *Oreochromis niloticus*; Palmetto bass, *Morone saxatilis* × *M. chrysops*; sunshine bass, *M. chrysops* × *M. saxatilis*; red drum, *Sciaenops ocellatus*; rainbow trout, *Oncorhynchus mykiss*.

[b]See specific references for detailed description of the diet ingredients.

[c]From Allan et al., 2000a; Stone et al., 2000.

[d]Cruz, 1975; Brown et al., 1985; Wilson and Poe, 1985.

[e]Hajen et al., 1993.

[f]Pongmaneerat and Watanabe, 1991.

[g]Nengas et al., 1995; Lupatsch et al., 1997.

[h]Popma, 1982.

[i]Sullivan and Reigh, 1995.

[j]Rawles and Gatlin, 2000.

[k]McGoogan and Reigh, 1996; Gaylord and Gatlin, 1996.

[l]Watanabe and Pongmaneerat, 1991; Smith et al., 1995; Bureau et al., 1999; Sugiura and Hardy, 2000.

(90 percent versus 83 percent) (Lupatsch et al., 1997). Hajen et al. (1993) reported a very low digestibility (29 percent) of blood meal protein by chinook salmon, *Oncorhynchus tshwaytscha.* The authors suggested that the extremely low digestibility was likely caused by excessive heating during processing of the blood meal, as indicated by a dark purple color and the presence of charred materials.

Amino Acid Availability

So far, amino acid availabilities of meatpacking products have been determined for only a few species (Tables 4.3 and 4.4). For Australian silver perch, amino acid availabilities for blood meal and M&B meal are lower than those for fish meal, with those for M&B meal being lower than those for blood meal (Allan et al., 2000a). Amino acids

TABLE 4.3. Apparent amino acid availabilities[a] (expressed as a percentage) for blood meal, meat meal, M&B meal, fish meal, and soybean meal determined for Australian silver perch, *Bidyanus bidyanus,* and channel *catfish, Ictalurus punctatus.*

	Australian silver perch[b]				Channel catfish[c]		
Amino acid	M&B meal	Blood meal	Fish meal	Soybean meal	M&B meal	Fish meal	Soybean meal
Arginine	75.4	93.3	95.0	97.7	86.1	89.2	95.4
Histidine	82.7	94.4	97.0	96.8	74.8	79.3	83.6
Isoleucine	80.2	80.0	95.4	95.9	77.0	84.8	77.6
Leucine	81.9	92.9	96.1	95.6	79.4	86.2	81.0
Lysine	80.2	92.9	96.5	96.8	81.6	82.5	90.9
Methionine	83.1	91.6	95.0	96.9	76.4	80.8	80.4
Cystine	74.9	87.0	96.2	94.1	–	–	–
Phenylalanine	79.2	92.9	94.8	96.5	82.2	84.1	81.3
Tyrosine	84.0	93.1	94.2	96.3	77.6	84.8	78.7
Threonine	80.1	93.8	95.3	95.5	69.9	83.3	77.5
Valine	78.9	92.2	95.4	95.7	77.5	84.0	75.5

[a]See specific references for detailed description of the diet ingredients.

[b]From Allan et al., 2000a.

[c]Wilson et al., 1981.

TABLE 4.4. Apparent amino acid availabilities[a] (expressed as a percentage) for blood meal, meat meal, M&B meal, fish meal, and soybean meal determined for gilthead seabream, *Sparus aurata,* and yellowtail, *Seriola quinqueradiata.*

Amino acid	Gilthead seabream[b]				Yellowtail[c]		
	Meat meal	Blood meal	Fish meal	Soybean meal	Meat meal	Fish meal	Soy protein
Arginine	89	97	93	96	82.2	92.5	89.9
Histidine	75	95	85	89	86.0	93.0	92.5
Isoleucine	82	80	88	89	75.9	90.2	87.9
Leucine	83	96	92	90	77.5	90.7	86.9
Lysine	86	94	91	92	85.0	93.1	91.2
Methionine	86	89	91	89	83.8	92.2	86.8
Cystine	–	–	–	–	43.8	90.3	87.2
Phenylalanine	80	95	88	89	78.4	88.8	88.9
Tyrosine	–	–	–	–	76.3	90.1	89.1
Threonine	88	96	93	91	73.8	88.9	83.0
Valine	81	96	90	89	72.3	85.7	79.7

[a]See specific references for detailed description of the diet ingredients.

[b]From Lupatsch et al., 1997.

[c]Masumoto et al., 1996.

in M&B meal are generally less available to channel catfish than those in fish meal (Wilson et al., 1981). Similarly, amino acids in meat meal are less available to yellowtail, *Seriola quinqueradiata* than those in fish meal (Masumoto et al., 1996). Generally, amino acids in blood meal are equally available to gilthead seabream as those in fish meal except isoleucine; however, amino acids in meat meal are less available to the fish than those in fish meal (Lupatsch et al., 1997).

Energy Digestibility

Digestibility of energy from meat meal appears to be approximately equal to that from fish meal for common carp (Pongmaneerat and Watanabe, 1991) and rainbow trout (Watanabe and Pongmaneerat, 1991) (Table 4.5). Energy digestibility of meat meal appears to be lower than that of fish meal for Australian silver perch (Stone et al., 2000). Nengas et al. (1995) reported that meat meal energy was less digestible

TABLE 4.5. Apparent energy digestibility coefficients of selected species. [a]

Feedstuff[b]	Australian silver perch[c]	Channel catfish[d]	Chinook salmon[e]	Common carp[f]	Gilthead seabream[g]	Nile tilapia[h]	Palmetto bass[i]	Sunshine bass[j]	Red drum[k]	Rainbow trout[l]
Meat meal	82	–	–	86	69, 78	–	–	–	85	–
M&B meal	76, 78	73, 81	–	73	23	69	80	92	54, 86	64, 73, 77
Blood meal	100	–	32	–	58, 83	–	–	–	58	67, 89
Fish meal	92	85, 94	89	88	80, 94	87	96	98	60, 94	87, 90
Soybean meal	78	56, 73	68	–	45, 72	73	55	56	38, 63	63

[a]See note below Table 4.2 for scientific names of fish species.

[b]See specific references for detailed description of the diet ingredients.

[c]From Allan et al., 2000a; Stone et al., 2000.

[d]Cruz, 1975; Wilson and Poe, 1985;

[e]Hajen et al., 1993.

[f]Pongmaneerat and Watanabe, 1991.

[g]Nengas et al., 1995; Lupatsch et al., 1997.

[h]Popma, 1982.

[i]Sullivan and Reigh, 1995.

[j]Rawles and Gatlin, 2000.

[k]McGoogan and Reigh, 1996; Gaylord and Gatlin, 1996.

[l]Watanabe and Pongmaneerat, 1991; Smith et al., 1995; Bureau et al., 1999.

to gilthead seabream than fish meal. However, for the same species, Lupatsch et al. (1997) found that energy digestibility of meat meal and fish meal was approximately the same. Similar to protein digestibility, energy digestibility coefficients for M&B meal determined for various species are generally lower than those for fish meal.

Energy from blood meal appears to be as easily digested as that from fish meal by Australian silver perch (Allan et al., 2000a), rainbow trout (Smith et al., 1995), and red drum (McGoogan and Reigh, 1996). Nengas et al. (1995) reported an energy digestibility of 58 percent for steam-dried blood meal as against 94 percent for fish meal by gilthead seabream. However, Lupatsch et al. (1997) found that energy digestibility coefficients for spray-dried blood meal and fish meal were similar for gilthead seabream. Hajen et al. (1993) reported a very low energy digestibility for blood meal by chinook salmon (digestibility was only 32 percent), which was likely caused by excessive heating during processing of the blood meal.

Lipid Digestibility

Lipid from meat meal is highly digestible by common carp (Pongmaneerat and Watanabe, 1991), gilthead seabream (Lupatsch et al., 1997), and rainbow trout (Watanabe and Pongmaneerat, 1991) with apparent digestibility coefficients of approximately 90 percent, almost the same as that of fish meal (Table 4.6). Lipid from blood meal appears to be highly digestible (94 percent versus 98 percent for fish meal) by rainbow trout (Bureau et al., 1999). However, lipid digestibility of M&B meal determined for several species appears to be lower than that of fish meal.

Phosphorus Availability

Phosphorus availabilities of meatpacking by-products have been determined for only a few species (Table 4.7). For channel catfish, phosphorus in M&B meal and M&B/blood meal mix appears to be equally digestible as that in fish meal (Lovell, 1978; Wilson et al., 1982; Li and Robinson, 1996; Buyukates et al., 2000). Rainbow trout appears to utilize phosphorus in meat meal equally well as that in fish meal; however, phosphorus availability for M&B meal appears to be slightly lower than that of fish meal and meat meal (Sugiura and Hardy,

TABLE 4.6. Apparent lipid digestibility coefficients of selected species.[a]

Feedstuff[b]	Apparent lipid digestibility coefficient (%)					
	Channel catfish[c]	Common carp[d]	Gilthead seabream[e]	Sunshine bass[f]	Rainbow trout[g]	Red drum[h]
Meat meal	–	91	88, 90	–	91	–
M&B meal	77	73	–	88	71, 81	67
Blood meal	–	–	–	–	94	–
Fish meal	97	88	95	95	98	77
Soybean meal	81	–	63	54	–	63

[a]See note below Table 4.2 for scientific names of fish species.

[b]See specific references for detailed description of the diet ingredients.

[c]From Cruz, 1975.

[d]Pongmaneerat and Watanabe, 1991.

[e]Nengas et al., 1995; Lupatsch et al., 1997.

[f]Rawles and Gatlin, 2000.

[g]Watanabe and Pongmaneerat, 1991; Bureau et al., 1999.

[h]Gaylord and Gatlin, 1996.

2000). In red drum, phosphorus utilization from M&B meal and fish meal is approximately the same (Gaylord and Gatlin, 1996).

UTILIZATION OF MEATPACKING BY-PRODUCTS BY FINFISH

Meatpacking by-products are an abundant protein source that may be used at various levels in fish diets. These products are high in protein and generally less expensive than fish meal, though protein quality is inferior to that of whole fish meal. They may be used to partially or totally replace fish meal in the diet to reduce diet cost and improve profitability. Amounts of meatpacking by-products that can be used in fish diets depend on fish species, product quality (essential amino acid and ash content), and cost per unit protein of these products (Tables 4.8-4.10).

TABLE 4.7. Apparent phosphorus availabilities[a] (expressed as a percentage) for blood meal, meat meal, M&B meal, fish meal, and soybean meal determined for channel catfish, *Ictalurus punctatus,* red drum, *Sciaenops ocellatus,* and rainbow trout, *Oncorhynchus mykiss.*

Feedstuff	Channel catfish[b]	Red drum[c]	Rainbow trout[d]
Meat meal	–	–	45
M&B meal	53, 53	66	28
Meat and bone and blood meal mix	84[e]	–	–
Fish meal	40, 46, 46, 75[e]	49	40
Soybean meal	29, 36, 36, 49, 54[e]	47	24

[a]See specific references for detailed description of the diet ingredients.

Sources: [b]Lovell, 1978; Wilson et al., 1982; Li and Robinson, 1996; Buyukates et al., 2000.

[c]Gaylord and Gatlin, 1996.

[d]Sugiura and Hardy, 2000.

[e]Values were not determined by a digestibility trial, rather based on weight gain of fish compared to that of fish fed a reference diet containing 0.4% available phosphorus from monobasic sodium phosphate (Li and Robinson, 1996). Weight gain appeared to be a more reliable indicator than did bone phosphorus concentrations.

Meat Meal and M&B Meal

Meat and bonemeal appears to be a good protein source for channel catfish. Mohsen and Lovell (1990) reported that 11.3 percent M&B meal could replace all menhaden fish meal in channel catfish (raised from approximately 3 to 15 g in aquaria) diets without affecting fish growth and feed efficiency. Pond-raised Australian silver perch may utilize 37 percent M&B meal to replace 80 percent Danish fish meal in the diet without any reduction in weight gain, feed efficiency, or protein efficiency ratio (PER) (Allan et al., 2000b).

Various dietary inclusion levels of M&B meal have been reported for common carp. The differences are likely caused by differences in the quality of M&B meal used and the experimental conditions. Zhu et al. (2002) reported that 4.5 percent M&B meal could replace 25 percent fish meal protein in the diet of common carp raised from 1.50 to 1.94 kg in tanks without affecting growth and feed efficiency.

TABLE 4.8. Inclusion levels of M&B meal in fish diets that do not affect growth and feed efficiency (values in parenthesis are fish meal replacement levels).

Species[a]	Level not affecting weight gain (%)	Level not affecting feed efficiency (%)	References
Australian silver perch	37 (80)	37 (80)	Allan et al. (2000a)
Channel catfish	11 (100)	11 (100)	Mohsen and Lovell (1990)
Common carp	4.5 (25)	4.5 (25)	Zhu et al. (2002)
	<3.8 (25)	<3.8 (25)	Liu et al. (2003)
	12 (100)	12 (100)	Wang et al. (2003a)
Gilthead seabream	28 (40)	28 (40)	Robaina et al. (1997)
Japanese flounder	9 (10)	36 (40)	Kikuchi et al. (1997)
	24 (60)	–	Zhu and Yu (2003)
Nile tilapia	40 (100)	<40 (100)	El-Sayed (1998)
	6 (100)	6 (100)	Wu et al. (1999)
	12 (67)	12 (67)	Xue et al. (2002)
Palmetto bass	<36 (100)	<36 (100)	Webster et al. (1997)
Sunshine bass	29 (100)	29 (100)	Webster et al. (1999)
	45 (100)	45 (100)	Bharawaj et al. (2002)
Rainbow trout	24 (32)	12 (16)	Bureau et al. (2000)
	19.5 (100)	<19.5 (100)	Yamamoto et al. (2002)
Red drum	<5.3 (17)	5.3 (17)	Kureshy et al. (2000)

[a]Scientific names: Japanese flounder, *Paralichthys olivaceus*. See note below Table 4.2 for scientific names of other fish species.

Higher levels of fish meal substitution by M&B meal (with lysine and methionine supplementation) reduced weight gain and feed efficiency. In contrast, Liu et al. (2003) found that a dietary level of 3.8 percent M&B meal replacing 25 percent fish meal protein reduced growth of common carp cultured from 28 to 135 g in ponds. In another study, Wang et al. (2003a) demonstrated that 12 percent M&B meal plus lysine and methionine supplementation could replace all fish meal in

TABLE 4.9. Inclusion levels of blood meal in fish diets that do not affect growth and feed efficiency (values in parenthesis are fish meal replacement levels).

Species[a]	Blood meal type	Level not affecting weight gain (%)	Level not affecting feed efficiency (%)	Reference
Channel catfish	Ring-dried	<3.4 (100)	<3.4 (100)	Mohsen and Lovell (1990)
Nile tilapia	Sun-dried	<30 (100)	<30 (100)	El-Sayed (1999)
Palmetto bass	Spray-dried	9.5 (25)	9.5 (25)	Gallagher and LaDouceur (1995)
Japanese eel	Hemoglobin powder	35.8 (75)	35.8 (75)	Lee and Bai (1997)
Japanese flounder	Type not specified	10 (47)	10 (47)	Bharawaj et al. (2002)

[a]Scientific names: Japanese eel, *Anguilla japonicus.* See note below Tables 4.2 and 4.8 for scientific names of other fish species.

TABLE 4.10. Inclusion levels of meat and bone/blood meal mix in fish diets that do not affect growth and feed efficiency (values in parenthesis are fish meal replacement levels).

Species[a]	M&B meal/ blood meal ratio	Level not affecting weight gain (%)	Level not affecting feed efficiency (%)	Reference
Australian silver perch	11:1	24 (63)	24 (63)	Allan et al. (2000b)
Channel catfish	3:2	10 (100)	10 (100)	Mohsen and Lovell (1990)
	3:2	8 (100)	8 (100)	Robinson (1992)
	3:2	5 (100)	5 (100)	Li et al. (2002)
Grouper	4:1	40 (80)	50 (100)	Millamena (2002)

[a]Scientific names: Grouper, *Epinephelus coioides.* See note below Table 4.2 for scientific names of other fish species.

the diet of common carp raised from 200 to 423 g in net cages without reduction in growth and feed efficiency.

Nile tilapia, *Oreochromis niloticus,* appear to utilize M&B meal well. El-Sayed (1998) reported that a dietary level of 40 percent M&B meal could completely replace herring fish meal in the diet for Nile tilapia without affecting growth or PER, but feed efficiency was reduced. Wu et al. (1999) found that at the 6 percent inclusion level, M&B meal could replace all menhaden fish meal in Nile tilapia diets without affecting fish performance. Xue et al. (2002) showed that 12 percent M&B meal could replace 67 percent fish meal protein without affecting weight gain and feed efficiency of Nile tilapia.

In hybrid striped bass, Webster et al. (1997) reported that palmetto bass fed a diet containing 36.5 percent M&B meal without fish meal had similar feed efficiency and PER, but lower weight gain, compared to fish fed diets containing 15 to 45 percent anchovy fish meal. A dietary level of 29 percent M&B meal, in combination with soybean meal and distiller's grains with solubles, may be used to totally replace menhaden fish meal in sunshine bass, *Morone chrysops* $\times$ *M. saxatilis,* diet without affecting growth, feed efficiency, PER, net protein utilization, or body composition (Webster et al., 1999). In another sunshine bass study, Bharadwaj et al. (2002) reported that M&B meal could be used at levels up to 45 percent in the diet (replacing all fish meal) without affecting growth, feed efficiency or PER; however, protein digestibility, and phosphorus and amino acid availabilities were lower compared with fish fed 30 percent or lower levels of M&B meal.

In rainbow trout, a dietary level of 12 percent M&B meal could substitute for 16 percent herring fish meal in the diet without adverse effects on growth, feed efficiency, and nitrogen and energy retention (Bureau et al., 2000). A dietary level of 24 percent M&B meal (replacing 32 percent fish meal) resulted in similar growth characteristics except for reduced feed efficiency. Yamamoto et al. (2002) found that 19.5 percent M&B meal in combination with extruded soybean meal, corn gluten meal, and supplemental essential amino acids could completely replace white fish meal in the diet without affecting growth, PER, but resulted in lower feed efficiency in rainbow trout. Fish did not grow well on diets without supplementation of essential amino acids.

Meat and bonemeal (28 percent of diet) may be used to replace up to 40 percent sardine fish meal in gilthead seabream diets without affecting weight gain, feed efficiency, PER, or body composition (Robaina et al., 1997). However, polarization and isolated necrosis in hepatocites were observed when the diet included 40 percent M&B meal. Replacing 30 percent or more fish meal with M&B meal resulted in increased fat deposit in liver tissues. Also, at the 20 percent fish meal replacement level, protein digestibility was reduced most likely because of the high ash content in the M&B meal.

An earlier study indicated that Japanese flounder, *Paralichthys olivaceus,* could only utilize approximately 9 percent M&B meal to replace 10 percent white fish meal protein without affecting growth (Kikuchi et al., 1997). A later study demonstrated that 24 percent M&B meal could be used to replace 60 percent fish meal protein in Japanese flounder diets without reducing weight gain (Zhu and Yu, 2003). Japanese eel, *Anguilla japonicus,* may utilize approximately 15 percent M&B meal to replace 23 percent fish meal protein without affecting growth and feed efficiency (Wang et al., 2003b). Red drum do not appear to utilize M&B meal well based on results reported by Kureshy et al. (2000), who found that a dietary level of 5.3 percent M&B meal (replacing 17 percent menhaden fish meal) resulted in a lower weight gain than fish fed a control diet, but this level did not affect feed efficiency and protein conversion efficiency.

Blood Meal

Blood meal appears to be of lower quality than M&B meal for channel catfish. Mohsen and Lovell (1990) reported that fingerling channel catfish fed a diet containing 3.4 percent ring-dried blood meal had a lower weight gain than fish fed a diet containing 5 percent menhaden fish meal. In Nile tilapia, total substitution of fish meal by sun-dried blood meal (30 percent of diet) reduces growth, feed efficiency, and PER of the fish (El-Sayed, 1998). Palmetto bass appear to utilize 9.5 percent spray-dried blood meal in combination with supplemental methionine to replace up to 25 percent herring fish meal without affecting growth, feed efficiency, or PER; however, replacing 50 percent fish meal with blood meal markedly reduces growth, feed efficiency, or PER (Gallagher and LaDouceur, 1995).

Adelizi et al. (1998) reported that rainbow trout fed a soybean meal and corn gluten meal-based diet (no fish meal) containing 10.6 percent blood meal had lower weight gain and feed efficiency than fish fed a commercial salmon diet. However, in commercial salmonid diets that typically contain relatively high levels of fish meal, blood meal is commonly included. For example, Pacific salmon, *Oncorhynchus* spp., diets typically contain approximately 2.5 to 5 percent blood meal and rainbow trout diets contain approximatelyt 9 to 10 percent blood meal (Hardy, 1998). Japanese flounder may utilize approximately 10 percent spray-dried blood meal in the diet without affecting growth and feed efficiency (Kikuchi, 1999). Hemoglobin powder (92 percent crude protein) may replace up to 50 percent of the fish meal in Japanese eel diets without supplementation of essential amino acids, and up to 75 percent fish meal with supplemental essential amino acids (Lee and Bai, 1997).

M&B Meal and Blood Meal Mix

Blended products made from good quality M&B meal and blood meal are well utilized by channel catfish. Mohsen and Lovell (1990) found that 10 percent M&B/blood meal mix (3:2 ratio) could replace all menhaden fish meal in the channel catfish diet without affecting fish growth and feed efficiency. Robinson (1992) found that growth and feed efficiency of pond-raised channel catfish fed 8 percent M&B/ blood meal mix or menhaden fish meal were essentially the same. Recently, Li et al. (2002) reported that at 5 percent dietary inclusion level, M&B/blood meal mix (3:2 ratio) could replace all menhaden fish meal for channel catfish fingerlings raised in aquaria. In contrast, Li et al. (2003) found that weight gain and feed efficiency were lower in fish fed 6 or 12 percent M&B/blood meal mix than those fed the same levels of menhaden fish meal, respectively. The relatively lower utilization of M&B/blood meal mix compared to that of fish meal in the later study was likely due to a difference in the quality of M&B/ blood meal mix used since the quality of these products varies considerably due to the raw materials used and processing methods. Use of M&B/blood meal mix in channel catfish diets does not appear to affect body composition of the fish.

El-Sayed (1998) reported that total substitution of fish meal by M&B/blood meal mix (4:3 ratio, 35 percent of diet) reduced growth, feed efficiency, and PER of Nile tilapia. Meat meal (15.4 percent of diet) plus blood meal (3.4 percent) may substitute for approximately 50 percent Danish fish meal without affecting weight gain, feed efficiency, PER, or protein and energy retention in juvenile Australian silver perch raised in tanks (Stone et al., 2000). Up to 80 percent Chilean fish meal may be replaced by M&B/blood meal mix

(4:1 ratio) without adverse effects on growth, survival, feed efficiency, or body composition for grouper, *Epinephelus coioides* (Millamena, 2002).

CONCLUSION

Research studies with various species have demonstrated that meat-packing by-products are suitable ingredients in aquaculture diets as a partial or total replacement of fish meal. Although protein quality of these sources is somewhat inferior to that of fish meal, a mixture of meat meal/M&B meal and blood meal may be used to improve the protein quality because of their complementary essential amino acids. Inclusion levels of these diet ingredients vary among fish species. In general, meatpacking by-products of good quality may be used at levels of 5 to 10 percent in fish diets provided that the diet is formulated to contain adequate amounts of essential amino acids.

Owing to concerns of bovine spongiform encephalopathy or "mad cow" disease, many countries prohibit the use of certain animal by-products in animal diets, especially diets for ruminant animals. The European Community has placed a total ban on the use of meat meal and M&B meal in diets for all farmed animals. Recently, the European Community also passed regulations on animal by-products and classified these products into three categories. Category 3 products—materials that are derived from healthy animals—may be allowed in animal diets (intra-species recycling is still prohibited) after proper treatment in approved processing plants in the future if the total ban on meat meal and M&B meal is ever lifted. The U.S. Food and Drug Administration (FDA) prohibits the use of meat meal and M&B meal and is considering to ban the use of blood meal from ruminant animals in diets for the ruminants. Some fish diet manufacturers in the United States have voluntarily discontinued the use of meat meal, M&B meal, and blood meal for precautionary measures. Owing to government regulations and public perception and safety concerns, there is uncertainty about the future use of these products in fish diets.

REFERENCES

Adelizi, P.D., R.R. Rosati, K. Warner, Y.V. Wu, T.R. Muench, M.R. White, and P.B. Brown (1998). Evaluation of fish meal-free diets for rainbow trout, *Oncorhynchus mykiss. Aquaculture Nutrition* 4:255-262.

Allan, G.L., S. Parkinson, M.A. Booth, D.A.J. Stone, S.J. Rowland, J. Frances, and R. Warner-Smith (2000a). Replacement of fish meal in diets for Australian silver perch, *Bidyanus bidyanus:* I. Digestibility of alterative ingredients. *Aquaculture* 186:293-310.

Allan, G.L., S.J. Rowland, C. Mifsud, D. Glendenning, D.A.J. Stone, and A. Ford (2000b). Replacement of fish meal in diets for Australian silver perch, *Bidyanus bidyanus:* V. Least-cost formulation of practical diets. *Aquaculture* 186:327-340.

Bharadwaj, A.S., W.R. Brignon, N.L. Gould, P.B. Brown, and Y.V. Wu (2002). Evaluation of meat and bone meal in practical diets fed to juvenile hybrid striped bass *Morone chrysops* × *M. saxatilis. Journal of the World Aquaculture Society* 33:448-457.

Brown, P.B., R.J. Strange, and K.R. Robbins (1985). Protein digestibility coefficients for yearling channel catfish fed high protein feedstuffs. *Progressive Fish-Culturist* 47:94-97.

Bureau, D.P., A.M. Harris, D.J. Bevan, L.A. Simmons, P.A. Azevedo, and C.Y. Cho (2000). Feather meals and meat and bone meals from different origins as protein sources in rainbow trout *(Oncorhynchus mykiss)* diets. *Aquaculture* 181:281-291.

Bureau, D.P., A.M. Harris, and C.Y. Cho (1999). Apparent digestibility of rendered animal protein ingredients for rainbow trout *(Oncorhynchus mykiss). Aquaculture* 180:345-358.

Buyukates, Y., S.D. Rawels, and D.M. Gatlin (2000). Phosphorus fractions of various feedstuffs and apparent phosphorus availability to channel catfish. *North American Journal of Aquaculture* 62:184-188.

Cruz, E.M (1975). Determination of nutrient digestibility in various classes of natural and purified feed materials for channel catfish. Ph.D. dissertation, Auburn University, AL.

Dale, N (1998). Ingredient analysis table. *Feedstuffs Reference Issue* 70 (30):24-30.

El-Sayed, A.F (1998). Total replacement of fish meal with animal protein sources in Nile tilapia, *Oreochromis niloticus* (L.) feeds. *Aquaculture Research* 29:275-280.

FAO (2003). Food Outlook No. 2. Economic and Social Department, FAO, Rome, Italy.

Gallagher, M.L. and M. LaDouceur (1995). The use of blood meal and poultry products as partial replacements for fish meal in diets for juvenile palmetto bass *(Morone saxatilis* × *M. chrysops). Journal of Applied Aquaculture* 5:57-65.

Gaylord, T.G. and G.M. Gatlin (1996). Determination of digestibility coefficients of various feedstuffs for red drum *(Sciaenops ocellatus). Aquaculture* 139:303-314.

Hajen, W.E., D.A. Higgs, R.M. Beames, and B.S. Dosanjh (1993). Digestibility of various feedstuffs by post-juvenile chinook salmon *(Oncorhynchus tshawytscha)* in sea water. 2. Measurement of digestibility. *Aquaculture* 112:333-348.

Hardy, R.W (1998). Feeding salmon and trout. In *Nutrition and Feeding of Feed, second edition,* T. Lovell (ed). Boston, MA: Kluwer Academic Publishers, pp. 175-192.

Kikuchi, K (1999). Use of defatted soybean meal as a substitute for fish meal in diets of Japanese flounder *(Paralichthys olivaceus). Aquaculture* 179:3-11.

Kikuchi, K., T. Sato, T. Furuta, I. Sakaguchi, and Y. Deguchi (1997). Use of meat and bone meal as a protein source in the diet of juvenile Japanese flounder. *Fisheries Sciences* 63:29-32.

Kureshy, N., D.A. Davis, and C.R. Arnold (2000). Partial replacement of fish meal with meat-and-bone meal, flash-dried poultry by-product meal, and enzyme-digested poultry by-product meal in practical diets for juvenile red drum. *North American Journal of Aquaculture* 62:266-272.

Lee, K.J. and S.C. Bai (1997). Hemoglobin powder as a dietary fish meal replacer in juvenile Japanese eel, *Anguilla japonica* (Temminck et Schlegeli). *Aquaculture Research* 28:509-516.

Li, M.H. and E.H. Robinson (1996). Phosphorus availability (digestibility) of common feedstuffs to channel catfish *Ictalurus punctatus* as measured by weight gain and bone mineralization. *Journal of the World Aquaculture Society* 27:297-302.

Li, M.H., E.H. Robinson, and B.B. Manning (2002). Comparison of various animal protein sources for growth, feed efficiency, and body composition of juvenile channel catfish *Ictalurus punctatus. Journal of the World Aquaculture Society* 33:489-493.

Li, M.H., D.J. Wise, B.B. Manning, and E.H. Robinson (2003). Effect of dietary total protein and animal protein on growth and feed efficiency of channel catfish *Ictalurus punctatus* and their response to *Edwardsiella ictaluri* challenge. *Journal of the World Aquaculture Society* 34:223-228.

Liu, T., X. Qian, and Y. Yu (2003). Replacement of fish meal, with or without the supplementation of synthetic amino acids, with meat and bone meal or poultry meal (pet food grade) on growth performance of juvenile common carp. Report No. 5, Aquaculture Research Reports (Asia), National Renderers Association, Inc., Hong Kong, China.

Lovell, R.T. (1978). Dietary phosphorus requirement of channel catfish *(Ictalurus punctatus). Transactions of the American Fisheries Society* 107:617-621.

Lupatsch, I, G.W. Kissil, D. Sklan, and E. Pfeiffer (1997). Apparent digestibility coefficients of feed ingredients and their predictability in compound diets for gilthead seabream, *Sparus aurata* L. *Aquaculture Nutrition* 3:81-97.

Masumoto, T., T. Ruchimat, Y. Ito, H. Hosokawa, and S. Shimeno (1996). Amino acid availability values for several protein sources for yellowtail *(Seriola quinqueradiata). Aquaculture* 146:109-119.

McGoogan, B.B. and R.C. Reigh (1996). Apparent digestibility of selected ingredients in red drum *(Sciaenops ocellatus)* diets. *Aquaculture* 141:233-244.

Millamena, O.M. (2002). Replacement of fish meal by animal by-product meals in a practical diet for grow-out culture of grouper *Epinephelus coioides. Aquaculture* 204:75-84.

Mohsen, A.A. and R.T. Lovell (1990). Partial substitution of soybean meal with animal protein sources in diets for channel catfish. *Aquaculture* 90:303-311.

Nengas, I., M.N. Alexis, S.J. Davies, and G. Petichakis (1995). Investigation to determine digestibility coefficients of various raw materials in the diets for gilthead sea bream *Sparus auratus,* L. *Aquaculture Research* 26:185-194.

Pongmaneerat, J. and T. Watanabe (1991). Nutritive value of protein of feed ingredients for carp *Cyprinus carpio. Nippon Suisan Gakkashi* 57:503-510.

Popma, T.J. (1982). Digestibility of selected feedstuffs and naturally occurring algae by tilapia. Ph.D. dissertation, Auburn University, Alabama.

Rawles, S.D. and D.M. Gatlin (2000). Nutrient digestibility of common feedstuffs in extruded diets for sunshine bass *Morone chrysops* 9 × *M. saxatilis 8. Journal of the World Aquaculture Society* 31:570-579.

Robaina, L., F.J. Moyano, M.S. Izquierdo, J. Socorro, J.M Vergara, and D. Montero (1997). Corn gluten meal and meat and bone meals as protein sources in diets for gilthead seabream *(Sparus aurata)*: Nutritional and histological implications. *Aquaculture* 157:347-359.

Robinson, E.H. (1992). Evaluation of MSM fish formula as a replacement for fish meal in catfish feeds. *The Catfish Journal* 6 (12):12-13.

Smith, R.R., R.A. Winfree, G.L. Rumsey, A. Allred, and M. Peterson (1995). Apparent digestion coefficients and metabolizable energy of feed ingredients for rainbow trout *Oncorhynchus mykiss. Journal of the World Aquaculture Society* 26:432-437.

Stone, D.J., G.L. Allan, S. Parkinson, and S.J. Rowland (2000). Replacement of fish meal in diets for Australian silver perch, *Bidyanus bidyanus:* III. Digestibility and growth using meat meal products. *Aquaculture* 186:311-326.

Sugiura, S.H. and R.W. Hardy (2000). Environmentally friendly feeds. In *Encyclopedia of Aquaculture,* R.R. Stickney (ed). New York, NY: Wiley & Sons, Inc., pp. 299-310.

Sullivan, A.J. and R.C. Reigh (1995). Apparent digestibility of selected feedstuffs in diets for hybrid striped bass *(Morone saxatilis 9 × Morone chrysops 8). Aquaculture* 138:313-322.

Wang, C., X. Qian, and Y. Yu (2003a). Replacement of fish meal, with or without the supplementation of synthetic amino acids, with meat and bone meal or poultry by-product meal (pet food grade) on growth performance of growing common carp. Report No. 16, Aquaculture Research Reports II, National Renderers Association, Inc., Hong Kong, China.

Wang, C., X. Qian, and Y. Yu (2003b). Replacement of fish meal with meat and bone meal or poultry by-product meal (pet food grade) on growth performance of Japanese eel. Report No. 20, Aquaculture Research Reports II, National Renderers Association, Inc., Hong Kong, China.

Watanabe, T. and J. Pongmaneerat (1991). Quality evaluation of some animal protein sources for rainbow trout *Oncorhynchus mykiss. Nippon Suisan Gakkashi* 57: 495-501.

Webster, C.D., L.G. Tiu, A.M. Morgan, A.M., and A. Gannam (1999). Effect of partial and total replacement of fish meal on growth and body composition of sunshine bass, *Morone chrysops × M. saxatilis,* fed practical diets. *Journal of the World Aquaculture Society* 30:443-453.

Webster, C.D., L.G. Tiu, and J.H. Tidwell (1997). Effects of replacing fish meal in diets on growth and body composition of palmetto bass *(Morone saxatilis ×* *M. chrysops)* raised in cages. *Journal of Applied Aquaculture* 7:53-68.

Wilson, R.P. and P.E. Poe (1985). Apparent digestible protein and energy coefficients of common feed ingredients for channel catfish. *Progressive Fish-Culturist* 47:154-158.

Wilson, R.P., E.H. Robinson, D.M. Gatlin, and W.E. Poe (1982). Dietary phosphorus requirement of channel catfish. *Journal of Nutrition* 112:1197-1202.

Wilson, R.P., E.H. Robinson, and W.E. Poe (1981). Apparent and true availability of amino acids from common feed ingredients for channel catfish. *Journal of Nutrition* 111:923-929.

Wu, Y.V., K.W. Tudor, P.B. Brown, and R.R. Rosati (1999). Substitution of plant proteins or meat and bone meal for fish meal in diets for Nile tilapia. *North American Journal of Aquaculture* 61:58-63.

Xue, M., Z. Zhou, C.He, Y. Yu, and Z. Ren (2002). Partial or total replacement of fish meal by meat and bone meal in practical diets for Nile tilapia, *Oreochromis niloticus*. Report No. 4, Aquaculture Research Reports I, National Renderers Association, Inc., Hong Kong, China.

Yamamoto, T., T. Shima, H. Furuita, and N. Suzuki (2002). Influence of feeding diets with and without fish meal by hand and by self-feeders on feed intake, growth and nutrient utilization of juvenile rainbow trout *(Oncorhynchus mykiss)*. *Aquaculture* 214:289-305.

Zhu, W., S. Lu, and Y. Yu (2002). Replacement of fish meal with meat and bone meal in common carp and gibel carp diets. Report No. 6, Aquaculture Research Reports II, National Renderers Association, Inc., Hong Kong, China.

Zhu, W. and Y. Yu (2003). Replacement of fish meal with meat and bone meal or poultry by-product meal (pet food grade) on performance of Japanese flounder. Report No. 18, Aquaculture Research Reports II, National Renderers Association, Inc., Hong Kong, China.

Chapter 5

Use of Fisheries Coproducts in Feeds for Aquatic Animals

Ian Forster

INTRODUCTION

Fish meal is used in the manufacture of diets for most aquatic species, primarily as a source of high-quality protein and for its attractant characteristics. Globally, most fish meals are manufactured from fish that are harvested for this purpose off the west coast of South America, Asia, and northern Europe. The supply of traditional sources of fish meal has been static over the past several years, and this condition is predicted to remain for the foreseeable future (see review by Barlow, 2003). As aquaculture has increased, so has the demand for fish meal, resulting in a shift of this commodity from use in diets for other animals, especially poultry (Hardy and Tacon, 2002). The relatively high cost and highly variable supply of fish meal of suitable quality for aquaculture has resulted in the search for replacements. Most of this effort has been directed to identifying and developing suitable plant protein sources. Recently, however, coproducts of the fish harvesting industry have shown promise as alternative protein sources for aquafeeds. A coproduct is a material that is produced for

This work was sponsored by the Agricultural Research Service (ARS) of the United States Department of Agriculture (USDA) under grant agreement No. 59-5325-9-216. Any opinions, findings, conclusions, or recommendations expressed in this publication are those of the author and do not necessarily reflect the view of the U.S. Department of Agriculture.

the general public's use and is ordinarily used in the form it is produced by the process; a by-product, in contrast, is a material that is not one of the primary products of a production process and is not solely or separately produced by the production process. Examples of coproducts include meals, hydrolysates, and stickwater.

Two of the most important challenges to incorporating these into mainstream aquaculture diets are the production of material with sufficiently high quality and the logistics and cost of transportation of material. The coproducts of the Alaska fisheries offer a good example (Smiley et al., 2003). The principal species of the Alaskan fisheries is Alaska walleye pollock *(Theragra calchogramma)*, at well in excess of 1 million tons harvested annually. All commercial landings in Alaska are for human food. The by-products (what remains after the principal product is obtained) of the processing industry are either disposed of into the ocean, or directed to the production of coproducts, primarily fish meal. Wet reduction is the method most commonly employed for fish meal production in Alaska. The products of this method are fish meal, fish oil, and condensed fish solubles (stickwater). All of these have value in aquaculture, if the quality is maintained sufficiently high. Screening the bones from the material used to manufacture these products (Babbitt et al., 1994) has resulted in higher protein, and lower ash fish meals, and these have been demonstrated to have excellent nutritional properties, allowing incorporation into aquatic diets for marine and freshwater fish and shrimp. Animals maintained on diets containing these fish meals have performed equally well as those maintained on a diet containing fish meal of established high quality.

This chapter describes some of the recent findings related to the use of fisheries by-products as protein sources in diets for aquatic animals and presents recommendations for further research and commercial development. It should be mentioned that the most direct utilization of fisheries by-products comes from improved efficiency of recovery for human food, including development of new food products. Although many of the examples in this paper come from the Alaska fisheries, the conclusions apply to almost all fisheries around the world. Although this discussion focuses on finfish, there are significant opportunities in other forms of marine life, notably crustaceans and mollusks (Nicol et al., 2000; Hardy et al., 2005).

SOURCES

There are two sources of raw material that can be used for the production of coproducts as protein sources for aquatic diets: harvested wild fish, either in directed fisheries or as by-catch, and fish processing by-products. Each of these sources has its own considerations that affect the effectiveness with which it can be used.

Almost all of the marine protein currently used in aquatic diets is contained in the meals manufactured from directed fisheries. Most of the species of fish in this group are "industrial fish"; fish for which there is not a great demand for direct human consumption. These include: herring, menhaden, capelin, anchovy, pilchard, and mackerel. These species contain high levels of oil throughout the flesh and are used whole as harvested. During the manufacture of fish meal, the oil from the fish must be greatly reduced to make a flowable product and to reduce susceptibility to oxidation. The protein content of these fish varies only slightly, although the oil content is considerably more variable. The level and quality of the protein in fish meal is highly dependant on the manufacturing methods used.

An alternative source of marine protein is the by-products of fish processed for human consumption, including material from capture fisheries and from aquaculture. The raw materials from these sources include trimmings, bone, viscera, heads, skin, and gonads. The fish in this category are generally lower in fat and store their body lipid in the liver. These fish are referred to as "lean fish." As the majority of the flesh is removed for human consumption, the by-products tend to contain higher levels of bone and extra effort is required to control this in the final product.

The size of the global capture fisheries harvest (marine and freshwater) is approximately 90 to 95 million tons annually (FAO, 2003). In addition, aquaculture production is approximately 37 million tons (FAO, 2003), including fish and shellfish. The proportion of fish that are retained for human food is highly variable by species and region, but generally ranges from 25 to 50 percent. Improvements are being made to maximize the recovery of products for human consumption. Disposal of the remaining parts after food portions are removed is a common problem for the processing industry. Frequently, these by-products are dumped into the ocean or landfill, but many jurisdictions restrict this practice, forcing the development of alternate disposal methods. Conversion of these fishery by-products into ingredients suitable for incorporation in aquatic diets presents a suitable solution, although several issues need to be considered.

PRODUCTS AND THEIR USE IN AQUATIC DIETS

Raw Fish or Fish Parts

The simplest method of using by-products of the fish processing industry is by direct incorporation into diets. Currently, a considerable proportion of aquaculture diets, particularly in Asia, are produced using whole fish or fish parts. This is done on the farm site where the raw product is loaded into a mixer, often along with other ingredients, and then ground, mixed, and formed into moist pellets, which are then fed directly to the animals. As the resulting diets often have poor stability, they tend to break apart in the water, resulting in diminished water quality and inefficiency of feeding. For this reason, there is a general movement away from this kind of use of fisheries products.

Fish Meal

Fish meal is the most common source of aquatic protein in diets for aquatic and terrestrial animals (primarily poultry), and to which by-products can be successfully diverted. A number of processes are involved with fish meal manufacture: grinding, cooking, oil removal, bone removal, and drying. After the raw material is ground and cooked, the oil is removed, usually by means of a press, and sometimes by centrifugation. The presscake, the solid material left from the press, is then dried and ground to produce fish meal. The liquid portion that is removed contains a mixture of oil, water, and the soluble proteins. The oil can be separated from the aqueous portion, which is known as stickwater. The stickwater contains water-soluble proteins and may be added to the presscake prior to drying. Fish meals made without addition of the stickwater are called presscake meals, and those with stickwater are whole fish meal. The issues involved in adding stickwater to fish meal will be discussed in a later section.

There are a number of factors influencing the value of the final product, including the quality of the starting material, the temperature and duration of cooking and drying, and the quality and quantity of stickwater added back to the presscake.

There are two advantages in using fish processing by-products over industrial fish, in terms of their suitability for fish meal production. As the fish used for human consumption tend to be lean (unless the livers are utilized), it is possible to eliminate the oil removal step in fish meal manufacture. Also, because fish intended for human consumption must be of the highest quality, the raw by-products from processing are also fresh. On the other hand, because the parts of the fish that are removed for food generally are the highest in protein, the by-products frequently contain considerable amount of bones, lowering their usefulness as ingredients. The bone content can be reduced by screening the cooked material (Babbitt et al., 1994), resulting in an increased protein content and reduced the mineral (ash) content.

During the manufacture of fish meal, heat is applied at two steps: cooking and drying. Cooking is an important step in liberating the oils to ease their removal and reduce the microbial level in the material. Cooking is done on the wet product and temperature is rarely a problem. There is far more potential for thermal damage during the drying process, as the temperature of the material surpasses the boiling point of water. It is desirable to reduce the moisture content in fish meals in order to prevent microbial growth (mold, bacteria, etc.) and to reduce the cost of shipping. Drying is generally achieved either by flame drying, in which hot air from a flue is brought into contact with moist presscake in a cylinder, or steam drying, in which the presscake is passed through a double jacketed drum with steam applied in the middle jacket. Flame drying is cheaper, but is often associated with lower-quality meals. Steam drying, although more costly, provides more even heating that prevents damage to the protein and minimizes oxidation of the lipids. Presscake contains varying levels of moisture (40 percent) and fish meals generally contain below 10 percent moisture. The most critical period where heat damage may occur is at the lowest level of moisture. More recently, a number of techniques have become available to reduce the moisture while minimizing the opportunity for heat damage. For example, flash-driers use hot air traveling at high velocity to reduce the moisture quickly, without damaging the product.

Fish meals produced from fisheries by-products have performed well in research trials designed to examine their nutritional quality. Rathbone et al. (2001) examined the nutritional quality of three rock sole *(Lepidopsetta bilineata)* processing coproducts: a meal made directly from processing by-products; a meal made from this material after passing through a Brown Refiner (Babbitt et al., 1994); and, a skin-and-bone meal. A series of diets were made containing these three meals as the principal protein source and these were fed along with a commercial feed to juvenile coho salmon *(Oncorhynchus kisutch)* for twelve weeks. At the end of this trial, it was found that the deboned fish meal was similar to the commercial diet in nutritional quality as measured by growth and protein utilization, although the whole fish meal was somewhat lower and the skin-and-bone meal was considerably poorer.

In a more extensive trial, Forster et al. (2004) examined eleven whitefish meals made from by-products of the Alaska pollock and cod fisheries. These meals were obtained from commercial sources in Alaska and had suitable composition for use in aquatic diets (Smiley et al., 2003). These meals were incorporated into diets as complete replacement of a high-quality fish meal (Norse LT-94®) and were fed to Pacific white shrimp *(Litopenaeus vannamei)* under intensive static-water culture conditions for eight weeks. Under these conditions, the shrimp in all the experimental treatments performed well as those in the control, illustrating the high quality of the by-product fish meals. Similar results were found using these same meals in a growth trial with Pacific threadfin *(Polydactylus sexfilis),* a marine warm water fish (Forster et al., 2005).

Hydrolysates

Hydrolysates are a class of materials in which some of the peptide bonds of proteins have been broken down. This can occur either chemically or with enzymatic mediation. Hydrolysates can have useful properties for use in aquatic diets, with particular relevance to fisheries by-product utilization.

Methods for production of protein hydrolysates from fish have been known for several decades (Tarr, 1948). In commercial production, hydrolysates are made by adding hydrolytic enzyme preparations

(Shahidi et al., 1995) or by utilizing naturally occurring enzymes or to accelerate the hydrolysis reaction (Wasson et al., 1992; Tschersich and Choudhury, 1998). Very low pH by itself is effective in inducing hydrolysis, and many hydrolytic enzymes are most active in mildly acidic conditions. The terms "silage" and "liquefied fish" refer to different kinds of hydrolysates (Stone and Hardy, 1986): silage is conventionally made by grinding fish by-products and adding organic or inorganic acids to a pH of 4.0, and liquefied fish uses endogenous enzymes in the by-product under conditions of physiological pH and mild heating (below 60°C).

Hydrolysates vary widely in their characteristics, and these are determined by the composition of the protein in the starting material, the degree of hydrolysis (the proportion of peptide bonds broken), and the location of the breaks (the amino acids that are part of the hydrolyzed peptide bonds). In practice, these are controlled by the conditions of production, especially the amount and type of hydrolytic enzymes used, the pH of the reaction medium, and the duration and temperature of the process. In general, increasing the degree of hydrolysis increases the solubility of the peptides. Also, under some conditions of production, hydrolysates develop a bitter taste that can limit their use (Daukšas et al., 2003). This bitterness is associated with the amount of hydrophobic amino acids at the terminal position of peptide. The degree of hydrolysis is also related to bitterness, with the highest risk occurring between 4 and 40 percent (Belitz and Wieser, 1976; Adler-Nissen, 1984).

Properly prepared fish hydrolysates have superior nutritional and storage characteristics (Hardy et al., 1983; Jackson et al., 1984; Stone and Hardy, 1986). In addition, these hydrolysates can have good attractant properties and can stimulate strong feeding response in aquatic animals. Some hydrolysates have functional properties. For example, hydrolysis of skin produces gelatine, which is a source of highly digestible protein and which is very useful as a binder in diets. It is claimed that some hydrolysates confer immunostimulant properties, but this has not been well established (an example using fish by-products is Murray et al., 2003).

Hydrolysis can improve the efficiency of recovery of protein from by-products, including reducing connective tissue and assisting the separation of the protein and bone fractions. An example of this is provided by Bechtel et al. (2003). In this work, pink salmon heads and viscera, by-products of the processing industry, were separately ground and mixed with papain at 61-65°C. Hydrolysis was terminated by elevating temperature to 85°C, followed by removal of bones by screening. The characteristics of the viscera meal were comparable to good quality fish meals (Table 5.1) with high protein and low ash content. The extra steps involved and the cost of the enzymes used in the production of hydrolysates increase their cost of production as compared to that of standard fish meals. However, the increased recovery efficiency and higher value of the resulting products can offset this increased expense.

Stickwater

Stickwater is the de-oiled liquid obtained during the manufacture of fish meal. Although stickwater is primarily composed of water, it contains an appreciable level of the solids in the source material. The actual amount of solids in stickwater is determined by the source of the raw material. The stickwater of fillet processing by-products contains approximately 8 to 10 percent solids, representing approximately 33 to 45 percent of the total raw solids, whereas stickwater from whole fish contains 5 to 6 percent, representing 16 to 22 of the incoming raw material solids (Pedersen et al., 2003).

Protein is the major component of stickwater solids, comprising approximately 56 to 70 percent (Bimbo, 2002). This protein may be more digestible and palatable than those left in the presscake, and if stickwater is added back to the presscake prior to final drying, the nutritional value (higher digestible protein content, more palatability) of the meal may be improved.

TABLE 5.1. The composition, degree of hydrolysis (DH) and pepsin digestibility of hydrolysates made from pink salmon processing by-products.

Source	Moisture (%)	Protein (%)	Lipid (%)	Ash (%)	DH (%)	Digestibility (%)
Head	6.6	59.6	29.8	5.5	12.0-14.5	>94
Viscera	7.4	74.6	12.8	6.8	16.3-16.9	>94

Source: Bechtel et al., 2003.

A limitation to stickwater utilization is the salt content. If salt water has been used to chill or pump harvested fish, the salt content of the resulting stickwater is elevated, reducing its suitability for addition back to the presscake. In offshore fish meal production, it may not be practical to carry sufficient quantities of freshwater to ensure that suitable stickwater is obtained. Membrane filtration technology has shown promise in reducing water and salt content of stickwater (Pedersen et al., 2003), and may enable development of higher quality fish meals from by-products without the need for costly evaporators.

In many cases, stickwater is disposed by dumping, despite the potentially higher nutritional benefits of whole fish meal relative to presscake fish meal. This occurs in areas where fishery by-product utilization is mandated by law and fish meal production is viewed as a "necessary evil" by harvesters and processors. In this case, quality is of little concern and stickwater is dumped. For industry to change its practices willingly, it is necessary to quantify the increase in nutritional value, which can be then realized in higher demand and price of including stickwater into fish meal production.

An example of the degree to which stickwater may improve the value of fish meal made from fisheries by-products is described by Forster et al. (2003). In this work, a series of meals were made from pollock *(Theragra calchogramma)* fishery by-products, incorporating stickwater at 0, 10, 20, and 40 percent on a protein basis. Diets specifically formulated for fish and for shrimp were produced using these meals as the main protein source. Separate diets using a high grade European fish meal (Norse LT94) were also produced for comparison. Each of these diets was then fed to groups of Pacific white shrimp *(Litopenaeus vannamei)* and Pacific threadfin *(Polydactylus sexfilis)* for several weeks under controlled conditions. The final weight of the shrimp fed the fish meal with the highest level of stickwater added was 38 percent greater than that of the shrimp fed the diet containing the presscake meal with no stickwater added. The final weight of the Pacific threadfin fed the diet with at least 20 percent stickwater added was 49 percent higher than those fed the diet with presscake meal. These results quantify the nutritional value of incorporating stickwater into fish meals made from fishery by-products for use in aquatic diets.

FACTORS TO CONSIDER IN THE USE
OF BY-PRODUCTS FOR AQUATIC DIETS

A number of factors affect the utilization of fishery by-products as ingredients for aquatic diets. These include logistics (the factors involved in collecting the by-products and bringing them to locations where they can be used), nutritional quality (the ability of products to supply suitable nutrients), and, safety (the absence of harmful elements, such as pesticides or heavy metals). Marketing and regulatory factors are also important to determining the success of these products.

Logistics

The locations that generate processing by-products (on-shore and off-shore) are not always amenable for large scale conversion of these to coproducts. In addition, most fisheries are seasonal, which can result in large volumes having to be processed in short periods of time. Alternatively, in some fisheries, processing takes place in relatively remote locations spread out over a wide area. The costs associated with transportation and storage of the material may preclude development of by-product utilization industry.

By locating aquaculture industries near to the locations of production, these costs can be at least partially offset, but this is not always possible. In Alaska, for example, where large amounts of fishery by-products are available, aquaculture is not legally permitted. One alternative approach to solving some of these issues is mobile processing plants. This has been attempted in Alaska and British Columbia (Canada).

Improving the storage characteristics of the wet material at above freezing temperatures can be achieved by ensiling, as described earlier (Stone and Hardy, 1986). This allows material generated remotely in relatively small batches to be accumulated and transported in bulk quantities to processing plants.

Nutritional Quality

Fish meal is the largest commodity of nonfood marine protein, and there is a wide spectrum of available qualities. Diet producers are careful to match the requirements of the species for which they are formulating with the type and quality of the meal that is available. Many higher value aquatic species are carnivorous and grow best when fed diets containing high-quality meals.

Fish meal quality is determined by several chemical and biological parameters, and by experience these have been reduced to a relatively small set of criteria for commercial purposes. Commercial specifications of fish meal (examples are found in Table 5.2) are used by diet manufacturers to predict the nutritional quality of the meals they use for their diets and as a guide to deciding the price they are willing to pay. High-quality fish meals contain higher protein levels and lower

TABLE 5.2. Specifications of some fish meals.

	FAQ[a]	Super prime	Norse LT-94[b]
Protein (%)	63-65	>68	68-71
Lipid (%)	<12	<10	≤11.5
Ash (%)		<17	≤14.0 (without salt)
Moisture (%)	<10	<10	6-10
Salt and sand (%)	<5	<4	
Sand	<2	<1	
Salt			≤3.5
TVN[c] (mg/100g)		<100	≤180
FFA[d] (%)		<7.5	
NH3-N[e] (%)			
Histamine (ppm)		<500	≤500
Cadaverine (ppm)			≤1,000
Water soluble protein (%)			18-32
Digestibility (Mink; %)			≥90
Salmonella			Not detected
Antioxidant			Ethoxyquin

Note: Note that some specifications and guarantees from different producers may vary.

[a]FAQ = fair average quality.

[b]From Norsildmel, Norway. http://www.norsildmel.no/

[c]TVN = total volatile nitrogen.

[d]FFA = free fatty acid.

[e]NH_3-N = ammonia nitrogen.

levels of extraneous elements (e.g., salt and sand), as well as lower substances associated with degradation (e.g., ammonia and biogenic amines such as cadaverine, histidine). Meals produced from by-products that meet higher standards will gain greater acceptance and command higher prices in the marketplace.

Safety

To be acceptable for use in aquaculture diets, protein products must be free of such compounds as biogenic amines, heavy metals, pesticides, PCBs, and dioxins. Although unusual, there are instances where fish meals have been found to contain some of these substances, and it is especially important to ensure that new sources of raw material are within acceptable guidelines. As mentioned, by-products of the seafood processing industry have the same quality in terms of freshness as portions used for food (Kilpatrick, 2003), and when these are used as the raw material for generation of protein sources for aquaculture, ensuring safety should not be difficult. It is important, however, to ensure that there is no generation of undesirable compounds during processing and to prevent contaminated material from entering the chain.

Marketing

Currently, the majority of fish meals used in aquatic diets are made from industrial fish produced in a few countries, notably, Peru, Chile, and Europe. The meals from these areas are well known and established in the market place. Alternative sources derived from fishery by-products are less well established among large end users, and there is some reluctance to use an unknown commodity. Diet manufacturers operate on small margins and rely heavily on the security of supply and quality of the ingredients that they use. In order to overcome this barrier, producers of ingredients from by-products need to establish and adhere to strict quality control parameters. To accomplish this requires information on all aspects, including nutritional quality, safety, standards, availability (security of supply), and price, in order to convince formulators to include by-product meals as a routine ingredient.

As mentioned, there are different quality-based grades of fish meal, depending on the raw material and the methods used in their production. There is an equally wide range of applications for fish meal in aquatic diets. Matching the quality of the product to the needs of the target species will maximize their utilization.

CONCLUSIONS, RESEARCH GAPS, AND RECOMMENDATIONS

Conclusions

- Coproducts (fish meals, hydrolysates, stickwater) recovered from fishery by-products are suitable for incorporation as protein sources into aquatic diets.
- Important issues concerning logistics, quality control, consistent availability, and marketing are yet to be addressed.

Research Gaps

- Cost-effective technology to maximize utilization of fisheries by-products in remote areas and to bring them to market.
- Information concerning the suitability of existing and novel coproducts (including blends with other ingredients) to meet the needs of different aspects of aquaculture.
- Standardization of products derived from multiple sources and species.
- Knowledge of the seasonal effects on quality of coproducts.
- Survey regarding safety of raw material and coproducts.

Recommendations

- Improve logistics of transportation of coproducts from remote locations.
- Develop new products to better meet the needs of aquaculture diets.
- Produce and disseminate information of the effectiveness and suitability of incorporating fish processing coproducts in aquafeeds.
- Develop and standardize methods for assessing the quality and freshness of fish processing coproducts.
- Develop programs to ensure the continuing safety (freedom from contaminants) of fish processing coproducts.

REFERENCES

Adler-Nissen, J. (1984). Control of the proteolytic reaction and of the level of bitterness in protein hydrolysis processes. *Journal of Chemical Technology and Biotechnology* 34B:215-222.

Babbitt, J.K., R.W. Hardy, K.D. Reppond, and T.M. Scott (1994). Processes for improving the quality of whitefish meal. *Journal of Aquatic Food Product Technology* 3:59-68.

Barlow, S. (2003). World market overview of fish meal and fish oil. In *Advances in Seafood Byproducts.* (P. Bechtel, ed.) Alaska Sea Grant College Program, University of Alaska Fairbanks, Fairbanks, Alaska. pp. 11-25.

Bechtel, P., S. Sathivel, A.C.M. Oliveira, S. Smiley, and J. Babbitt (2003). Properties of hydrolysates from pink salmon heads and viscera. *Proceedings of the First Joint Trans-Atlantic Fisheries Technology Conference 2003.* June 11-14, 2003, Reykjavik, Iceland. pp. 284-285.

Belitz, H.D. and H. Wieser (1976). Steric arrangement of sweet and bitter taste of amino-acids and peptides. *Z. Lebensm Unters Forsch* 160:251-253.

Bimbo, A.P. (2002). Pollution prevention and control in the seafood industry and particularly for small and medium sized fish meal plants. *US EPA/CEPIS Seminar: Prevención de la Contaminacion en la Pequeña y Mediana Industria,* Pan American Health Organization, Lima Peru. http://www.cepis.ops-oms.org/muwww/fulltext/epa/pcsi/pcsi.html

Daukšas, E., R. Šlizyte, T. Rustad, and I. Storrø (2003). Bitterness in fish protein hydrolysates: Origin and method for removal. *Journal of Aquatic Food Product Technology* 13:101-114.

FAO. Food and Agriculture Organization (2003). Fisheries Global Information Service. http://www.fao.org/fi/figis/index/jsp

Forster, I., J.K. Babbitt, and S. Smiley (2003). Nutritional quality of Alaska white fish meals made with different levels of hydrolyzed stickwater for Pacific threadfin *(Polydactylus sexfilis).* In *Advances in Seafood Byproducts.* (P. Bechtel, ed.) Alaska Sea Grant College Program, University of Alaska Fairbanks, Fairbanks, Alaska. pp. 169-174.

Forster, I.P., J.K. Babbitt, and S. Smiley (2004). Nutritional quality of fish meals made from by-products of the Alaska fishing industry in diets for Pacific white shrimp *(Litopenaeus vannamei). Journal of Aquatic Food Product Technology* 13:115-123.

Forster, I., J. Babbitt, and S. Smiley (2005). Comparison of the nutritional quality of fish meals made from byproducts of the Alaska fishing industry in diets for Pacific threadfin *(Polydactylus sexfilis). Journal of the World Aquaculture Society* 36:530-537.

Hardy, R.W., W.M. Sealey, and D.M. Gatlin (2005). Fisheries by-catch and by-product meals as protein sources for rainbow trout *Oncorhynchus mykiss. Journal of the World Aquaculture Society* 36:393-400.

Hardy, R.W., K.D. Shearer, F.E. Stone, and D.H. Wieg (1983). Fish silage in aquaculture diets. *Journal of the World Mariculture Society* 14:695-703.

Hardy, R.W. and A.G.J. Tacon (2002). Fish meal: Historical uses, production trends and future outlook for supplies. In *Responsible Marine Aquaculture.* (R.R. Stickney and J.P. McVey, eds.) CABI Publishing, New York, NY. pp. 311-325

Jackson, A.J., A.K. Kerr, and C.B. Cowey (1984). Fish silage as a dietary ingredient for salmon. I. Nutritional and storage characteristics. *Aquaculture* 38:211-220.

Kilpatrick, J.S. (2003). Fish processing wastes: Opportunity or liability. In *Advances in Seafood Byproducts.* (P. Bechtel, ed.) Alaska Sea Grant College Program. University of Alaska Fairbanks, Fairbanks, Alaska. pp. 1-10.

Murray, A.L., R.J. Pascho, S.W. Alcorn, W.T. Fairgrieve, K.D. Shearer, and D. Roley (2003). Effects of various feed supplements containing fish protein hydrolysate or fish processing by-products on the innate immune functions of juvenile coho salmon *(Oncorhynchus kisutch). Aquaculture* 220:643-653.

Nicol, S., I. Forster, and J. Spence (2000). Products derived from krill. In *Krill Biology, Ecology and Fisheries.* (I. Everson, ed.) Blackwell Science Ltd., Oxford, United Kingdom. pp. 262-283.

Pedersen, L.D., C. Crapo, J. Babbitt, and S. Smiley (2003). Membrane filtration of stickwater. In *Advances in Seafood Byproducts.* (P. Bechtel, ed.) Alaska Sea Grant College Program. University of Alaska Fairbanks, Fairbanks, Alaska. pp. 359-369.

Rathbone, C.K., J.K. Babbitt, F.M. Dong, and R.W. Hardy (2001). Performance of juvenile coho salmon *Oncorhynchus kisutch* fed diets containing meals from fish wastes, deboned fish wastes, or skin-and-bone by-product as the protein ingredient. *Journal of the World Aquaculture Society* 32:21-29.

Shahidi, F., X.-Q. Han, and J. Synowiecki (1995). Production and characteristics of protein hydrolysates from capelin *(Mallotus villosus). Food Chemistry* 53: 285-293.

Smiley, S., J. Babbitt, S. Divakaran, I. Forster, and A. de Oliveira (2003). Analysis of groundfish meals made in Alaska. In *Advances in Seafood Byproducts.* (P. Bechtel, ed.) Alaska Sea Grant College Program. University of Alaska Fairbanks, Fairbanks, Alaska. pp. 431-454.

Stone, F.E. and R.W. Hardy (1986). Nutritional value of acid stabilized silage and liquefied fish protein. *Journal of the Science of Food and Agriculture* 37:797-803.

Tarr, H.L.A. (1948). Possibilities in developing fisheries by-products. *Food Technology* 2:268-277.

Tschersich, P. and G.S. Choudhury (1998). Arrowtooth flounder *(Atheresthes stomias)* protease as a processing aid. *Journal of Aquatic Food Product Technology* 7:77-89.

Wasson, D.H., J.K. Babbitt, and J.S. French (1992). Characterization of a heat stable protease from arrowtooth flounder *(Atheresthes stomias). Journal of Aquatic Food Product Technology* 1:167-182.

Chapter 6

Use of Animal By-Products in Crustacean Diets

Shi-Yen Shiau

INTRODUCTION

In less than three decades, commercially farmed shrimp have become the world's top aquaculture commodity by value, and shrimp aquaculture is one of the world's fastest growing food production systems. This sector has grown at an average annual rate of 16 percent per year since 1984, with total farmed shrimp landings valued at US$6.2 billion (Tacon and Foster, 2001).

The protein requirement of crustaceans is higher than that of terrestrial animals. Fish meal is a high-protein-content feedstuff and is also an excellent source of essential amino acids. Shrimp diets have long been formulated with fish meal (FM) as the primary source of dietary protein (Hughes, 1990). Diet is the major cost variable in aquaculture, representing up to 60 percent of total costs (Akiyama et al., 1992; Sarac et al., 1993), and most of the actual diet cost can be attributed to the protein fraction. Although there is no doubt of the high nutritional value of FM, the best quality meals are expensive, and according to some projections, their availability is expected to decline and the price will dramatically increase (Dong et al., 1993). Therefore, there is a need to identify and utilize less expensive and more sustainable animal by-products within crustacean feeds.

I wish to thank my graduate student, Yu-Hung Lin, for his valuable assistance during preparation of this chapter.

FISHERIES BY-PRODUCTS

Fish Meal

The principle in FM processing is the separation of the solids from water and oils. The quality of FM depends on several factors, which include: (1) temperature at the time the fish are caught; (2) temperature at which the fish is stored prior to processing; (3) length of storage prior to processing; (4) type or composition of fish catch; and (5) method of catching the fish. Fish meal has to be processed as soon as the fish is caught in the ship offshore or at shore. A by-product of FM manufacture is fish oil. Fish meal is a high-quality protein feedstuff. The chemical composition, particularly the protein content, varies widely and depends on fish species used for FM manufacture (Table 6.1), season, and the latitude where the fish are caught. Fish meal is also an excellent source of essential amino acids. The lipid content and fatty acid composition of FM is species-specific.

Fish meal is commonly used as protein source in shrimp diet. In postlarvae (2.0 to 3.0 g) tiger prawn, *(Penaeus monodon),* the highest survival rate was attained with FM as the sole protein source, whereas the best growth rate obtained with the combination of FM and shrimp head meal (1:1) (Piedad-Pascual and Destajo, 1979).

The chemical composition of FM varies widely and depends on the species used to make the FM. In Great Britain, "white fish meal" is defined as containing not more than 6 percent oil and 4 percent salt. In Norway, distinction is made between herring meal and FM. The former is only herring, whereas the latter is from cod and fish offal and is of lower quality (Baelum, 1962). Teshima et al. (1986) compared the dietary value of six different protein sources for larvae of Kuruma prawn *(Marsupenaeus japonicus)* and reported that white FM, casein, and casein-gelatin (3:1) had superior nutritive values compared to gelatin, egg albumin, and an amino acid mixture approximating the composition of casein. Commercial diets for *M. japonicus* in Japan are costly due to the inclusion of expensive protein source such as squid meal. Teshima et al. (1991) indicated that brown FM-based diets supplemented with crystalline amino acids of arginine and lysine can sustain good growth and high survival of larval *M. japonicus* as comparable to the live food (control). Koshio et al. (1992) reported that

TABLE 6.1. Chemical composition of fish meal (% dry matter).[a]

Meal	Dry matter	Crude protein	Crude fat	Ash	Crude fiber	N-free extract
Fish meal from defined species						
Anchovy (true)	92.0	70.7	5.3	16.9	–	7.1
Anchovy (false)	93.0	78.0	9.0	12.5	–	0.5
Bream	93.0	63.2	10.3	25.2	–	1.3
Capelin	91.1	72.6	9.3	10.6	4.7	2.8
Chilean hake	81.6	83.3	2.4	14.3	–	–
Cod	89.7	68.6	3.8	26.0	–	1.6
Croaker	94.0	63.1	10.9	20.2	0.9	4.9
Haddock	93.4	65.9	4.7	26.5	–	2.9
Halibut	–	53.2	13.1	32.8	–	0.9
Herring	90.0	74.4	9.0	15.0	–	1.6
Horse Mackerel	95.4	70.9	13.7	–	–	–
Mackerel	92.0	66.4	10.3	21.1	–	2.2
Menhaden	92.6	66.6	11.1	20.9	–	1.4
Pilchard	91.8	66.5	7.6	20.4	–	5.5
Pollack	94.8	65.5	17.7	14.1	–	2.7
Sandeel	91.0	72.6	8.3	10.6	1.0	8.5
Sardine	93.0	65.2	5.0	19.8	–	9.0
Shark	92.0	72.3	17.9	–	–	–
Tuna	–	64.0	10.1	23.6	–	2.3
Fish meal from unspecified fish						
White fish meal	91.5	65.8	8.5	19.5	1.4	4.8
Brawn fish meal	91.3	69.0	6.0	14.8	–	10.2
Fish waste meal	90.0	49.2	9.0	34.4	–	7.4

[a]These data were adapted from Cho et al. (1985); Evans (1985); Feltwell and Fox (1978); Friesecke (1984); Murayama et al. (1962); NRC (1983); New (1987); Park (1989); Tacon (1993); Watanabe et al. (1988); and Wöhlbier and Jäger (1977).

the addition of both white FM and brown FM to crab protein concentrate (CPC) improved *M. japonicus* larval performance compared to the use of CPC alone. The growth performance of Chinese shrimp, *Fenneropenaeus chinensis,* was better when fed diets containing high-quality FM than shrimp fed diets with a low-quality FM (Liang and Anders, 2001).

Tuna meal, shrimp meal, and tilapia meal in combination with soybean meal (SBM) and copra (coconut) meal as protein sources in diets were fed to the freshwater prawn *Macrobrachium rosenbergii* (Balazs and Ross, 1976). After 167 days on these diets, growth was better in prawn fed diets with SBM and tuna meal, or soybean, tuna, and shrimp meal combinations than prawn fed diets with SBM and tilapia meal, or copra meal and tilapia meal.

Fish meals vary widely in their protein quality and nutrient composition depending on the freshness and type of the raw material, state of residual lipid, and processing temperature exposure (drying process). Fish freshness is affected by conditions and length of storage before processing. From the time of catching, fish undergo changes brought about by the action of the enzymes of the fish (autolysis) and also from the action of bacteria present on the surface of the fish and in the gut. The rate at which raw material spoils depends on the species of fish, storage time and temperature, and degree of microbiological contamination (Aksnes and Brekken, 1988; Haaland et al., 1990). Raw material freshness is an important criterion of FM quality for shrimp. In a study conducted by Ricque-Marie et al. (1998), three batches of anchovy meal were produced in a commercial low-temperature processing plant, from a unique source of raw fish, either fresh (FR, twelve hours postcapture), moderately fresh (MF, twenty-five hours postcapture), or stale (ST, thirty-six hours postcapture). Freshness was assessed through the total volatile nitrogen content in fish before process (TVN 14, 30, and 50 mg N/100 g, respectively), and biogenic amines in FM (histamine 28; 1,850; and 4,701 mg/kg, respectively, and also with increasing content of cadaverine, putrescine, and tyramine). Samples of the three FMs were incorporated at levels of 30 or 40 percent into isoenergetic diets fed ad libitum to Pacific white shrimp, *Litopenaeus vannamei* (0.9, 1.5, and 7.6 g body wt), tiger prawn, *Penaeus monodon* (2.5 g body wt), and blue shrimp, *Litopenaeus stylirostris* (8.4 g body wt). Results indicated that weight gain of *L. stylirostris* fed diets with ST and MF and *P. monodon* fed diet with ST were significantly lower than shrimp fed diet with FR. For *L. vannamei*, small shrimp (0.9 g) fed diets with ST and MF gained significantly less body weight than shrimp fed diets with FR. However, this difference was not observed in larger shrimp (i.e., 1.5 g and 7.6 g).

Shrimp Meal

Shrimp by-products, such as heads and shells, are produced either on the fishing ships or in the coastal shrimp processing plants in large amounts during the shrimp fishing season. Shrimp heads alone represent 35 to 45 percent of the total shrimp production (Meyers, 1986). Shrimp by-products are made into a meal and have been used widely as ingredients in shrimp diets. The chemical composition of shrimp meal varies widely depending on the source (whether whole shrimp or heads) (Table 6.2).

A comparative study of the nutritional value of four proteins (casein, egg, feather, and shrimp) was studied for juvenile American lobster, *Homarus americanus* (Boghen and Castell, 1981). The shrimp protein diet produced significantly better lobster growth and survival. It was suggested that the superiority of the shrimp-based diet may be a result of factors other than the amino acid composition of the diet.

Several reports have shown that the growth of shrimp and freshwater prawns fed with FM-based diets is significantly improved if the diets used contain shrimp meal (Colvin, 1976; Ali, 1982; Ravishankar and Keshavanath, 1986). Shrimp meal as the sole source of protein in the diet for *P. monodon* juveniles did not perform as well as the combination of shrimp meal and molluscan meal (Ogle and Beaugez, 1991). The combination of shrimp head meal and FM at the ratio of 1:1 or 2:1 in a diet produced significantly better growth in post-larval *P. monodon* compared to a diet which had shrimp head meal at a level of 60 percent (Piedad-Pascual and Destajo, 1979).

Various combinations of shrimp head meal, scallop waste meal, lobster waste meal, and sardine meal in diets were evaluated for juvenile *P. monodon* (Sudaryono et al., 1995). The diet combining scallop waste and shrimp head meals as major protein sources was superior to other diets for the shrimp, followed by the diets containing sardine FM and shrimp head meal, and sardine FM and lobster waste meal. The commercial anchovy FM and shrimp head meal gave the poorest growth performance. The finding of the poorest growth performance in shrimp fed diet with a combination of commercial anchovy fish and shrimp head meal was in agreement with Cruz-Suárez et al. (1993) who also found that *L. vannamei* fed a diet containing commercial FM had reduced growth compared to that of shrimp fed diets containing

TABLE 6.2. Chemical composition of fisheries by-product (% dry matter).[a]

	Dry matter	Crude protein	Crude fat	Ash	Crude fiber	N-free extract
Shrimp meal						
Whole	–	70.9	3.3	18.3	3.1	4.4
Head	–	43.2	5.6	33.0	15.8	2.4
Waste	–	31.2	4.1	28.9	20.0	15.8
Process residue	–	46.0	3.5	31.0	14.6	4.9
Acetes	–	72.0	3.8	15.9	3.1	5.2
Shrimp head silage dried	–	74.2	7.4	18.4	–	–
Squid meal						
Squid meal	95.3	80.5	4.0	6.4	–	–
Squid liver meal	88.8	50.8	17.2	7.6	–	–
Squid viscera (fresh)	17.8	75.3	7.4	6.8	–	–
Squid viscera (silage)	19.9	60.8	7.3	9.5	–	–
Crab meal						
Crab meal	92.7	32.2	2.8	39.4	10.6	29.3
Crab protein concentrate	90.0	60.5	0.4	6.1	–	–
Krill meal	92.0	58.8	9.2	13.6	6.4	3.7
Mollusc						
Blue mussel	–	60.4	8.5	11.6	–	19.5
Green mussel	–	52.5	15.4	9.3	3.1	19.7
Brown mussel	14.4	62.7	7.9	5.1	11.6	12.7
Clam	–	56.5	8.8	9.6	–	25.1
Oyster	16.9	48.0	10.8	10.8	3.8	24.6
Scallops	21.7	83.5	1.9	14.4	–	–

[a]These data were adapted from Ali (1992); Anonymous (1992); Carver et al. (1989); Chou (1993); Christians and Leinemann (1980); Deshimaru (1981); Djunaidah (1993); FDS (1994); Food and Nutrition and Research Institute (1968); Guillame et al. (1990); Johnson (1988); Joseph et al. (1984); Kling and Wöhlbier (1977); Lim et al. (1979); Marsden et al. (1992); Nandeesha (1993); NRC (1977, 1981, 1983, 1991, 1993); Peñaflorida (1989); Piedad-Pascual (1993); Piedad-Pascual and Destajo (1979); Primavera et al. (1979); Steffens (1980); Steffens and Albrecht (1981); Tacon (1987); van Lunen and Anderson (1990); Watt and Merrill (1963).

shrimp by-product meals. The inferior growth performance obtained with this diet could have been due to differences in protein quality related to species or source (origin; age/size), or to differences in processing techniques. For example, the degree of heat treatment can affect bioavailability of essential amino acids, particularly lysine and methionine (Knipfel, 1981; Hurrel and Finot, 1985), the digestibility of feed ingredients (McDonald et al., 1988), or the level of the anti-nutritional factor, thiaminase (NRC, 1983).

Shrimp by-product meals from the Mexican Gulf coast and from the Pacific coast were evaluated as partial protein source replacement for FM and SBM-based diets for juvenile *L. vannamei* (Cruz-Suárez et al., 1993). The shrimp by-product meals were included at dietary levels of 3, 6, and 18 percent. The addition of shrimp by-product meals at graded levels up to 18 percent in shrimp diets resulted in a positive dose-response relationship. At the 18 percent inclusion level, the growth rate of the shrimp was significantly higher than the control group regardless of the source of shrimp by-product meals. The improvement in feed conversion ratios compensated for the increase in cost of the diets due to the use of shrimp by-product meal.

Shrimp head meals obtained by either drying the raw material in a solar simulator, drying in an oven, or passing the raw material through a commercial meat/bone separator first, followed by drying the meat fraction (MBDD) were studied on diet utilization by juvenile *P. monodon* (Fox et al., 1994). Experimental diets were prepared incorporating either 54 percent FM, or 31 percent of the solar-dried, oven-dried, or MBDD shrimp head meals, with enough FM added to make isonitrogenous diets. Generally, shrimp fed diets containing shrimp head meal performed better in terms of final weight, feed conversion rate, and production than shrimp fed the 54 percent FM-based diet. The differences among the three processing methods were not significant among each other.

The effects of supplementary shrimp head meal (SHM) contaminated with white spot syndrome virus (WSSV-SHM) in the diet on detection of WSSV in *P. monodon* were investigated (Pongmaneerat et al., 2001). The results indicated that using the cooked WSSV-SHM (steamed WSSV-SHM and oven-dried WSSV-SHM), the uncooked WSSV-SHM (freeze-dried WSSV-SHM and raw-fresh WSSV-SHM), and the commercial SHM in the diets did not induce WSSV infection in the shrimp.

Squid Meal

The mantle, head, and tentacles of squid are used mainly for human consumption. The squid waste, which usually includes viscera, may also contain head and tentacles, fin, skin, and pen amounting to approximately 52 percent of the whole-body weight (Joseph et al., 1984). Squid meal has been found to be an effective protein source for many penaeids. Squid meal has high protein values, from 70 percent to almost 90 percent (Table 6.2). The crude protein content of squid liver meal is lower because it is not pure squid; however, the amino acid profile is well balanced.

In *M. japonicas,* Deshimaru and Shigueno (1972) found an excellent nutritive value of squid meal and stated it was due to the essential amino acid composition of squid protein, which was similar to that of shrimp protein. High levels of squid meal (more than 45 percent) were also used in diets reported as being efficient (Deshimaru and Shigueno, 1972). For brown shrimp, *Farfentepenaeus aztecus,* an optimum level of squid meal (between 5 and 15 percent) was reported (Fennuci and Zein-Eldin, 1976). In the case of *P. monodon,* Lim et al. (1979) evaluated the efficiency of various dietary protein sources and found that squid meal was the best source for growth, feed conversion, and protein efficiency ratio. In white shrimp, *Litopenaeus setiferus,* and blue shrimp, *L. stylirostris,* diets containing squid meal produced better growth than those without this ingredient; the inclusion of 5 to 6 percent squid meal being beneficial (Fennuci et al., 1980).

To assess the reference dietary amino acid profiles for juvenile *M. japonicus,* a feeding trial was conducted by Alam et al. (2002) using six semipurified diets containing casein-gelatin and precoated supplemental crystalline amino acids (CAA), and a control diet containing intact protein (casein-gelatin). Diets were supplemented with precoated CAA to obtain dietary amino acid profiles similar to those of the prawn egg protein (PEP), prawn larvae whole-body protein (PLP), prawn juvenile whole-body protein (PJP), squid meal protein (SMP), short-necked clam protein (SNP), and brown fish meal protein (BFP). The result showed that *M. japonicus* juveniles are capable of utilizing the precoated CAA, and higher growth performances were observed in the groups fed the PJP, SMP, and the control diets compared to shrimp fed the PLP, SNP, BFP, and PEP diets. This suggests that squid meal protein would be suitable as a reference dietary amino acid profile for *M. japonicus.*

A protein fraction extracted from squid (squid protein fraction, SPF) was tested in juvenile *M. japonicus* (Cruz-Suárez et al., 1987). Squid protein fraction was substituted for purified fish protein in a mixed diet at a level ranging from 1.5 to 16 percent. Weight gain and number of molts of the shrimp increased with SPF level and reached a plateau when SPF was added at > 6 percent. Growth improvement was also observed in *L. stylirostris, L. vannamei,* and *P. monodon* fed diets containing SPF (Cruz-Ricque et al., 1987).

In field experiments (brackishwater ponds), *P. monodon* fed diets with squid meal as the main ingredient showed higher growth rates compared to shrimp fed diets with FM or SBM as the main ingredients (Rajyalakshmi et al., 1982). The use of squid meal (10 percent) in extruded diets substantially improved yields of *P. monodon* under semi-intensive pond production conditions (Cruz-Suárez et al., 1992). Squid meal incorporation (5 to 10 percent) in shrimp diet formulation is recommended, not only for the very high-quality diets commonly used in intensive systems, but for a wide range of culture conditions, especially in semi-intensive systems.

Crab Meal

Boghen et al. (1982) first tested purified protein concentrates from a number of marine organisms and reported that diets containing rock crab, *Cancer irroratus,* protein gave superior growth and survival among juvenile lobsters. Thereafter, several investigations have also demonstrated that CPC derived from rock crab is a good protein source for the growth and survival of crustaceans (Kean et al., 1985; Brodner, 1989; Castell et al., 1989a; Morrissy, 1989; Reed and D'Abramo, 1989; Koshio et al., 1990, 1992). The chemical composition of crab meal is rather heterogeneous due to the variability of the raw material. The average crude protein is about 32.2 percent with a wide variation (Table 6.2). Crab protein concentrate on the other hand has a crude protein content almost double that of crab meal. The high ash content of crab meal is due to the exoskeleton of crabs, which is particularly high in calcium.

Crab meal is produced by drying the waste from crab processing and/or whole crabs. The nutritive value of the CPC varies depending upon how the crab meat is processed. For example, lysine can become nutritionally unavailable during processing in the presence of reducing sugars because of the Maillard reaction. Castell et al. (1989b) conducted a feeding trial with juvenile lobster *(H. americanus)* fed diets containing CPC that was produced by several different purification process. Among the differently derived purified rock CPC products, the precooked, isopropanol-purified was the most effective dietary ingredient (Figure 6.1).

Crab protein concentrate used as a sole protein source in microbound diets (MBD) fed to larval *M. japonicus* was found not nutritious enough to sustain larval growth and survival. The poor performance of larvae is thought to be due to CPC's physical properties, which may not allow the binder to give proper pellet stability, and/or an unsatisfactory amino acid profile of CPC, since growth and survival were improved when 20 to 40 percent of CPC in the diets was replaced by casein (Koshio et al., 1989). Therefore, it is likely that other protein sources, besides casein, are potentially good additions to CPC in the MBD. A series of studies were conducted by Koshio et al. (1992) to search for an optimal combination of several protein sources with CPC for larval *M. japonicus* fed MBD diets. Although the addition of squid meal, brown FM, white FM, and krill meal to CPC improved larval

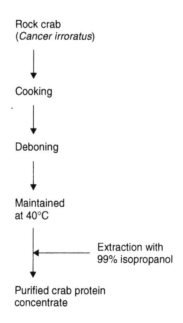

FIGURE 6.1. Flow diagram for processing of CPC. *Source:* Castell et al., 1986b.

performance compared to the single use of CPC, the magnitude of improvement by them was not as great as that by casein and soybean protein (SBP). In the diet containing both CPC and squid meal as the protein source, more CPC in the diet produced a poorer growth performance of larvae, particularly at low levels of protein. Besides casein, the best combination for larval development was CPC and SBP suggesting that plant proteins such as SBP rather than squid meal may be suitable for use as additional protein sources to CPC for larval *M. japonicus*.

Krill Meal

Crustacean meals are usually the preferred protein sources in formulating prepared diets for other carnivorous crustacean species because of the close resemblance in their amino and fatty acid profiles. Krill, *Euphasia* sp., a crustacean found in large quantities in the Pacific and Atlantic oceans, and a primary consumer in the ocean food chain, is currently used as a nutrient-rich additive in many commercial aquaculture diets. In contrast to traditional FMs used in crustacean diets, it contains high levels of carotenoid pigments and attractants, and has amino and fatty acid profiles that more closely resemble the profiles of the farmed crustaceans. The chemical composition of krill meal (Table 6.2) is affected by the season, availability of food, the area where krill is caught, the age, and sex of the animals.

Not much is known on krill meal feeding to crustaceans. Recently, the feasibility of soy-based diets for juvenile American lobster, *H. americanus,* was investigated using diets of SBM containing various proportions of freeze-dried krill hydrolysate (FDKH) at 0, 12.5, 25, 50, 75, and 100 percent of dietary protein. A diet of 100 percent of protein from FM was included for comparison. Results indicated that lobster fed all krill-containing diets and a 100 percent FM diet had significantly higher weight gain, higher survival percentages, and shorter molting cycles than those fed 100 percent SBM diet. It was also to note that the weight gain was higher in the 100 percent FDKH group than that in the 100 percent FM group (Floreto et al., 2001).

Mollusc Products (Mussels, Clams, Oysters, and Scallops)

For processing of many molluscs, steaming will release the meat from the shells without individual shucking (Quayle and Newkirk,

1989). Although meat of molluscs can be boiled, dried, and ground into a meal, the procedure is tedious. In addition, the live or frozen meat is preferred by shrimp particularly in larval and broodstock diets. The protein content of mollusc meal varies widely. It is highest for scallop meat and lowest for oyster meat (Table 6.2). Mollusc meat is a good source of essential amino acids.

The gut contents of wild *P. monodon* contained 76 percent remains of molluscs and bivalves which formed 31 percent of total food (Marte, 1980). Fresh or frozen mussel meat is used in feeding *P. monodon* broodstock, postlarvae, and as a food supplement in grow-out pond (Lim et al., 1979; Piedad-Pascual and Destajo, 1979; Primavera et al., 1979). In one study, brookstock of ablated *P. monodon* maintained under laboratory conditions and allowed to spawn in tanks were fed 3.0 percent of body weight frozen mussel meat in the morning and formulated pellets in the afternoon (Primavera et al., 1979). It was reported that the mussel-pellets combination followed by mussel-mussel and squid-pellet combinations gave the better reproductive performance than the pellet-pellet feeding regime.

The positive effect on ovarian maturation of penaeid shrimps in captivity of fresh mussel meat in combination with formulated pelletized diets was reported in pond-grown *P. monodon* (Quinitio et al., 1994), *L. vannamei,* and *L. stylirostris* (Galgani and Aquacop, 1989).

Banana prawn *(Fenneropenaeus merguiensis)* grew significantly better on fresh mussel meat than on formulated feed (Sedgewick, 1980). Fresh mussel meat is also comparable to shrimp meal as the sole source of protein when fed to *P. monodon* (Lim et al., 1979).

The short-necked clam *(Venerupis philippinarum)* is a very important species in the development of practical diets for the *M. japonicus.* Its chemical composition was the basis of the basal prepared diet for determining nutrient requirements of *M. japonicus* prior to the formulation of practical diets (Deshimaru, 1981). It was also used in the fresh form, minced, and fed to *M. japonicus* larvae in the early years of prawn culture in Japan (Shigueno, 1975).

Fresh frozen clams *(Mercenaria mercenaria)* were fed to *L. vannamei* and *L. stylirostris,* or in a composite diet consisting of adult and juvenile squid (*Loligo* sp. and *Lolloiguncula* sp.), blood worms *(Glycera dibranchiata),* and penaeid shrimps (primarily brown shrimp, *F. aztecus*). With regard to ovarian maturation, spawning, molting rate, and survival, the diet that contained the most nutritious single food ingredient was squid. Diets in which clam meat comprised the single protein source gave the poorest results (Chamberlain and Lawrence, 1981).

Juveniles of Indian white shrimp *(Fenneropenaeus indicus)* were fed various animal- and plant-protein sources. A diet containing clam meat *(Sunneta scripta)* gave higher weight gain for shrimp compared to a diet containing FM. The feed/gain ratio was the same as the FM diet; however, net protein utilization (NPU), biological value (BV), and survival of shrimp fed the diet containing clam meat were lower than values of shrimp fed the diet containing FM (Ali, 1992).

POULTRY BY-PRODUCTS

Poultry By-Product Meal

Sources of terrestrial animal protein are primarily rendered by-products, such as poultry by-product meal and meat and bonemeal, which are often good sources of indispensable amino acids. The quality of these meals depends on both the quality of raw ingredients and the type of processing. One of the more promising alternative ingredients is poultry by-product meal (Dong et al., 1993), which is an ingredient from by-products of the poultry industry. Currently, few studies have evaluated the use of poultry by-product meals in practical shrimp diets. Poultry by-product meal is a rather heterogeneous material. The protein is predominantly from connective tissue. The mean crude protein content of 61 percent ranges from 56.4 to 84.2 percent. "Whole poultry meal" is much lower in protein and almost equal to its fat content (Table 6.3). Poultry by-product meal is a good source of essential amino acids.

For juvenile *L. vannamei,* the replacement of 40 to 80 percent of the FM in a basal diet with flashed-dried poultry by-product meal resulted in an increase in weight gain and feed efficiency in shrimp, as compared to the control group, suggesting that the FM content of practical diets can be reduced from 30 percent to 6 percent (Davis and Arnold, 2000). It has also been reported that up to 67 percent of the FM could be replaced by regular poultry by-product in *L. vannamei* diets without a reduction in shrimp growth (Cheng et al., 2002). A study

TABLE 6.3. Chemical composition of meat and poultry by-product (% dry matter).[a]

	Dry matter	Crude protein	Crude fat	Ash	Crude fiber	N-free extract
Poultry by-product meal						
Whole poultry meal	98.3	45.2	44.5	8.4	–	0.2
Poultry by-product meal	92.4	61.0	17.5	13.9	2.1	3.5
Feather meal	93.3	86.9	3.6	3.1	0.8	0.6
Meat by-product meal						
Meat meal	94.1	56.9	5.6	21.6	2.4	17.7
Meat and bonemeal	93.3	50.4	9.8	31.3	2.6	2.5
Blood meal						
Fresh blood	–	90.8	0.5	5.8	–	2.9
Blood meal	90.3	92.5	1.2	5.3	0.9	3.3

[a]These data were adapted from Anonymous (1987); Bath et al. (o/w year); Cho and Slinger (1979); Cho et al. (1985); De Boer and Bickel (1988); Dupree and Huner (1984); Evans (1985); Feltwell and Fox (1978); Friesecke (1984); Just et al. (1982); Meyer and Heckötter (1986); New (1987); NRA (1993); NRC (1981, 1983); Pesti and Dubuc (1986); Schulz (1995); Stählin (1957); Steffens (1985); Wöhlbier (1977); Wöhlbier and Tran Thu (1977).

by Tan et al. (2003) indicated that up to 80 percent of the FM in a diet can be replaced by poultry by-product meal without adverse effect on growth, survival, and feed conversion ratio of *L. vannamei.*

Feather Meal

At present, there are few examples in the literature concerning the successful formulation of diets containing varying percentage of feather meal for crustaceans. Among the reasons for the limited use of feather meal in diets for shrimp are its poor digestibility and essential amino acid profile, attributed to the natural structure of keratin and/or to commercial steam pressure-processing condition. The crude protein content of feather meal as a dry matter basis averages 86.9 percent with a variation of more than 20 percent (Table 6.3).

In a semipurified diet for *L. vannamei,* 2.5, 5.0, and 10.0 percent of the diet's protein were replaced by feather meal. Although there was no difference in survival rate between the control group and the group fed feather meal, growth for shrimp fed the FM diet was higher than that of shrimp fed the diet in which feather meal partially replaced FM (Lawrence and Castille, 1991).

The use of a commercial steam-processed-feather meal (SPFM) and feathers enzymatically hydrolyzed for 60 or 120 minutes (EHF60 and EHF120) as substitutes for FM in diets for *L. vannamei* have been investigated (Mendoza et al., 2001). Enzymatically hydrolyzed feathers or SPFM were blended through an extruder with SBM in a 1:1 ratio (EHF-SBM, SPFM-SBM). Results indicate that *L. vannamei* can be fed a practical diet containing 20 percent EHF-SBM (2:1) without impairing growth or food conversion. The use of 20 percent EHF-SBM (2:1) allowed the FM protein to be reduced by nearly 55 percent.

MEAT BY-PRODUCTS

Meat by-products are offal of abattoirs, such as material confiscated and not fit for human consumption, scrap from the meat processing and canning industry, and livestock casualties processed by the rendering industry.

Meat and Bonemeal

The chemical composition of meat by-product meal has extreme variations, and it depends on the quality of the available raw material (Table 6.3). The amino acid profile of meat and bonemeal (MBM) varies from that of meat meal because the protein properties of bones and meat differ from each other as reflected in the amino acid profile. The essential amino acids of meat by-product meals have limitations. Isoleucine and methionine + cystine are particularly limiting amino acids when compared to the amino acid profile of whole egg protein (New, 1987).

Meat meal can replace about half of the digestible crude protein in a shrimp diet without significantly affecting biological performance of *P. monodon* (Smith et al., 2001). It has also been demonstrated that processed (minced, pressed, and sterilized) by-product, in the form of meat solubles from authorized slaughter houses in Denmark, can be a cost effective replacement for FM in *P. monodon* diets (Millamena et al., 2000).

Three samples of MBM were evaluated as ingredients in diets for *L. vannamei* (Forster et al., 2003). The dry matter digestibilities of the three sample (A, B, and C) were 74.25, 68.08, and 63.60 percent, respectively; while the crude protein digestibilities were 87.84, 81.08, and 81.04 percent, respectively. In the feeding trial, the experimental diets were formulated to contain each protein source (i.e., A, B, and C) at 25, 50, and 75 percent level of replacement of FM. The results of the growth trial provide evidence that MSMs can replace as much as 75 percent of FM in diet for *L. vannamei,* although there was considerable variation in nutritive value among species.

As noted, MBMs are manufactured from a variety of materials, and the specific combination used influences the nutritional quality of the final product. Standardization of component mix, manufacturing methodology, and final composition will go a long way toward improving the usefulness of MBM for commercial shrimp diets. Since the relative rankings of the three MBMs were different based on digestibility and growth performance data, the previously mentioned study also suggests that digestibility itself is insufficient to explain the reduction in growth response relative to dietary FM.

Another growth trial was conducted by Tan and Yu (2003a) to estimate the replacement of FM with MBM on growth performance of *L. vannamei.* The control diet contained 40 percent FM, which was replaced on a weight basis by MBM at the levels of 0, 20, 30, 40, 50, 60, and 80 percent with adjustments in amino acids for the experimental diets. Results showed that MBM can replace 60 percent of FM in a shrimp diet without affecting the weight gain, feed efficiency, and flavor of the tail muscle.

Blood Meal

Blood products are protein feedstuffs. Blood from slaughtered food animals is used for human consumption, but large quantities are still discarded. The nutrient content of blood meal varies widely (Table 6.3). The fat and carbohydrate content is rather low but blood meal is a rich source of leucine, but a poor source of methionine and isoleucine (Hardy, 1989).

A feeding study by Dominy and Ako (1988) suggest that blood meal products can replace marine proteins in grow-out rations for medium sized (3 to 4 g) *L. vannamei.* Reigh and Ellis (1994) indicated that both FM and MBM are superior to a 60:40 MBM/blood meal mixture. However, MBM and MBM/blood meal mixtures could be useful as lower-cost alternative to FM in diets for pond-raised crayfish.

DIGESTIBILITY

A formulated diet may appear from its chemical composition to be an excellent source of nutrients but will be of little value unless it can be digested and absorbed. An effective diet can be prepared when information on digestibility of nutrients in feedstuffs has been considered in its formulation.

Fisheries By-Products

Apparent digestibilities of dry matter and crude protein of several fisheries by-products are summarized in Table 6.4. Akiyama et al. (1989) evaluated apparent digestibility of several feedstuffs, including FM, shrimp meal, and squid meal by *L. vannamei.* Each experimental diet contained 88 percent of the tested feedstuff, 4 percent attractant, 7 percent binder mixture, and 1 percent chromic oxide. The apparent dry matter digestibility and apparent protein digestibility of the three fisheries by-products were: FM, 64.3 and 80.7 percent, respectively; shrimp meal, 56.8 and 74.6 percent, respectively; and squid meal, 68.9 and 79.7 percent, respectively.

Apparent digestibility of nutrients in Peruvian FM, squid meal, and planktonic shrimp (*Acetes* sp.) by mud crab *(Scylla serrata)* were measured using a reference diet (RF) with 1 percent chromic oxide as an external indicator (Catacutan et al., 2003). A test diet consisted of 70 percent RF and 30 percent of the feedstuff to be tested. Apparent dry matter digestibility was highest in squid meal (93.6 percent), followed by Peruvian FM (89.9 percent), and then *Acetes* sp. (88.3 percent). Apparent crude protein digestibility was higher in squid meal (97.6 percent) than in Peruvian FM (94.8 percent) and *Acetes* sp. (94.9 percent).

TABLE 6.4. Apparent digestibility of crustacean fed diets contained fisheries by-product meals.

Species	Dry matter digestibility (%)	Crude protein digestibility (%)	Reference
Fish meal			
L. setiferus	59	75.85	Brunson et al. (1997)
L. vannamei	64.3	80.7	Akiyama et al. (1989)
Mud crab, S. serrata	89.9	94.8	Catacutan et al. (2003)
Red swamp crayfish, P. clarkii	71.6	80.7	Reigh et al. (1990)
Shrimp meal			
L. vannamei	56.8	74.6	Akiyama et al. (1989)
Squid meal			
L. setiferus	63.15	81.78	Brunson et al. (1997)
L. vannamei	68.9	79.9	Akiyama et al. (1989)
Mud crab, S. serrata	93.6	97.6	Catacutan et al. (2003)
Crab meal			
L. setiferus	22.22	59.09	Brunson et al. (1997)
Red swamp crayfish, P. clarkii	61.3	76.5	Reigh et al. (1990)
Crayfish meal			
Australian freshwater crayfish, C. destructor	89.31	94.05	Jones and De Silva (1997)
Acetes sp.			
Mud crab, S. serrata	88.3	94.9	Catacutan et al. (2003)

Apparent crude protein digestibility coefficient by *L. setiferus* for shrimp meal and crab meal (58 to 59 percent) were significantly lower than those for menhaden FM (76 percent) and squid liver powder (82 percent) (Brunson et al., 1997; with chromic oxide as indicator, digestibility was determined based upon the 70:30 ratio of RF to each test ingredient in the test diet).

Apparent dry matter and crude protein digestibilities coefficient by red swamp crayfish *(Procambarus clarkii)* for menhaden FM (71.7 and 80.7 percent, respectively) were significantly higher than those for crab meal (61.3 and 76.5 percent, respectively) (Reigh et al., 1990).

Jones and De Silva (1997) indicated that the apparent dry matter and crude protein digestibilities coefficient by the freshwater crayfish *(Cherax destructor)* for crayfish meal were 89.31 and 94.05 percent, respectively.

MBM

Apparent digestibilities of dry matter and crude protein of several meat by-products are summarized in Table 6.5. For MBM, apparent dry matter and crude protein digestibilities coefficient by *L. setiferus* were 46.41 and 58.57 percent, respectively; apparent dry matter and crude protein digestibilities coefficient by mud crab *(S. serrata)* were 85.2 and 95.0 percent, respectively; apparent dry matter and crude protein digestibilities coefficient by *P. clarkii* were 61.3 and 76.5 percent, respectively (Reigh et al., 1990); apparent dry matter and crude

TABLE 6.5. Apparent digestibility of crustacean fed diets contained meat and bonemeal, poultry by-product meal and snail meal.

Species	Dry matter digestibility (%)	Crude protein digestibility (%)	Reference
Meat and bonemeal			
L. setiferus	46.41	58.57	Brunson et al. (1997)
L. vannamei	63.60-74.25	81.04-87.84	Forster et al. (2003)
	69	82	Tan and Yu (2003a)
Mud crab, *S. serrata*	85.2	95.0	Catacutan et al. (2003)
Red swamp crayfish, *P. clarkii*	61.3	76.5	Reigh et al. (1990)
Australian freshwater crayfish, *C. destructor*	84.83	90.60	Jones and De Silva (1997)
Poultry by-product meal			
L. vannamei	76	90	Tan and Yu (2003b)
Snail meal			
Australian freshwater crayfish, *C. destructor*	80.13	88.27	Jones and De Silva (1997)

protein digestibilities coefficient by *C. destructor* were 84.83 and 90.60 percent, respectively (Jones and De Silva, 1997).

Tan and Yu (2003b) determined the digestibility of FM, MBM, and poultry by-product meal (pet food grade, FGPBM) by *L. vannamei*. Test ingredients were mixed with a commercial FM-based RF at 30:70 ratio prior to pelleting. Results showed that crude protein digestibility was higher for FGPBM (90 percent) than MBM (82 percent), with a similar trend for dry matter digestibility (76 percent for FGPBM and 69 percent for MBM).

Jones and De Silva (1997) indicated that the apparent dry matter and crude protein digestibilities coefficient by *C. destructor* for snail meal were 84.83 and 90.60 percent, respectively.

Combined Ingredients

Apparent digestibility of four practical diets containing different combinations of marine animal-protein sources prepared from fisheries by-products were determined for *P. monodon* (Sudaryono et al., 1996). The diet formulations tested contained mollusc and crustacean waste meals (13.0 percent scallop and 7.8 percent shrimp head, diet D1) and fish and crustacean waste meals (14.7 percent sardine and 8.3 percent lobster, diet D2; 14.7 percent sardine and 7.8 percent shrimp head, diet D3). Diet D4 contained commercially available fish and crustacean meals (15.0 percent anchovy and 10.0 percent shrimp). The apparent dry matter and protein digestibility values of the three diets, D1, D2, and D4 were similar with a range of 75.8 to 78.1 percent and 92.0 to 92.8 percent, respectively, and were significantly higher than D3 (53.2 to 85.4 percent) (Table 6.6).

CONCLUSION

Crustaceans require higher dietary protein levels than most terrestrial food animals. Fisheries by-products are often utilized in shrimp diets because they are excellent sources of indispensable amino acids, essential fatty acids, and generally enhance palatability. The quality of fisheries by-products depends largely on the source of the materials, freshness, and processing method.

TABLE 6.6. Apparent digestibility of *P. monodon* fed diets containing combined fisheries by-products.

Combined food	Dry matter digestibility (%)	Crude protein digestibility (%)	Reference
Scallop waste meal + shrimp head meal	75.8	92.0	Sudaryono et al. (1996)
Sardine meal + lobster waste meal	78.1	92.5	
Sardine meal + shrimp head meal	53.2	85.4	
Commercial fish meal + commercial shrimp meal	77.8	92.8	

Fish meal and marine by-products are commodities for which supplies are limited and demand is expected to continue to increase. Hence, maintenance of the economical viability of commercial aquaculture will require the replacement of expensive marine proteins, with lower-cost ingredients such as meat by-products and poultry by-products for which production is not limited. It should be noted, however, that concerns related to the spread of bovine spongiform encephalitis have restricted the use of MBM in animal diets. Therefore, for the control of disease and epidemics collection of these materials is a must.

REFERENCES

Akiyama, D.M., S.R. Coelho, A.L. Lawrence, and E.H. Robinson (1989). Apparent digestibility of feedstuffs by the marine shrimp *Penaeus vannamei* Boone. *Nippon Suisan Gakkaishi* 55: 91-98.

Akiyama, D.M., W.G. Dominy, and A.L. Lawrence (1992). Penaeid shrimp nutrition. In *Marine Shrimp Culture: Principles and Practices,* A.W. Fast and L.J. Lester (Eds.). Elsevier Science Publishers B.V., Amsterdam, The Netherlands, pp. 535-568.

Aksnes, A. and B. Brekken (1988). Tissue degradation, amino acid liberation and bacterial decomposition of bulk stored capelin. *Journal of the Science of Food and Agriculture* 45: 53-60.

Alam, M.S., S.I. Teshima, D. Yaniharto, M. Ishikawa, and S. Koshio (2002). Dietary amino acid profiles and growth performance in juvenile kuruma shrimp *Marsupenaeus japonicus. Comparative Biochemistry and Physiology* 133B: 289-297.

Ali, S.A. (1982). Relative efficiencies of pelletized feeds compounded with different animal proteins and the effect of protein level on the growth of the prawn *Penaeus indicus*. *Proceedings of the Symposium of Coastal Aquaculture* 1: 321-325.

Ali, S.A. (1992). Evaluation of some animal and plant protein sources in the diet for the shrimp *Penaeus indicus*. *Asian Fisheries Science* 5: 277-289.

Anonymous (1987). By-product meals may have place in turkey diets. *Feedstuffs* 59: 13-14.

Anonymous (1992). Innovations. *Infofish International* 6: 53.

Baelum, J. (1962). El pescado y los productors pesqueros en la alimentacion de las aces. In *Fish in Nutrition*, E. Heen and B. Kruezer (Eds.). FAO, Rome, Italy.

Balazs, G.H. and E. Ross (1976). Effect of protein source and level on growth and performance of the captive freshwater prawn, *Macrobrachium rosenbergii*. *Aquaculture* 7: 299-313.

Boghen, A.D. and J.D. Castell (1981). Nutritional value of different dietary proteins to juvenile lobsters, *Homarus americanus*. *Aquaculture* 22: 343-351.

Boghen, A.D., J.D. Castell, and D.E. Conklin (1982). In search of a reference protein to replace "vitamin-free casein" in lobster nutrition studies. *Canadian Journal of Zoology* 60: 2033-2038.

Brodner, C.E. (1989). A standard reference diet for crustacean nutrition research. V. Growth and survival of juvenile dungeness crabs *Cancer magister*. *Journal of the World Aquaculture Society* 20: 118-121.

Brunson, J.F., R.P. Romaire, and R.C. Reigh (1997). Apparent digestibility of selected ingredients in diets for white shrimp *Penaeus setiferus* L. *Aquaculture Nutrition* 3: 9-16.

Carver, L.A., D.M. Akiyama, and W.G. Dominy (1989). Processing of wet shrimp heads and squid viscera with soy meal by a dry extrusion process. In *Proceedings of the World Congress on Vegetable Protein Utilization in Human Foods and Animal Feedstuffs*. American Oil Chemists Society, Champaign, IL, pp. 167-170.

Castell, J.D., J.C. Kean, L.R. D'Abramo, and D.E. Conklin (1989a). A standard reference diet for crustacean nutrition research. I. Evaluation of two formulations. *Journal of the World Aquaculture Society* 20: 93-99.

Castell, J.D., J.C. Kean, D.G.C. McCann, A.D. Boghen, D.E. Conklin, and L.R. D'Abramo (1989b). A standard reference diet for crustacean research. II. Selection of purification procedures for production of the rock crab *(Cancer irroratus)* protein ingredient. *Journal of the World Aquaculture Society* 20: 100-106.

Catacutan, M.R., P.S. Eusebio, and S.I. Teshima (2003). Apparent digestibility of selected feedstuffs by mud crab, *Scylla serrata*. *Aquaculture* 216: 253-261.

Chamberlain, G.W. and A.L. Lawrence (1981). Maturation, reproduction, and growth of *Penaeus vannamei* and *Penaeus stylirostris* fed natural diets. *Journal of the World Mariculture Society* 12: 209-224.

Cheng, Z.J., K.C. Behnke, and W.G. Dominy (2002). Effects of poultry by-product meal as a substitute for fish meal in diets on growth and body composition of juvenile Pacific white shrimp, *Litopenaeus vannamei*. *Journal of Applied Aquaculture* 12: 70-83.

Cho, C.Y., C. Cowey, and T. Watanabe (1985). Finfish nutrition in Asia: Methodological approaches to research and development. IDRC, 233e. Otawa, Canada.

Cho, C.Y. and S.J. Slinger (1979). Apparent digestibility measurement in feedstuffs for rainbow trout. In *Proceedings of World Symposium on Finfish Nutrition and Fishfeed Technology,* Vol. II. Heenemann Verlagsgesell, Berlin, Germany. pp. 239-247.

Chou, R. (1993). Aquafeeds and feeding strategies in Singapore. In *Proceedings of FAO/AADCP Regional Expert Consultation on Farm-Made Aquafeeds,* December 14-18, 1992, Bangkok/Thailand. FAO-RAPA/AADCP, Bangkok, Thailand, pp. 354-364.

Christians, O. and M. Leinemann (1980). Untersuchungen über Fluor im Krill (*Euphausia superba* Dana). *Information Fischwirtschaft* 28: 254-260.

Colvin, P.M. (1976). Nutritional studies on penaeid prawns: Protein requirements in compounded diets for juvenile *Penaeus indicus* (Milne Edwards). *Aquaculture* 7: 315-326.

Cruz-Ricque, L.E., J. Guillaume, and G. Cuzon (1987). Squid protein effect on growth of four penaeid shrimp. *Journal of the World Aquaculture Society* 18: 209-217.

Cruz-Suárez, L.E., J. Guillaume, and A. van Wormhoudt (1987). Effect of various levels of squid protein on growth and some biochemical parameters of *Penaeus japonicus* juveniles. *Nippon Suisan Gakkaishi* 53: 2083-2088.

Cruz-Suárez, L.E., D. Ricque, and Aquacop (1992). Effect of squid meal on growth of *Penaeus monodon* juveniles reared in pond pens and tanks. *Aquaculture* 106: 293-299.

Cruz-Suárez, L.E., D. Ricque, J.A. Martinez-Vega, and P. Wesche-Ebeling (1993). Evaluation of two shrimp by-product meals as protein sources in diets for *Penaeus vannamei. Aquaculture* 115: 53-62.

Davis, A.D. and C.R. Arnold (2000). Replacement of fish meal in practical diets for the Pacific white shrimp, *Litopenaeus vannamei. Aquaculture* 185: 291-298.

De Boer, F. and H. Bickel (1988). *Livestock Feed Resources and Feed Evaluation in Europe.* Elsevier Sciences Publishers, Amsterdam, The Netherlands.

Deshimaru, O. (1981). Studies on nutrition and diet for prawn, *Penaeus japonicus. Memoirs of Kagoshima Prefectural Fisheries Experiment Station* 12: 188.

Deshimaru, O. and K. Shigueno (1972). Introduction to the artificial diet for prawn *Penaeus japonicus. Aquaculture* 1: 115-133.

Djunaidah, I.S. (1993). Aquafeeds and feeding strategies in Singapore. In *Proceedings of FAO/AADCP Regional Expert Consultation on Farm-Made Aquafeeds,* December 14-18, 1992, Bangkok/Thailand. FAO-RAPA/AADCP, Bangkok, Thailand, pp. 255-281.

Dominy, W.C. and H. Ako (1988). The utilization of blood meal as a protein ingredient in the diet of the marine shrimp *Penaeus vannamei. Aquaculture* 70: 289-299.

Dong, F.M., R.W. Hardy, N.F. Haard, F.T. Barrows, B.A. Rasco, W.T. Fairgrieve, and I.P. Foster (1993). Chemical composition and digestibility of poultry by-product meals for salmonid diets. *Aquaculture* 116: 149-158.

Dupree, H.K. and J.V. Huner (1984). Third report to the fish farmers. In *Publication of U.S. Fish and Wildlife Service.* Washington, DC, 270 pp.

Evans, M. (1985). *Nutrient Composition of Feedstuff for Pig and Poultry.* Queensland Department of Primary Industries, Information Series Q185001, Brisbane, Australia.

Feed Development Section (FDS) (1994). *Feeds and Feeding of Milkfish, Nile Tilapia, Asian Sea Bass, and Tiger Shrimp.* Extension Manual No. 21, SEAFDEC Aquaculture Department, Tigbauan, Iloilo, Philippines.

Feltwell, R. and S. Fox (1978). *Practical Poultry Feeding.* Faber and Faber, Boston, MA.

Fennuci, J.L. and Z.P. Zein-Eldin (1976). Evaluation of squid mantle meal as a protein source in penaeid nutrition. In *Advances in Aquaculture,* T.V. Pillay and W.A. Dill (Eds.). FAO Technical Conference on Aquaculture, Kyoto, Japan, pp. 601-605.

Fennuci, J.L., Z.P. Zein-Eldin, and A.L. Lawrence (1980). The nutritional response of two penaeid species to various levels of squid meal in prepared feed. *Proceedings of the World Mariculture Society* 11: 403-409.

Floreto, E.A.T., P.B. Brown, and R.C. Bayer (2001). The effects of krill hydrolysate-supplemented soya-bean based diets on the growth, colouration, amino and fatty acid profiles of juvenile American lobster, *Homarus americanus. Aquaculture Nutrition* 7: 33-43.

Food and Nutrition and Research Institute (1968). *Food Composition Table Handbook I.* National Science Development Board, Manila, Philippines, 134 pp.

Forster, I.P., W. Dominy, L. Obaldo, and A.G.J. Tacon (2003). Rendered meat and bone meals as ingredients of diets for shrimp *Litopenaeus vannamei* (Boone, 1931). *Aquaculture* 219: 655-670.

Fox, C.J., P. Blow, J.H. Brown, and I. Watson (1994). The effect of various processing methods on the physical and biochemical properties of shrimp head meals and their utilization by juvenile *Penaeus monodon* Fab. *Aquaculture* 122: 209-226.

Friesecke, H. (1984). *Handbuch der praktischen Fütterung.* BLV Verlagsgesellschaft München, Germany.

Galgani, M.L. and Aquacop (1989). Influence du regime alimentaire sur la reproduction en captivite *Penaeus vannamei* et *Penaeus stylirostris. Aquaculture* 80: 97-109.

Guillame, J., E. Cruz-Ricque, E. Cuzon, A.V. Wormhoudt, and A. Revol (1990). Growth factors in penaeid shrimp feeding. In *Advances in Tropical Aquaculture,* Tahiti, February 20 to March 4, 1989. Aquacop Ifremer Actes de Colloque, pp. 327-338.

Haaland, H., E. Arnesen, and L.R. Njaa (1990). Amino acid composition of whole mackerel *(Scomber scombrus)* stored anaerobically at 20°C and at 2°C. *International Journal of Food Science and Technology* 25: 82-87.

Hardy, R.W. (1989). Diet preparation. In *Fish Nutriton,* J.E. Halver (Ed.). Academic Press, San Diego, pp 41-72.

Hughes, S. (1990). Feather meal can displace fishmeal in aquaculture rations. *Feed International* 15: 13-15.

Hurrel, R.F. and P.A. Finot (1985). Effects of food processing on protein digestibility and amino acid availability. In *Digestibility and Amino Acid Availability in*

Cereals and Oilseeds, J.W. Finley and D.T. Hopkins (Eds.). American Association of Cereal Chemists, St. Paul, MN, pp. 233-246.

Johnson, D. (1988). Crab meal as feed for hogs. Prince Edward Isl. Agriculture and Food Development Sub-Agreement Report, Technology Development, pp. 1-35.

Jones, P.L. and S.S. De Silva (1997). Apparent nutrient digestibility of formulated diets by the Australian freshwater crayfish *Cherax destructor* Clark (Decapoda, Parastacidae). *Aquaculture Research* 28: 881-891.

Joseph, J., V. Prabhu, and P. Madhavan (1984). Utilization of squid waste as meal. *Progressive Technology of Society,* Cochin 24: 41-43.

Just, A., J.A. Fernandez, and H. Jørgensen (1982). Kødbenmels værdi til svin. *Bert. Statens Husdyrbrugsfor* 525: 1-52.

Kean, J.C., A.D. Boghen, L.R. D'Abramo, and D.E. Conklin (1985). Re-evaluation of the lecithin and cholesterol requirements of juvenile lobster *(Homarus americanus)* using crab protein-based diets. *Aquaculture* 47: 143-149.

Kling, M. and W. Wöhlbier (1977). *Handels-Futtermittel.* Verlag Eugen Ulmer, Stuttgart, Germany.

Knipfel, J.E. (1981). Nitrogen and energy availability in foods and feeds subjected to heating. In *Maillard Reaction in Food,* C. Eriksson (Ed.). *Progressive of Food and Nutrition Science* 5(1-6), 177-192.

Koshio, S., A. Kanazawa, and S.I. Teshima (1992). Search for effective protein combination with crab protein for the larval kuruma prawn *Penaeus japonicus. Nippon Suisan Gakkaishi* 58: 1083-1089.

Koshio, S., A. Kanazawa, S. Teshima, and J.D. Castell (1989). Nutritive evaluation of crab protein for larval *Penaeus japonicus* fed microparticulate diets. *Aquaculture* 81: 145-154.

Koshio, S., R.K. O'Dor, and J.D. Castell (1990). The effect of varying dietary energy levels on growth and survival of eyestalk ablated and intact juvenile lobster, *Homarus americanus. Journal of the World Aquaculture Society* 21: 160-169.

Lawrence, A.K. and F. Castille (1991). Nutritive response of a western hemisphere shrimp *Penaeus vannamei,* to meat and bone, feather and poultry by-product meal. *Director's Digest,* No. 215, Fat and Protein Research Foundation, Bloomington, IL.

Liang, M. and A. Anders (2001). Influence of fish meal quality on growth, feed conversion rate and protein digestibility in shrimp *(Penaeus chinensis)* and red seabream *(Pagrosomus major). Marine Fisheries Research* 22: 75-79.

Lim, C., P. Suraniranat, and R. Platon (1979). Evaluation of various protein sources for *Penaeus monodon* post-larvae. *Kalikasan, Philippine Journal of Biology* 8: 29-36.

Marsden, G., J.M. McGuren, H.Z. Sarac, A.R. Neill, I.J. Brock, and C.L. Palmer (1992). Nutritional composition of some natural marine feeds used in prawn maturation. In *Proceedings of Aquaculture Nutrition Workshop,* Salamander Bay, April 15-17, 1991, pp. 82-86.

Marte, C. (1980). The food and feeding habit of *Penaeus monodon* Fabricus collected from Makato River, Aklan, Philippines (Decapoda Natantia). *Crustaceana* 38: 225-235.

McDonald, P., R.A. Edwards, and J.F.D. Greenhalgh (1988). *Animal Nutrition (Evaluation of Food)*. 4th edn. John Wiley & Sons, Inc., New York, NY, pp. 200-215.

Mendoza, R., A. De Dios, C. Vazquez, E. Cruz, D. Ricque, C. Aguilera, and J. Montemayor (2001). Fishmeal replacement with feather-enzymatic hydrolyzates co-extruded with soya-bean meal in practical diets for the Pacific white shrimp *(Litopenaeus vannamei)*. *Aquaculture Nutrition* 7: 143-151.

Meyer, S.P. (1986). Utilization of shrimp processing wastes. *Infofish Marketing Digest* 4 (86): 18-19.

Meyer, S.P. and E. Heckötter (1986). *Futterwerttabelle für Hunde und Katzen.* Schlütersche Verlagsanstalt, Hannover, Germany.

Millamena, O.M., N.V. Golez, J.A.J. Janssen, and M. Peschcke-Koedt (2000). Processed meat solubles, protamino aqua, used as an ingredient in juvenile shrimp feeds. *The Israeli Journal of Aquaculture-Bamidgeh* 52: 91-97.

Morrissy, N.M. (1989). A standard reference diet for crustacean nutrition research. IV. Growth of freshwater crayfish *Cherax tenuimanus. Journal of the World Aquaculture Society* 20: 114-117.

Murayama, S., M. Yanase, and Y. Masaki (1962). Nutritive constituents of fish meal and fish solubles. In *Fish in Nutrition,* E. Heen and B. Kruezer (Eds.). FAO, Rome, Italy, pp. 320-323.

Nandeesha, M.C. (1993). Aquafeeds and feeding strategies in India. In *Farm-Made Aquafeeds Proceedings of FAO/AADCP Regional Expert Consultation on Farm-Made Aquafeeds,* M.B. New, A.G.J. Tacon, and I. Csavas (Eds.), December 14-18, 1992, Bangkok/Thailand. FAO-RAPA/AADCP, Bangkok, Thailand, pp. 213-254.

National Renderers Association (NRA) (1993). Pocket Information Manual—A Buyers Guide to Rendered Products. National Renderers Association, Inc., Alexandria, Virginia.

National Research Council (NRC) (1977). *Nutrient Requirements of Warmwater Fishes. Nutrient Requirements of Domestic Animals.* National Academic Press, Washington, DC.

National Research Council (NRC) (1981). *Nutrient Requirements of Coldwater Fishes.* National Academic Press, Washington, DC.

National Research Council (NRC) (1983). *Nutrient Requirements of Warmwater Fishes and Shellfish.* National Academic Press, Washington, DC.

National Research Council (NRC) (1991). *Nutrient Requirement of Coldwater Fishes. Nutrient Requirements of Domestic Animals.* National Academic Press, Washington, DC.

National Research Council (NRC) (1993). *Nutrient Requirements of Fish. Nutrient Requirements of Domestic Animals.* National Academic Press, Washington, DC.

New, M.B. (1987). Feed and Feeding of Fish and Shrimps—A Manual on the Preparation and Presentation of Compound Feeds for Shrimps and Fish in Aquaculture. UNDP/FAO/ADCP/REP/87/26, Rome, Italy.

Park, C.W. (1989). Effect of sardine meal on growth and mineral contents of red sea bream, *Pagrus major* and Japanese eel, *Anguilla japonicus. Ocean Research Korea* 11: 9-13.

Peñaflorida, V.D. (1989). An evaluation of indigenous protein sources as potential component in the diet formulation for tiger prawn, *Penaeus monodon,* using amino acid index (EAAI). *Aquaculture* 83: 319-330.

Pesti, G.M. and P.G. Dubuc (1986). The relationship between poultry by-product meal content and market values. In *Proceedings of Arkansas Nutrition Conference,* Little Rock, Arkansas.

Piedad-Pascual, P. (1993). Aquafeeds and feeding strategies in Philippines. In *Farm-Made Aquafeeds Proceedings of FAO/AADCP Regional Expert Consultation on Farm-Made Aquafeeds,* M.B. New, A.G.J. Tacon, and I. Csavas (Eds.), December 14-18, 1992, Bangkok/Thailand. FAO-RAPA/AADCP, Bangkok, Thailand, pp. 317-353.

Piedad-Pascual, F. and W.H. Destajo (1979). Growth and survival of *Penaeus monodon* postlarvae fed shrimp head meal and fish meal as primary animal sources of protein. *Fisheries Research Journal, Philippine* 4: 29-36.

Pongmaneerat, J., J. Kasornchandra, S. Boonyaratpalin, and M. Boonyaratpalin (2001). Effect of dietary shrimp head meal contaminated with white spot syndrome virus (WSSV) on detection of WSSV in black tiger shrimp (*Penaeus monodon* Fabricius). *Aquaculture Research* 32: 383-387.

Primavera, J.H., C. Lom, and E. Borlongan (1979). Feeding regimes in relation to reproduction and survival of abalated *Penaeus monodon. Kalikasan Philippine Journal of Biology* 8: 227-235.

Quayle, D.B. and G.F. Newkirk (1989). Farming bivalve molluscs: Methods for study and development. *Advances in World Aquaculture,* Vol. 1. Canada, World Aquaculture Soc: In association with the Int'l Dev. Res. Center.

Quinitio, E.T., F.D. Parado-Estepa, O.M. Millmena, and H. Biona (1994). Reproduction performance of captive *Penaeus monodon* fed various sources of carotenoids. In *National Seminar-Workshop on Fish Nutrition and Feeds,* SEAFDEC, Tigbiuan, Iloilo, Philippines.

Rajyalakshmi, T., S.M. Pillar, A.K. Roy, and P.U. Verghese (1982). Studies on rearing of *Penaeus monodon* Fabricius in brackishwater ponds using pelleted feeds. *Journal of Inland Fisheries Society India* 14: 28-35.

Ravishankar, A.N. and P. Keshavanath (1986). Growth response of *Macrobrachium rosenbergii* (de Man) fed on four pelleted feeds. *Indian Journal of Animal Science* 56: 110-115.

Reed, L. and L.R. D'Abramo (1989). A standard reference diet for crustacean nutrition research. III. Effects on weight gain and amino acid composition of whole body and tail muscle of juvenile prawns *Macrobrachium rosenbergii. Journal of the World Aquaculture Society* 20: 107-113.

Reigh, R.C., S.L. Braden, and R.J. Craig (1990). Apparent digestibility coefficients for common feedstuffs in formulated diets for red swamp crayfish, *Procambarus clarkii. Aquaculture* 84: 321-334.

Reigh, R.C. and S.C. Ellis (1994). Utilization of animal-protein and plant-protein supplements by red swamp crayfish *Procambarus clarkii* fed formulated diets. *Journal of the World Aquaculture Society* 25: 541-552.

Ricque-Marie, D., Ma.I. Adbo-de La Parra, L.E. Cruz-Suarez, M. Cuzon, and I.H. Pike (1998). Raw material freshness, a quality criterion for fish meal fed to shrimp. *Aquaculture* 165: 95-109.

Sarac, H.Z., M. Grvel, J. Saunders, and S. Tabrett (1993). Evaluation of Australian protein sources for diets of the black tiger prawn *(Penaeus monodon)* by proximate analysis and essential amino acid index. In *Abstracts of the International Conference World Aquaculture '93,* May 26-28, 1993, M. Carrillo, L. Dahle, J. Morales, P. Sorgeloos, N. Svennevig, and J. Wyban (Eds.). Special Publication, No. 19, European Aquaculture Society, Torremolinos, Spain, Oostende, Belgium.

Schulz, E. (1995). *Nährstoff- und Energiegehalt in deutschen Tiermehlen.* Die Fleischmehl-Industrie, Washington, DC.

Sedgewick, R.W. (1980). The requirements of *Penaeus merguiensis* for vitamin and mineral supplements in diets based on freeze-dried *Mytilus edulis* meal. *Aquaculture* 19: 127-137.

Shigueno, K. (1975). *Shrimp Culture in Japan.* Association International Technical Promotion, Tokyo, Japan.

Smith, D.M., G.L., Allan, K.C. Williams, and C.G. Barlow (2001). Fishmeal replement research for shrimp feed in Australia. In *Aquaculture 2001, Book of Abstracts.* World Aquaculture Society, Baton Rouge, LA, Vol. 31, 598 pp.

Stählin, A. (1957). *Methodenbuch, Beurteilung der Futtermittel,* Vol. XII. Die Beurteilung der Futtermittel. Neumann Verlag, Radebeul and Berlin, Germany.

Steffens, W. (1980). Krillmehl als Eiweißquelle im Fischfutter. 1. Mitt. Biologie and Nährstoffzusammensetzung des Krills. *Z. Binnenfischerei DDR* 27: 182-186.

Steffens, W. (1985). Grundlagen der Fischernährung. In *VEG Gustay Fishcher Verlag,* Jena, Germany.

Steffens, W. and M.L. Albrecht (1981). Krillmehl als Eiweißquelle im Fischfutter. 3. Mitt Vollständiger Ersatz von tierischem Protein durch Krillmehl ´im Forellenfutter. *Z. Binnenfischerei DDR* 28: 178-184.

Sudaryono, A., M.J. Hoxey, S.G. Kailis, and L.H. Evans (1995). Investigation of alternative protein sources in practical diets for juvenile shrimp, *Penaeus monodon. Aquaculture* 134: 313-323.

Sudaryono, A.E., Tsvetnenko, and L.H. Evans (1996). Digestibility studies on fisheries by-product based diets for *Penaeus monodon. Aquaculture* 143: 331-340.

Tacon, A.G.J. (1987). The Nutrition and Feeding of Farmed Fish and Shrimp. 2. A Training Manual on Nutrient Sources and Composition. Field Document 5/E GCP/RLA/075/ITA, FAO, Rome, Italy, 129 pp.

Tacon, A.G.J. (1993). Feed Ingredients for Crustaceans Natural Foods and Processed Feedstuffs. FAO Fisheries Circular No. 866, FAO, Rome, Italy.

Tacon, A.G.J. and I.P. Foster (2001). Global trends and challenges to aquaculture and aquafeed development in the new millennium. In *International Aquafeed Directory and Buyer's Guide.* Pickmansworth, Turret Rai Group, Pickmansworth, UK, pp. 4-25.

Tan, B. and Y. Yu (2003a). Replacement of Fish Meal with Meat and Bone Meal on Growth Performance of White Leg Shrimp (*P. vannamei*). Research Reports No. 24, National Renderers Association, Inc., Causeway Bay, Hong Kong.

Tan, B. and Y. Yu (2003b). Digestibility of Fish Meal, Meat and Bone Meal, and Poultry Byproduct Meal (Pet Food Grade) by White Leg Shrimp (*P. vannamei*). Research Reports No. 24, National Renderers Association, Inc., Causeway Bay, Hong Kong.

Tan, B., S. Zheng, H. Yu, and Y. Yu (2003). Growth, Feed Efficiency of Juvenile *Litopenaeus vannamei* Fed Practical Diets Containing Different Levels of Poultry By-Product Meal. Research Reports No. 24, National Renderers Association, Inc., Causeway Bay, Hong Kong.

Teshima, S., A. Kanazawa, and M. Yamashita (1986). Dietary value of several proteins and supplemental amino acids for larvae of the prawn *Penaeus japonicus. Aquaculture* 51: 225-235.

Teshima, S., A. Kanazawa, M. Yamashita, and S. Koshio (1991). Rearing of larval prawn *Penaeus japonicus* with brown fish meal based diets. *Nippon Suisan Gakkaishi* 57: 175.

van Lunen, T.A. and D.M. Anderson (1990). Crab meal. In *Nontraditional Feed Sources for Used in Swine Production,* P.A. Thacker and R.N. Kirkwood (Eds.). Butterworths Publishers, Stoneham, MA, pp. 153-159.

Watanabe, T., S. Satoh, and T. Takeuchi (1988). Availability of minerals in fish meal to fish. *Asian Fisheries Science* 1: 175-195.

Watt, B.K. and A.L. Merrill (1963). *Composition of Foods.* Agricultural Handbook No. 8. United States Department of Agriculture, Washington DC, 190 pp.

Wöhlbier, W. (1977). Geflügelschlachtabfälle. In *Handelsfuttermittel,* M. Kling and W. Wöhlbier (Eds.). Verlag Eugen Ulmer, Stuttgart, Germany.

Wöhlbier, W. and F. Jäger (1977). Futtermittel aus Meerestieren. In *Handelsfuttermittel,* M. Kling and W. Wöhlbier (Eds.). Verlag Eugen Ulmer, Stuttgart, Germany.

Wöhlbier, W. and D. Tran Thu (1977). Tiermehle. In *Handelsfuttermittel,* M. Kling and W. Wöhlbier (Eds.). Verlag Eugen Ulmer, Stuttgart, Germany.

Chapter 7

Use of Plant Protein Sources in Crustacean Diets

Jesus A. Venero
D. Allen Davis
Chhorn Lim

INTRODUCTION

As part of the global growth of aquaculture, the world production of crustaceans has experienced a steady increase that is expected to continue as world population increases and demand for quality seafood continues to rise (FAO, 2000; Tacon et al., 2000). Although crustaceans contributed only 5.10 percent of the world aquaculture production in 2003, it represented 19.80 percent of the total production by value (FAO, 2005a). In 2003, marine shrimp were the third most important world aquaculture species with a value of US$9.32 billion (FAO, 2005a). In addition, marine shrimp production represented 64.65 percent of the total crustacean production. Other crustaceans groups produced included crab and sea spiders, 6.55 percent, freshwater crustaceans, 24.65 percent, lobster and spiny-rock lobster, 0.3 percent, and other marine crustaceans, 4.15 percent (FAO, 2005a).

Paralleling the growth of the industry has been an expansion in diet production (Tacon et al., 2000). The majority of finfish and crustaceans species fed with these commercial diets are omnivorous or carnivorous (Lee et al., 1980; Lee and Lawrence, 1997; Tacon and Akiyama, 1997). It is generally felt that these groups demand high-quality protein and special organoleptic properties in their diets. Fish

Alternative Protein Sources in Aquaculture Diets
© 2008 by The Haworth Press, Taylor & Francis Group. All rights reserved.
doi:10.1300/5892_07

meal, as well as other marine animal meals such as krill, shrimp, squid, and scallop waste are often included in aquatic diets as they are considered an excellent source of high-quality proteins, highly unsaturated fatty acids, minerals, and attractants (Tacon and Akiyama, 1997). Owing to these properties, fish meal has become one of the primary components of commercial diet formulations. The demand for fish meal in aquatic diets has been estimated to account for 31 to 42.5 percent of total world fish meal production (Tacon and Barg, 1998). Of this marine shrimp consumed approximately 17.6 percent of the total amount.

Even though fish meal and other marine meals are excellent sources of protein and other essential nutrients for aquaculture diets, their demand is subject to competition from other sectors of the agriculture industry. Also, the supply, quality, and price often fluctuate from year to year due to both market constraints and natural phenomena. There are also environmental concerns, in terms of pollution and overfishing, as well as ethical considerations for the use of fish products that could be used directly to feed humans (Chamberlain, 1993; Tacon and Akiyama, 1997; Tacon and Forster, 2000). On account of these constraints, replacement of fish meal and other marine proteins with alternative sources of proteins such as those derived from terrestrial animals or plant proteins have been encouraged (Tacon and Akiyama, 1997).

The animal by-product meals include meat and bonemeal, blood meal, poultry by-product meal, feather meal (e.g., hydrolyzed or enzyme treated), and specialized protein blends (Tacon and Forster, 2000). These feedstuffs are characterized by higher production than fish meals and less cost, but they are of variable composition, their nutritional quality is lower than fish meals (e.g., imbalances in essential amino acids (EAAs), high ash content, and low nutrient digestibility among others), they have possible palatability problems, and possible microbial contamination (Tacon, 1994; Tacon and Forster, 2000). Other sources of protein for aquatic diets are plant proteins (Li et al., 2000). From a nutritional standpoint, fish meal, and most of the marine meals, can be replaced totally or almost completely by a single or a combination of plant protein sources (Viola et al., 1982, 1988; Tidwell, Webster, Clark, and D'Abramo, 1993; Tidwell, Webster, Yancey, and D'Abramo, 1993; Sudaryono et al., 1995; Webster et al.,

1995; Wu et al., 1995; Davis and Arnold, 2000; Davis et al., 2004; Samocha et al., 2004) Plant proteins are produced in larger quantities than fish meals, their production is more stable, they are often less expensive, and their expanded use does not threaten overexploitation of a limited resource as can occur with fisheries products. Among different sources of vegetable proteins, soybean meal (SBM) is the most commonly used source as a replacement or complement to marine proteins (Hertrampf and Piedad-Pascual, 2000; Olvera-Novoa and Olivera-Castillo, 2000). Other sources of plant proteins, such as cottonseed meal, peanut meal, canola meal, distillers grain with solubles, and some legume meals, are commonly utilized (Li et al., 2000). Usually the use of plant proteins shows some limitation due to a variety of factors including a deficiency or imbalance of EAAs, presence of antinutritional factors or toxins, and decreased palatability. Many of these limitations can be overcome through the use of proper combinations of different types of plant proteins to balance essential nutrient profiles (e.g., amino acids and fatty acids); through developing specific processing procedures to inactivate, reduce, or eliminate antinutritional factors (e.g., heat treatment to inactivate heat labile components); and/or through limiting their inclusion in the diet to a level that does not influence animal performance (Li et al., 2000).

In this chapter we will present and discuss published information on research on the use of plant proteins in crustacean diets as partial or total replacement of fish meal and other marine products. The information provided will be mainly on marine and freshwater shrimps, due to their commercial importance, as well as the availability of research information.

PLANT PROTEIN SOURCES

Protein ingredients are defined as those feedstuffs used in animal diets whose levels of crude protein (CP) are equal to or higher than 20 percent (NRC, 1969). Different plant ingredients are grouped in this category and have been commonly used as main or complementary sources of protein in animal diets. Many of these protein sources are by-products of other industries, such as the oilseed meals obtained after extracting oils from the seeds. The leaves or seeds of plants can be processed directly as plant protein meals. Protein concentrates are also used and are obtained after removing some of the nonprotein components from protein meals (Li et al., 2000; Hertrampf and Piedad-Pascual, 2000). Based on their origin we could classify the primary plant protein sources as follows:

Oilseeds

1. SBM *(Glycine max)*
2. Cottonseed meal *(Gossypum hirsitum)*
3. Rapeseed/canola meal *(Brassica napus* and *Brassica campestris)*
4. Peanut meal *(Arachis hypogaea)*
5. Sunflower meal *(Helianthus annuus)*
6. Linseed meal/flax *(Linum usitatissimum)*
7. Sesame meal *(Sesamum indicum)*

Other Leguminous Seeds

1. Lupins: *Lupinus* sp.
2. Field/feed pea meal *(Pisum sativum)*
3. Others: Cowpea (*Vigna unguiculata, Canavalia ensiformis, Phaseolus* sp.)

Leguminous Leaf Meals

1. *Leucaena leucocephala*
2. Alfalfa *(Medicago sativa)*

By-Products of the Brewery Industry

1. Distillers' dried grains
2. Distillers' dried grains with solubles

Protein Isolates or Concentrates

1. Corn gluten meal
2. Wheat gluten
3. Soy, canola, potato, and pea protein concentrates

INGREDIENT SELECTION CRITERIA

When considering the replacement of fish meal (or any ingredient), a number of factors should be considered. These include the availability of the product, nutritional content and quality, ingredient cost, and the effects of inclusion on processing and physical quality of the final product.

Availability

The worldwide production of oilseed meals and plant proteins during 2002 was estimated to be 194.5 mmt (FAO, 2005b). Of this, SBM represented 66.6 percent, rapeseed/canola meal 10.1 percent, and cottonseed meal 6.3 percent (Table 7.1). These protein sources are readily available in the market and their amounts easily surpass the total amount of fish meal that is produced each year (FAO, 2003).

Nutritional Considerations

To replace fish meal with plant ingredients, the nutrient requirement of the species to be fed, as well as the nutrient composition and biological availability of the feedstuffs, should be known (NRC, 1993). Examples of the nutrient compositions of common plant protein

TABLE 7.1. Production in metric tons (mt) of plant protein meals in 2002.

Plant protein	World production	Percentage
SBM	129,645,082	66.6
Rapeseed/canola meal	19,652,653	10.1
Cottonseed meal	12,281,397	6.3
Sunflower seed meal	9,273,780	4.8
Ground nut meal	6,753,445	3.5
Palm kernel meal	3,753,695	1.9
Coconut meal	1,821,560	0.9
Sesame seed meal	837,474	0.43
Other oilseed meals	10,561,440	5.2
Total	194,580,526	100

Data adapted from FAO (2005b).

sources as well as that of the fish meal (herring meal) are presented in Table 7.2.

Information on nutrient requirements is available for a few species of crustaceans, especially that of penaeid shrimps (Lim and Akiyama, 1995; Guillaume, 1997; Shiau, 1998). This information is more limited than that established for finfish species due to a variety of factors common to working with aquatic species and also to the difficulty in determining nutrient intake in crustaceans, which often externally masticate the diet (Tacon and Akiyama, 1997). For example, the determination of EAA requirements in shrimp has been difficult due to the high leaching rate of crystalline amino acids, differential rates of absorption of a mixture of the amino acids as compared to those from intact proteins, and the feeding habits of crustaceans. Consequently, limited and/or variable success has been reported for the supplementation of crystalline amino acids to shrimp diets (for review see Guillaume, 1997). More recently, procedures to encapsulate the amino

TABLE 7.2. Average CP, ether extract (EE), ash, and crude fiber of fish meal and common protein ingredients used in crustacean diets.

Protein ingredient	Composition (% as fed; dry matter ~90-91%)			
	CP	EE	Ash	Crude fiber
Fish meal (herring)	67	8.1	13.5	—
Canola meal (solvent extracted)	37.3	1.9	7.2	11.4
Cottonseed meal (decorticated)	41.1	1	5.4	8.2
Peanut meal (decorticated)	46.5	1	5.4	8.2
SBM (full-fat)	37.4	18.9	5.5	5.4
SBM (expeller extracted)	43.5	5.6	6.2	5.4
SBM (solvent extracted)	48.5	0.9	5.8	3.2
Field pea meal	23.7	1.7	3.5	6.8
Lupin meal (L. albus)	34.5	6.1	3.7	15.5
Corn gluten meal	59.9	3.6	2.5	2.4
Linseed meal	35	2	6.2	9.2
Sunflower seed meal (corticated expeller)	35.1	8.2	6.5	25.6

Data adapted from NRC (1993) and Hertrampf and Piedad-Pascual (2000).

acids and prevent leaching to the water have been utilized. Millamena et al. (1996, 1997, 1998, 1999) and Chen et al. (1992) determined several EAA requirements for *Penaeus monodon* using a procedure of microencapsulation. For *Litopenaeus vannamei* only the lysine (Fox et al., 1995) and arginine (Colvin and Brand, 1977) requirements have been reported. Similarly for *Litopenaeus stylirostris,* only the lysine and threonine requirements have been reported (Colvin and Brand, 1977). The ten EAA requirements for *Marsupenaeus japonicus* were estimated based on the daily increase of each amino acid in the body when the prawn were fed a nutritionally complete diet (Teshima et al., 2002). The requirement of arginine by the dose-response method using coated amino acid was determined also for *M. japonicus* (Alam et al., 2004).

When information on the nutrient requirement of a species is not available, some nutrient requirements can be predicted from body composition (e.g., Wilson and Poe, 1985) and growth (e.g., Lupatsch et al., 1998). Since protein is the main component in a crustacean body, especially decapods, the profile of amino acids, particularly the EAA, can be used to assess the nutritional quality of protein ingredients. Protein sources with EAA profiles similar to those of the animal body are of high nutritional value (Guillaume, 1997). The essential amino acid index (EAAI) is a common coefficient used to evaluate the suitability of a protein to an animal species (Philipps and Brockway, 1956; Deshimaru and Shigeno, 1972). It reflects the proportion of EAA between a feedstuff and an animal body. The EAAI has been used successfully to formulate diets for finfish species whose requirements have not been determined (Cowey and Tacon, 1983; Murai et al., 1984; Wilson and Poe, 1985). Based on Oser's method (Oser, 1959) a protein with an EAAI equal to or higher than 0.9 is considered of good quality, useful when it is around 0.8, and inadequate when it is lower than 0.7. Peñaflorida (1989) estimated the EAAI of some diet ingredients for *P. monodon* (Table 7.3). We included additional calculations to compare the EAAI of fish meal with some plant protein sources using ingredient compositions from the NRC (1993). In general, SBM and canola meal show the EAA profile that best matches the body composition of shrimp (Table 7.3). However, SBM seems to be deficient in the sulfur-containing amino acids (methionine and cysteine) and lysine, whereas canola meal is low in lysine.

TABLE 7.3. Essential amino acid index (EAAI) of various feed ingredients for *P. monodon*.

Protein source	EAAI
Fish meal (herring meal)	0.95[a]-0.97
Shrimp meal (*Acetes* sp.)	0.98
Fish meal (anchovy)	0.97
SBM (solvent)	0.87[a]-0.95
Canola meal	0.95
Peanut meal	0.84
Cottonseed meal	0.89
Acacia meal	0.53[a]
Leucaena leaf meal	0.5[a]

[a]Peñaflorida (1989).

Other values calculated by the authors using data from NRC (1993) and the body composition of *P. monodon* from Peñaflorida (1989).

The most limiting EAAs in plant proteins are the sulfur-containing amino acids (methionine-cysteine), lysine, threonine, and arginine (Deshimaru et al., 1985; NRC, 1993). With the exception of gluten (corn or wheat) which has a high level of methionine, all plant proteins have lower levels of these EAAs than fish meal (Table 7.4).

The C18 series polyunsaturated fatty acids (PUFA) (linoleic, 18:2 n-6 and linolenic, 18:3 n-3) as well as n-3 and n-6 highly unsaturated fatty acids (HUFA) (eicosapentaenoic acid, EPA; decosahexanoic acid, DHA; and arachidonic acid, AA), are dietary essentials in shrimp and other crustaceans (Kanazawa et al., 1977; Martin, 1980; Fenucci et al., 1981; Read, 1981; Shiau, 1998). It has been demonstrated that these essential fatty acids (EFA) are required in the diets of penaeid shrimp at a level between 0.5 and 1 percent (for review see Lim and Akiyama, 1995). Some authors have found that shrimp fed diets that contain vegetable oils high in linolenic acid exhibit better growth and survival (Guary et al., 1976). This response was probably due to n-3/n-6 PUFA ratio. It has been suggested that a n-3/n-6 ratio that resembles that of the crustaceans body, especially that of the reproductive tissues, promotes better growth and reproductive performance of crustaceans (Floreto et al., 2000; Millamena and Quinitio, 2000).

TABLE 7.4. Essential amino acid content (percentage of protein) in fish meal and selected plant protein sources.[a]

Amino acid	Soybean meal	Field pea meal[b]	Peanut meal	Cottonseed meal	Canola meal	Fish meal (herring)
Cysteine	1.56	0.96	1.23	1.09	1.24	1.03
Methionine	1.27	1.22	1.02	1.21	1.84	2.89
Threonine	3.97	3.65	3.47	2.48	4.50	4.03
Tryptophan	1.43	0.83	1.00	1.02	1.16	1.07
Valine	4.51	4.57	3.91	4.08	5.11	5.97
Isoleucine	4.53	4.78	3.66	2.79	3.97	4.35
Leucine	7.79	7.83	6.92	4.37	6.97	7.21
Tyrosine	3.50	3.17	4.64	1.94	2.45	3.06
Phenylalanine	4.96	4.26	5.18	5.10	4.00	3.76
Lysine	6.36	7.26	3.56	4.59	5.97	7.74
Histidine	2.66	3.13	2.77	2.01	2.82	2.29
Arginine	7.57	10.04	12.25	9.64	6.11	6.31
Crude protein (%)	44.8	23	48.1	41.2	38	72

[a]Data adapted from NRC (1993).

[b]Data adapted from Pulse Canada (2005).

The oils present in plant protein sources are higher in linoleic acid than linolenic acid (except for the linseed oil) and do not contain n-3 HUFA. In contrast, the oil in fish meal has a higher level of n-3 HUFA and is very low in n-3 and n-6 PUFA (Table 7.5). As dietary lipids influence the lipid composition of the crustacean body (Sandifer and Joshep, 1976; D'Abramo and Sheen, 1993), replacement of fish meal by plant protein sources may result in an increase in the levels of PUFA, especially of linoleic acid (n-6), and a reduction in the n-3/n-6 ratio as well as n-3 HUFA. Floreto et al. (2000) found that the n-3 HUFA and the n-3/n-6 ratio of juvenile lobster *(Homarus americanus)* were lower at higher levels of replacement of fish meal by SBM, which was associated with longer molting cycles. Addition of fish oil, or supplementation with EFA to achieve an appropriate n-3/n-6 ratio, should be considered when using plant proteins and oils in crustacean diets. Complete replacement of fish products in the diet of *L. vannamei*

TABLE 7.5. Linoleic acid, linolenic acid, arachidonic acid (AA), eicosapentaenoic acid (EPA), and decosahexanoic acid (DHA) content (percentage total fatty acids) of selected oils.

Oil	Linoleic acid	Linolenic acid	AA	EPA	DHA
Herring	0.6	0.4	0.4	8.1	4.8
Soybean	51	6.8	0	0	0
Cottonseed	51.5	0.2	0	0	0
Canola	20.2	12	0	0	0
Peanut	32	0	0	0	0
Linseed	12.7	53.3	0	0	0

Data adapted from NRC (1993).

by co-extruded soybean and poultry by-product meal or SBM is possible (without affecting growth and other production parameters) when appropriate supplementation of algae meals containing high levels of DHA and EAA is incorporated (Davis et al., 2004).

The levels of certain nutritionally important minerals are often lower or less available in plant proteins than in fish meal (Table 7.6). Crustaceans have a requirement for certain minerals, such as phosphorus that must be provided by the diet (Kitabayashi et al., 1971; Deshimaru and Yone, 1978; Kanazawa, 1984; Davis et al., 1993a, 1993b). Quite often, the bioavailability of minerals is lower in plants. For example, approximately 67 percent of the phosphorus (P) in common plant protein sources is in the form of phytate that has limited biological availability to crustaceans (Akiyama, 1991). The use of phytase (an enzyme that degrades phytate) has shown to increase P availability for some fish (Ketola, 1994; Jackson et al., 1996; Hughes and Soares, 1998; Forsters et al., 1999; Sugiura et al., 1999) and shrimp (Civera and Guillaume, 1989; Davis et al., 2000; Cruz-Suarez et al., 2004).

The nutrient composition of a feedstuff does not provide complete information or the suitability of a diet ingredient for an animal species. Another important consideration is the availability or digestibility of the nutrients from the diet ingredients (NRC, 1993). As the protein from plant sources may be less digestible than that from fish meal (Guillaume, 1997), it is critical to understand nutrient availability.

TABLE 7.6. Percentage of phosphorous and magnesium in fish meal and selected plan protein sources (as-fed basis).

Feedstuff	Phosphorous	Magnesium
Fish meal (herring)	1.67	0.14
Full-fat SBM	0.65	0.29
Cottonseed meal	1.17	0.41
Peanut meal	0.61	0.27
Corn gluten meal	0.44	0.07
Distillers' grains	0.51	0.15

Data adapted from NRC (1993).

Digestibility varies due to a variety of factors such as species, age, diet, particle size, feeding management, method of feces collection, and diet processing temperature (Lee and Lawrence, 1997; Tacon and Akiyama, 1997). Digestibility for selected plant protein ingredients has been determined for different species of crustaceans. Akiyama et al. (1989) provide information on digestibility of EAAs of thirteen feedstuffs for *L. vannamei*. Selected values of digestibility reported in the literature for different plant ingredients are summarized in Table 7.7. As there are limited reports on the digestibility of amino acids, amino acid availability data for various ingredients is one of the primary limitations to more accurate feed formulations.

Costs

The cost of the different sources of plant proteins, on unit protein basis, is quite often lower than that of fish meal (Table 7.8). However, prices will vary depending on the region, shipping costs as well as import tariffs. Even if prices are equal, plant proteins are often less variable in terms of nutrient profiles, more shelf-life stable, and easier to handle and incorporate into diet formulations.

PLANT PROTEIN USE IN CRUSTACEAN DIETS

The use of plant proteins in crustacean diets has centered on the most available protein source (SBM); however, there is also

TABLE 7.7. Apparent digestibility coefficients (ADC) of the dry matter (DM) and the crude protein (CP) of selected plant protein ingredients used in crustacean diets.

Species (reference)	Diet ingredient	ADC (%)	
		DM	CP
P. monodon (Smith et al., 2000)	Canola (solvent)	49	79
	Soybean (solvent)	67	92
	Lupin (whole)	39	88
	Lupin (dehulled)	67	94
	Field pea	72	89
	Gluten (wheat)	96	100
L. vannamei (Divakaran et al., 2000)	Soybean (solvent)	61.2-84.7	
L. vannamei (Ezquerra et al., 1997)	Soybean (solvent)		90.97
L. vannamei (Akiyama et al., 1989)	SBM (SBM)	55.9	89.9
	Gluten (wheat)	85.4	98.0
L. vannamei (Davis et al., 2002)	Whole raw field pea (WRA)	72.3	90.4
	Whole extruded field pea (WEX)	86	111.3±13.1
	Dehulled raw field pea (DRA)	88.4	93.5
	Dehulled extruded field pea (DEX)	94.4	96.7
	Whole micronized field pea (WMI)	94.1	117
		73.4	96.7
L. stylirostris (McCallum et al., 2000)	WRA		79.1
	WEX		83.4
	DRA		85.4
	DEX		82.7
	WMI		80.0
L. stylirostris (Cruz-Suarez et al., 2001)	WRA	89	79.8
	WEX	91.3	83
	DRA	87.6	85.4
	DEX	92.4	82.7
	WMI	80.7	80
L. vannamei (Davis et al., 2002)	Basal diet (BS) + WRA	72.6	77.4
	BS + WEX	76.7	81.6
	BS 1 DRA	77.3	78.1
	BS 1 DEX	79.1	79.0

TABLE 7.7 *(continued)*

| Species (reference) | Diet ingredient | ADC (%) | |
		DM	CP
	BS + WMI	78.9	83.3
	BS + Canola (extruded)	72.9	81.3
P. monodon	SBM, 16 ppt salinity	77.8	84.8
(Shiau et al., 1992)	SBM, 32 ppt salinity	71.2	78.4
P. monodon	SBM (defatted)		90.99
(Eusebio, 1991)	Whole cowpea meal		90.83
	Dehulled Cowpea meal		91.03
	Whole rice bean meal		88.52
	Dehulled rice bean meal		90.98
P. setiferus	SBM	67	94.63
(Brunson et al., 1997)	Cottonseed meal	53.64	82.85
	Wheat gluten	100.97	109.76
L. vannamei	SBM	55.9	89.9
(Akiyama et al., 1989)	Soy protein	84.1	96.4
	Wheat gluten	85.4	98
P. japonicus	94% SBM in the diet	63.6	90.1
P. monodon		60.1	90.4
L. vannamei		61.2	91.6
(Akiyama, 1988)			

TABLE 7.8. Costs of fish meal (anchovy) and some plant protein sources in 2007.

Protein source	Cost ($/MT)[a]	U.S.$/kg protein	Relative (%)
Anchovy meal, Peru (67%)	1015.00	0.854-0.862	100
Meat and bonemeal (50%)	290.00	0.49-0.51	59.2
SBM (48%)	249.00	0.408	47.3
Cottonseed meal (41%)	163.00	0.390	45.2
Peanut meal (41.2%)	166.00	0.340	39.4
Sunflower (32%)	125.00	0.316	36.7
Canola meal (34%)	193.50	0.363	42.1
Distillers' dried grains (29.5%)	139.00	0.339	39.3
Corn gluten (60%)	403.00	0.442-0.475	55.1

[a]Regional average prices for June 2007.

considerable information with regard to other plant proteins, such as cottonseed meal and canola meal. A brief discussion of the plant protein feedstuffs, as well as listing or description of research conducted on those ingredients, is presented in the following sections.

USE OF OILSEED MEALS IN CRUSTACEAN DIETS

SBM

This ingredient is fairly palatable, but is often considered bland for many fish and crustacean species, may be low in EAA (methionine, lysine, and tryptophan), and low in essential minerals (e.g., P, Se). Soybean meal is the most commonly used ingredient in aquatic diets and demand is expected to continue to increase. This may bring about future problems in terms of availability and cost.

A number of studies have been conducted that evaluate the influence of various inclusion levels of SBM on growth and survival of crustacean species. Table 7.9 summarizes some of the more important studies published in the literature. The appropriate level of inclusion of SBM in the diet, without affecting growth, seems to be related more to the species and age of the animals. For instance, *Macrobrachium rosenbergii* and *L. vannamei* can tolerate higher levels of SBM in their diet than *M. japonicus* (Table 7.9). Koshio et al. (1992) reported that *M. rosenbergii* can grow well when offered diets with soybean at levels up to 53 percent. A level of up to 60 percent SBM had no effect on the growth of *L. vannamei* (Fernandez and Lawrence, 1988a) and levels between 50 percent (Akiyama, 1990) and 55 percent (Piedad-Pascual et al., 1990) have been utilized in *P. monodon* diets. However, Kanazawa (1992) found that diets with 35 percent of SBM are appropriate for *M. japonicus*.

Studies to evaluate the effect of replacement of fish meal with SBM (17.8 percent of the diet) and SBM co-extruded with poultry by-product meal in practical diets of *L. vannamei* demonstrated that fish meal can be totally replaced by a proper combination of protein sources (Davis and Arnold, 2000; Samocha et al., 2004). In a pond experiment with *M. rosenbergii,* prawn did not show significant differences in growth when all the fish meal in a 32 percent CP diet was replaced with 26 percent of SBM and 40 percent of distiller grain

TABLE 7.9. Research on the use of SBM in crustacean diets.

Species	Protein g/100g (culture system) CP (%)	SBM inclusion (%)	Response
F. aztecus, F. dorarum, L. stylirostris, L. schmitti, L. vannamei, L. setiferus (Lawrence et al., 1986)	25 and 35 (tanks)	15-75	Sig. D. depending on species, size, and level of protein 20 to 50% of SBM in the feed does not affect growth in tanks
			N.S.D. in L. vannamei fed commercial diets with 10 and 40% SBM
P. monodon (Sudaryono et al., 1995)	41 (tanks)	41	N.S.D. weight gain than diet in which 71% of lupin seed meal had replaced all the SBM (Sig. higher FCR and lower PER)
P. monodon (Akiyama, 1990)	42 (tanks)	20,30,40,50 with white fish meal at 18, 12, 6, and 0%, respectively	N.S.D. on weight gain, survival, and FCR
P. monodon (Piedad-Pascual et al., 1990)	40 (cages in ponds)	15, 25, 35, 45, and 55	N.S.D. in weight gain
P. monodon (Piedad-Pascual and Catacutan, 1990)	42-48 (tanks)	15 and 35	N.S.D. in survival and growth
L. vannamei (Fernandez and Lawrence, 1988a)	2535	53 45-60 75	N.S.D. in growth N.S.D. in growth slightly decreased growth at 75%
L. vannamei (Lim and Dominy, 1990)	35	0,14, and 28 42, 56, and 70	N.S.D in weight gain S. lower weight gain at higher SBM levels

TABLE 7.9 *(continued)*

Species	Protein g/100g (culture system) CP (%)	SBM inclusion (%)	Response
L. vannamei (Sessa and Lim, 1991)	35	Toasted soy flour (S.F), lightly toasted S.F, and modified S.F (replaced 40% of animal protein)	N.S.D. in weight gain
		Three other soybean protein concentrates	Sig. lower weight gain than the three first diets
		0.64 to 6.17 mg/g diet of trypsin inhibitor	N.S.D.
L. vannamei (Lim and Dominy, 1992)	35-37	Full-fat double extruded from 0 to 35.8% replaced 0 to 28% solvent ext. SBM	N.S.D. in weight gain and other parameters
M. rosenbergii (Teichert-Coddington and Rodriguez, 1995)	30 (ponds)	30	Good weight gain
M. rosenbergii (Balazs and Ross, 1976)	15, 25, and 35 (tanks)	29% + tuna fish meal 21% + tuna fish meal + shrimp meal	N.S.D. in weight gain
M. rosenbergii (Tidwell, Webster, Yancey, and D'Abramo.)	32 (pond)	25, 15, and 26.25 of SBM and 15, 7.5, and 0% of fish meal, respectively with 40% distillers' grains with solubles in the 2nd and the 3rd diet	N.S.D. in survival, weight gain, and FCR

Species (Reference)		Diet / Treatment	Results
M. rosenbergii (Du and Niu, 2003)	30	0, 21, 42, 62.5, and 84 (100% of total protein) with 62.5, 47, 31.2, 15.5, and 0% of fish meal	Sig. diff. between the 0% SBM and the SBM containing diets. Growth rate decreased in proportion to the increase in SBM
M. rosenbergii (Koshio et al., 1992)	30-55	27,36,44, and 53 of soybean protein concentrate (SBP)	N.S.D. when SBP was replaced by crab protein conc. except for the diet at 47% CP
C. quadricarinatus (Webster et al., 1994)	25, 35, 45, and 55 (tanks)	3, 15, 28, and 30.5, respectively Compared to a commercial diet 45% CP	N.S.D. among diets. Commercial diet was Sig. lower than diets 35, 45, and 55% CP
C. quadricarinatus (Muzinic et al., 2004)	40 (tanks)	Exp. 1: 25, 10, and 0% of fish meal(FM) and 35, 46.8, and 79.8 of SBM, respectively. Also 0, 30, and 5% of brewers grain with yeast (BGY), respectively	N.S.D. growth and survival
		Exp. 2 and 3: 24, 0, 0, 0% of FM and 23, 56, 47, and 40% of SBM, respectively. Also 0, 10, 20, and 30% of BGY, respectively	N.S.D. growth and survival
P. clarkii (Reigh and Ellis, 1994)	30 (tanks)	41% SBM and 18% fish meal SBM was replaced by 44.66% of cottonseed meal	Sig. higher weight gain, FCE, PER than the other diets with cottonseed meal, MB, or MBB) Sig. lower weight gain
P. clarkia (Reigh and Ellis, 1994)	30 (tank)	68% plus phosphorous supplementation	Sig. higher than control (10% shrimp meal, 15% SBM and 25% peanut meal)

TABLE 7.9 *(continued)*

Species	Protein g/100g (culture system) CP (%)	SBM inclusion (%)	Response
H. americanus (Floreto et al., 2000)	40	0 (salted fish and fish racks), 20.8, 41.7, 62.5, 72.9, and 83.3 EAA supplementation (arg, met, leu, tryp)	Sig. lower than 45% SBM plus 15% fish meal N.S.D. weight gain and survival Sig. lower weight gain. Early mortality at 100% SBM Sig. higher weight gain with EAA supplementation

with solubles (Tidwell, Webster, Yancey, and D'Abramo, 1993). The successful replacement of all the fish meal with vegetable protein in this experiment was probably due in part to the contribution of natural productivity from the pond as has been noted by Tacon and Akiyama (1997). In a tank study, levels of 27 to 53 percent of SBM have shown not to affect growth of this species (Koshio et al., 1992). However, more recently, Du and Niu (2003) reported that replacement of fish meal with SBM depressed growth rate and feed efficiency ratio (FER) even at a level of 21 percent SBM and 47 percent fish meal. In this experiment, fish meal and SBM or a combination of both were used as the protein sources.

Quite often when replacing protein sources, changes in the EAA profiles are overlooked. An appropriate balance of EAAs can be reached either by combining different sources of protein with complementary amino acids or by adding a mixture of crystalline amino acids to compensate for the deficient EAA. Peñaflorida (2002) observed higher weight gain of post-larvae *Penaeus margulensis* fed with diets that contained fish meal, SBM, and yeast as the main sources of protein, compared to post-larvae fed diets that only contained fish meal and SBM. However, the same author did not find significant differences in the growth rate of *P. indicus* fed either with a diet containing 21.6 percent of fish meal and 34 percent SBM or a diet that contained 21.6 percent fish meal, 17 percent SBM, and 15.2 percent of yeast. Muzinic et al. (2004) did not observe significant differences in growth of juvenile Australian red craw crayfish *(Cherax quadricarinatus)* when a diet with 25 percent fish meal and 35 percent SBM was replaced by a diet that contained 79.8 percent SBM and 5 percent brewer's grain with yeast. In two more experiments, the same authors did not find significant differences in the growth rate between animals fed with a diet with fish meal, SBM, and brewer's grain with yeast and diets in which all the fish meal had been replaced by SBM and brewer's grain with yeast. They reported that the EAAI was very similar between the fish meal diet and the plant protein diets.

The supplementation with crystalline amino acids of crustacean diets has improved the nutritive value of diets containing SBM as the main source of protein. Floreto et al. (2000) obtained similar growth of juvenile American lobster *(H. americanus)* fed a diet that contained 41.7 percent SBM and a mix of the crystalline amino acids, arginine, leucine, methionine, and tryptophan, and a diet that included blue mussel as the main source of protein. They observed significant lower growth rate when the same diet was fed without the amino acid supplementation.

Another potential disadvantage of SBM is that if it is not properly processed or used in its raw state, it contains trypsin inhibitors. The presence of antinutrients, such as trypsin inhibitors, does not seem to affect crustaceans as much as finfish species. Sessa and Lim (1991) did not observe significant differences in weight gain of *L. vannamei* fed with diets containing from 0.64 to 6.17 mg/g of diet of trypsin inhibitor.

Another common problem encountered when replacing fish meal and/or other animal by-product meals with high levels of SBM is that P availability generally decreases due to the presence of phytin-bound P. More than 70 percent of the soybean P is unavailable (NRC, 1993; Hertrampf and Piedad-Pascual, 2000). Owing to the low biological availability of phytin-bound P, P released into the environment may increase. Hence, supplementation of P as well as other minerals, such as selenium, may be necessary as the animal meal is replaced with plant proteins. This bioavailability issue could be reduced through the use of the enzyme phytase to the diet to breakdown phytate which in turn releases P and prevents the sequestering of other minerals.

Cottonseed Meal

Cottonseed meal has not been broadly used in crustacean diets. The presence of free gossypol, and the low levels and reduced availability of the amino acid lysine, limit the amount of this feedstuff that can be used in the diet. Lim (1996) included cottonseed meal in diets of *L. vannamei* at levels of 0, 13.3, 26.5, 39.8, 53, and 66.3 percent of the diet as substitutes of an animal protein mix (menhaden meal, shrimp meal, and squid meal). It was reported that the three lowest levels (0, 13.3, and 26.5 percent) yielded similar weight gain, diet consumption, and survival. However, growth performance was affected at levels of cottonseed meal higher than 26.5 percent of the diet and free gossypol levels higher than 1,100 ppm. The two highest levels of cottonseed meal diets (53 and 66.3 percent) caused reduced weight gain at week 4 and high mortality between weeks 6 and 8. The shrimp body tended to show a yellow-green coloration, probably due to accumulation of free gossypol. During processing of cottonseed meal, the free form of gossypol binds to protein, especially the epsilon carbon of lysine, reducing the availability of this amino acid (Robinson and Li, 1995). Lim (1996) reported that the observed growth depression of *L. vannamei* was probably due to free gossypol toxicity rather than lysine deficiency.

The level of dietary protein, diet composition, size, and species may affect the level of cottonseed meal that can be incorporated into the diet. Fernandez and Lawrence (1988b) observed that *L. vannamei* and *Penaeus stylirostris* weight gain and survival was not affected in a 30 percent-protein diet that contained 20 percent cottonseed meal; however, growth depression was observed in *Penaeus setiferus* fed the same diet. When the level of protein was reduced to 20 percent, the inclusion of 10 percent of cottonseed meal in the diet did not affect growth of *L. vannamei* and *P. stylirostris,* but levels of dietary cottonseed meal higher than 5 percent affected growth of *P. setiferus.* Chen et al. (1985) reported that replacement of fish meal by fish hydrolysate, SBM, and cottonseed meal, reduced the growth of juvenile *P. setiferus* and *L. vannamei* at three body sizes. They observed that growth depression was more pronounced for smaller shrimp at a higher plant-protein-to-animal-protein ratio (1.5:1) than a lower ratio (1:1). Such responses are often due to nutritional imbalances or antinutrients, which produce a more pronounced effect in the small, faster growing shrimp. Lawrence and Castille (1989) showed that cottonseed meal could partially substitute animal protein (fish meal and shrimp head meal) in diets of penaeid shrimp, but at lower levels than SBM.

The effect of cottonseed meal in the diet has also been tested in red swamp crayfish *(Procambarus clarkii).* For this species, an adverse effect on growth performance was observed when SBM (41.07 percent of the diet) was replaced by cottonseed meal (44.66 percent of the diet) in a 30 percent CP isoenergetic diet containing a mixture of plant and animal protein (ratio of 65:35; Reigh and Ellis, 1994). These authors did not report a negative effect of cottonseed meal on palatability of the diets.

Based on the previous findings, a maximum level of inclusion of 25 percent of cottonseed meal in crustacean diets should be considered. Although cottonseed meal is the third most abundant plant protein source in the market (Table 7.1) and its price is quite competitive

(Table 7.8), inclusion levels are limited mainly due to the presence of free gossypol, low lysine availability, and high levels of crude fiber (Hertrampf and Piedad-Pascual, 2000). Crude fiber content is approximately 10 percent in decorticated meal and can be higher than 20 percent in corticated meal. The amount of cottonseed meal that can be incorporated in crustacean diets can be increased by using glandless (gossypol-free) varieties. Currently, production of these varieties is limited due to undesirable characteristics, such as reduced insect resistance, low lint levels, and inferior fiber quality (Robinson and Li, 1995). In terrestrial animals, the toxicity of gossypol may be overcome by adding iron salts to the diet (e.g., ferrous sulfate) (Hertrampf and Piedad-Pascual, 2000). Addition of iron salts to the diet with this purpose has not been reported for crustaceans.

Pellet quality and water stability are very important for crustaceans, especially for shrimp species, which are slow feeders. The high level of crude fiber in cottonseed meal could affect the water stability of the diet when this ingredient is used at high concentration. High levels of fiber in cottonseed meal also limit its incorporation in other aquatic species diets. Robinson and Li (1995) pointed out that pellet quality and water stability is not affected by inclusion of cottonseed meal. However, they indicated that the physical characteristics and the high abrasiveness of cottonseed meal slow down the diet production process and increase the wear in the equipments, thus limiting the amount of this ingredient to a level of 10 to 15 percent.

Rapeseed/Canola Meal

Based on the EAAI (Table 7.3) and the profile of EAA, rapeseed/canola meal is one of the most complete ingredients among the protein sources. Canola meal is a particularly rich plant source of the sulfur-containing amino acids, methionine and cysteine (Table 7.4). Even though canola meal is produced from low erucic acid and low glucosinolate rapeseed varieties of *B. napus* and *B. campestris,* these antinutrients may still be present and limit the level of inclusion of canola meal in animal diets, and probably in crustacean diets. The utilization of high levels of canola meal in crustacean diets has resulted in reduced consumption and growth of penaeid shrimp (Lim et al., 1997). The high level of fiber of this ingredient could affect water stability and diet consumption when included at high levels. Lim et al. (1997) evaluated the effect of different levels of high-fiber and low-fiber canola meal in diets of *L. vannamei*. They found that levels of 14 and 28 percent of a high-fiber canola meal did not affect growth of juvenile shrimp. However, growth was depressed when the level of canola meal was increased to 42 percent replacing all the menhaden fish meal in the diets. Using low-fiber canola meal, decreased growth of *L. vannamei* was observed for diets that contained 24 and 35 percent of canola meal. For all the diets containing canola meal, a reduction in diet consumption was observed when compared with the basal diet that contained a mixture of animal proteins. It was also reported that there was a decrease of water stability of the diets containing canola meal, especially when using the high-fiber canola meal at the higher inclusion levels. The diets with the highest levels of canola meal had detectable levels of glucosinolates and probably other antinutritional factors, such as tannins and low digestible carbohydrates (e.g., stachyose and raffinose) that could have affected consumption and growth performance of the animals. Lower survival was also reported for shrimp fed diets containing 28 and 42 percent of the high-fiber canola as well as those containing 24 and 35 percent of the low-fiber canola.

The presence of indigestible components in the diet, as well as some antinutrients, appears to limit the use of canola meal in crustacean diets. Inclusion of higher levels of canola meal in marine shrimp diets has been shown to be possible by addition of digestive enzymes to the diet or by special processing of the canola meal before adding it to the diet (e.g., extrusion). Buchanan et al. (1997) obtained the same growth rate for *P. monodon* fed either a basal diet containing only squid meal as the source of protein, or a diet containing 64 percent of canola meal supplemented with 0.25 percent of an enzyme mixture (porzyme, heat stabilized). When the enzyme mixture was not added, the growth performance of the shrimp was not affected when the diet contained 20 percent canola meal, but it was reduced when the diet contained 64 percent canola meal. The enzyme mixture apparently improved the utilization of indigestible carbohydrates present in the canola meal, thus increasing the available energy and improved the utilization of the dietary protein. In another study, substitution of a portion of SBM, fish meal, and wheat (1:2:3) by 30 percent of extruded canola meal in a diet of juvenile *L. stylirostris* did not affect growth performance (Cruz-Suarez et al., 2001). It seems that the high temperatures during the extrusion process deactivated or reduced the activity of some of the antinutrients and/or enhanced the availability of carbohydrates present in the canola meal.

Specialized processing has the potential to improve the quality of canola meal. For example dehulling before further processing or removal of hulls by air classification of defatted meal, or milling moisture-adjusted meals before sieving have all been utilized. Boiling of dehulled seeds (to deactivate myrosinase), followed by washes in water and hexanes to remove undesirable carbohydrates and phenolic compounds, result in rapeseed/canola protein concentrates. As with other plant protein sources, 60 to 90 percent of total P is found in the form of phytic acid (3.1 to 3.7 percent, mostly unavailable) which has very low P availability. Further processing allows one to obtain dephytinized canola protein concentrates. A recent study showed that dephytinized canola protein concentrate can replace up to 50 percent of the fish meal protein in the diet without affecting the growth of juvenile *L. vannamei* (Rique-Marie et al., 2005). Replacement of 75 and 100 percent of the fish meal protein reduced growth due presumably to leaching of the supplemental free, noncoated amino acids lysine and methionine.

Peanut Meal

Research on the use of peanut meal for crustacean diets is limited. The low levels of the amino acids lysine, methionine, and threonine seem to limit the use of this plant protein ingredient in crustacean diets (Tables 7.3 and 7.4). An adverse effect on diet palatability with high inclusion levels of peanut meal also appears to limit the amount that can be incorporated in the diet. Lim (1997) substituted an animal protein mixture (53 percent menhaden meal, 34 percent shrimp waste meal, and 13 percent squid meal) by peanut meal in diets of *L. vannamei*. He used peanut meal at 0, 11.7, 23.4, 35.1, 46.8, and 58.5 percent that substituted 0, 20, 40, 60, 80, and 100 percent of the animal proteins. He observed no effect on weight gain and dry matter intake when peanut meal was included in the diet at levels up to 11.7 percent. However, weight gain was reduced when included at 23.4 percent or higher levels, probably due to an adverse effect of the peanut meal on diet palatability. He reported that the inclusion of 0, 11.7, 23.4, and

35.1 percent peanut meal in the diets resulted in similar feed conversion ratio (FCR), protein efficincey ratio (PER), and apparent protein utilization, but these values decreased significantly for 46.8 and 58.5 percent peanut meal diets, probably due to the low levels of lysine and sulfur-containing amino acids. He recommended a level of 12 percent of peanut meal in diets of penaeid shrimp as a replacement of 20 percent of the animal protein mix. However, he pointed out that if palatability could be improved, the level of peanut meal could be elevated to 35 percent to replace 60 percent of the animal protein mix.

Studies with other crustacean species have also shown that the use of peanut meal in the diet affects the animal growth. Lochmann et al. (1992) reported that *P. clarkii* fed a diet containing 25 percent of peanut meal and a mixture of shrimp meal (10 percent) and SBM (15 percent) yielded lower weight gain than a diet containing 35 percent SBM and 10 percent shrimp meal. They did not observe differences in diet consumption among the crayfish fed both diets, but reported a lower percentage of lysine, threonine, and methionine in the diet containing 25 percent peanut meal. Supplementation with crystalline amino acids did not improve growth. The amino acid supplemented diet containing peanut meal did not improve the growth of shrimp to a level comparable to that of the control diet, in which shrimp tail muscle was the only source of protein. Freshwater prawns *(M. rosenbergii)* seem to tolerate higher levels of peanut meal in their diets. Hari and Madhusoona-Kurup (2003) tested 30 percent CP diets for *M. rosenbergii* with plant:animal protein ratios of 1:2, 1:1, 1:1.5, and 2:1 with levels of peanut meal of 7, 11.5, 18.5, and 30 percent, respectively and did not find significant differences in growth rate, PER, and survival.

Sunflower Seed Meal

The protein content of this feedstuff ranges between 25 percent (whole diet) and 50 percent (dehulled) and it is widely available for incorporation in aquatic diets (Tables 7.1 and 7.2). Sunflower seed meal is the fourth most abundant plant protein source with annual worldwide production between 9 and 15 mmt (Hertrampf and Piedad-Pascual, 2000; FAO, 2005b). Sunflower seed meal has been evaluated in fish diets (Tacon et al., 1984; Cardenete et al., 1993), but reports on its use in crustacean diets are limited. This feedstuff is low in lysine, but its levels of methionine and arginine are higher than those of SBM (Hertrampf and Piedad-Pascual, 2000).

Linseed Meal

The use of linseed meal in crustacean diets has not been reported. Compared to SBM, linseed meal is low in lysine and sulphur amino acids, and high in crude fiber. Also, the presence of mucilage, a sticky water-dispersible carbohydrate, has been shown to reduce the digestibility of carbohydrates by monogastric animals (Hertrampf and Piedad-Pascual, 2000). Similar effect may also occur in crustacean diets. Hertrampf and Piedad-Pascual (2000) recommended a level between 3 and 7 percent of linseed meal in diets of aquatic herbivorous/omnivorous species, and up to 5 percent for carnivorous species.

USE OF OTHER LEGUMINOUS SEED MEALS IN CRUSTACEAN DIETS

Leguminous seeds include beans (*Phaseolus* sp.), lupins (*Lupinus* sp.), peanuts *(Arachis hypogaea),* peas *(Pisum sativum),* vigna *(Vigna ungiculata),* canavalia *(Canavalia ensiformis),* alfalfa *(Medicago sativa),* and ipil-ipil *(Leucaena leucocephala).* They are generally low in sulfur-containing amino acids and tryptophan and contain antinutritional factors and indigestible components, such as oligosacharides (Chow and Halver, 1980). They are good sources of protein, but also a potential energy source. The use of modern processing technologies, such as extrusion processing, can make the energy from carbohydrates more available (Hertrampf and Piedad-Pascual, 2000). Untreated leguminous feedstuffs are not recommended in aquatic diets. Hertrampf and Piedad-Pascual (2000) recommended that aquatic diets should contain no more than 15 percent treated leguminous seeds. Some of the *Phaseolus* beans and cow peas are low in crude fiber, making them more suitable for aquatic diets.

Lupins

Five species of lupins are cultivated for human food and animal diets. They are low in lysine and methionine, contain some carbohydrates of low digestibility (e.g., galactosydes), as well as high levels of hy-drosoluble alkaloids (e.g., quinolizadine). Some varieties have reduced levels of these alkaloids. Lupin meal has been evaluated as partial replacement for fish meal and/or SBM in shrimp diets in Australia (Smith et al., 1999) and Indonesia (Sudaryono et al., 1999). A more detailed discussion on the use of lupins in aquaculture diets is provided in Chapter 14 of this book.

Feed Peas

Feed peas are relatively high in lysine content (>7 percent of protein) and are often used in combination with other plant proteins deficient in this amino acid (e.g., soybean or canola meal). They are low in protein (Table 7.2), but are a rich source of energy in the form of carbohydrates. Feed peas can be used to replace both more expensive sources of protein, such as fish meal and SBM, and less-digestible carbohydrate sources, such as grains or grain by-products.

Like other plant protein sources, the amount of feed pea meal that can be used in crustacean diets is limited by the presence of indigestible carbohydrates, antinutritional factors, and high levels of fiber. Removal of the seed coat, where some antinutrients (e.g., tannins) tend to accumulate, has shown to improve the nutritional properties of some leguminous seeds (Eusebio, 1991). However, McCallum et al. (2000) did not observe any difference in growth of *L. vannamei* and *L. stylirostris* when whole feed pea meal was replaced by dehulled (decorticated) meal. The same results were observed by Davis et al. (2002) and Cruz-Suarez et al. (2001) with *L. vannamei* and *L. styliro-stris*, respectively. However, processing feed pea meal by extrusion or by micronization (infrared heat treatment) had an effect on growth performance of the shrimp in some cases. McCallum et al. (2000) observed no significant effect of extrusion on growth of *L. stylirostris*, but obtained an improvement in the growth rate of *L. vannamei*. In other experiments, extrusion did not show a significant effect on growth of *L. stylirostris* (Cruz-Suarez et al., 2001) and *L. vannamei* (Davis et al., 2002).

Micronization of fed pea meal increased the growth rate of *L. stylirostris* because of an increase in food consumption (McCallum et al., 2000; Cruz-Suarez et al., 2001). However, there was no such effect for *L. vannamei* (Davis et al., 2002). Davis et al. (2002) reported an increase in apparent digestibility of the diets containing extruded or micronized feed peas. This is likely the result of increasing carbohydrate digestibility due to gelatinization. It has also been suggested that heat treatment by micronization enhances palatability or destroys certain substances that affect palatability.

Normally it is not recommended to use more than 15 percent of treated leguminous seeds in aquatic diets (Hertrampf and Piedad-Pascual, 2000). However, higher levels of feed pea meal have been incorporated in diets of shrimp without observing negative effects on growth performance. Davis et al. (2002) compared the growth of shrimp fed with 5, 10, and 20 percent of feed pea meal in their diets and did not observe any significant difference among the treatments. Similarly, Bautista-Teruel et al. (2003) did not find significant differences in growth of *P. monodon* fed 40 percent CP diets containing 0 (basal diet) to 42 percent feed pea meal.

Methods to concentrate the pea protein flour, such as by air classification, have been suggested to improve the nutritional quality and reduce the indigestible fraction. However, use of these methods results in an increase in the concentration of phytate-bound P. However, Cruz-Suarez et al. (2004) increased the digestibility of phytate-bound P by 180 percent by adding phytase to a *L. vannamei* diet that contained 40 percent of air-classified pea protein flour.

USE OF LEGUMINOUS LEAF MEALS IN CRUSTACEAN DIETS

There are a number of leaf meals (leucaena and alfalfa) that are commonly utilized by a number of animal production industries. Leucaena leaf meal is often used to feed ruminants. It contains high levels of the nonprotein amino acid, mimosine, and low levels of lysine and methionine. Soaking the leaf for forty-eight hours reduced the levels of mimosine making this feedstuff less toxic for *P. monodon* (Piedad-Pascual and Catacutan, 1990). Another common leaf meal is alfalfa, which is commonly used to feed ruminants and is occasionally used in aquatic animal diet. In addition to containing saponines and having a high fiber level, alfalfa meal contains trypsin inhibitors. It is also relatively low in lysine and methionine. Although, there is limited information on the use of these meals in aquatic diets, they are not likely to be used at high levels in shrimp diets.

USE OF BY-PRODUCTS OF THE BREWERY INDUSTRY IN CRUSTACEAN DIETS

There are four types of grain by-products (distillers' dried grains, distillers' dried grain with solubles, distillers' dried solubles, and condensed distillers' solubles) that are obtained from brewing or fermentation processes. Distillers' dried grains (DDG) and distillers' dried grains with solubles (DDGS) are the most commonly evaluated brewery by-products in aquatic diets. Their composition varies with brewing protocols and type of grain used, but generally is low in lysine and contains approximately 28 percent protein and 12 percent fiber.

Different levels of DDGS have been evaluated in diets (30 percent CP) of juvenile *M. rosenbergii* grown in ponds. Prawns fed diets containing 0, 20, and 40 percent of DDGs, in combination with 19 percent SBM, did not show significant differences in weight gain, survival, and FCR. (Tidwell, Webster, Clark, and D'Abramo, 1993). For juvenile *L. vannamei,* replacement of animal protein (69 percent shrimp meal, 21 percent fish meal, 10 percent squid meal) with 0 and 16 percent of DDGs did not produce significant differences in weight gain. However, shrimp showed lower growth at animal protein replacement levels of 32 and 48 percent (Molina-Poveda and Morales, 2002). Distillers' dried grains with solubles have been successfully used in combination with other plant proteins in crustacean diets, especially with SBM. A more detailed discussion on the use of DDG and DDGS can be found in Chapter 16 of this book.

USE OF PROTEIN ISOLATES/CONCENTRATES IN CRUSTACEAN DIETS

Quite often, plant proteins contain antinutritional factors and carbohydrates that are poorly digested by monogastric animals, including crustaceans. A number of processing technologies have been developed to remove antinutrients and carbohydrates leading to increase in protein content of the products. Protein concentrates have been made from a wide variety of plant meals such as soybean, canola, corn, wheat, and pea. However, the most commonly used concentrates are the gluten from corn and wheat. Corn gluten meal is the residue from corn after removing all the starch and germ, and after separation of the bran. Two corn gluten products are generally available, a 41 percent CP and 4 percent fiber meal and a 60 percent CP and 2.5 percent fiber meal. Corn gluten meal is considered a good source of methionine, but a relatively poor source of lysine. Wheat gluten has similar properties to those of corn gluten. Although wheat gluten is a better binder it is often more expensive and not as readily available. Wheat gluten is commonly used in shrimp diets to increase binding properties of the diets and as a methionine source, with inclusion levels generally being less than 8 percent of the diet.

USE OF ALGAE MEALS IN CRUSTACEAN DIETS

The algae, or kelp, meals are generally not included as protein sources in crustacean diets. They have been used as sources of pigments, essential nutrients (EFA and minerals), binding and texturizing agents, or attractants to improve the palatability of the diet (Cruz-Suarez et al., 2000). An immunostimulating effect in fish and shrimp has also been suggested for some kelp (Fujiki et al., 1994; Takahashi et al., 1998). Kelps include pheophyte algae of the order Fucales (i.e., *Ascopphyllum nodosum, Sargassum* sp., and *Palvetia* sp.) and Laminariales (i.e., *Laminaria hyperborea, Macrocystis pyrifera,* and *Nereocystis luetkaena*). There is limited information on the use of kelp in crustacean diets. Cruz-Suarez et al. (2000) reported the use of kelp as a binder and immunostimulant in growth-out diets for *L. vannamei* in some commercial diets in Mexico, noticing an improvement in growth, food conversion, and survival. In a laboratory-controlled experiment, Cruz-Suarez et al. (2000) found that incorporation of kelp meal *(M. pyrifera)* at levels between 2 and 4 percent in *L. vannamei* diets increased diet consumption, growth, and yield of juvenile shrimp. This response was probably due to an improvement in the palatability, texture, and high stability of the pellets in the water. It was suggested that improved growth was due to a synergistic effect of the kelp meals with other sources of proteins in the diet, especially animal proteins like fish meal. They reported maximal growth by feeding a 25 percent CP diet with 3.2 percent of kelp meal.

Other algae have also been used in aquatic diets to improve their nutritional quality. The algae *Crypthecodinium* sp. has been grown under heterotrophic conditions to produce a supplement for aquatic diets high in HUFA, especially DHA. These products could be considered as potential sources of essential nutrients to balance crustacean diets when fish meal and fish oil are replaced with plant proteins and oils that contain lower levels of EFA.

CONCLUSION

There are a number of major plant protein sources, but their use in crustacean diets varies depending on their availability, price, farmer preferences or tradition, undesirable factors present in the ingredients, and lack of research data. Since plant proteins are generally cheaper than proteins derived from animal sources, replacement of animal proteins by plant proteins would lead to reduction of diet cost. Earlier research results showed that, provided that the diets are nutritionally balanced, high levels of plant sources of proteins can be included in crustacean diets without affecting diet intake and growth performance. It has been reported that a number of crustacean species, especially those having low food chain such as the omnivorous *L. vannamei,* performed well on diets with minimal level of or containing no marine animal proteins. Currently, tests are being conducted using a diet without animal proteins or fish meal (Davis et al., 2004).

Nutritionally balanced diets containing high levels of alternative protein sources can be formulated by the use of a combination of ingredients with complementary nutrients or by supplementation of the deficient nutrients. This, however, requires information on nutrient requirements, nutrient content, and bioavailability. A good understanding of the palatability of plant feedstuffs and possible enhancement with the appropriate use of attractants is the key area of interest that will lead to increased use of plant feedstuffs in crustacean diets.

Currently, different processing methods to improve the nutrient content and nutritional value of plant products are being investigated (Barrows et al., 2005). For example, a mechanical process, such as air classification, allows the development of protein concentrates. In addition to increasing the protein content, subjecting the ingredient to further processing reduces or eliminates antinutritional factors, thus improving the product quality allowing higher levels of dietary inclusion. Biological modification of ingredients through the use of microorganisms to remove undesirable components/antinutrients has also been used to improve the nutritional quality of plant ingredients. Improvement of existing varieties and development of new strains through breeding, genetic selection, and molecular techniques (i.e., high lysine corn, low glucosinolate rapeseed, and soybean low in trypsin inhibitors) will also lead to the development of better quality alternative protein sources.

REFERENCES

Akiyama, D.M. (1988). Soybean meal utilization by marine shrimp. *AOCS World Congress on Vegetable Protein Utilization in Human Food and Animal Feedstuffs*. Singapore: AOCS, 27 pp.

Akiyama, D.M. (1990). The use of soybean meal to replace white fish meal in commercially processed *Penaeus monodon* feeds in Taiwan. In *Proceeding from the 3rd International Symposium on Feeding and Nutrition in Fish: The Current Status of Fish Nutrition in Aquaculture,* M. Takeda and T. Watanable (eds.). Toba, Japan, pp. 289-299.

Akiyama, D.M. (1991). Soybean meal utilization by marine shrimp. In *Proceedings of the Aquaculture Feed Processing and Nutrition Workshop,* D.M Akiyama and R.K.H. Tan (eds.). Singapore: American Soybean Association, pp. 207-225.

Akiyama, D.M., S.R Coelho, A.L Lawrence, and E.H. Robinson (1989). Apparent digestibility of feedstuffs by the marine shrimp *Penaeus vannamei* Boone. *Bulletin of the Japanese Society of Scientific Fisheries* 55:91-98.

Alam, M.S., S. Teshima, M. Ishykawa, D. Hasegawa, and S. Koshio (2004). Dietary arginine requirement of juvenile kuruma shrimp *Marsupenaeus japonicus* (Bate). *Aquaculture Research* 35:842-849.

Balazs, G.H. and E. Ross (1976). Effect of protein source and level on growth and performance of the captive freshwater prawn, *Macrobrachium rosenbergii. Aquaculture* 7:299-313.

Barrows, F., C. Bradley, D. Sands, A. Pilgeram, and M. Kirkpatrick (2005). *Methods for Improving the Nutrient Profile of Plan-Derived ingredients.* Aquaculture America 2005. Baton Rouge, LA: World Aquaculture Society.

Bautista-Teruel, M.N., P.S. Eusebio, and T.P. Welsh (2003). Utilization of feed pea, *Pisum sativum,* meal as a protein source in practical diets for juvenile tiger shrimp, *Penaeus monodon. Aquaculture* 225:121-131.

Brunson, J.F., R.P. Romaire, and R.C. Reigh (1997). Apparent digestibility of selected ingredients in diets for white shrimp *Penaeus setiferus* L. *Aquaculture Nutrition* 3:9-16.

Buchanan, J., H.Z. Sarac, D. Poppi, and R.T. Cowan (1997). Effects of enzyme addition to canola meal in prawn diets. *Aquaculture* 151:29-35.

Cardenete, G., A.E. Morales, M. de la Higuera, and A. Sanz (1993). Nutritive evaluation of sunflower meal as a protein source for rainbow trout. In *Proceeding 4th International Symposium on Fish Nutrition and Feeding*. Paris, France: Institut de National de la Recherche Agronomique, pp. 927-931.

Chamberlain, G.W. (1993). Aquaculture trends and feed projections. *World Aquaculture* 24(1):19-29.

Chen, H.Y., Y.T. Leu, and I. Roelants (1992). Quantification of arginine requirements on juvenile marine shrimp, *P. monodon* using microencapsulated arginine. *Marine Biology* 114:229-233.

Chen, H.Y., Z.P. Zein-Eldin, and D.V. Aldrich (1985). Combined effects of shrimp size and dietary protein source on the growth of *Penaeus setiferus* and *P. vannamei*. *Journal of the World Mariculture Society* 16:288-296.

Chow, K.E. and J.E. Halver (1980). Carbohydrates. In *Feed Fish Technology*, ADCP (eds.). Rome, Italy: Food and Agriculture Organization, 395 pp.

Civera, R. and J. Guillaume (1989). Effect of sodium phytate on growth and tissue mineralization of *Penaeus japonicus* and *Penaeus vannamei* juveniles. *Aquaculture* 77:145-156.

Colvin, L.B. and C.W. Brand (1977). The protein requirement of penaeid shrimp at various life cycle stages in controlled environment systems. *Proceeding of the World Mariculture Society* 8:821-840.

Cowey, C.B. and A.G.J. Tacon (1983). Fish nutrition-relevance to invertebrates. In *Proceeding of the Second International Conference on Aquaculture Nutrition: Biochemical and Physiological Approaches to Shellfish Nutrition*, G.D. Pruder, C.J. Langdon, and D.E. Conklin (eds.). Baton Rouge, LA: Louisiana State University, Division of Continuing Education, 444 pp.

Cruz-Suarez, L.E., D. Ricque-Marie, M. Tapia-Salazar, I.M. McCallum, and G. Barbosa (2000). Uso de harina de kelp *(Macrocystes pyrifera)* en alimentos para camarón. In *Avances en Nutrición Acuícola V. Memorias del V Symposium Internacional de Nutrición Acuícola*, L.E. Cruz-Suárez, D. Ricque-Marie, M. Tapia-Salazar, M.A. Olvera-Novoa, and R. Civera-Cerecedo (eds.). Mérida, Yucatan: Universidad Autónoma de Nuevo León, pp. 227-266.

Cruz-Suarez, L.E., D. Ricque-Marie, M. Tapia-Salazar, I.M. McCallum, and D. Hickling (2001). Assessment of differently processed feed pea *(Pisum sativum)* meals and canola meal *(Brassica* sp.) in diets for blue shrimp *(Litopenaeus stylirostris)*. *Aquaculture* 196:87-101.

Cruz-Suarez, L.E., B.M. Zabala-Chavez, M. Nieto-Lopez, C. Guajardo, D. Ricque-Marie, M. Tapia-Salazar, and I.M. McCallum (2004). Effect of a phytase product on protein and phosphorus digestibility in shrimp *Litopenaeus vannamei* fed an air classified pea protein flour (ppf) based diet. *Annual Meeting of the World Aquaculture Society*. Baton Rouge, LA: World Aquaculture Society.

D'Abramo, L.R. and S.S. Sheen (1993). Polyunsaturated fatty acid nutrition in juvenile freshwater prawn *Macrobrachium rosenbergii*. *Aquaculture* 115:63-86.

Davis, D.A. and C.R. Arnold (2000). Replacement of fish meal in practical diets for the Pacific white shrimp, *Litopenaeus vannamei*. *Aquaculture* 185:291-298.

Davis, D.A., C.R. Arnold, and I.M. Mc Callum (2002). Nutritional value of feed peas *(Pisum sativum)* in practical diet formulations for *Litopenaeus vannamei*. *Aquaculture Nutrition* 8:87-94.

Davis, D.A., W.L. Johnston, and C.R. Arnold (2000). El uso de suplementos enzimáticos en dietas para camarón. In *Avances en Nutrición Acuícola. Memorias del IV simposio Internacional de Nutrición Acuícola*, R.Civera-Cerecedo, C.J. Pérez-Estrada, D. Ricque-Marie, and L.E Cruz-Suárez (eds.). Mérida, Yucatan: Universidad Autonoma de Nuevo León, pp. 19-22.

Davis, D.A., A.L. Lawrence, and D.M. Gatlin III (1993a). Dietary copper requirement of *Penaeus vannamei*. *Nippon Suisan Gakkaishi* 59:117-122.

Davis, D.A., A.L. Lawrence, and D.M. Gatlin III (1993b). Response of *Penaeus vannamei* to dietary calcium, phosphorus and calcium: Phosphorus ratio. *Journal of the World Aquaculture Society* 24:504-515.

Davis, D.A., T.M. Samocha, R.A. Bullis, S. Patnaik, C. Browdy, A. Stokes, and H. Atwood (2004). Practical diets for *Litopenaeus vannamei* (Boone, 1931): Working toward organic and/or all plant production diets. In *Avances en Nutrición Acuícola VII. Memorias del VII Simposium Internacional de Nutrición Acuícola*, L.E. Cruz Suárez, D. Ricque Marie, M.G. Nieto López, D.Villarreal, U. Scholz, and M. González (eds.). Hermosillo, Sonora, Mexico: Universidad Autónoma de Nuevo León, pp. 202-214.

Deshimaru, O., K. Kuroki, M.A. Mazid, and S. Kitamura (1985). Nutritional quality of compounded diets for prawn *Penaeus monodon*. *Bulletin Japanese Society of Scientific Fisheries* 51:1037-1044.

Deshimaru, O. and K. Shigeno (1972). Introduction to the artificial diet for prawn *Penaeus japonicus*. *Aquaculture* 1:115-133.

Deshimaru, O. and Y. Yone (1978). Requirements of prawn for dietary minerals. *Bulletin Japanese Society of Scientific Fisheries* 44:907-910.

Divakaran, S., M. Velasco, E. Beyer, I. Forster, and I. Tacon (2000). Soybean meal apparent digestibility for *Litopenaeus vannamei*, including a critique of methodology. In *Avances en Nutrition Acuícola V. Memorias del V Symposium Internacional de Nutrición Acuícola*, L.E. Cruz-Suarez, D. Ricque-Marie, M. Tapia-Salazar, M.A. Olvera-Novoa, and R. Civera-Cerecedo (eds.). Mérida, Yucatán: Universidad Autónoma de Nuevo León, pp. 267-276.

Du, L. and C.J. Niu (2003). Effect of dietary substitution of soya bean meal for fish meal on consumption, growth, and metabolism of juvenile giant freshwater prawn, *Macrobrachium rosenbergii*. *Aquaculture Nutrition* 9:139-143.

Eusebio, P.S. (1991). Effect of dehulling on the nutritive value of some leguminous seeds as protein sources for tiger prawn, *Penaeus monodon*, juveniles. *Aquaculture* 99:297-308.

Ezquerra, J.M., L. Fernando, F.L. Garcia-Carrefio, R. Civera, and N.F. Haard (1997). pH-stat method to predict protein digestibility in white shrimp *(Penaeus vannamei)*. *Aquaculture* 157:249-260.

FAO (2000). *Yearbook of Fishery Statistics 1998. Vol. 86/2. Aquaculture Production.* FAO Statistics Series N. 154 and Fisheries Series N. 56, Rome, Italy: FAO, 182 pp.

FAO (2003). *Overview of Fish Production, Utilization, Consumption and Trade Based on 2001 Data.* Rome, Italy: FAO, 3 pp.

FAO (2005a). *FAOSTAT Data for Agriculture, Aquaculture Production 2003,* http:// www.fao.org/fi/statist/fisoft/FISHPLUS.asp, Rome, Italy: FAO.

FAO (2005b). *FAOSTAT Data for Agriculture, 2005-Crops, Primary Equivalents,* http://faostat.fao.org/faostat. Rome, Italy: FAO.

Fenucci, J.J., A.L. Lawrence, and Z.P. Zein-Eldin (1981). The effects of fatty acids and shrimp meal composition of prepared diets on growth of juvenile shrimp, *Penaeus stylirostris. Journal of the World Mariculture Society* 12(1):315-324.

Fernandez, R.R. and A.L. Lawrence (1988a). Nutritional responses of postlarval penaeus vannamei to different soybean levels. Abstracts of the 19th Annual Meeting of the World Aquaculture Society. *Journal of the World Aquaculture Society* 19(1):30-A.

Fernandez, R.R and A.L. Lawrence (1988b). The nutritional responses of three species of postlarval penaeid shrimp to cottonseed meal. Abstracts of the 19th Annual Meeting of the World Aquaculture Society. *Journal of the World Aquaculture Society* 19(1):30-A.

Floreto, E.A., R.C. Bayer, and P.B. Brown (2000). The effects of soybean-based diets, with and without amino acid supplementation, on growth and biochemical composition of juvenile American lobster, *Homarus americanus. Aquaculture* 189:211-235.

Forster, I., D.A. Higgs, B.S. Dosanjh, M. Rowshandeli, and J. Parr (1999). Potential for dietary phytase to improve the nutritive value of canola protein concentrate and decrease phosphorus output in rainbow trout *(Onchorhynchus mykiss)* held in 11°C freshwater. *Aquaculture* 179:109-125.

Fox, J.M, A.L. Lawrence, and E. Li-Chan (1995). Dietary requirement for lysine by juvenile *Penaeus vannamei* using intact and free amino acid sources. *Aquaculture* 131:279-290.

Fujiki, K., H. Matsuyama, and T. Yano (1994). Protective effect of sodium alginate against bacterial infection in common carp, *Cyprinus carpio* L. *Journal of Fish Diseases* 17(4):349-355.

Guary, J.C., M. Kayama, and H.J. Ceccaldi (1976). The effects of a fat-free diet and compounded diets supplemented with various oils on moult, growth and fatty acid composition of prawn, *Penaeus japonicus* Bate. *Aquaculture* 7:245-254.

Guillaume, J. (1997). Protein and amino acids. In *Advances in World Aquaculture 6: Crustacean Nutrition,* L.R. D'Abramo, D.E. Conklin, and D.M. Akiyama (eds.). Baton Rouge, LA: World Aquaculture Society, pp. 26-50.

Hari, B. and B. Madhusoodna-Kurup (2003). Comparative evaluation of dietary protein levels and plant-animal protein ratios in *Macrobrachium rosenbergii* (de Man). *Aquaculture Nutrition* 9:131-137.

Hertrampf, J.W. and F. Piedad-Pascual (2000). *Handbook on Ingredients for Aquaculture Feeds.* Dordrecht, The Netherlands: Kluwer Academic Publishers, 573 pp.

Hughes, K.P. and J.H. Soares Jr. (1998). Efficacy of phytase on phosphorus utilization in practical diets fed to striped bass *Morone sexatilis*. *Aquaculture Nutrition* 4:133-140.

Jackson, L.S., M.H. Li, and E.H. Robinson (1996). Use of microbial phytase in channel catfish *Ictalurus punctatus* diets to improve utilization of phytate phosphorus. *Journal of the World Aquaculture Society* 27:309-313.

Kanazawa, A. (1984). Feed formulation for penaeid shrimp, sea bass, grouper, and rabbitfish culture in Malaysia. In *Coastal Aquaculture Development. FAO Project: Mal/77/008-Coastal Aquaculture Demonstration and Training*. Rome, Italy: FAO, pp. 62-78.

Kanazawa, A.S. (1992). Utilization of soybean meal and other non-marine protein sources in diets for penaeid prawns. In *Proceeding of the Aquaculture Nutrition Workshop. Program and Abstracts*, G.L. Allan (ed.). Salamander Bay, Australia: NSW Fisheries.

Kanazawa, A., S. Tokiwa, M. Kayama, and M. Hirata (1977). Essential fatty acids in the diet of prawn: I. Effects of linoleic and linolenic acids on growth. *Bulletin of the Japanese Society of Scientific Fisheries* 43:1111-1114.

Ketola, H.G. (1994). Use of enzymes in diets of trout to reduce environmental discharge of phosphorus. *World Aquaculture Society, Book of Abstracts*. Baton Rouge, LA: World Aquaculture Society, 211 pp.

Kitabayashi, K., H. Kurata, K. Shudo, K. Nakamura, and S. Ishikawa. (1971). Studies on formula feed for kuruma prawn—I. On the relationship among glycocyamine, phosphorus and calcium. *Bulletin of the Tokai Regional Fisheries Research Laboratory*, Tokyo 65:91-108.

Koshio, S., A. Kanazawa, and S. Teshima (1992). Nutritional evaluation of dietary soybean protein for juvenile freshwater prawn *Macrobrachium rosenbergii*. *Nippon Suisan Gakkaishi* 58(5):965-970.

Lawrence, A.L. and F.L. Castille Jr. (1989). Use of soybeans, cottonseed, meat and blood meal and dry blood in shrimp feeds. *Journal of the American Oil Chemists Society* 66:464.

Lawrence, A.L., F.L. Castille Jr., L.N. Sturmer, and D.M. Akiyama (1986). *Nutritional response of marine shrimp to different levels of soybean meal in feeds. USA-ROC and ROC-USA Economic Councils' Tenth Anniversary*. Republic of Singapore: American Soybean Association, 12 pp.

Lee, P.G., N.J. Blake, and G.E. Rodrick (1980). A quantitative analysis of digestive enzymes for the freshwater prawn, *Macrobrachium rosenbergii*. *Proceedings of the World Mariculture Society* 11:392-402.

Lee, P.G. and A.L. Lawrence (1997). Digestibility. In *Advances in World Aquaculture 6: Crustacean Nutrition*, L.R. D'Abramo, D.E. Conklin, and D.M. Akiyama (eds.). Baton Rouge, LA: World Aquaculture Society, pp. 194-260.

Li, M.H., E.H. Robinson, and R.W. Hardy (2000). Protein sources for feed. In *Encyclopedia of Aquaculture*, R.R. Stickney (ed.). New York, NY: John Wiley & Sons, Inc., pp. 688-695.

Lim, C. (1996). Substitution of cottonseed meal for marine animal protein in diets for *Penaeus vannamei*. *Journal of the World Aquaculture Society* (27)4:402-409.

Lim, C. (1997). Replacement of marine animal protein with peanut meal in diets for juvenile white shrimp, *Penaeus vannamei*. *Journal of Applied Aquaculture* 7(3):67-78.

Lim, C. and D.M. Akiyama (1995). Nutrient requirements of penaeid shrimps. In *Nutrition and Utilization Technology in Aquaculture,* C.E. Lim and D.J. Sessa (eds.). Champaign, Illinois: AOCS Press, pp. 60-73.

Lim, C., R.M. Beames, J.G. Eales, A.F. Prendergast, J.M. Mcleese, K.D. Shearer, and D.A Higgs (1997). Nutritive values of low and high fiber canola meals for shrimp *(Penaeus vannamei)*. *Aquaculture Nutrition* 3:269-279.

Lim, C. and W. Dominy (1990). Evaluation of soybean meal as a replacement for marine animal protein in diets for shrimp *(Penaeus vannamei)*. *Aquaculture* 87:53-63.

Lim, C. and W. Dominy (1992). Substitution of full-fat soybeans for commercial soybean meal in diets for shrimp, *Penaeus vannamei*. *Journal of Applied Aquaculture* 1(3):35-45.

Lochmann, R., W.R McClain, and D.M. Gatlin, III (1992). Evaluation of practical feed formulations and dietary supplements for red swamp crayfish. *Journal of the World Aquaculture Society* (23)3:217-227.

Lupatsch, I., G. Wm. Kissil, D. Sklan, and E. Pfeffer (1998). Energy and protein requirements for maintenance and growth in gilthead seabream *(Sparus aurata* L.). *Aquaculture Nutrition* 4:165-173.

Martin, B.J. (1980). Croissance et acides gras de la crevette *Palaemon serratus* (Crustacea, Decapoda) nourrie avec des aliments composés contenant différentes proportions d'acide linoleique et linolénique: Growth and fatty acids of *Palaemon serratus* (Crustacea, Decapoda) fed with compounded diets containing different proportions of linoleic and linolenic acids. *Aquaculture* 19:325-337.

McCallum, I.M., I.W. Newell, L.E. Cruz-Suárez, D.R. Ricque-Marie, M. Tapia-Salazar, A. Davis, D. Thiessen et al. (2000). Uso de arvejón (feed pea, chicharo) *Pisum sativum* en alimentos para camarones *(Litopenaeus stylirostris* y *L. vannamei),* tilapia *(Oreochromis niloticus)* y trucha *(Oncorhynchus mykiss).* In *Avances en Nutrición Acuícola V. Memorias del V Symposium Internacional de Nutrición Acuícola,* L.E. Cruz-Suárez, D. Ricque-Marie, M. Tapia-Salazar, M.A. Olvera-Novoa, and R. Civera-Cerecedo (eds.). Mérida, Yucatán: Universidad Autónoma de Nuevo León, pp. 193-215.

Millamena, O.M., M.N. Bautista-Teruel, O.S. Reyes, and A. Kanazawa (1996). Methionine requirement of juvenile tiger shrimp, *Penaues monodon* Fabricius. *Aquaculture* 143:403-410.

Millamena, O.M., M.N. Bautista-Teruel, O.S. Reyes, and A. Kanazawa (1997). Threonine requirement of juvenile marine shrimp *Penaeus monodon. Aquaculture* 151:9-14.

Millamena, O.M., M.N. Bautista-Teruel, O.S. Reyes, and A. Kanazawa (1998). Requirements of juvenile marine shrimp, *Penaues monodon* (Fabricius) for lysine and arginine. *Aquaculture* 164:95-104.

Millamena, O.M. and E. Quinitio (2000).The effects of diets and reproductive performance of eyestalk ablated and intact mud crab *Scylla serrata. Aquaculture* 181:81-90.

Millamena, O.M., M.B. Teruel, A. Kanazawa, and S. Teshima (1999). Quantitative dietary requirements of postlarval tiger shrimp, *Penaues monodon*, for histidine, isoleucine, leucine, phenylalanine and tryptophen. *Aquaculture* 179:169-179.

Molina-Poveda, C. and M.E. Morales (2002). Utilization of dry distillers' grains derived from barley-based brewer production as an alternative protein source in diets of *Litopenaeus vannamei*. In *Avances en nutrición acuícola VI. Memorias del VI simposium internacional de nutrición acuícola (abstracts)*, L.E. Cruz-Suárez, D. Ricque-Marie, M. Tapia-Salazar, M.G. Gaxiola-Cortez, and S. Nuno (eds.). Monterrey, Nuevo León, México: Universidad Autónoma de Nuevo León.

Murai, T., T. Akiyama, and T. Nose (1984). Effect of amino acid balance on efficiency and utilization of diet by fingerling carp. *Bulletin of the Japanese Society of Scientific Fisheries* 50:893-897.

Muzinic, L.A., K.R. Thompson, A. Morris, C.D. Webster, D.B. Rouse, and L. Manomaitis (2004). Partial of total replacement of fish meal with soybean meal and brewer's grains with yeast in practical diets for Australian red craw crayfish *Cherax quadricarinatus. Aquaculture* 230:359-376.

National Research Council (NRC) (1969). *United States-Canadian Table of Feed Composition. National Academy of Sciences.* Washington, DC: National Academy Press, 92 pp.

National Research Council (NRC) (1993). *Nutrient Requirements of Fish.* Washington, DC: National Academy Press, 114 pp.

Olvera-Novoa, M.A. and L. Olivera-Castillo (2000). Potencialidad del uso de las leguminosas como fuente proteica en alimentos para peces. In *Avances en Nutrición Acuícola. Memorias del IV simposio Internacional de Nutrición Acuícola,* R.Civera-Cerecedo, C.J. Perez-Estrada, D. Ricque-Marie, and L.E Cruz-Suárez (eds.). Mérida, Yucatán, México: Universidad Autónoma de Nuevo León, pp. 327-348.

Oser, B.L. (1959). Protein and amino acids. In *Protein and Amino Acid Nutrition,* A.A. Albanese (ed.). New York, NY: Academic Press, pp. 281-295.

Philipps, A.M. Jr. and D.R. Brockway (1956). The nutrition of trout II: Protein and carbohydrate. *The Progressive Fish Culturist* 18:159-164.

Peñaflorida, V.D. (1989). An evaluation of indigenous protein sources as potential component in the diet formulation for tiger prawn, *Penaeus monodon,* using essential amino acid index (EAAI). *Aquaculture* 83:319-330.

Peñaflorida, V.D. (2002). Evaluation of plan proteins as partial replacement for animal protein in diets for *Penaeus indicus* and *P. merguiensis* juveniles. *The Israeli Journal of Aquaculture* 54(3):116-124.

Piedad-Pascual, F. and M. Catacutan (1990). Defatted soybean meal and leucaena leaf meal as protein sources in diets for *Penaeus monodon* juveniles. In *The Second Asian Fisheries Forum,* R. Hirano and I. Hanyu (eds.). Manila, Philippines: Asian Fisheries Society, 991 pp.

Piedad-Pascual, F., E.M. Cruz, and A. Sumalangcay Jr. (1990). Supplemental feeding of *Penaeus monodon* juveniles with diets containing various levels of defatted soybean meal. *Aquaculture* 89:183-191.

Pulse Canada (2005). *Feed Peas.* In www.pulse.ab.ca/marketing/Feed%Peas% Sheet%English.pdf. Manitoba, Canada.

Read, G.H.L. (1981). The response of *Penaeus indicus* (Crustacea: Penaeidea) to purified and compounded diets of varying fatty acid composition. *Aquaculture* 24:245-256.

Reigh, R.C. and S.C. Ellis (1994). Utilization of animal protein and plant protein supplements by red swamp crayfish *Procambarus clarkii* fed formulated diets. *Journal of the World Aquaculture Society* (25)4:541-552.

Ricque-Marie, D., M. Nieto-Lopez, B. Zabala-Chavez, C. Guajardo-Barboza, U. Scholz, M.L. Locatelli, D. Thiessen et al. (2005). *Massive Improvement in* Litopenaeus vannamei *Growth by Replacing 50% Fish Meal with Canola Protein Concentrate in a 30% Crude Protein Diet.* Aquaculture America 2005. Baton Rouge, LA: World Aquaculture Society.

Robinson, E.H. and M.H. Li (1995). Use of cottonseed meal in aquaculture feeds. In *Nutrition and Utilization Technology in Aquaculture,* C.E. Lim and D.J. Sessa (eds.). Champaign, IL: AOCS Press, pp. 157-165.

Samocha, T.M., D.A. Davis, I.P. Saoud, and K. DeBault (2004). Substitution of fish meal by co-extruded soybean poultry by-product meal in practical diets for the Pacific white shrimp, *Litopenaeus vannamei. Aquaculture* 231:197-203.

Sandifer, P.A. and J.D. Joshep (1976). Growth responses and fatty acid composition of juvenile prawns *(Macrobrachium rosenbergii)* fed a prepared ration augmented with shrimp head oil. *Aquaculture* 8:129-138.

Sessa, D.J. and C. Lim (1991). Effect of feeding soy products with varying trypsin inhibitor activities on growth of shrimp. *Journal of the America Oil Chemists Society* 69(3):209-212.

Shiau S.Y. (1998). Nutrient requirements of penaeid shrimps. *Aquaculture* 164:77-93.

Shiau, S.Y., K.P. Lin, and C.L. Chiou (1992). Digestibility of different protein protein sources by *Penaeus monodon* raised in brackish water and in sea water. *Journal of Applied Aquaculture* 1(3):47-53.

Smith, D.M., G.L. Allan, K.C. Williams, and C.G. Barlow (2000). Fishmeal replacement research for shrimp feed in Australia. In *Avances en Nutrición Acuícola V. Memorias del V Symposium Internacional de Nutrición Acuícola,* L.E. Cruz-Suárez, D. Ricque-Marie, M. Tapia-Salazar, M.A. Olvera-Novoa, and R. Civera-Cerecedo (eds.). Mérida, Yucatán: Universidad Autónoma de Nuevo León, pp. 277-286.

Smith, D.M., S.J. Tabrett, and H.Z. Sarac (1999). Fish meal replacement in the diet of the prawn *Penaeus monodon. Book of Abstracts of the World Aquaculture Society.* Baton Rouge, LA: World Aquaculture Society.

Sudaryono, A., M.J. Hoxey, S.G. Kailis, and L.H. Evans (1995). Investigation of alternative protein sources in practical diets for juvenile shrimp *Penaeus monodon. Aquaculture* 134:313-323.

Sudaryono, A.E., J. Tsvetnenko, S. Hutabarat, and L.H. Evans (1999). Lupin ingredients in shrimp *(Penaeus monodon)* diets: Influence of lupin species and types of meals. *Aquaculture* 171:121-133.

Sugiura, S.H, F.M. Dong, and R.W. Hardy (1999). Availability of phosphorus and trace elements in low-phytate varieties of barley and corn for rainbow trout *(Oncorhynchus mykiss). Aquaculture* 170:285-296.

Tacon, A.G.J. (1994). Feed ingredients for carnivorous fish species: Alternatives to fishmeal and other fishery resources. *FAO Fisheries Circular No. 881.* Rome: FAO, 35 pp.

Tacon, A.G.J. and D.M. Akiyama (1997). Feed ingredients. In *Advances in World Aquaculture 6: Crustacean Nutrition,* L.R. D'Abramo, D.E. Conklin, and D.M. Akiyama (eds.). Baton Rouge, LA: World Aquaculture Society, pp. 411-472.

Tacon, A.G.J. and U.C. Barg (1998). Major challenges to feed development for marine and diadromous finfish and crustaceans species. In *Tropical Mariculture,* S.S. De Silva (ed.). New York, NY: Academic Press, pp. 171-207.

Tacon, A.G.J., W.G. Dominy, and G.D. Pruder (2000). Tendencias y retos globales de los alimentos para el camarón. In *Avances en Nutrición Acuícola IV. Memorias del IV Symposium Internacional de Nutrición Acuícola,* R. Civera-Cerecedo, C.J. Pérez Estrada, D. Ricque-Marie, and L.E. Cruz-Suárez (eds.). La Paz, México: Universidad Autónoma de Nuevo León, pp. 1-27.

Tacon, A.G.J. and I.P. Forster (2000). Trends and challenges to aquaculture and aquafeed development in the new millennium. In *Avances en Nutrición Acuícola V. Memorias del V Symposium Internacional de Nutrición Acuícola,* L.E. Cruz-Suárez, D. Ricque-Marie, M. Tapia-Salazar, M.A. Olvera-Novoa, and R. Civera-Cerecedo (eds.). Mérida, Yucatán: Universidad Autónoma de Nuevo León, pp. 1-12.

Tacon, A.G.J., J.L. Webster, and C.A. Martinez (1984). Use of solvent extracted sunflower meal in complete diets for fingerling rainbow trout (*Salmo gairdneri* Richardson). *Aquaculture* 43:381-389.

Takahashi, Y., K. Uehra, R. Watanable, T. Okumara, T. Yamashita, H. Omura, T. Kawano et al. (1998). Efficacy of oral administration of fucoidan, a sulphated polysaccharide, in controlling white spot syndrome in kuruma shrimp in Japan. In *Advances in Shrimp Biotechnology,* T.W. Flegel (ed.). Bangkok, Thailand: National Center for Genetic Engineering and Biotechnology. pp. 171-174.

Teichert-Coddington, D.R. and R. Rodriguez (1995). Semi-intensive commercial growth-out of *Penaeus vannamei* fed diets containing differing levels of crude protein during wet and dry seasons in Honduras. *Journal of the World Aquaculture Society* 26(1):72-79.

Teshima, S., M.S. Alam, S. Koshio, M. Ishikawa, and A. Kanazawa (2002). Assestment of requirement values for essential amino acids in the prawn *Marsupenaeus japonicus* (Bate). *Aquaculture Research* 33:395-402.

Tidwell, J.H., C.D. Webster, J.A. Clark, and L.R. D'Abramo (1993). Evaluation of distiller dried grains with solubles as an ingredient in diets for pond culture of the freshwater prawn *Macrobrachium rosenbergii. Journal of the World Aquaculture Society* 24(1):66-70.

Tidwell, J.H., C.D. Webster, D.H. Yancey, and L.R. D'Abramo (1993). Partial and total replacement of fish meal with soybean meal and distiller by-products in diets for pond culture of the freshwater prawn *Macrobrachium rosenbergii. Aquaculture* 118:119-130.

Viola, S., Y. Arieli, and G. Zohar (1988). Animal-protein-free feeds for hybrid tilapia *(Oreochromis niloticus × O. aureus)* in intensive culture. *Aquaculture* 75:115-125.

Viola, S., S. Mokady, U. Rappaport, and Y. Arieli (1982). Partial and complete replacement of fishmeal by soybean meal in feeds for intensive culture of carp. *Aquaculture* 26:223-236.

Webster, C.D., L.S. Goodgame-Tiu, J.H. Tidwell, and D. Rouse (1994). Evaluation of practical feed formulations with different protein levels for juvenile red crayfish *(Cherax quadricarinatus)*. *Transactions of the Kentucky Academy of Science* 55(34):108-112.

Webster, C.D., D.H. Yancey, and J.H. Tidwell (1995). Effect of partially or totally replacing fish meal with soybean meal on growth of blue catfish *(Ictalurus furcatus)*. *Aquaculture* 103:141-152.

Wilson, R.P. and W.E. Poe (1985). Relationship of whole body and egg essential amino acid patterns to amino acid requirement patterns in channel catfish, *Ictalurus punctatus. Comparative Biochemistry and Physiology* 80B:385-388.

Wu, Y.V., R. Rosati, D.J. Sessa, and P. Brown (1995).Utilization of corn gluten feed by Nile tilapia. *The Progressive Fish Culturist* 57:305-309.

Chapter 8

Protein Feedstuffs Originating from Soybeans

Paul B. Brown
Sadasivam J. Kaushik
Helena Peres

INTRODUCTION

The soybean (*Glycine* sp.) is a leguminous plant native to Asia that has been introduced to every nonpolar continent. Agricultural production of soybean seeds was the largest global oilseed crop in 2003 (Table 8.1; United Nations, Food and Agriculture Organization, 2004 database). Soybean seeds are classified as an oilseed because of the relatively high concentration of lipid. Oilseeds can be used as is, also known as full-fat, or after lipid extraction. Lipid is extracted from oilseeds by a variety of methods, most often solvent extraction, and the remaining cake can be further processed into various meals for use in human or animal foods. In 2003, a significant portion of global soybean production was used for extraction of oil resulting in a relatively large global supply of high-protein soybean cake (Table 8.1). The global supply of soybean cake is approximately 65 percent of the total soybean production. Rapeseed and safflower cakes are approximately 49 and 43 percent, respectively, of total soybean production, other percentages are less.

Alternative Protein Sources in Aquaculture Diets
© 2008 by The Haworth Press, Taylor & Francis Group. All rights reserved.
doi:10.1300/5892_08

TABLE 8.1. Global harvest of some oilseed crops in 2003 and volume of processed products (cakes) after extraction of oils (metric tons).[a]

Crop	Total harvest	Cake
Soybeans	189,523,638	123,327,242
Oil, palm fruit	139,148,600	3,727,703
Seed cotton	56,969,044	12,388,472
Coconuts	50,227,217	1,902,410
Ground nuts in shell	37,057,652	7,416,247
Rapeseed	35,931,652	17,760,805
Sunflower	26,085,901	9,267,370
Dry beans	18,886,344	N/A[b]
Sesame	2,765,419	736,057
Linseed	2,054,195	861,863
Lupins	1,596,444	N/A
Castor beans	1,153,768	N/A
Safflower	731,425	318,700

[a]Values from United Nations Food and Agriculture (http://faostat.fao.org/).

[b]Data not available.

NUTRITIONAL CONCENTRATIONS

After harvesting, the raw soybean can be processed into a variety of products (Figure 8.1). The most commonly used products in aquaculture are the toasted soybean meals (SBM). Two forms are commonly available, dehulled (dh) soybean meal (usual protein content of ~49 percent, as is basis), and a product in which the hulls have been added back to the toasted meal resulting in a lower protein concentration (~44 percent, as is basis). For the remainder of this document, SBM will indicate the dehulled, solvent-extracted toasted soybean meal. Raw soybean seeds, soy flours, soybean meals, protein concentrates, and protein isolates have been evaluated in a variety of animal species. There are limitations on use of raw soybeans in diets fed to aquatic animals because of the antinutritional factors (ANF) present in seeds. Exposing the seeds to heat treatment diminishes the concentrations of ANF. Heated soybean seeds can be classified into two categories,

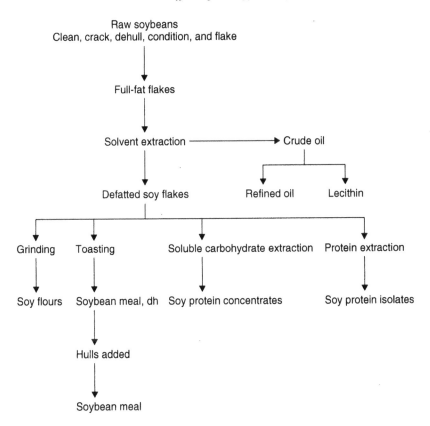

FIGURE 8.1. Schematic representation of soybean processing resulting in soy flours and meal, protein concentrates, protein isolates, oil, and lecithin.

depending on the temperature used for heating. Heated seeds are typically exposed to temperatures of 225°F (107°C), roasted seeds are exposed to higher temperatures and exposed for a longer period of time. Temperatures used for roasting can be as high as 400°F (204°C), but the most commonly used is around 300°F (149°C). Heating or roasting can improve utility of full-fat soybeans in diets fed to some species. Protein concentrates and isolates are more expensive than SBM because of the additional processing steps. Expected macronutrient concentrations are presented in Table 8.2 for various soybean products.

Agronomists classify soybeans as oilseeds, but nutritionists consider soybean seeds as high lipid and high protein feedstuffs. After

TABLE 8.2. Expected macronutrient concentrations in heated soybean seeds, soybean meal (SBM), soy concentrate, and soy isolate compared to a standard fish meal (%, dry matter basis).

Macronutrient	Seeds, heated[a]	SBM[a]	Concentrate[b]	Isolate[b]	Fish meal[a,c]
Crude protein[d]	42.2	53.9	65-72	90-92	78.3
Crude fat	20.0	1.1	0.5-1.0	0.5-1.0	9.1
Crude fiber	5.5	4.3	3.5-5.0	0.1-0.2	0.6
Ash	5.0	5.4	4.0-6.5	4.0-5.0	11.3
NFE[e]	27.3	35.3	20-22	3.0-4.0	0.7

[a]Values from NRC (1994).

[b]Values from Endres (2001).

[c]Herring fish meal.

[d]$N \times 6.25$.

[e]Nitrogen-free extract, calculated by difference.

lipid extraction, the resulting meal is clearly a protein feedstuff. As soybean products are processed into meals, concentrates, and isolates, the crude protein concentrations approach and exceed those values found in standard fish meals. However, the crude fat and ash concentrations of all solvent-extracted soy products are lower than values found in fish meals, but the crude fiber and nitrogen-free extract concentrations are higher than in fish meals. The differences in energy-yielding macronutrients (fat and carbohydrate) are problematic in diets fed to some species of fish. Lower ash, and therefore mineral concentrations, can be overcome by addition of relatively inexpensive inorganic minerals. The fiber concentrations in soy products are not high enough to warrant concern in most dietary formulations. Each of the macronutrients can be further subdivided into their constituent components.

Concentrations of the ten essential amino acids plus cystine and tyrosine in soybean seeds, meals, concentrates, and isolates are presented in Table 8.3 and compared to herring fish meal. Concentrations of all amino acids presented are lower in SBM than in fish meal with the exception of cystine. Of particular concern are lysine, methionine, and

TABLE 8.3. Amino acid concentrations of heated soybean seeds, soybean meal (SBM), soy concentrate. and soy isolate compared to a standard fish meal (%, dry matter basis).[a]

Amino acid	Seeds, heated	SBM	Concentrate	Isolate	Fish meal[b]
Arginine	2.9	3.9	6.4	7.5	4.3
Histidine	1.1	1.4	2.0	2.4	1.6
Isoleucine	1.8	2.4	3.7	4.6	3.1
Leucine	3.1	4.2	5.9	7.2	5.6
Lysine	2.5	3.3	4.7	5.7	5.9
Methionine	0.6	0.7	1.0	1.1	2.2
Cystine	0.6	0.8	1.1	1.3	0.7
Phenylalanine	2.0	2.6	3.8	4.7	2.9
Tyrosine	1.5	2.2	2.8	3.4	2.3
Threonine	1.6	2.1	3.1	3.4	3.2
Tryptophan	0.5	0.8	1.0	1.2	0.8
Valine	1.9	2.5	3.8	4.6	3.7

[a]Values from NRC (1998).
[b]Herring fish meal.

threonine. The lysine concentration is 56 percent of the concentration found in herring fish meal, the concentration of methionine is 32 percent of the concentration in fish meal, and the threonine concentration is 66 percent of the value reported for fish meal. These three essential amino acids tend to be limiting in soy-based diets fed to aquatic animals. Dietary cystine can spare approximately 50 percent of the dietary methionine requirement in fish (Harding et al., 1977; Moon and Gatlin, 1991; Griffin et al., 1994; Twibell et al., 2000) and a more appropriate comparison would be the total values for methionine plus cystine (total sulfur amino acid concentrations). Adding both values and expressing those as a function of the values in fish meal yields a comparative value of 52 percent of the total sulfur amino acid concentration in SBM compared to herring fish meal. As soy protein is further purified by processing into soy concentrates, the amino acid concentrations increase. In a typical soy concentrate, the lysine concentration is 85 percent of the concentration found in herring fish meal, the total sulfur amino acid concentrations are 95 percent of the values in fish meal and the threonine concentration is higher than that in fish meal. Amino acid concentrations in soy isolates are higher than those in fish meal with the exception of lysine (5.7 percent in soy isolates compared to 5.9 percent in herring fish meal).

The lipid fraction (crude fat) of soy is a consideration only in the full-fat heated seeds (Lim and Akiyama, 1992). Crude soybean oil contains 12.2 percent palmitic acid (16:0), 26.0 percent oleic acid (18:1n-9), 51.5 percent linoleic acid (18:2n-6), and 7.0 percent linolenic acid (18:3n-3) (expressed as a percentage of the total fatty acids; NRC, 1994). A full-fat heated seed meal contains 10.3 percent 18:2n-6 and 1.4 percent 18:3n-3 (expressed as percentage of the dry meal). The essential fatty acids of fishes are the n-6 and n-3 families and range from 0.5 to 1.7 percent of the diet (NRC, 1993). Thus, full-fat soybean meal at 50 percent incorporation into the diet can provide the essential fatty acid needs of those fishes requiring n-6 fatty acids and a significant portion of the fatty acid requirements in those fishes requiring n-3s. The crude fat concentration of solvent-extracted soy products is low. For example, SBM contains 0.52 percent 18:2n-6 and 0.08 percent 18:3n-3 (dry matter basis; NRC, 1994). Thus, the contribution of SBM, concentrates, and isolates to the essential fatty needs of most fishes is not significant.

The carbohydrate fraction of soybeans, presented as the nitrogen-free extract in Table 8.2 and often referred to as the soluble carbohydrate fraction, contains sucrose, raffinose, and stachyose as the primary oligosaccharides. Sucrose is generally available to animals, but raffinose and stachyose are not as animals lack the α-galactosidase necessary to catabolize the complex sugar (Snyder and Kwon, 1987).

Mineral concentrations of SBM, concentrates, and isolates are presented in Table 8.4 along with values for herring fish meal. Concentrations of calcium, phosphorus, sodium, chlorine, sulfur, zinc, and selenium are lower in most soy products compared to fish meal. The only exception is the concentration of sulfur in soy isolates, which is similar to the concentration found in fish meal. As stated previously, minerals can be easily added to diets in a premix and they are relatively inexpensive. Common additions to mineral premixes in diets fed to fish include zinc and selenium as well as other minerals considered limiting. Furthermore, aquatic animals have multiple mechanisms for

TABLE 8.4. Mineral concentrations (dry matter basis) of heated soybean seed, soybean meal (SBM), concentrates. and isolates compared to a standard fish meal.[a]

Mineral	Seeds, heated	SBM	Concentrate	Isolates	Fish meal[b]
Calcium (%)	0.28	0.38	0.39	0.16	2.58
Phosphorus (%)	0.65	0.77	0.90	0.71	1.89
Sodium (%)	0.03	0.02	0.05	0.08	0.65
Chlorine (%)	0.03	0.05	–	0.02	1.20
Potassium (%)	1.89	2.38	2.44	0.29	1.09
Magnesium (%)	0.31	0.33	0.35	0.09	0.19
Sulfur (%)	0.33	0.49	–	0.77	0.74
Copper (mg/kg)	17.8	22.2	14.4	14.1	6.4
Iron (mg/kg)	88.9	195.5	122.2	148.9	194.6
Manganese (mg/kg)	33.3	40.0	–	5.4	8.6
Zinc (mg/kg)	43.3	61.1	33.3	36.9	141.9
Selenium (mg/kg)	0.12	0.30	–	0.15	2.07

[a]Values from NRC (1998).

[b]Herring fish meal.

obtaining minerals from the environment. There is significant ion exchange at the gill in fresh- and saltwater animals in an effort to maintain osmotic balance. Calcium, sodium, chloride, and potassium are readily uptaken from the environment. Marine fish also drink copious amounts of water, usually measured in L/hour/unit body weight, and obtain minerals directly from their water intake. There is increasing data supporting mineral uptake through the skin as well. Phosphorus (P) deserves some attention as the form of P is different in oilseed crops from that in fish meal.

The majority of P (60 to 80 percent of total) in plant tissues occurs in the form of myoinositol 1, 2, 3, 4, 5, 6-hexa kis dihydrogen phosphate, also known as phytic acid (Erdman, 1979). This is a storage form of P in plants and a nutritionally unavailable form of P in most animals. Bacteria possess phytase and can cleave the phosphate groups from the inositol ring, but vertebrates do not possess phytase in their gastrointestinal tract. Ruminant animals are capable of using some of the P from phytate, but the passage of food through the gastrointestinal tract of fish is more rapid than in ruminants and the contribution of nutrients from bacterial action to fishes is not well understood. Phytate also interacts with cations, reducing their availability. There are six charged phosphate groups on the myoinositol ring that can and will bind cations in the gastrointestinal tract. Once these cations are bound, availability is reduced. Cations that can bind to phytate include calcium, magnesium, iron, zinc, copper, manganese, molybdenum, and cobalt (Erdman, 1979; Cheryan, 1980). That topic is considered in more detail in subsequent chapters.

Vitamin concentrations in heated seeds, SBM, soy concentrate, and isolate are presented in Table 8.5 along with values for herring fish meal. Soy products contain higher concentrations of biotin, folic acid,

TABLE 8.5. Vitamin concentrations (mg/kg, dry matter basis, unless otherwise denoted) of heated soybean seeds, soybean meal (SBM), soy concentrate, and soy isolate compared to herring fish meal.

Vitamin	Seeds, heated	SBM[a]	Concentrate[b]	Isolate[a]	Fish meal[a,c]
Biotin	0.27	0.29	0.3	0.3	0.14
Choline	2563	3034	2.2	2.2	5705
Folic acid	4.0	1.54	2.8	2.7	0.4
Niacin	24.4	24.4	6.7	6.5	100.0
Pantothenic acid	16.7	16.7	4.7	4.6	18.3
Riboflavin	2.9	3.4	1.3	1.8	10.6
Thiamin	12.2	3.5	0.2	0.3	0.4
Pyridoxine	12.0	7.1	6.0	5.9	5.2
Vitamin B_{12} (μg/kg)	0	0	–	0	433.3
Vitamin E	20.1	2.5	–	–	16.1
β-Carotene[d]	2.1	0.22	–	–	–

[a]Values from NRC (1998).

[b]Values from NRC (1994).

[c]Herring fish meal.

[d]Conversion of β-carotene to vitamin A in swine – 1 mg of all-trans β-carotene = 267 IU vitamin A, 80 μg retinol, or 92 μg retinyl acetate (NRC, 1998).

thiamin, and pyridoxine, but lower concentrations of choline, niacin, pantothenic acid, riboflavin, vitamin B_{12}, vitamin E, and vitamin A compared to fish meal. There are few vitamin availability data from feedstuffs fed to fish and most practical diets are formulated assuming low or negligible vitamin availability. As SBM is processed into concentrates and isolates, most of the vitamin concentrations decrease. The concentrations of biotin, folic acid, and pyridoxine remain at similar or higher concentrations compared to SBM.

A new soybean product was developed in the mid-1990s that deserves mention. Expelled soybean meal is a product of extruding raw soybeans, pressing the extruded beans, and removing a smaller fraction of the lipid than in solvent extraction. Resulting nutritional concentrations are 45 percent crude protein, 6.5 percent fat, 5.5 percent fiber, 0.26 percent calcium, 0.63 percent phosphorus, 0.3 percent sulfur, 0.7 percent methionine, 1.35 percent total sulfur amino acids, 2.75 percent lysine, 0.7 percent tryptophan, 1.8 percent threonine, and 2.0 percent isoleucine. The expelled meal apparently is beneficial for young swine and dairy cattle (Producers' Natural Processing, 2004; http://www.pnpi.com/ExpMeal.htm).

ANTINUTRITIONAL FACTORS

Soybeans contain a diverse array of biologically active compounds, often referred to as antinutritional factors (ANF). These compounds tend to disrupt various aspects of nutrient absorption. There are several mechanisms for diminishing the concentrations of ANF in soybeans, including traditional breeding programs, genetic modification of seeds, heating and processing of harvested seeds. Extrusion of fish feeds also reduces the concentrations of some ANF. Concentrations of ANF in soybean products are presented in Table 8.6. Processing of soybeans to SBM and soy concentrates tends to decrease concentrations of ANF.

Trypsin Inhibitors

Trypsin inhibitors (TI) are commonly associated with soybeans. They are proteins found in soybeans that bind with trypsin and chymotrypsin in the gastrointestinal tract of animals and effectively block

TABLE 8.6. Concentrations of the predominant ANFs in various soybean products.[a]

Component	Raw soybeans	Soybean meal	Soy concentrate
Trypsin inhibitor, mg TI/g	45-50	5-8	<4
Lectin, μg/g	3,600	10-200[b]	<0.1
Saponin, %	0.5	0.6	0
Glycinin antigen, mg/kg	184,000	66,000	<30
β-Conglycinin antigen, mg/kg	>69,000	16,000	<10

[a]Data from Russett (2002).

[b]Range dependent on processing and growing conditions.

the enzymatic activity of both enzymes. Trypsin and chymotrypsin are two of the most important enzymes involved in breakdown of proteins in the gastrointestinal tract of animals. As these proteins can bind with other enzymes, they are also referred to as proteinase inhibitors (Ikenaka and Odani, 1977; Hymowitz, 1985) and serine proteinase inhibitors, as they contain relatively high concentrations of serine. Inhibitor levels in soybean seed is influenced by growing location and genotype (Kumar et al., 2003).

There are two main categories of TI in soybeans, the Bowman-Birk proteinase inhibitors and Kunitz trypsin inhibitor (Hymowitz, 1985), and they are in nearly equal concentrations (Pusztai et al., 1991). There are several Bowman-Birk proteins in soybeans, and they have the ability to bind two gastrointestinal tract enzymes simultaneously, a characteristic referred to as double-headed. Bowman-Birks can bind both trypsin and chymotrypsin at the same time (Hymowitz, 1985), two trypsins, or trypsin and elastase (Laskowski and Kato, 1980). Substrate specificity is a function of amino acid sequences. The Kunitz trypsin inhibitor binds trypsin preferentially, but not exclusively. Rendering trypsin and chymotrypsin inactive in the gastrointestinal tract leads to suppression of the feedback inhibition typically present, elevated levels of cholecystokinin to overcome the apparent deficiency, and hypersecretion of both enzymes (Liener, 2002). This situation results in enlargement of pancreatic tissue. Elevated levels of cholecystokinin and hypersecretion of trypsin and chymotrypsin occurs regardless of whether the TI are introduced into the gastrointestinal tract as free or previously complexed forms (Pusztai et al., 1997). This finding suggests that the TI may be the compound detected leading to hypersecretion of enzymes, not a decrease in protease activity. Both trypsin and chymotrypsin contain relatively high levels of sulfur containing amino acids and those are the limiting amino acids in soy products. Thus, hypersecretion of proteases may exacerbate the limitation of sulfur amino acids in animals fed high levels of TI in soy products (Liener, 1994).

There are conflicting data on resistance to hydrolysis of TI in the gastrointestinal tract of animals. In vitro data indicate the Kunitz inhibitors are inactivated and Bowman-Birks remain active, while in vivo data indicate the opposite (Lajalo and Genovese, 2002). It is clear that heat inactivates both protease inhibitor types from soybeans. Typical SBM is toasted to reduce or eliminate TI activity and feed manufacturing for fish generally incorporates heat that should further reduce TI activity.

Germplasm lines of soybeans lacking the Kunitz inhibitor have been identified and registered (Orf and Hymowitz, 1979; Bernard and Hymowitz 1986a,b).

Lectin

Soybeans contain glycoproteins referred to as lectins or phytohemagglutinins because of their ability to agglutinate, or cause clumping of, red blood cells. Concentration of lectins in soy products varies widely. Liener (1979) reported values of 3 percent, while more recent values are in the range of 10 to 200 μg/g (Russett, 2002). Soybean lectins can bind with the intestinal villi, leading to hypertrophy and hyperplasia, reduced nutrient uptake, and reduction in gut enzyme activities (Lajalo and Genovese, 2002). The net effect is a lengthening and thickening of the gastrointestinal tract. Lectins also stimulate hypersecretion of enzymes from the pancreas, but this effect appears to be only partially mediated through cholecystokinin (Lajalo and Genovese, 2002). Short-term effects of lectin binding to cell surface sites in the gastrointestinal tract appear reversible if inputs are stopped (Bardocz et al., 1995).

Lectins are resistant to degradation in the gastrointestinal tract (Bardocz et al., 1995; Hajos and Gelencser, 1995), but are generally thought to be inactivated by heat. Armour et al. (1998) reported that lectin concentrations remained unaltered when seeds were heated to 60°C, but destroyed by five minute exposures to 100°C in an aqueous medium. Extrusion of dry beans *(Phaeolus vulgaris)* did not inactivate lectins. This deserves additional research in fish as most diets are extruded. Soybean germplasm lines that do not express lectin have been identified (Orf et al., 1978).

Saponins

These compounds are glucosides and five different forms have been identified. The carbohydrate fraction can be glucose, glucuronic acid, galactose, rhamnose, xylose, or arabinose (Hymowitz, 1985). The saponins are considered harmless by some authors (Birk, 1969), not by other authors (Cheeke, 1996). Saponins are poorly absorbed and their effects are largely in the gastrointestinal tract. The effects are thought to be associated with the carbohydrate moiety, although details of the binding of the carbohydrate with intestinal cells have not been determined. Saponins are irritants and they promote their own absorption (Kingsbury, 1980) by increasing the permeability of intestinal mucosal cells and diminishing active uptake of other nutrients (Russett, 2002). Saponins significantly alter cell morphology of the gastrointestinal tract (Gee and Johnson, 1988).

Antigenic Proteins

Globular storage proteins comprise approximately 80 percent of the protein found in defatted soybean flakes (Plumb et al., 1994). Two of these proteins, glycinin (or the 11-12S protein) and conglycinin (or the 7-8S protein) are the predominant forms. There are at least three forms of conglycinin with the β form usually found in the highest concentrations. There is considerable variation in the ratios of glycinin to β-conglycinin, ranging from 1:1 to 3:1 (D'Mello, 1991). The globular proteins, or more specifically, various peptides resulting from partial catabolism in the gastrointestinal tract, cause an allergic reaction in pre-ruminant calves, young pigs (Barratt et al., 1978), and rats (Yamauchi and Suetsuna, 1993). These peptides also display antioxidative (Chen, H.-M et al., 1995) and immunostimulative properties (Chen, J.-R. et al., 1995). The globular proteins are complex molecules with individual protein strands synthesized by as many as five genes that are post-translationally bound by several methods including disulfide bonds (Nielsen et al., 1989). Processing conditions affect configuration of the protein and the interaction with animals (Sissons et al., 1982; Sorgentini et al., 1995).

Other ANFs

Goitrogens were identified in early soy-based diets for rats and poultry (Hymowitz, 1985). The effects were countered with addition of iodine to diets. Little work has been conducted with these compounds since that time and the specific causes have not been identified.

Hydrolysis of glucosides may yield hydrocyanic acid. Honig et al. (1983) reported very low levels in soybean meal and cyanogens have not been considered a significant problem since.

Soybeans contain factors referred to as anti-vitamin in nature. The anti-vitamin factors have not been fully characterized, but they appear to be specific for vitamins A, B_{12}, and D (Rackis, 1972). The significance of these anti-vitamin factors remains unknown in fish.

Urease is present in soybeans and will liberate ammonia if urea is used in diets. However, urea is not a common ingredient in diets fed to fish.

Phytoestrogens, or soy isoflavones, are considered ANFs by some authors, but their beneficial effects elucidated in recent years suggest this may be a misnomer. The phytoestrogens are structurally similar to mammalian estrogens and can bind to estrogen receptors. There are three major isoflavones in soybeans, daidzin, genistin, and glycitin. These native forms do not exhibit estrogenic activity, but hydrolyzed versions, daidzein, genistein, and glycitein, display estrogenic activity (Magee and Rowland, 2004). Soy isoflavones act as estrogenic antagonists and there is increasing evidence they can protect against hormone-dependent neoplasias as well as atherosclerotic lesions (Barnes, 1998). Table 8.7 lists isoflavone concentrations in a variety of raw soybeans as well as processed soy co-products. There is considerable variation in the major isoflavone concentrations in soybeans from various geographic locations. Processed soybean products generally contain similar concentrations of isoflavones except for the alcohol washed soy protein concentrate. These data indicate that isoflavones are soluble in alcohol.

TABLE 8.7. Isoflavone concentrations (mg/100 g) of various soybean products.[a]

Soybean product	Daidzein	Genistein	Glycitein	Total
Soybeans, raw				
Brazil	20.2	67.5	N/R[b]	87.6
Japan	34.5	64.8	13.8	118.5
Korea	72.7	72.3	N/R	145.0
Taiwan	28.2	31.5	N/R	59.7
US, food quality	46.6	73.8	10.9	128.3
US, commodity grade	52.2	91.7	12.1	153.4
Soybeans, roasted	52.0	65.9	13.4	128.3
Soy flakes, full fat	48.2	80.0	1.6	129.0
Soy flakes, defatted	37.0	85.7	14.2	125.8
Soy flour	57.5	71.2	7.6	131.2
Soy protein concentrate[c]	43.0	55.6	5.2	102.1
Soy protein concentrate[d]	6.8	5.3	1.6	12.5
Soy protein isolate	33.6	59.6	9.5	97.4

[a]Data from http://www.nal.usda.gov/fnic/foodcomp/Data/isoflav/isfl_ref.pdf

[b]Not reported.

[c]Aqueous washed.

[d]Alcohol extraction.

Soy sterols include β-sitosterol, campesterol, stigmasterol, and di-hydrobrassicasterol. Plant sterols and stanols (saturated plant sterols) are an active area of investigation as they contribute to a decrease in serum cholesterol in humans by decreasing absorption of cholesterol (de Jong et al., 2003). Plant sterols also interfere with absorption of carotinoids and vitamin E, but have no effect on plasma retinol or vitamin D concentrations (Plat and Mensink, 2000).

SOYBEAN QUALITY

Protein quality is usually discussed as a function of essential amino acid balance relative to the dietary essential amino acid requirements of the target species. Soy proteins are generally lacking in lysine, methionine, and threonine, with methionine and threonine being first limiting in diets fed to pigs and chicks (Emmert and Baker, 1995).

Soybean quality can also be assessed by chemical methods. Urease activity is measured by the change in pH when soybean products are placed in water. Urease is a common measure of soybean quality, but is limited to detecting undercooking of soybean meal (Parsons et al., 1991). Solubility of protein in a weak solution of potassium hydroxide is a second measure and that measure appears to coincide well with in vivo results (Parsons et al., 1991). Finally, there are two methods that rely on solubility of protein from soybeans placed in water, the protein dispersibility index and nitrogen water solubility index (Dudley-Cash, 1999).

CONCLUSIONS AND RECOMMENDATIONS

Increases in aquaculture production are occurring on every heavily inhabited continent. Most of the increase is occurring with intensively raised fish requiring a formulated diet. Thus, the demand for feedstuffs is increasing and global. Soybeans are grown on every nonpolar continent, the annual soybean crop is the largest among the oilseeds, and the resulting lipid-free cake used to make various soybean meals is the largest supply available among these crops. Thus, based on volume of product and global availability, soybeans are a logical protein source to consider for use in diets fed to aquatic animals.

Nutritional concentrations of soy meals place them in the protein feedstuff category. However, disproportionately high concentrations of carbohydrate and low concentrations of lipid in traditional SBM pose problems with some species of fish. There are three essential amino acids that appear limiting in SBM, lysine, methionine, and threonine. Mineral and vitamin limitations also exist, and data on micronutrient availability have only recently become the focus of research efforts.

Biologically active compounds, or ANFs, are present in soybeans and they may be affecting utility of the various soybean products in diets fed to fish. Beneficial compounds are also present in soy products and their effects on fish need to be understood.

Despite the nutrient limitations, soybean products are often one of the first protein sources evaluated with new aquaculture species, probably because of price and availability. In the subsequent chapters, research results will be summarized on use of soy products in diets.

REFERENCES

Armour, J.C., R.L. Chanaka Perera, W.C. Buchan, and G. Grant (1998). Protease inhibitors and lectins in soya beans and effects of aqueous heat-treatment. *Journal of Science and Food Agriculture* 78:225-231.

Bardocz, S., G. Grant, S.W.B. Ewen, T.J. Duguid, D.S. Brown, K. Englyst, and A. Pusztai (1995). Reversible effect of phytohemagglutinin on the growth and metabolism of rat gastrointestinal tract. *Gut* 37:353-360.

Barnes, S. (1998). Evolution of the health benefits of soy isoflavoneṣ. *Proceedings of the Society for Experimental Biology and Medicine* 217:386-392.

Barratt, M.E., P.J. Strachan, and P. Porter (1978). Antibody mechanisms implicated in digestive disturbances following ingestion of soya protein in calves and piglets. *Clinical and Experimental Immunology* 31:305-312.

Bernard, R.L. and T. Hymowitz (1986a). Registration of L81-4590, L81-4871, and L83-4387 soybean germplasm lines lacking the kunitz trypsin inhibitor. *Crop Science* 26:650.

Bernard, R.L. and T. Hymowitz (1986b). Registration of L82-2024 and L82-2051 soybean germplasm lines with kunitz trypsin inhibitor variants. *Crop Science* 26:651.

Birk, Y. (1969). Saponins. In *Toxic Constituents of Plant Foodstuffs*, I.E. Liener (ed). New York, NY: Academic Press, pp 169-210.

Cheeke, P.R. (1996). Biological effects of feed and forage saponins and their impacts on animal production. In *Saponins Used in Food and Agriculture*, G.R. Wallen and K. Yamasaki (eds.). New York, NY: Plenum Press, pp. 377-385.

Chen, H.-M., K. Muramoto, and F. Yamauchi (1995). Structural analysis of antioxidative peptides from soybean ß-conglycinin. *Journal of Agriculture and Food Chemistry* 43:574-578.

Chen, J.-R., K. Suetsuna, and F. Yamauchi (1995). Isolation and characterization of immunostimulative peptides from soybean. *Nutritional Biochemistry* 6:310-313.

Cheryan, M. (1980). Phytic acid interactions in food systems. *CRC Critical Reviews in Food Science and Nutrition* 20:297-335.

de Jong, A., J. Plat, and R.P. Mensink (2003). Metabolic effects of plant sterols and stanols. *Journal of Nutritional Biochemistry* 14:362-269.

D'Mello, J.P.F. (1991). Antigenic proteins. In *Toxic Substances in Crop Plants*, J.P.F. D'Mello, C.M. Duffus, and J.H. Duffus (eds.). Cambridge, UK: The Royal Society of Chemistry, pp. 107-125.

Dudley-Cash, W.A. (January 4, 1999). Methods for determining quality of soybean meal protein important. *Feedstuffs* 71:10-11.

Emmert, J.L. and D.H. Baker (1995). Protein quality assessment of soy products. *Nutrition Research* 15:1647-1656.

Endres, J.G. (2001). *Soy Protein Products: Characteristics, Nutritional Aspects and Utilization, Revised and Expanded Edition*. Champaign, IL: AOCS Press, 53 pp.

Erdman, J.W., Jr. (1979). Oilseed phytates: Nutritional implications. *Journal of the American Oil Chemists' Society* 56:736-741.

Gee, J.M. and I.T. Johnson (1988). Interactions between hemolytic saponins, bile salts and small intestinal mucosa in the rat. *Journal of Nutrition* 118:1391-1397.

Griffin, M.E., M.R. White, and P.B. Brown (1994). Total sulfur amino acid requirement and cysteine replacement value for juvenile hybrid striped bass *(Morone saxatilis × M. chrysops)*. *Comparative Biochemistry and Physiology* 108A:423-429.

Hajos, G. and G. Gelencser (1995). Biological effects and survival of trypsin inhibitors and the agglutinin from soybeans in the small intestine of the rat. *Journal of Agricultural and Food Chemistry* 43:165-170.

Harding, D.E., O.W. Allen, Jr., and R.P. Wilson (1977). Sulfur amino acid requirement of channel catfish: L-methionine and L-cystine. *Journal of Nutrition* 107: 2031-2035.

Honig, D.H., M.E. Hockridge, R.M. Gould, and J.J. Rackis (1983). Determination of cyanide in soybeans and soybean products. *Journal of Agricultural and Food Chemistry* 31:272-275.

Hymowitz, T. (1985). Anti-nutritional factors in soybeans: Genetics and breeding. In *World Soybean Research Conference: Proceedings,* Richard Shibles (ed.). London, UK: Westview Press, pp. 368-373.

Ikenaka, T. and S. Odani (1977). Diversity of legume proteinase inhibitors. In *Proceedings of the Symposium on Evolution of Protein Molecules,* H. Matsubara and T. Yamanaka (eds.). Tokyo, Japan: Japan Scientific Societies Press, pp. 287-296.

Kinsgbury, J.M. (1980). Phytotoxicology. In *Toxicology: The Basic Science of Poisons,* J. Doull, C.D. Klaassen and M.O. Amdur (eds.). New York, NY: Macmillan Publishing Company, Inc., pp. 578-590.

Kumar, V., A. Rani, C. Tindwani, and M. Kain (2003). Lipoxygenase isozymes and trypsin inhibitor activities in soybean as influenced by growing location. *Food Chemistry* 83:79-83.

Lajalo, F.M. and M.I. Genovese (2002). Nutritional significance of lectins and enzyme inhibitors from legumes. *Journal of Agricultural and Food Chemistry* 50:6592-6598.

Laskowski, M. and I. Kato (1980). Protein inhibitors of proteinases. *Annual Review of Biochemistry* 49:593-626.

Liener, I.E. (1979). Significance for humans of biologically active factors in soybeans and other feed legumes. *Journal of the American Oil Chemists' Society* 56:121-129.

Liener, I.E. (1994). Implications of antinutritional components in soybean foods. *Critical Reviews in Food Science and Nutrition* 34:31-67.

Liener, I.E. (2002). A trail of research revisited. *Journal of Agricultural and Food Chemistry* 50:6580-6582.

Lim, C. and D.M. Akiyama (1992). Full-fat soybean meal utilization by fish. *Asian Fisheries Science* 5:181-197.

Magee, P.J. and I.R. Rowland (2004). Phyto-oestrogens, their mechanism of action: Current evidence for a role in breast and prostate cancer. *British Journal of Nutrition* 91:513-531.

Moon, H.Y. and D.M. Gatlin (1991). Total sulfur amino acid requirement of juvenile red drum, *Sciaenops ocellatus. Aquaculture* 95:97-106.

National Research Council (NRC) (1993). *Nutrient Requirements of Fish*. Washington, DC: National Academy Press, 114 pp.

National Research Council (NRC) (1994). *Nutrient Requirements of Poultry, NInth Revised Edition*. Washington, DC: National Academy Press, 155 pp.

National Research Council (NRC) (1998). *Nutrient Requirements of Swine, Tenth Rrevised Edition*. Washington, DC: National Academy Press, 188 pp.

Nielsen, N.C., C.D. Dickinson, T.-J. Cho, V.H. Thanh, B.J. Scallon, R.L. Fischer, T.L. Sims et al. (1989). Characterization of the glycinin gene family in soybean. *Plant Cell* 1:313-328.

Orf, J.H. and T. Hymowitz (1979). Genetics of the Kunitz trypsin inhibitor: An antinutritional factor in soybeans. *Journal of the American Oil Chemists' Society* 56:722-726.

Orf, J.H., T. Hymowitz, S.P. Pull, and S.C. Pueppke (1978). Inheritance of a soybean seed lectin. *Crop Science* 18:899-900.

Parsons, C.M., K. Hashimoto, K.J. Underwood, and D.H. Baker (1991). Soybean protein solubility in potassium hydroxide: An *in vitro* test of *in vivo* protein quality. *Journal of Animal Science* 69:2918-2924.

Plat, J. and R.P. Mensink (2000). Effects of diets enriched with two different plant sterol mixtures on plasma ubiquinol-10 and fat-soluble antioxidant concentrations. *Metabolism* 50:520-529.

Plumb, G.W., E.N. Clare Mills, M.J. Tatton, C.C.M. D'Ursel, N. Lambert, and M.R.A. Morgan (1994). Effect of thermal and proteolytic processing on glycinin, the 11S globulin of soy *(Glycine max):* A study utilizing monoclonal and polyclonal antibodies. *Journal of Agriculture and Food Chemistry* 42:834-840.

Pusztai, A., G. Grant, S. Bardocz, K. Baintner, E. Gelencser, and S.W.B. Ewen (1997). Both free and complexed trypsin inhibitors stimulate pancreatic secretion and change duodenal enzyme levels. *American Journal of Physiology* 35:G340-350.

Pusztai, A., W.B. Watt, and J.C. Stewart (1991). A comprehensive scheme for the isolation of trypsin inhibitors and the agglutinin from soybean seeds. *Journal of Agricultural and Food Chemistry* 39:862-866.

Rackis, J.J. (1972). Biologically active components. In *Soybeans: Chemistry and Technology, Volume 1, Proteins*, A.K. Smith and S.J. Circle (eds.). Westport, CT: AVI Publishing Company, pp. 158-202.

Russett, J.C. (2002). *Soy Protein Concentrate for Animal Feeds*. Specialty Products Research Notes SPC-T-47, Ft. Wayne, IN: Central Soya Company, Inc.

Sissons, J.W., A. Nyrup, P.J. Kilshaw, and R.H. Smith (1982). Ethanol denaturation of soya bean protein antigens. *Journal of Science for Food and Agriculture* 33:706-710.

Snyder, H.E. and T.W. Kwon (1987). *Soybean Utilization*. New York, NY: Van Nostrand Reinhold Publishing Co., 239 pp.

Sorgentini, D.A., J.R. Wagner, and M.C. Anon (1995). Effects of thermal treatment of soy protein isolate on the characteristics and structure-function relationship of soluble and insoluble fractions. *Journal of Agriculture and Food Chemistry* 43:2471-2479.

Twibell, R.G., K.A. Wilson, and P.B. Brown (2000). Dietary sulfur amino acid requirement of juvenile yellow perch fed the maximum cystine replacement value for methionine. *Journal of Nutrition* 130:612-616.

Yamauchi, F. and K. Suetsuna (1993). Immunological effects of dietary peptide derived from soybean protein. *Journal of Nutritional Biochemistry* 4:450-457.

Chapter 9

Utilization of Soy Products in Diets of Freshwater Fishes

Paul B. Brown

INTRODUCTION

Soybean seeds and resulting meals from those seeds have become an important source of crude protein and essential amino acids (EAA) in diets fed to freshwater fishes. There are numerous soybean feedstuffs available for use in diets and most have been evaluated. This chapter will focus on nutrient availability data for the various freshwater fishes, maximum levels of incorporation, supplementary nutrients needed in high soybean diets, and the known effects of antinutritional factors (ANF).

NUTRIENT AVAILABILITY

Macronutrients

It is not the intent of this chapter to critique methods used to determine nutrient digestibility values, but to simply report those that have been published. Undoubtedly, different methods contributed to some of the variation in values. Specific indicator of digestibility used in thev arious published studies was not reported as most used chromic oxide. Sintayehu et al. (1996) used a form of acid-insoluble ash as the indicator, but this was the only departure from the use of chromic

Alternative Protein Sources in Aquaculture Diets
© 2008 by The Haworth Press, Taylor & Francis Group. All rights reserved.
doi:10.1300/5892_09

oxide. Availability of dry matter, crude protein, nitrogen-free extract (NFE), and energy for freshwater fishes fed various soybean products is presented in Table 9.1. To date, dry matter digestibility has been evaluated in seventeen species, crude protein digestibility (CPD) of soybean products has been evaluated in twenty-seven species or hybrids raised in aquaculture, digestibility of NFE in nine, and energy digestibility has been evaluated in twenty-one species or hybrids. The number of species in which soybean have been evaluated is an indication of the importance of this ingredient in diets fed to freshwater fishes. Furthermore, many of those evaluations are relatively recent.

Dry matter digestibility values range from a low of 40 and 44 percent for the striped bass *(Morone saxatilis)* and hybrid striped bass *(M. saxatilis* × *M. chrysops),* respectively, fed soybean meal (SBM) to a high of 91 percent for striped bass fed soy protein isolate (SPI). Within each species, there is generally good agreement among values despite varying methodology and size of fish. The exception is with the striped bass, whose values vary from 40 to 72 percent despite similar methodology. Based on these data, it appears members of the genus *Morone* (striped bass and its hybrid) digest the dry matter from soybean products inefficiently compared to other species tested. There appears to be differences among species, but the current data set is relatively small to make many conclusions.

Crude protein digestibility values range from a low of 69 percent in the jelawat, or sultan fish *(Leptobarbus hoeveni)* fed SBM to a high of 100 percent in common carp *(Cyprinus carpio)* fed the low-protein SBM with hulls. As with the dry matter digestibility values, the within species variation is relatively low, although there appear to be differences among species. The values, as a group, are relatively high with most values over 80 percent and approximately one half of the values are greater than 90 percent. Thus, the CPD values indicate the crude protein fraction from soy products is highly available, despite the variability in species, soy product tested, methodology, and age and size of fish. The range of products tested include heated, full-fat soybeans, both low- and high-protein SBM, soybean hulls, soy flours, protein isolates, and expelled SBM.

Digestibility of NFE varies from a low of 51 percent in rohu *(Labeo rohita)* fed soybean hulls to 89 percent in Nile tilapia *(Oreochromis niloticus)* fed SBM. Most of these values are low relative to CPD and

TABLE 9.1. Apparent macronutrient digestibility (%) of soybean feedstuffs by freshwater fishes.

Species	Soybean product[a]	Size of fish (g)	Temperature (°C)	Fecal collection[b]	Formulation[c]	Digestibility	Reference
African catfish *C. isheriensis*	SBM	47.5-51.2	28	Dissection	Reference (70:30)	67	Fagbenro (1996)
Asian redtail catfish[d]	SBM	35.5	25-28.4	Post-excretion	Reference (70:30)	92	Khan (1994)
Australian shortfin eel	SBM	39.6	21	Post-excretion	Reference (70:30)	82	De Silva et al. (2000)
Black carp	SBM	5	25-27	Post-excretion	Reference (70:30)	79	You et al. (1993)
Brown discuss	SBM	3.5-5.0	N/R[e]	Post-excretion	Reference (70:30)	66	Chong et al. (2002)
Channel catfish	SBM	22	25	Dissection	Blend (45)	76	Peres et al. (2003)
Channel catfish	RSBM$_{ht}$	22	25	Dissection	Blend (45)	76	Peres et al. (2003)
Common carp	SBM$_{hu}$	45.8	20	Post-excretion	Blend (10-40)	79-84	Appleford and Anderson (1997)
Grass carp	SBM	15.3	N/R	Post-excretion	Reference (70:30)	82	Law (1986)
Hybrid striped bass	SBM	50	N/R	Stripping	Reference (70:30)	44	Sullivan and Reigh (1995)
Hybrid striped bass	SBM	50	28	Stripping	Reference (70:30)	51	Rawles and Gatlin (2000)

Dry matter

TABLE 9.1 *(continued)*

Species	Soybean product[a]	Size of fish (g)	Temperature (°C)	Fecal collection[b]	Formulation[c]	Digestibility	Reference
Indian carps							
Catla catla	Hulls	1.5–2.5	N/R	Post-excretion	Reference (70:30)	88	Erfanullah (1998)
Labeo rohita	Hulls	1.5–2.6	N/R	Post-excretion	Reference (70:30)	87	Erfanullah (1998)
Cirrhinus mrigala	Hulls	1.6–2.8	N/R	Post-excretion	Reference (70:30)	87	Erfanullah (1998)
Murray cod	SBM	100.4	21	Post-excretion	Reference (70:30)	80	De Silva et al. (2000)
Pacu	SBM	370	27	Post-excretion	Reference (70:30)	84	Fernandes et al. (2004)
Silver perch	SBM	N/R	23–28	Post-excretion	Reference (69.3/29.7)	75	Allan et al. (2000)
Silver perch	SBM$_{ex}$	N/R	23–28	Post-excretion	Reference (69.3/29.7)	84	Allan et al. (2000)
Silver perch	SBM$_{deff}$	N/R	23–28	Post-excretion	Reference (69.3/29.7)	77	Allan et al. (2000)
Striped bass	SBM	N/R	23	Stripping	Reference (60:40)	40	Papatryphon and Soares (2001)
Striped bass	SPI	N/R	23	Stripping	Reference (75:25)	91	Papatryphon and Soares (2001)
Striped bass	SBM	150	23	Stripping	Single (75.5)	72	Small et al. (1999)

Species	Ingredient		Crude protein	Method		ADC	Reference
Tilapia							
O. niloticus	SBM_hu	13	22-26	Dissection	Single (66.5)	80	Lorico-Querijero and Chiu (1989)
O. niloticus	SBM	93	26.5	Post-excretion	Reference (80:20)	84	Sintayehu et al. (1996)
O. niloticus	FFSB_ht	48-53	28	Dissection	Reference (70:30)	66	Fagbenro (1998a)
O. niloticus	SBM	25.2	26	Post-excretion	Reference (70:30)	89	Furuya et al. (2001)
African catfish							
C. gariepinus	SBM_ck	47.5-51.2	28	Dissection	Reference (70:30)	87	Fagbenro (1998b)
C. isheriensis	SBM	47.5-51.2	28	Dissection	Reference (70:30)	90	Fagbenro (1996)
Asian redtail catfish[d]	SBM	35.5	25-28.4	Post-excretion	Reference (70:30)	86	Khan (1994)
Australian shortfin eel	SBM	39.6	21	Post-excretion	Reference (70:30)	92	De Silva et al. (2000)
Ayu[f]	SBM_hu	1.4-46	20-25	Post-excretion	Reference (80:20)	85-89	Watanabe et al. (1996)
Black carp	SBM	5	25-27	Post-excretion	Reference (70:30)	97	You et al. (1993)
Brown discuss	SBM	3.5-5.0	N/R	Post-excretion	Reference (70:30)	83	Chong et al. (2002)
Channel catfish	SBM	N/R	N/R	Dissection	Single (99.5)	72	Hastings (1967)
Channel catfish	SBM	22	25	Dissection	Blend (45)	90	Peres et al. (2003)

229

TABLE 9.1 *(continued)*

Species	Soybean product[a]	Size of fish (g)	Temperature (°C)	Fecal collection[b]	Formulation[c]	Digestibility	Reference
Channel catfish	RSBM_ht	22	25	Dissection	Blend (45)	86	Peres et al. (2003)
Channel catfish	SBM	500-1,000	27	Dissection	Reference (70:30)	93-97	Wilson and Poe (1985)
Channel catfish	SBM	212	28	Catheterization	Single (35.7)	85	Brown et al. (1985)
Channel catfish	SBM	480-721	23-27	Dissection	Single (94)	77 (79)[g]	Cruz (1975)
Channel catfish	SBM	480-721	23-27	Catheterization	Single (78.5)	84 (85)	Cruz (1975)
Chinese long-snout catfish	SBM	15	27	Post-excretion	Reference (69.3:29.7)	76	Wu et al. (1996)
Common carp	SBM	120-130	25	Stripping	Single (20, 40)	83-86	Eid and Matty (1989)
Common carp	SBM	3.2	22	Post-excretion	Single (75)	91	Pongmaneerat and Watanabe (1993)
Common carp	SBM	3-295	N/R	N/R	N/R	94-96	Takeuchi et al. (2002)
Common carp	SBM	N/R	25	Catheterization	Blend (72.5)	70	Degani et al. 1997a
Common carp	SBM_hu	15.4	25	Post-excretion	Single (75)	90 (92)	Yamamoto et al. (1998)
Common carp	SBM_ex	15.4	25	Post-excretion	Single (73.9)	94 (96)	Yamamoto et al. (1998)

Species	Ingredient	Size	Temp	Method	Diet	Digestibility	Reference
Common carp	SBM$_{hu}$	45.8	20	Post-excretion	Blend (10-40)	98-100	Appleford and Anderson (1997)
Common carp	SBM$_{hu}$	3-295	15-25	Post-excretion	Reference (80:20)	94-96	Watanabe et al. (1996)
Grass carp	SBM	15.3	N/R	Post-excretion	Reference (70:30)	96	Law (1986)
Hybrid striped bass	SBM	50	N/R	Stripping	Reference (70:30)	80	Sullivan and Reigh (1995)
Hybrid striped bass	SBM	50	28	Stripping	Reference (70:30)	77	Rawles and Gatlin (2000)
Hybrid sturgeon	SBM	250-400	20	Stripping	Reference (50:50)	82	Degani (2002)
Indian carps							
Catla catla	Hulls	1.5-2.5	N/R	Post-excretion	Reference (70:30)	86	Erfanullah (1998)
Catla catla	SBM	0.8-1.2	27	Post-excretion	Reference (70:30)	97	Jafri and Anwar (1995)
Labeo rohita	Hulls	1.5-2.6	N/R	Post-excretion	Reference (70:30)	86	Erfanullah (1998)
Labeo rohita	FFSBM	5.2-6.5	26-28	Post-excretion	Single (60.65)	84 (89)	Hossain et al. (1997)
Labeo rohita	SBM	1.5-2.0	27	Post-excretion	Reference (70:30)	94	Jafri and Anwar (1995)
Cirrhinus mrigala	Hulls	1.6-2.8	N/R	Post-excretion	Reference (70:30)	86	Erfanullah (1998)
Cirrhinus mrigala	SBM	1.2-1.9	27	Post-excretion	Reference (70:30)	93	Jafri and Anwar (1995)
Jelawat[h]	SBM	12-26 cm	25-27	Post-excretion	Reference (70:30)	69	Law (1984)
Milkfish[i]	SBM	60-175	25.5-28.5	Dissection	Single (45)	70-90[h]	Ferraris et al. (1986)

TABLE 9.1 (continued)

Species	Soybean product[a]	Size of fish (g)	Temperature (°C)	Fecal collection[b]	Formulation[c]	Digestibility	Reference
Murray cod	SBM	100.4	21	Post-excretion	Reference (70:30)	94	De Silva et al. (2000)
Pacu	SBM	370	27	Post-excretion	Reference (70:30)	76	Fernandes et al. (2004)
Silver perch	SBM	N/R	23-28	Post-excretion	69.3/29.7	95	Allan et al. (2000)
Silver perch	SBM_{ex}	N/R	23-28	Post-excretion	69.3/29.7	96	Allan et al. (2000)
Silver perch	SBM_{deff}	N/R	23-28	Post-excretion	69.3/29.7	92	Allan et al. (2000)
Striped bass	SBM	N/R	23	Stripping	Reference (60:40)	84	Papatryphon and Soares (2001)
Striped bass	SPI	N/R	23	Stripping	Reference (75:25)	92	Papatryphon and Soares (2001)
Striped bass	SBM	150	23	Stripping	Single (75.5)	93	Small et al. (1999)
Tilapia							
O. mossambicus	FFSBM	6.2-6.8	N/R	Post-excretion	Single (60.65)	85 (88)	Hossain et al. (1992)
O. niloticus	SBM_{hu}	34	26	Post-excretion	Single (84)	91	Hanley (1987)
O. niloticus	Soy flour	22	28	Post-excretion	Blend (75:25)	75	Sadiku and Jauncey (1995)

Species	Ingredient	Value	Temp	Method	Design	Digestibility	Reference
O. niloticus	SBM$_{hu}$	13	22-26	Dissection	Single (66.5)	80 (93)	Lorico-Querijero and Chiu (1989)
O. niloticus	SBM	93	26.5	Post-excretion	Reference (80:20)	93	Sintayehu et al. (1996)
O. niloticus	SBM$_{hu}$	2.7-204	15-25	Post-excretion	Reference (80:20)	89-91	Watanabe et al. (1996)
O. niloticus	SBM	7.2	29.5-30.5	Post-excretion	Reference (70:30)	99	Wu et al. (2000)
O. niloticus	FFSB$_{ht}$	7.2	29.5-30.5	Post-excretion	Reference (70:30)	93	Wu et al. (2000)
O. niloticus	FFSB$_{ht}$	48-53	28	Dissection	Reference (70:30)	88	Fagbenro (1998)
O. niloticus	SBM	25.2	26	Post-excretion	Reference (70:30)	93	Furuya et al. (2001)
O. aureus	SBM	30-60	29	Post-excretion	Single (77.8)	94	Popma (1982)
O. aureus × *O. niloticus*	SBM	250-400	23	Stripping	Reference (52:48)	95	Degani et al. (1997b)
O. niloticus × *O. aureus*	SBM	4.5	26	Post-excretion	Blend (67:33)	82	Shiau et al. (1989)
Nitrogen-free extract[i]							
Black carp	SBM	5	25-27	Post-excretion	Reference (70:30)	67	You et al. (1993)
Common carp	SBM	N/R	25	Catheterization	Blend (72.5)	62	Degani et al. 1997[a]
Grass carp	SBM	15.3	N/R	Post-excretion	Reference (70:30)	63	Law (1986)
Indian carps							
Catla catla	Hulls	1.5-2.5	N/R	Post-excretion	Reference (70:30)	60	Erfanullah (1998)

TABLE 9.1 *(continued)*

Species	Soybean product[a]	Size of fish (g)	Temperature (°C)	Fecal collection[b]	Formulation[c]	Digestibility	Reference
Labeo rohita	Hulls	1.5-2.6	N/R	Post-excretion	Reference (70:30)	51	Erfanullah (1998)
Cirrhinus mrigala	Hulls	1.6-2.8	N/R	Post-excretion	Reference (70:30)	56	Erfanullah (1998)
Jelawat[g]	SBM	12-26 cm	25-27	Post-excretion	Reference (70:30)	74	Law (1984)
Tilapia							
O. aureus	SBM	30-60	29	Post-excretion	Single (77.8)	54[k]	Popma (1982)
O. niloticus	SBM	93	26.5	Post-excretion	Reference (80:20)	89	Sintayehu et al. (1996)
				Energy			
African catfish							
C. gariepinus	SBM$_{ck}$	47.5-51.2	28	Dissection	Reference (70:30)	76	Fagbenro (1998b)
C. isheriensis	SBM	47.5-51.2	28	Dissection	Reference (70:30)	85	Fagbenro (1996)
Asian redtail catfish[d]	SBM	35.5	25-28.4	Post-excretion	Reference (70:30)	68	Khan (1994)
Australian shortfin eel	SBM	39.6	21	Post-excretion	Reference (70:30)	56	De Silva et al. (2000)
Black carp	SBM	5	25-27	Post-excretion	Reference (70:30)	81	You et al. (1993)
Channel catfish	SBM	500-1,000	27	Dissection	Reference (70:30)	72-73	Wilson and Poe (1985)
Channel catfish	SBM	480-721	23-27	Dissection	Single (94)	56	Cruz (1975)
Channel catfish	SBM	480-721	23-27	Catheterization	Single (78.5)	57	Cruz (1975)

Species	Ingredient	Value	Temp	Method	Diet type	%	Reference
Chinese long-snout catfish	SBM	15	27	Post-excretion	Reference (69.3:29.7)	61	Wu et al. (1996)
Common carp	SBM	3.2	22	Post-excretion	Single (60)	85	Pongman-eerat and Watanabe (1993)
Common carp	SBM	3-295	N/R	N/R	N/R	76-80	Takeuchi et al. (2002)
Common carp	SBM	N/R	25	Catheterization	Blend (72.5)	75	Degani et al. (1997a)
Common carp	SBM_hu	45.8	20	Post-excretion	Blend (10-40)	80-87	Appleford and Anderson (1997)
Grass carp	SBM	15.3	N/R	Post-excretion	Reference (70:30)	83	Law (1986)
Hybrid striped bass	SBM	50	N/R	Stripping	Reference (70:30)	55	Sullivan and Reigh (1995)
Hybrid striped bass	SBM	50	28	Stripping	Reference (70:30)	56	Rawles and Gatlin (2000)
Hybrid sturgeon	SBM	250-400	20	Stripping	Reference (50:50)	88	Degani (2002)
Indian carps							
Catla catla	Hulls	1.5-2.5	N/R	Post-excretion	Reference (70:30)	74	Erfanullah (1998)
Labeo rohita	Hulls	1.5-2.6	N/R	Post-excretion	Reference (70:30)	70	Erfanullah (1998)
Cirrhinus mrigala	Hulls	1.6-2.8	N/R	Post-excretion	Reference (70:30)	72	Erfanullah (1998)
Jelawat[g]	SBM	12-26 cm	25-27	Post-excretion	Reference (70:30)	59	Law (1984)
Murray cod	SBM	100.4	21	Post-excretion	Reference (70:30)	58	De Silva et al. (2000)

TABLE 9.1 *(continued)*

Species	Soybean product[a]	Size of fish (g)	Temperature (°C)	Fecal collection[b]	Formulation[c]	Digestibility	Reference
Pacu	SBM	370	27	Post-excretion	Reference (70:30)	51[1]	Fernandes et al. (2004)
Silver perch	SBM	N/R	23-28	Post-excretion	Reference (69.3/29.7)	78	Allan et al. (2000)
Silver perch	SBM$_{ex}$	N/R	23-28	Post-excretion	Reference (69.3/29.7)	84	Allan et al. (2000)
Silver perch	SBM$_{deff}$	N/R	23-28	Post-excretion	Reference (69.3/29.7)	80	Allan et al. (2000)
Tilapia							
O. niloticus	SBM$_{hu}$	34	26	Post-excretion	Single (84)	57	Hanley (1987)
O. niloticus	SBM	93	26.5	Post-excretion	Reference (80:20)	77	Sintayehu et al. (1996)
O. niloticus	FFSB$_{ht}$	48-53	28	Dissection	Reference (70:30)	76	Fagbenro (1998)
O. niloticus	SBM	25.2	26	Post-excretion	Reference (70:30)	77	Furuya et al. (2001)
O. aureus	SBM	30-60	29	Post-excretion	Single (77.8)	72	Popma (1982)
O. aureus × *O. niloticus*	SBM	250-400	23	Stripping	Reference (52:48)	81	Degani et al. (1997)

[a]Soybean products are: SBM, solvent-extracted, dehulled, toasted meal; SBM$_{hu}$, solvent-extracted, toasted meal with hulls added back after toasting; SBM$_{ex}$, expelled SBM; SBM$_{deff}$, dehulled, full-fat meal; and, RSBM$_{ht}$, full-fat meal, heat treated.

[b]Fecal collection methods were: Dissection, collecting samples by excising the gastrointestinal tract and removing ingesta from the posterior portion of the tract; stripping, massaging the gastrointestinal tract to expel samples; post-excretion, allowing

defacation into water and collecting samples from the component that settles within the tank; and catheterization, inserting canula into rectum, applying negative pressure, and withdrawing samples.

[c]Formulation approaches used were: single (x), as the sole source of macronutrient tested and (concentration of soy in diet); reference (70:30), 30% substitution of test ingredient into a reference diet such that the final ratios are 70% reference and 30% test ingredient; and, blend (x:y), soybean product tested in a blend of other ingredients (soybean concentration:other ingredients).

[d]Asian redtail catfish, *Mystus nemurus*, family Bagridae, Order Siluriformes.

[e]Not reported.

[f]Ayu, *Plecoglossus altivelis*, family Plecoglossidae, Order Osmeridae.

[g]Parenthetical values are true digestibility.

[h]Jelawat, or sultan fish, *Leptobarbus hoeveni*, family Cyprinidae.

[i]Milkfish raised in freshwater.

[j]Nitrogen-free extract is largely the carbohydrate fraction of soy products and is calculated by difference (NFE = 100 – (Σ crude protein + fat + ash + fiber)).

[k]Starch concentrations in feed and feces used to calculate available NFE coefficients.

[l]Digestible energy value expressed as 2,382 kcal/kg. That value was expressed as a function of gross energy (4650 kcal/kg; NRC, 1977).

dry matter digestibility values. The NFE fraction is largely the soluble carbohydrate portion of the test feedstuff and these data, along with some of the dry matter digestibility values, seem to indicate the NFE fraction may not be as available as other components of soy products.

Apparent digestible energy (ADE) values range from a low of 55 to 57 percent for channel catfish *(Ictalurus punctatus),* hybrid striped bass, and Nile tilapia fed high- and low-protein SBM to a high of 85 to 87 percent in common carp fed the same products. Published ADE values tend to be lower than values for CPD, again emphasizing the difficulty freshwater fish have in digesting the carbohydrate fraction of SBMs.

Micronutrients

Availability of amino acids from soybean products fed to freshwater fishes is presented in Table 9.2. As with the macronutrient data in Table 9.1, values for fish fed soy products are relatively high. Most of the published apparent and true amino acid availability values are more than 90 percent. The limiting amino acids tend to be highly available to fish with the exception of cystine which, in all species tested to date, is the least available of all amino acids. There have been two published comparisons of apparent and true availability of amino acids and in both cases, corrections to true availability were in the range of 0 to 5 percent from apparent values.

Availability of phosphorus (P), magnesium, choline, and niacin from soybean products is presented in Table 9.3. Li and Robinson (1996) used weight gain and bone mineralization as indications of P availability, Zhang and Wilson (1999) used liver choline concentrations as indictors of choline availability, and Ng et al. (1998) used liver nicotine adenine dinucleotide as indicator of niacin status. All other data were determined by standard digestibility experiments. Availability of total P from SBM fed to channel catfish or striped bass varies from 29 to 59 percent. Buyukates et al. (2000) separated soluble reactive phosphorus (SRP) from total in feed and feces and reported complete availability of SRP in catfish fed SBM. Zhang and Wilson (1999) reported 73 percent available choline from SBM fed to catfish, but

TABLE 9.2. Availability of amino acids (%) from soybean products fed to freshwater fishes.

Species	Arg	Cys	His	Ile	Leu	Lys	Met	Phe	Thr	Trp	Tyr	Val	Reference
Australian shortfin eel[a]													
Apparent	92	N/R[b]	89	89	87	66	91	87	89	N/R	N/R	85	De Silva et al. (2000)
Channel catfish[c]													
Apparent	95	N/R	84	78	81	91	80	81	82	N/R	79	75	Wilson et al. (1981)
True[d]	97	N/R	88	80	83	94	85	84	82	N/R	83	78	Wilson et al. (1981)
Common carp													
Apparent[e]	96	88	92	90	92	93	91	93	87	91	92	90	Yamamoto et al. (1998)
True	96	89	93	92	93	94	93	94	90	N/R	93	91	Yamamoto et al. (1998)
Apparent[f]	98	92	95	94	95	95	95	97	93	93	95	94	Yamamoto et al. (1998)
True	98	93	96	96	97	97	97	98	96	N/R	97	96	Yamamoto et al. (1998)
Murray cod[g]													
Apparent	91	N/R	90	89	85	76	90	84	88	N/R	N/R	84	De Silva et al. (2000)
Silver perch													
Apparent[h]	98	94	97	95	95	97	96	96	95	N/R	96	95	Allan et al. (2000)
Apparent[i]	98	94	97	97	96	97	98	97	95	N/R	96	96	Allan et al. (2000)

TABLE 9.2 (continued)

Species	Arg	Cys	His	Ile	Leu	Lys	Met	Phe	Thr	Trp	Tyr	Val	Reference
Apparent[j]	95	96	94	92	92	94	96	93	94	N/R	94	91	Allan et al. (2000)
Striped bass[k]													
True	97	70	92	94	95	95	92	N/R	91	N/R	N/R	93	Small et al. (1999)
Tilapia													
Apparent[l]	99	87	98	94	96	96	95	96	97	N/R	94	94	Wu et al. (2000)
Apparent[m]	96	79	95	93	90	89	93	92	92	N/R	92	89	Wu et al. (2000)
Apparent[n]	93	N/R	92	91	95	93	93	91	93	67	N/R	96	Fagbenro (1998)

[a]Soybean meal evaluated contained 47.8% crude protein and 2.9% lipid (dry matter basis), fish weighed 39.6 g, water temperature was 21°C, chromic oxide was used as the indicator, fecal collection was post-excretion, and the test feedstuff was substituted into a reference diet (70:30, reference:test).

[b]Not reported.

[c]Solvent-extracted, toasted, dehulled SBM was used in the evaluation, fish weighed 500-1,000 g, test water temperature was not reported, chromic oxide was used as the indicator, fecal samples were collected by dissection, and the test ingredient was incorporated into the diet as the sole source of crude protein (incorporation level not provided).

[d]Apparent availability values do not account for endogenous losses of amino acids, true availability values account for endogenous losses by feeding an amino acid-free diet and quantifying excretion.

eSoybean meal used in this evaluation was not characterized, fish weighed 7.2 g, water temperature was 29.5-30.5°C, chromic oxide was used as the indicator, feces were collected post-excretion, and the test feedstuff was evaluated in a reference diet (70:30).

fSoybean used was described as "Soybean cooked." All other parameters of the evaluation were as described under note e.

gSoybean meal evaluated contained 47.8% crude protein and 2.9% lipid (dry matter basis), mean fish weight was 100.4 g, water temperature was 21°C, chromic oxide was used as the indicator, fecal collection was post-excretion and the test feedstuff was substituted into a reference diet (70:30, reference:test).

hSoybean meal evaluated contained 47.8% crude protein and 3.7% fat, fish weight was not reported, water temperature was 23.2-28.0°C, chromic oxide was used as the indicator, feces were collected post-excretion and the test ingredient was substituted into a reference diet (71.3:29.7, reference:test ingredient).

iWhole, expeller SBM evaluated contained 47.5% crude protein and 6.4% fat. All other conditions of the test were as described under note h.

jDehulled, full-fat SBM evaluated contained 35.8% crude protein and 19.5% fat. All other conditions of the test were as described under note h.

kSoybean meal used in this evaluation was not characterized, fish weighed 150 g, water temperature was 23°C, chromic oxide was used as the indicator, feces were collected by stripping, and the test ingredients was evaluated as the sole source of protein incorporated into the diet at 75.5%.

lSoybean meal was not characterized, mean fish weight was 7.2 g, water temperature was 29.5-30.5°C, chromic oxide was used as the indicator, feces were collected post-excretion, and the test ingredient was substituted into a reference diet (70:30, reference:test ingredient).

mTest ingredient was described as "Soybean cooked." All other conditions of the test were as described under note l.

nFull-fat, heated soybean was used in the test, fish weight was 48-53 g, water temperature was 28°C, chromic oxide was used as the indicator, feces were collected by dissection, and the test ingredients was substituted into a reference diet (70:30, reference:test ingredient).

241

TABLE 9.3. Mineral and vitamin availabilities (%) from soybean products fed to freshwater fishes.

Species	Soybean product[a]	Size of fish (g)	Temperature (°C)	Fecal collection[b]	Formulation[c]	Availability (%)	Reference
Phosphorus							
Black carp	SBM	5	25-27	Post-excretion	Reference (70:30)	64	You et al. (1993)
Channel catfish	SBM	280	26	Catheterization	Single (60)	54	Lovell (1978)
Channel catfish	SBM_hu	280	26	Catheterization	Single (60)	50	Lovell (1978)
Channel catfish	SBM	500-1000	27	Dissection	Single (61)	29	Wilson et al. (1982)
Channel catfish	SBM	7.8	30	N/A[d]	Single (59.3)	45-56	Li and Robinson (1996)
Channel catfish	SBM	150-200	28	Stripping	Reference (70:30)	35	Buyukates et al. (2000)
Channel catfish	SBM	150-200	28	Stripping	Reference (70:30)	100	Buyukates et al. (2000)
Grass carp	SBM	63-118	22-25	Post-excretion	Single (99)	67	Huang and Liu (1990)
Striped bass	SBM	N/R	23	Stripping	Reference (60:40)	59	Papatryphon and Soares (2001)

242

Striped bass	SPI	N/R	23	Stripping	Reference (75:25)	52	Papatryphon and Soares (2001)
Tilapia, O. niloticus	SBM	25.2	26	Post-excretion	Reference (70:30)	47	Furuya et al. (2001)
Magnesium							
Grass carp	SBM	63-118	22-25	Post-excretion	Single (99)	80	Huang and Liu (1990)
Choline							
Channel catfish	SBM	6.6	28	N/A	Single (7.55)	73	Zhang and Wilson (1999)
Niacin							
Channel catfish	SBM	10.9	28	N/A	Single (45.6)	57	Ng et al. (1998)

[a]Soybean products tested were solvent extracted, toasted dehulled SBM, SBM with hulls added back (SBM$_{hu}$), and SPI.

[b]Fecal collection methods were: dissection, collecting samples by excising the gastrointestinal tract, and removing ingesta from the posterior portion of the tract; stripping, massaging the gastrointestinal tract to expel samples; and catheterization, inserting a canula into rectum, applying negative pressure, and withdrawing samples.

[c]Formulations used for test diets were evaluation of single test feedstuff (single (level of soybean incorporation)), or substituted into a reference diet (percentage reference diet:percentage soybean feedstuff).

[d]Not applicable.

the level of SBM inclusion was 7.55 percent of the diet, well below current concentrations of SBM in diets fed to catfish.

UTILIZATION OF SOYBEAN FEEDSTUFFS

The first multi-ingredient, formulated, pelleted diets for fish contained animal by-product meals, most often fish meal, as the primary source of crude protein (Barrows and Hardy, 2001). The channel catfish industry in the United States is an example of dietary evolution away from fish meal–based diets to the use of alternative protein sources. Hastings and Dupree (1969) were apparently the first to conduct systematic evaluations of alternative protein sources in catfish with a focus on SBM. Working with a basal formulation containing 20 percent SBM and 12 percent fish meal, they removed all fish meal, increased SBM to 35 percent in test diets, and evaluated supplemental lysine, methionine, and dried fish solubles in practical diets. Supplementation with amino acids improved response of catfish grown in ponds. However, all values were significantly lower than in fish fed the basal diet. Andrews and Page (1974) replaced fish meal with SBM in diets fed to catfish and systematically added methionine and cystine as well as lipid and unidentified nonlipid components to the diet. Fish fed SBM exhibited significantly lower weight gain than fish fed fish meal. Supplemental EAA did not improve response parameters. In a series of studies, Dr. R.T. Lovell and students at Auburn University evaluated soy products as a replacement for fish meal in practical diets (Cruz, 1975; Saad, 1979; Leibowitz, 1981; Murray, 1982). Initial results indicated decreases in weight gain and increased feed conversion ratios as SBM inclusion increased and fish meal concentration decreased.

Research with catfish, and almost simultaneously with tilapia (Jackson et al., 1982; Viola and Arieli, 1983a,b; Viola et al., 1986; Viola, Arieli et al., 1988) and common carp (Viola et al., 1982; Kim and Oh, 1985; Viola, Zohar et al., 1988) indicated additional lipid and P were required in diets containing low levels of fish meal. In addition, supplemental amino acids (lysine and/or methionine) were needed in diets for carp (Viola et al., 1982), but not for hybrid tilapia reared in ponds (Viola et al., 1988). As these results became available and modified formulations tested, catfish formulations quickly changed, and solvent-extracted, dehulled, toasted SBM became the dominant source of crude protein in what became the catfish formulation, also known in some parts of the world as the carp or tilapia diet, or as the more general omnivorous or warmwater fish diet. Several factors contributed to this shift in ingredient use with catfish as well as the carp and tilapia.

As species specific aquaculture industries increase annual production, price to producers tends to decrease. As diet costs are one of the highest annual variable costs in most aquaculture facilities, less expensive diets become an important research topic. The need for lower diet costs in the catfish industry stimulated nutrition research. In addition, as the EAA requirements for channel catfish were quantified, use of ingredients to meet those requirements became possible. In 1977, the recommended maximum level of inclusion of SBM in diets for channel catfish was less than 20 percent (Robinette, 1977). At that time, only the dietary requirements for methionine and lysine had been quantified. By 1984, all ten EAA requirements had been quantified (Wilson, 1984) and a model catfish diet contained 37 to 48 percent SBM (Robinette, 1984). This line of research designed to diminish use of fish meal facilitated expansion of the catfish industry (Tucker, 2003; E.H. Robinson, Mississippi State University, personal communication), and also resulted in a fish diet formulation that was not predominantly fish meal.

As new aquaculture candidates are evaluated and as production in older industries intensifies with the use of formulated diets, the fundamental choice in diets are the salmonid diets that still contain relatively high level of fish meal or catfish diets that contain low levels of fish meal and high levels of SBM. As there is usually a significant difference in price, soy-based diets, formulated along the lines of catfish formulations, are often evaluated with new culture species (Hancz, 1993), replacement of fish meal by soy products is usually an early study in nutrition of the respective species (Xie et al., 1998), and those formulations have been the focus of efforts to intensify production in extensive aquaculture industries in Asia (Cremer et al., 2000, 2001). It is probably safe to speculate that SBM, along with fish meal, has been fed to virtually all species of freshwater fish raised in intensive commercial aquaculture. However, there are relatively few formal evaluations of soy products as replacement for fish meal.

Protein Quality

Soybean protein is considered a high-quality plant protein source in diets fed to freshwater fish, but contains lower concentrations of methionine, lysine, and threonine compared to fish meal. Thus, protein quality is generally thought to be inferior to fish meal. However, there are relatively few evaluations of protein quality in fishes. Stuart and Hung (1989) documented that a soybean protein concentrate (SPC) was of poorer quality than casein, defatted shrimp meal, and a defatted herring fish meal, but of better quality than egg white in diets fed to white sturgeon *(Acipenser transmontanus)*. There were no clear indications why the SPC was of poorer quality, but the implication is lower concentrations of EAA. Dabrowski and Kozak (1979) also evaluated single sources of protein (fish meal or SBM) and reported decreased responses in grass carp *(Ctenopharyngodon idellus)* fry even when the dietary crude protein concentrations were similar.

This classic approach to determination of protein quality (evaluating single protein feedstuffs in otherwise nutritionally complete diets) has not been routinely conducted with cultured fishes. A modified approach was used by Davis and Stickney (1978). Using fish meal, SBM, or a mixture of the two sources, they established diets containing graded levels of crude protein and reported no significant differences in growth of tilapia when the total dietary crude protein concentration was 36 percent. At lower dietary crude protein concentrations, significant differences existed between SBM and fish meal, suggesting inferior protein quality when using only SBM as the protein source. A similar finding was reported by Lovell et al. (1974) in one of the first formal attempts designed to develop a fish meal–free diet for channel catfish. They reported similar responses in catfish fed a standard fish meal–based diet compared to fish fed a fish meal–free diet, but only at higher protein concentrations. Thus, high dietary crude protein concentrations can alleviate an apparent deficiency in EAAs.

Contrary to these findings, Shiau et al. (1987) reported similar weight gains in hybrid tilapia fed a 70:30 mix of fish meal and SBM at 24 percent dietary crude protein and decreased weight gains in fish fed the same mixture at 32 percent dietary crude protein. Furthermore, there was a positive response to supplemental methionine in diets at 32 percent crude protein, but not in fish fed 24 percent crude protein.

Reasons for the differences in these studies are not clear. However, crude protein is relatively expensive, and the general desire is to incorporate a minimum concentration in practical diets. As the EAA requirements were quantified for freshwater fishes, those values were used to maintain some minimum concentrations in experimental diets. However, it is not clear if the EAA requirements for all freshwater fishes are the same.

Traditional SBM is probably not adequate as the sole source of protein and EAA in diets fed to fish. Complimentary proteins are needed to meet the EAA requirements. Several complimentary plant proteins have been evaluated and some new feedstuffs are available that have SBM as a constituent. Many of the diets for channel catfish are currently mixtures of SBM and cottonseed meal. Robinson and Daniels (1987) initially evaluated replacement of SBM with cottonseed meal and that line of research led to combinations of the two products that were cost-effective and promoted the same weight gains as fish fed SBM/fish meal combinations (Robinson and Brent, 1989; Robinson and Li, 1994). Webster et al. (1992a) evaluated a combination of SBM and distillers' grains in diets fed to channel catfish and reported no significant differences in response of catfish fed this combination. There was a small decrease in weight gain in fish fed 49 percent SBM and 35 percent distillers' grains, but that decrease was alleviated in fish fed a diet containing 48.5 percent SBM and 35 percent distillers' grains with supplemental lysine and methionine. Common carp fed a combination of SBM and corn gluten meal gained less weight than fish fed fish meal (Pongmaneerat et al., 1993). Weight gain was significantly improved with supplemental lysine, methionine, and threonine, but did not equal the response of fish fed the fish meal control diet. Fagbenro et al. (1994) developed a mixture of SBM and lactic acid fermented fish silage as a replacement for fish meal in diets fed to tilapia and African catfish *(Clarias gariepinus),* and reported no significant differences in response of either species. Fish meal concentrations were reduced from 40 and 53 percent to 10 and 13.5 percent, respectively.

Another method of combining ingredients is coextrusion prior to mixing into diets. This approach has potential with products that have a high moisture content. Robinson et al. (1985) were apparently the first to examine coextrusion of SBM with other ingredients. Channel catfish were fed an ingredient that contained SBM, full-fat soybeans, and hydrolyzed or nonhydrolyzed catfish offal. Weight gain of fish fed the various coextruded products was significantly higher than fish fed a standard SBM/fish meal diet. Carver et al. (1989) provided recommendations for manufacturing coextruded SBM and shrimp or squid waste products.

An alternative to complimentary proteins is supplementation of diets with feed-grade EAA or their chemical analogs, which have become readily available in the global marketplace. As soybean evaluations increased, supplementation with lysine, methionine, and threonine became commonplace. Initially, Andrews and Page (1974) could not demonstrate a beneficial effect of supplemental methionine, cystine, or lysine in diets fed to channel catfish. Later Murai et al. (1982) demonstrated a beneficial effect of various methionine products into catfish and common carp diets. Shun (1989) supplemented soy-based diets for common carp with methionine and reported no decreases in weight gain. Supplemental amino acids are now routinely used in a wide variety of diets.

Webster et al. (1992b) evaluated replacement of fish meal with SBM in diets fed to blue catfish *(Ictalurus furcatus)*. Beginning with a basal diet containing 48 percent SBM and 13 percent fish meal, they systematically lowered fish meal to 0 percent and increased SBM to 69 percent. There were no other complimentary protein sources or supplemental EAA in the diets. Chemical analyses of diets indicated they met or exceeded the EAA requirements for channel catfish, but there was a significant reduction in weight gain in fish fed any SBM concentration more than 48 percent. Thus, the authors speculated the methionine requirement, which appeared to be first limiting, may be higher in blue catfish than in channel catfish.

Brown et al. (1997) evaluated raw soybean seeds ground into a meal, roasted soybean seeds ground into a meal, and SBM in diets fed to hybrid striped bass. Any level of raw soybeans resulted in depressed responses of fish. Response of fish fed 20 percent roasted soybeans was not significantly depressed compared to fish fed 0 percent, but at the next levels of incorporation (40 percent), weight gain and feed efficiency was significantly reduced. Weight gain of fish fed 45 percent SBM was significantly lower then fish fed 30 percent or lower concentrations. When a complete mineral premix was incorporated into the diet, there were no significant decreases in response of fish fed 20, 25, 30, 35, or 40 percent SBM. Thus, it appears ANFs present in raw soybeans limit its use in diets fed to hybrid striped bass. When soybeans are roasted, some of the limitation is removed, most likely denatured, and SBM can be used at relatively high levels in the diet when the mineral concentrations are adequate. Gallagher (1994) also replaced fish meal in diets fed to hybrid striped bass and reported significant reductions in weight gain in fish fed 34 percent SBM.

Khan et al. (2003) conducted a series of studies with rohu designed to identify potential protein supplying ingredients followed by a systematic evaluation of SBM use as a replacement for fish meal. Experimental diets contained 14, 26, 37, 49, or 61 percent SBM, with fish meal concentrations decreasing from 50 to 0 percent. In diets containing higher levels of SBM (26 percent of the diet through 61 percent of the dry diet), supplemental methionine and lipid were added. There were no significant decreases in response of rohu fed up to 61 percent SBM.

Antinutritional Factors

Heat treatment of soybeans has been evaluated in several species, including African catfish (Balogun and Ologhobo, 1989; Sadiku and Jauncey, 1998), common carp (Viola et al., 1983; Abel et al., 1984), channel catfish (Wilson and Poe, 1985b; Peres et al., 2003), Indian carps (Garg et al., 2002; Maity and Patra, 2002), and tilapia (Wee and Shu, 1989). The intent is to decrease the concentration of heat labile ANF such as trypsin inhibitors (TI). Heated raw or processed soy products resulted in improved responses in most fish tested. Shiau et al. (1990) reported similar response in hybrid tilapia fed SBM or full-fat soybeans at 17 percent of the diet, but this appears to be the only report of no significant effect caused by full-fat soybeans. Both experimental diets also contained 25 percent fish meal. In the evaluations conducted to date, treatment temperatures and duration of heating varied, but most treatment conditions were within the realm of normal extrusion conditions. However, definitive studies have not been conducted. Soybeans contain several ANFs and several are heat labile. Although the correlations between improved response and decreased TI activity are beneficial, research with purified sources of ANF is the only means by which effects can be quantified.

Lopez et al. (1999), working with in vitro systems, documented that TI significantly decreased activity of digestive proteases. Furthermore, they reported species specific effects with the Nile tilapia being more sensitive than either the seabream *(Sparus aurata)* or African sole *(Solea senegalensis)*. Escaffre et al. (1997) are apparently the only group to have fed purified TI and documented decreased trypsin activity. Using an SPC, with and without supplemental methionine, increased TI in the diets of common carp larvae and resulted in decreased trypsin activity in whole larvae, but not a depression in growth. Thus, the authors concluded that other ANFs were contributing to the depressed weight gain in carp larvae seen in fish fed 60 and 70 percent SPC.

There have been several evaluations of saponins in diets fed to freshwater fishes. Hossain et al. (2001) fed an alcohol extract from sesbania *(Sesbania aculeate,* Leguminosae) to common carp. There was a significant decrease in weight gain in fish fed the alcohol extract at 2,400 mg/kg, but not at lower concentrations. Francis et al. (2001, 2002a,b) used a purified source of saponins from the soapbark tree *(Quillaja saponaria,* Rosaceae). They reported an improvement in weight gain in adult Nile tilapia fed 150 and 300 mg/kg diet, but a progressive inhibition of reproduction as dietary saponin concentrations increased (Francis et al., 2001). They also reported an improvement in weight gain in juvenile common carp fed 150 mg/kg saponins, but a decrease in fish fed 300 mg/kg diet (Francis et al., 2002a). Finally, they reported that continuous feeding of common carp juveniles resulted in better performance than fish fed the same levels (150 mg/kg diet) during alternate weeks (Francis et al., 2002b). Purified soy saponins have apparently not been evaluated in nonsalmonid freshwater fishes.

Phytic acid concentrations in SBM have the potential to impact availability of minerals as they do in terrestrial animals and the decreased mineral availability may lead to depressions in overall response of fish. Usmani and Jafri (2002) documented a reduction in weight gain and specific growth rate in mrigal *(Cirrhinus mrigala)* fed 1 percent phytic acid as well as significant changes in proximate composition of whole fish. Purified phytate sources fed to fishes resulted in significant reductions in bone zinc concentrations in channel catfish (Satoh et al., 1989), and bone and scale zinc concentrations in blue tilapia *(Oreochromis aureus)* (McClain et al., 1988), and weight gain, apparent crude protein digestibility, serum mineral concentrations, and liver and kidney mineral concentrations in common carp (Hossain and Jauncey, 1991). Concentrations of phytate eliciting negative responses were 1.0 to 2.2 percent in those studies.

Gatlin and Phillips (1989) reported a dietary zinc requirement of 200 mg/kg diet in the presence of phytic acid and relatively high levels of dietary calcium, while the dietary Zn requirement in catfish fed purified diets is 20 mg/kg diet (Gatlin and Wilson, 1983). More recently, Sa et al. (2004) reported a requirement of 44 to 79 mg zinc/kg diet for tilapia in practical diets containing 65 percent SBM. The cation deficiencies that develop when feeding SBM can be alleviated by simply increasing supplementation with higher levels of targeted minerals or by using certain forms of minerals that are not as susceptible to binding with phytic acid as others. For example, Paripatananont and Lovell (1995) reported that zinc-methionine was approximately four to five times more available than zinc sulfate in practical catfish diets.

The enzyme phytase can also diminish the negative effects of phytic acid in diets. Jackson et al. (1996) documented the beneficial effects of phytase fed to channel catfish on growth and bone mineral concentrations, then subsequently Li and Robinson (1997) reported that 250 units of phytase/kg diet improved P availability to the point no supplemental P was needed in the diet. Yan et al. (2002) analyzed contents from the gastrointestinal tract of catfish and reported the various partial breakdown products of phytate as dietary phytase increased. Phytase addition to practical diets containing 41 percent SBM improved weight gain, feed conversions, and tissue mineral concentrations in striped bass (Papatryphon et al., 1999).

CONCLUSIONS

Soy protein products have become an important ingredient in diets fed to many species of freshwater fish, even though formal evaluations are rare. This alternative protein source was one of the first fish meal replacements evaluated and remains a desirable ingredient for a variety of reasons. Use of soy products requires supplementation with other macro- and micronutrients, such as lipid, inorganic P and EAA in some species, but it is clear that soy has become the focus of research efforts in fish diets. There have been numerous publications on use of soy since the year 2000, and others are underway. The research studies conducted with channel catfish, tilapia, and common carp define the series of studies that should be conducted with new culture species and point us toward rapid evaluations of alternative protein sources. If these evaluations can be conducted with knowledge of the EAA requirements of the targeted species and knowledge of the optimum protein:nonprotein energy ratios (Twibell et al., 2003), the evaluations will proceed more efficiently than simply evaluating protein replacement. Rapid development of diets for new culture species should be of paramount importance for aquaculture in the twenty-first century.

REFERENCES

Abel, H.J., K. Becker, C. Meske, and W. Friedrich (1984). Possibilities of using heat-treated full-fat soybeans in carp feeding. *Aquaculture* 42:97-108.

Allan, G.L., S. Parkinson, M.A. Booth, D.A.J. Stone, S.J. Rowland, J. Frances, and R. Warner-Smith (2000). Replacement of fish meal in diets for Australian silver perch, *Bidyanus bidyanus:* I. Digestibility of alternative ingredients. *Aquaculture* 186:293-310.

Andrews, J.W. and J.W. Page (1974). Growth factors in the fish meal component of catfish diets. *Journal of Nutrition* 104:1091-1096.

Appleford, P. and T.A. Anderson (1997). Effect of inclusion level and time on apparent digestibility of solvent-extracted soybean meal for common carp (*Cyprinus carpio*, Cyprinidae). *Asian Fisheries Science* 10:65-74.

Balogun, A.M. and A.D. Ologhobo (1989). Growth performance and nutrient utilization of fingerling *Clarias gariepinus* (Burchell) fed raw and cooked soybean diets. *Aquaculture* 76:119-126.

Barrows, F.T. and R.W. Hardy (2001). Nutrition and feeding. In *Fish Hatchery Management,* second edition, G.A. Wedemeyer (ed.). Bethesda, MD: American Fisheries Society, pp. 483-558.

Brown, P.B., R.J. Strange, and K.R. Robbins (1985). Protein digestibility coefficients for yearling channel catfish fed various high-protein feedstuffs. *Progressive Fish-Culturist* 47:94-97.

Brown, P.B., R. Twibell, Y. Jonker, and K.A. Wilson (1997). Evaluation of three soybean products in diets fed to juvenile hybrid striped bass Morone Saxatillis × M. chrysops. *Journal of the World Aquaculture* 28:215-223.

Buyukates, Y., S.D. Rawles, and D.M. Gatlin, III (2000). Phosphorus fractions of various feedstuffs and apparent phosphorus availability to channel catfish. *North American Journal of Aquaculture* 62:184-188.

Carver, L.A., D.M. Akiyama, and W.G. Dominy (1989). Processing of wet shrimp heads and squid viscera with soy meal by a dry extrusion process. In *Proceedings of the People's Republic of China Aquaculture and Feed Workshop,* D.M. Akiyama (ed.). Singapore: American Soybean Association, pp. 416-428.

Chong, A.S.C., R. Hashim, and A.B. Ali (2002). Assessment of dry matter and protein digestibilities of selected raw ingredients by discuss fish *(Symphysodon aequifasciata)* using *in vivo* and *in vitro* methods. *Aquaculture Nutrition* 8:229-238.

Cremer, M.C., J. Zhang, and E. Zhou (2000). *Growth Performance of Grass Carp Fed a Low-Fat, High Fiber Feed Formulated with Soybean Meal and Soy Hulls as the Primary Protein and Fiber Sources.* Beijeng, China: American Soybean Association.

Cremer, M.C., J. Zhang, and E. Zhou (2001). *Mirror Carp Fry to Fingerling Growth Performance in Ponds in Harbin with Soymeal-Based Feeds.* Beijeng, China: American Soybean Association.

Cruz, E.M. (1975). Determination of nutrient digestibility in various classes of natural and purified feed materials for channel catfish. Ph.D. Dissertation, Auburn University, Auburn, AL, 82 pp.

Dabrowski, K. and B. Kozak (1979). The use of fish meal and soyabean meal as a protein source in the diet of grass carp fry. *Aquaculture* 18:107-114.

Davis, A.T. and R.R. Stickney (1978). Growth responses of *Tilapia aurea* to dietary protein quality and quantity. *Transactions of the American Fisheries Society* 107:479-483.

Degani, G. (2002). Availability of protein and energy from three protein sources in hybrid sturgeon *Acipenser guldenstadti* × *A. bester. Aquaculture Research* 33:725-727.

Degani, G., S. Viola, and Y. Yehuda (1997a). Apparent digestibility coefficient of protein sources for carp, *Cyprinus carpio L. Aquaculture Research* 28:23-28.

Degani, G., S. Viola, and Y. Yehuda (1997b). Apparent digestibility of protein and carbohydrate in feed ingredients for adult tilapia (*Oreochromis aureus* × *O. niloticus*). *Israeli Journal of Aquaculture—Bamidgeh* 49:115-123.

De Silva, S.S., R.M. Gunasekera, and G. Gooley (2000). Digestibility and amino acid availability of three protein-rich ingredient-incorporated diets by Murray cod *Maccullochella peelii peelii* (Mitchell) and the Australian shortfin eel *Anguilla australis* Richardson. *Aquaculture Research* 31:195-205.

Eid, A.E. and A.J. Matty (1989). A simple *in vitro* method for measuring protein digestibility. *Aquaculture* 79:111-119.

Erfanullah, and A.K. Jafri (1998). Evaluation of digestibility coefficients of some carbohydrate-rich feedstuffs for Indian major carp fingerlings. *Aquaculture Research* 29(7):511-519.

Escaffre, A.M., J.L.Z. Infante, C.L. Caho, M. Mabrini, P. Bergot, and S.J. Kaushik (1997). Nutritional value of soy protein concentrate for larvae of common carp

(Cyprinus carpio) based on growth performance and digestive enzyme activities. *Aquaculture* 153:63-80.

Fagbenro, O.A. (1996). Apparent digestibility of crude protein and gross energy in some plant and animal-based feedstuffs by *Clarias isheriensis* (Siluriforms: Clariidae) (Sydenham 1980). *Journal of Applied Ichthyology* 12:67-68.

Fagbenro, O.A. (1998a). Apparent digestibility of various legume seed meals in Nile tilapia diets. *Aquaculture International* 6:83-87.

Fagbenro, O.A. (1998b). Apparent digestibility of various oilseed cakes/meals in African catfish diets. *Aquaculture International* 6:317-322.

Fagbenro, O., K. Jauncey, and G. Haylor (1994). Nutritive value of diets containing dried lactic acid fermented fish silage and soybean meal for juvenile *Oreochromis niloticus* and *Clarias gariepinus*. *Aquatic Living Resources* 7:79-85.

Fernandes, J.B.K., R. Lochmann, and F.A. Bocanegra (2004). Apparent digestible energy and nutrient digestibility coefficients of diet ingredients for pacu *Piaractus brachypomus*. *Journal of the World Aquaculture Society* 35:237-244.

Ferraris, R.P., M.R. Catacutan, R.L. Mabelin, and A.P. Jazul (1986). Digestibility in milkfish, *Chanos chanos* (Forsskal): Effects of protein source, fish size, and salinity. *Aquaculture* 59:93-105.

Francis, G., H.P.S. Makkar, and K. Becker (2001). Effects of *Quillaja* saponins on growth, metabolism, egg production and muscle cholesterol in individually reared Nile tilapia *(Oreochromis niloticus)*. *Comparative Biochemistry and Physiology* 129C:105-114.

Francis, G., H.P.S. Makkar, and K. Becker (2002a). Dietary supplementation with a *Quillaja* saponin mixture improves growth performance and metabolic efficiency in common carp (*Cyprinus carpio* L.). *Aquaculture* 203:311-320.

Francis, G., H.P.S. Makkar, and K. Becker (2002b). Effects of cyclic and regular feeding of a *Quillaja* saponin supplemented diet on growth and metabolism of common carp (*Cyprinus carpio* L.). *Fish Physiology and Biochemistry* 24:343-350.

Furuya, W.M., L.E. Pezzato, E.C. de Miranda, V.R.B. Furuya, and M.M. Barros (2001). Apparent digestibility coefficients of energy and nutrients of some ingredients for Nile tilapia, *Oreochromis niloticus* (L.) (Thai strain). *Acta Scientiarum* 23:465-469.

Gallagher, M.L. (1994). The use of soybean meal as a replacement for fish meal in diets for hybrid striped bass *(Morone saxatilis* × *M. chrysops)*. *Aquaculture* 126:119-127.

Garg, S.K., A. Kalla, and A. Bhatnagar (2002). Evaluation of raw and hydrothermically processed leguminous seeds as supplementary feed for the growth of two Indian major carp species. *Aquaculture Research* 33:151-163.

Gatlin, D.M. III and H.F. Phillips (1989). Dietary calcium, phytate and zinc interactions in channel catfish. *Aquaculture* 79:259-266.

Gatlin, D.M. III and R.P. Wilson (1983). Dietary zinc requirement of fingerling channel catfish. *Journal of Nutrition* 114:630-635.

Hancz, C. (1993). Performance of the Amazonian tambaqui, *Colossoma macropomum*, in pond polyculture. *Aquacultural Engineering* 12:245-254.

Hanley, F. (1987). The digestibility of foodstuffs and the effects of feeding selectivity on digestibility determinations in tilapia, *Oreochromis niloticus* (L). *Aquaculture* 66:163-179.

Hastings, W.H. (1967). *Feeds. Progress in Sport Fishery Research 1966. Bureau of Sport Fisheries and Wildlife,* Resource Publication 39. Washington, DC: U.S. Government Printing Office, pp. 137-139.

Hastings, W. and H.K. Dupree (1969). Formula feeds for channel catfish. *Progressive Fish-Culturist* 31:187-196.

Hossain, M.A., U. Focken, and K. Becker (2001). The effect of purified alcohol extract from sesbania seeds on the growth and feed utilization in common carp, *Cyprinus carpio* L. *Journal of Aquaculture in the Tropics* 16:401-412.

Hossain, M.A. and K. Jauncey (1991). The effects of varying dietary phytic acid, calcium and magnesium levels on the nutrition of common carp, *Cyprinus carpio*. In *Fish Nutrition in Practice*, S.J. Kaushik and P. Luquet (eds.). Paris, France: Institut National de la Recherche Agronomique, Colloquim number 61, pp. 705-715

Hossain, M.A., N.M. Kamal, and M.N. Islam (1992). Nutrient digestibility coefficients of some plant and animal proteins for tilapia *(Orechromis mossambicus). Journal of Aquaculture in the Tropics* 7:257-266.

Hossain, M.A., N. Nahar, and M. Kamal (1997). Nutrient digestibility coefficients of some plant and animal proteins for rohu *(Labeo rohita). Aquaculture* 151:37-45.

Huang, Y. and Y. Liu (1990). Availabilities of Ca and P in nutritive salts and of Ca, P, Mg, Fe in feeds for grass carp *Ctenopharyngodon idellus. Acta Hydrobiologica Sinica* 14:145-152.

Jackson, A.J., B.S. Capper, and A.J. Matty (1982). Evaluation of some plant proteins in complete diets for the tilapia *Saratherodon mossambicus. Aquaculture* 27:97-109.

Jackson, S., M.H. Li, and E.H. Robinson (1996). Use of microbial phytase in channel catfish *Ictalurus punctatus* diets to improve utilization of phytate phosphorus. *Journal of the World Aquaculture Society* 27:309-313.

Jafri, A.K. and M.F. Anwar (1995). Protein digestibility of some low-cost feedstuffs in fingerling Indian major carps. *Asian Fisheries Science* 8:47-53.

Khan, M.A., A.K. Jafri, N.K. Chadra, and N. Usmani (2003). Growth and body composition of rohu *(Labeo rohita)* fed diets containing oilseed meals: Partial or total replacement of fish meal with soybean meal. *Aquaculture Nutrition* 9:391-396.

Khan, M.S. (1994). Apparent digestibility coefficients for common feed ingredients in formulated diets for tropical catfish, *Mystus nemurus* (Cuvier and Valenciennes). *Aquaculture and Fisheries Management* 25:167-174.

Kim, I.-B. and J.-K. Oh (1985). The effect of phosphorus supplementation to 40% soybean meal substituted diet for common carp. *Bulletin of the Korean Fisheries Society* 18:491-495.

Law, A.T. (1984). Nutritional study of jelawat, *Leptobarbus hoeveni* (Bleeker), fed on pelleted feed. *Aquaculture* 41:227-233.

Law, A.T. (1986). Digestibility of low-cost ingredients in pelleted feed by grass carp *(Ctenopharyngodon idella* C. et V.). *Aquaculture* 51:97-103.

Leibowitz, H.E. (1981). Replacing fish meal with soybean meal in practical catfish diets. M.S. Thesis, Auburn University, Auburn, AL, 56 pp.

Li, M.H. and E.H. Robinson (1996). Phosphorus availability of common feedstuffs to channel catfish *Ictalurus punctatus* as measured by weight gain and bone mineralization. *Journal of the World Aquaculture Society* 27:297-302.

Li, M.H. and E.H. Robinson (1997). Microbial phytase can replace inorganic phosphorus supplements in channel catfish *Ictalurus punctatus*. *Journal of the World Aquaculture Society* 28:402-406.

Lopez, F.J.M., I.M. Diaz, M.D. Lopez, and F.J.A. Lopez (1999). Inhibition of digestive proteases by vegetable meals in three fish species; seabream *(Sparus aurata)*, tilapia *(Oreochromis niloticus)* and African sole *(Solea senegalensis)*. *Comparative Biochemistry and Physiology* 122B:327-332.

Lorico-Querijero, B.V. and Y.N. Chiu (1989). Protein digestibility studies in *Oreochromis niloticus* using chromic oxide indicator. *Asian Fisheries Science* 2:177-191.

Lovell, R.T. (1978). Dietary phosphorus requirement of channel catfish *(Ictalurus punctatus)*. *Transactions of the American Fisheries Society* 107:617-621.

Lovell, R.T., E.E. Prather, J. Tres-Dick, and L. Chhorn (1974). Effects of addition of fish meal to all-plant feeds on the dietary protein needs of channel catfish in ponds. *28th Annual Conference of the Southeastern Association of Game and Fish Commissioners*, pp. 222-228.

Maity, J. and B.C. Patra (2002). Effect of feeding raw or heated soybean meal on the growth and survival of *Cirrhinus mrigala* Ham. fingerlings. *Applied Fisheries and Aquaculture* 2:19-24.

McClain, W.R. and D.M. Gatlin, III (1988). Dietary zinc requirement of *Oreochromis aureus* and effects of dietary calcium and phytate on zinc bioavailability. *Journal of the World Aquaculture Society* 19:103-108.

Murai, T., H. Ogata, and T. Nose (1982). Methionine coated with various materials supplemented to soybean meal diet for fingerling carp *Cyprinus carpio* and channel catfish *Ictalurus punctatus*. *Bulletin of the Japanese Society of Scientific Fisheries* 48:85-88.

Murray, M.G. (1982). Replacement of fish meal with soybean meal in diets fed to channel catfish in ponds. M.S. Thesis, Auburn, University, Auburn, AL, 40 pp.

National Research Council (NRC) (1977). *Nutrient requirements of warmwater fishes*. Washington, DC: National Academy Press, 78 pp.

Ng, W.K., C.N. Keembiyehetty, and R.P. Wilson (1998). Bioavailability of niacin from feed ingredients commonly used in feeds for channel catfish, *Ictalurus punctatus*. *Aquaculture* 161:391-402.

Papatryphon, E., R.A. Howell, and J.H. Soares, Jr. (1999). Growth and mineral absorption by striped bass *Morone saxatilis* fed a plant feedstuff based diet supplemented with phytase. *Journal of the World Aquaculture Society* 30:161-173.

Papatryphon, E. and J.H. Soares, Jr. (2001). The effect of phytase on apparent digestibility of four practical plant feedstuffs fed to striped bass, *Morone saxatilis*. *Aquaculture Nutrition* 7:161-167.

Paripatananont, T. and R.T. Lovell (1995). Chelated zinc reduces the dietary zinc requirement of channel catfish, *Ictalurus punctatus*. *Aquaculture* 133:73-82.

Peres, H., C. Lim, and P.H. Klesius (2003). Nutritional value of heat-treated soybean meal for channel catfish *(Ictalurus punctatus)*. *Aquaculture* 225:67-82.

Pongmaneerat, J. and T. Watanabe (1993). Nutritional evaluation of soybean meal for rainbow trout and carp. *Nippon Suisan Gakkaishi* 59:157-163.

Pongmaneerat, J., T. Watanabe, T. Takeuchi, and S. Satoh (1993). Use of different protein meals as partial or total substitution for fish meal in carp diets. *Nippon Suisan Gakkaishi* 59:1249-1257.

Popma, T.J. (1982). Digestibility of selected feedstuffs and naturally occurring algae by Tilapia. Ph.D. Dissertation, Auburn University, Auburn, AL, 78 pp.

Rawles, S.D. and D.M. Gatlin, III (2000). Nutrient digestibility of common feedstuffs in extruded diets for sunshine bass *Morone chrysops* E × *M. saxatilis* F. *Journal of the World Aquaculture Society* 31:570-579.

Robinette, H.R. (1977). Diet formulations. In *Nutrition and Feeding of Channel Catfish, Southern Cooperative Series Bulletin 218*, R.R. Stickney and R.T. Lovell (eds.). Auburn, AL: Alabama Agricultural Experiment Station, pp. 38-43.

Robinette, H.R. (1984). Feed formulation and processing. In *Nutrition and Feeding of Channel Catfish (Revised), Southern Cooperative Series Bulletin 296*, E.H. Robinson and R.T. Lovell (eds.). College Station, TX: Texas Agricultural Experiment Station, pp. 29-33.

Robinson, E.H. and J.R. Brent (1989). Use of cottonseed meal in channel catfish feeds. *Journal of the World Aquaculture Society* 20:250-255.

Robinson, E.H. and W.H. Daniels (1987). Substitution of soybean meal with cottonseed meal in pond feeds for channel catfish reared at low densities. *Journal of the World Aquaculture Society* 18:101-106.

Robinson, E.H. and M.H. Li (1994). Use of plant proteins in catfish feeds: replacement of soybean meal with cottonseed meal and replacements of fish meal with soybean meal and cottonseed meal. *Journal of the World Aquaculture Society* 25:271-276.

Robinson, E.H., J.K. Miller, V.M. Vergara, and G.A. Ducharme (1985). Evaluation of dry extrusion-cooked protein mixes as replacements for soybean meal and fish meal in catfish diets. *Progressive Fish-Culturist* 47:102-109.

Sa, M.V.C., L.E. Pezzato, M.M.B.F. Lima, and P.M. Padilha (2004). Optimum zinc supplementation level in Nile tilapia *Oreochromis niloticus* juveniles diets. *Aquaculture* 238:385-401.

Saad, C.R.B. (1979). Use of full-fat roasted soybeans in practical catfish diets. M.S. Thesis, Auburn University, Auburn, AL, 30 pp.

Sadiku, S.O.E. and K. Jauncey (1995). Digestibility, apparent amino acid availability and waste generation potential of soybean flour: Poultry meat meal blend based diets for tilapia, *Oreochromis niloticus* (L.), fingerlings. *Aquaculture Research* 26:651-657.

Sadiku, S.O.E. and K. Jauncey (1998). Utilisation of enriched soybean flour by *Clarias gariepinus*. *Journal of Aquaculture in the Tropics* 13:1-10.

Satoh, S., W.E. Poe, and R.P. Wilson (1989). Effect of supplemental phytate and/or tricalcium phosphate on weight gain, feed efficiency and zinc content in vertebrae of channel catfish. *Aquaculture* 80:155-161.

Shiau, S.-Y., J.-L. Chuang, and C.-L. Sun (1987). Inclusion of soybean meal in tilapia *(Oreochromis niloticus × O. aureus)* diets at two protein levels. *Aquaculture* 65:251-261.

Shiau, S.-Y., C.-C. Kwok, J.-Y. Hwang, C.-M. Chen, and S.-L. Lee (1989). Replacement of fishmeal with soybean meal in male tilapia *(Oreochromis niloticus × O. aureus)* fingerling diets at a suboptimal protein level. *Journal of the World Aquaculture Society* 20:230-235.

Shiau, S.-Y., S.-F. Lin, S.-L. Yu, A.-L. Lin, and C.-C. Kwok (1990). Defatted and full-fat soybean meal as partial replacements for fishmeal in tilapia *(Oreochromis niloticus × O. aureus)* diets at low protein level. *Aquaculture* 86:401-407.

Shun, T.J. (1989). The utilization of soybean as a main protein source for carp. In *Proceedings of the People's Republic of China Aquaculture and Feed Workshop,* D.M. Akiyama, (ed.). Singapore: American Soybean Association, pp. 133-142

Sintayehu, A., E. Mathies, K.-H. Meyer-Burgdirff, H. Rosenow, and K.-D. Gunther (1996). Apparent digestibilities and growth experiments with tilapia *(Oreochromis niloticus)* fed soybean meal, cottonseed meal and sunflower seed meal. *Journal of Applied Ichthyology* 12:125-130.

Small, B.C., R.E. Austic, and J.H. Soares, Jr. (1999). Amino acid availability of four practical feed ingredients fed to striped bass *Morone saxatilis. Journal of the World Aquaculture Society* 30:58-64.

Stuart, J.S. and S.S.O. Hung (1989). Growth of juvenile white sturgeon *(Acipenser transmontanus)* fed different proteins. *Aquaculture* 76:303-316.

Sullivan, J.A. and R.C. Reigh (1995). Apparent digestibility of selected feedstuffs in diets for hybrid striped bass *Morone saxatilis* E × *Morone chrysops* F. *Aquaculture* 138:313-322.

Takeuchi, T., S. Satoh, and V. Kiron (2002). Common carp, *Cyprinus carpio.* In *Nutrient Requirements and Feeding of Finfish for Aquaculture,* C.D. Webster and C. Lim (eds.). New York, NY: CABI Publishing, pp. 245-261

Tucker, C. (2003). Channel catfish. In *Aquaculture: Farming Aquatic Animals and Plants,* J.S. Lucas and P.C. Southgate (eds.). Oxford, United Kingdom: Blackwell Publishing, pp. 346-381

Twibell, R.G., M.E. Griffin, B. Martin, J. Price, and P.B. Brown (2003). Predicting dietary essential amino acid requirements for hybrid striped bass. *Aquaculture Nutrition* 9:373-382.

Usmani, N. and A.K. Jafri (2002). Influence of dietary phytic acid on the growth, conversion efficiency and carcass composition of mrigal *Cirrhinus mrigala* (Hamilton) fry. *Journal of the World Aquaculture Society* 33:199-204.

Viola, S. and Y. Arieli (1983a). Nutrition studies with tilapia *(Sarotherodon):* 1. Replacement of fishmeal by soybeanmeal in feeds for intensive tilapia culture. *Bamidgeh* 35:9-17.

Viola, S. and Y. Arieli (1983b). Nutrition studies with tilapia hybrids: 2. The effects of oil supplements to practical diets for intensive aquaculture. *Bamidgeh* 35:44-52.

Viola, S., Y. Arieli, and G. Zohar (1988). Animal-protein-free feeds for hybrid tilapia *(Oreochromis niloticus × O. aureus)* in intensive culture. *Aquaculture* 75:115-125.

Viola, S., S. Mokady, and Y. Arieli (1983). Effects of soybean processing methods on the growth of carp *(Cyprinus carpio). Aquaculture* 32:27-38.

Viola, S., S. Mokady, U. Rappaport, and U. Arieli (1982). Partial and complete replacement of fish meal by soybean meal in feeds for intensive culture of carp. *Aquaculture* 26:223-236.

Viola, S., G. Zohar, and Y. Arieli (1986). Phosphorus requirements and its availability from different sources for intensive pond culture species in Israel. *Bamidgeh* 38:3-12.

Viola, S., G. Zohar, and Y. Arieli (1988). Requirements of phosphorus and its availability from different sources for intensive pond culture species in Israel. Part II. Carp culture. *Bamidgeh* 40:44-54.

Watanabe, T., T. Takeuchi, S. Satoh, and V. Kiron (1996). Digestible crude protein in various feedstuffs determined with four freshwater fish species. *Fisheries Science* 62:278-282.

Webster, C.D., J.H. Tidwell, L.S. Goodgame, D.H. Yancey, and L. Mackey (1992a). Use of soybean meal and distillers grains with solubles as partial or total replacement of fish meal in diets for channel catfish, *Ictalurus punctatus. Aquaculture* 106:301-309.

Webster, C.D., D.H. Yancey, and J.H. Tidwell (1992b). Effect of partially or totally replacing fish meal with soybean meal on growth of blue catfish *(Ictalurus furcatus). Aquaculture* 103:141-152.

Wee, K.L. and S.-W. Shu (1989). The nutritive value of boiled full-fat soybean in pelleted feed for Nile tilapia. *Aquaculture* 81:303-314.

Wilson, R.P. (1984). Proteins and amino acids. In *Nutrition and Feeding of Channel Catfish (Revised), Southern Cooperative Series Bulletin 296,* E.H. Robinson and R.T. Lovell (eds.). College Station, TX: Texas Agricultural Experiment Station, pp. 5-11.

Wilson, R.P. and W.E. Poe (1985a). Apparent digestible protein and energy coefficients of common feed ingredients for channel catfish. *Progressive Fish-Culturist* 47:154-158.

Wilson, R.P. and W.E. Poe (1985b). Effects of feeding soybean meal with varying trypsin inhibitor activities on growth of fingerling channel catfish. *Aquaculture* 46:19-25.

Wilson, R.P., E.H. Robinson, D.M. Gatlin, III and W.E. Poe (1982). Dietary phosphorus requirement of channel catfish. *Journal of Nutrition* 112:1197-1202.

Wilson, R.P., E.H. Robinson, and W.E. Poe (1981). Apparent and true availability of amino acids from common feed ingredients for channel catfish. *Journal of Nutrition* 111:923-929.

Wu, J.-K., W.-Y. Yong, W.-Z. You, H. Wen, and C.-X. Liao (2000). Nutritional value of proteins in 13 feed ingredients for *Oreochromis niloticus. Journal of Fishery Sciences of China/ Zhongguo Shuichan Kexue* 7:37-42.

Wu, L., Y. Yunxia, and H. Xiqin (1996). Apparent digestibility coefficient of crude protein and energy for six commercial feed ingredients for Chinese long-snout catfish. *Acta Hydrobiologica Sinica* 20:113-118.

Xie, S., X. He, and Y. Yang (1998). Effects on growth and feed utilization of Chinese long-snout catfish *Leiocassis longirostris* Gunther of replacement of dietary fishmeal by soybean cake. *Aquaculture Nutrition* 4:187-192.

Yamamoto, T., A. Akimoto, S. Kishi, T. Unuma, and T. Akiyama (1998). Apparent and true availabilities of amino acids from several protein sources for fingerling rainbow trout, common carp and red sea bream. *Fisheries Science* 64:448-458.

Yan, W., R.C. Reigh, and Z. Xu (2002). Effects of fungal phytase on utilization of dietary protein and minerals, and dephosphorylation of phytic acid in the alimentary tract of channel catfish *Ictalurus punctatus* fed an all-plant protein diet. *Journal of the World Aquaculture Society* 33:10-22.

You, W., W. Yong, D. Wu, J. Wu, C. Liao, and H. Wen (1993). Evaluaiton of nutritive value of 11 feed ingredients for black carp *(Mylophayngodon piceus)*. *Freshwater Fisheries/Danshui yuye* 23:8-12.

Zhang, Z. and R.P. Wilson (1999). Reevaluation of the choline requirement of fingerling channel catfish *(Ictalurus punctatus)* and determination of the availability of choline from common feed ingredients. *Aquaculture* 180:89-98.

Chapter 10

Soybean Products in Salmonid Diets

INTRODUCTION

Salmonids are one of the most important groups of finfish farmed under controlled conditions and totally relying on formulated diets. The growth of the salmon industry across continents has seen a dramatic increase over the past decade, with global production close to 1.7 million tons. Atlantic salmon *(Salmo salar)* is the most widely farmed salmonid species. Production of rainbow trout *(Oncorhynchus mykiss)* amounts to about 500,000 tons and the other Pacific salmon to a much smaller extent (Table 10.1). To feed all these salmonids cultured globally, it would require at least 2 million tons of diets with an average protein content of 40 percent. Thus, 800,000 tons of protein would be needed to feed these fish.

The worldwide availability of soybean products in sufficient quantities makes them definitely a reliable and viable alternative to fish meal in salmonid diets. Current estimates are that globally, soybean meal (SBM) use in salmon diets is between 100,000 and 110,000 mt (Hardy, 2002). Although research on the use of soybean products as a protein source started when salmonid aquaculture was still developing, and soybean products were included in salmonid diets at varying levels, a number of questions remain. Given its lower protein content, SBM cannot be considered as a potent total substitute for fish meal. But, novel treated high-protein soy products definitely hold good potential.

Alternative Protein Sources in Aquaculture Diets
© 2008 by The Haworth Press, Taylor & Francis Group. All rights reserved.
doi:10.1300/5892_10   *261*

TABLE 10.1. Production of farmed salmonids (FAO, 2002).

Salmonid species	Production (tons)
Atlantic salmon *(Salmo salar)*	1,025,300
Rainbow trout *(Oncorhynchus mykiss)*	510,100
Coho salmon *(Oncorhynchus kisutch)*	151,400
Chinook salmon *(Oncorhynchus tshawytscha)*	23,300

NUTRIENT AVAILABILITY

Measurement of nutrient availability from ingredients requires a formulation of a basal diet and inclusion of varying proportions of this diet to the test ingredient (generally seven to three) and measure digestibility of the test and the basal diets (NRC, 1993). Such systematic studies on all soybean products with all salmonids have not been made. Very early studies by Cho et al. (1974) with rainbow trout or Lall and Bishop (1977) with the Atlantic salmon showed that apparent digestibility coefficient (ADC) of proteins from SBM (86 percent) was lower than that of casein (94 percent).

In the Atlantic salmon, most of the data available on digestibility coefficients relate to whole diets containing soybean by-products, but not on the digestibility of the different ingredients themselves (Olli et al., 1994; Refstie et al., 1999, 2001; Storebakken, Shearer, Baeverfjord, et al., 2000; Sajjadi and Carter, 2004). It is nevertheless clear from these studies that heat treatment of soybean reduces trypsin inhibitor activity and improves protein digestibility, and carbohydrate availability of soybean products in all salmonid species studied so far. Data on rainbow trout are presented in Table 10.2. Similar observations have also been made on Atlantic salmon (Refstie et al., 1998) or coho salmon (Arndt et al., 1999) with diets containing different soy by-products. In the case of chinook salmon, Hajen et al. (1993) attempted to measure ADC of soy products, but could not do so because of palatability problems.

Amino Acid Availability

Data on essential amino acid (EAA) requirements of salmonids along with data on amino acid composition of SBM are summarized in Table 10.3.

TABLE 10.2. Apparent digestibility coefficients (ADC), digestible protein (DP), and digestible energy (DE) of different soybean products in rainbow trout.[a]

Soy product	Dry matter	Protein	Energy	DP (% DM)	DE (kJ/g DM)
	ADC (%) of				
SBM	68.8 ± 7.1	92.8 ± 1.6	76.8 ± 5.1	46.5	13.5
FFSB[b] (1 ext)	82.1 ± 1.6	97.7 ± 0.7	85.1 ± 2.1	34.9	20.0
FFSB[b] (2 ext)	79.3 ± 1.0	97.2 +0.4	86.7 ± 1.7	35.3	19.0
SF[c]	73.5 ± 2.2	95.1 ± 1.2	80.7 ± 1.6	49.4	14.4
SPC[c]	74.3 ± 2.2	96.1 ± 1.9	83.3 ± 2.5	69.1	15.5

[a]From Kaushik et al. (1995).

[b]Full-fat SBM, single or twice extruded.

[c]From SOGIP, France.

TABLE 10.3. Essential amino acid requirements of salmonids, in comparison with average composition of SBM. Data are expressed as % dry matter and within brackets as % crude protein.

	Rainbow trout[a]	Pacific salmon[a]	Atlantic salmon[b]	Soybean meal[c]
Crude protein	40	40	45	48
Arginine	1.5 (3.8)	2.0 (5.0)	1.9-2.3 (4.2-5.1)	3.9 (8.1)
Histidine	0.7 (1.8)	0.6 (1.5)	0.9 (2.0)	1.4 (5.0)
Isoleucine	0.9 (2.3)	0.8 (2.0)	1.4 (3.1)	2.4 (5.0)
Leucine	1.4 (3.5)	1.3 (3.3)	2.3 (5.1)	4.2 (8.8)
Lysine	1.8 (4.5)	1.7 (4.3)	1.8-2.7 (4.0-6.0)	3.3 (6.9)
Methionine + Cystine	1.0 (2.5)	1.4 (3.5)	1.4 (3.1)	1.5 (3.1)
Phenyllanine + Tyrosine	1.8 (4.5)	1.7 (4.3)	2.6 (5.8)	4.8 (10.0)
Threonine	0.8 (2.0)	0.8 (2.0)	1.4 (3.1)	2.1 (4.4)
Tryptophan	0.2 (0.5)	0.2 (0.5)	–	0.8 (1.7)
Valine	1.2 (3.0)	1.1 (2.8)	1.8 (4.0)	2.5 (5.2)

[a]NRC (1993).

[b]Lall and Anderson (2005).

[c]NRC (1988).

Apparent and true availabilities of amino acids of SBMs (Table 10.4) are high, above 90 percent, in rainbow trout (Yamamoto et al., 1998), but significantly lower, 70 and 84 percent respectively for apparent and true availabilities, in Atlantic salmon (Anderson et al., 1992). It is, however, difficult to state whether such differences are due to differences between species, or due to inherent differences in the products tested, or to the methodologies used. It is interesting to note that extrusion processing improves amino acid availability of SBM

TABLE 10.4. Apparent amino acid availability (AAAA, %) and true amino acid availability (TAAA, %) from SBMs.

| Species product | Atlantic salmon[a] | | Rainbow trout[b] | | | |
| | SBM | | SBM | | Extruded SBM | |
	AAAA	TAAA	AAAA	TAAA	AAAA	TAAA
Protein	71.3	88.3	93.4	94.8	96.9	98.2
Arg	76.6	86.7	96.5	96.9	98.7	99.1
Cys	–	–	90.8	92.5	93.2	95.0
His	77.4	86.4	95.1	95.9	97.1	97.9
Ile	67.3	79.2	93.2	94.2	95.8	96.9
Leu	69.6	75.9	93.1	93.8	97.4	98.1
Lys	67.2	83.6	95.5	96.1	97.7	98.4
Met	70.6	94.0	95.9	96.7	97.5	98.4
Phe	70.1	78.7	94.2	94.8	98.1	98.7
Thr	62.0	84.5	94.2	95.8	96.5	98.2
Trp	45.4	50.3	94.0	–	94.9	–
Tyr	68.2	83.0	95.1	95.9	97.6	98.5
Val	66.5	77.3	93.9	94.8	97.3	98.2
Ala	74.1	87.0	94.3	95.2	97.1	98.1
Asp	67.1	80.8	94.4	95.1	97.1	98.8
Glu	74.8	81.3	95.9	96.2	98.2	98.5
Gly	63.3	94.6	93.8	94.8	96.1	97.1
Pro	77.4	90.4	94.0	94.9	96.8	97.8
Ser	72.9	87.8	93.5	94.5	96.6	97.8

[a]Anderson et al. (1992).

[b]Yamamoto et al. (1998).

(Table 10.4) as also shown by Cheng and Hardy (2003) with regard to full-fat SBM (Figure 10.1).

Macro Mineral and Trace Element Availability

Availability of phosphorus can vary depending on species and source (NRC, 1993; Riche and Brown, 1996). Compared to fish meal, mineral concentrations are low in soybean, except for iron, manganese, copper, and potassium. Soy by-products contain phosphorus (P) in the range of 0.6 to 0.9 percent dry matter, which is about three times lower than what is found in fish meal. This is an advantage in that reducing fish meal levels in a diet by incorporation of low P plant products can reduce P loads into the environment. But, the P in soy by-products is mainly in the phytic form, and since salmonids do not have phytase, the availability of phytic P from soybean products is

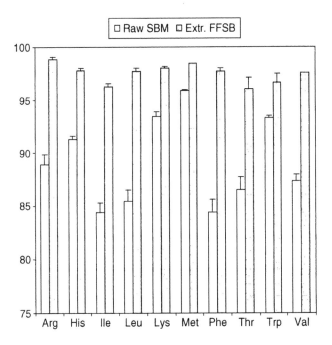

FIGURE 10.1. Apparent amino acid availabilities from full-fat soybean meal (FFSB) as affected by extrusion processes in the rainbow trout. *Source:* Adapted from Cheng and Hardy, 2003.

limited. Satoh et al. (2002) found that absorption of P also varies depending on body size of rainbow trout, but in general, the availability of P from extruded SBM was much higher (25 to 58 percent) than that from untreated SBM (3 to 37 percent).

Suguira et al. (1998) measured the nutrient availability of several ingredients in rainbow trout and coho salmon. Yamamoto et al. (1997) measured mineral availability in rainbow trout fed high levels of SBM with different mineral premixes. Data related to the availability of protein, some minerals, and trace-elements from SBM in these two species are presented in Table 10.5.

Addition of phytase to the diet, or pretreatment of soybean products with phytase improves the availability of phytic P (from 30 to 49 percent) in Atlantic salmon. Treatment with phytase improves not only the availability of phosphorus but also leads to a small, but significant, rise in the ADC of proteins in Atlantic salmon fed SPC-based diets (Storebakken et al., 1998). Furthermore, there was an increased availability of calcium, magnesium, and zinc in phytase-treated diets. Similar data were also obtained by Cheng and Hardy (2003), who showed that addition of phytase at 400 FTU/kg diet significantly increased the ADCs of magnesium, total-P, phytate-P, manganese, and zinc in rainbow trout. Even top-spraying diets with phytase led to increased availability of P, decrease of phytic P in the fecal matter in a significant manner and increase of the vertebral ash content in rain-

TABLE 10.5. Mineral availability from SBM in rainbow trout and coho salmon.

Element	Coho salmon	Rainbow trout	
Protein	93.0	90.1	94.0
Phosphorus	28.4	22.0	14-36
Magnesium	73.8	51.3	24-38
Potassium	97.6	98.9	–
Iron	11.4	–	19-40
Copper	96.4	84.3	74-83
Zinc			64-74
Manganese			15-21
Reference	Suguira et al. (1998)	Suguira et al. (1998)	Yamamoto et al. (1997)

bow trout (Vielma et al., 2000, 2002, 2004). Similarly, pretreatment of SBM with phytase (dephytinization) was found to improve P utilization by rainbow trout (Vielma et al., 2002). The question regarding whether the presence of phytic acid will affect protein digestibility or amino acid availability has also been raised. Addition of phytic acid in an otherwise complete diet appears to reduce protein digestibility, but it can be improved by addition of phytase (Sajjadi and Carter, 2004). Phytase, however, does not seem to improve lysine utilization (Vielma et al., 2004).

EFFECTS OF ANTINUTRITIONAL FACTORS FOUND IN SOYBEAN BY-PRODUCTS

All the potential antinutritional factors identified in soy by-products seem to exert their effects in salmonids. These include nonstarch polysaccharides (NSP), trypsin inhibitors, lectins or hemagglutinins, saponins, antigenic proteins, and isoflavones. Problems related to palatability of SBM have also been encountered, especially with chinook salmon (Hajen et al., 1993; Bureau et al., 1998).

Arnesen et al. (1989) observed that the poor utilization of nutrients by Atlantic salmon was probably related to the nature and level of carbohydrates supplied. Recently, Reftsie et al. (1998) also suggested that this could be linked to the level of NSP. However, compared to an SPC, SBM with reduced oligosaccharide contents still had lower digestibility. Detailed information on the possible implications of NSPs from soy by-products is needed. Storebakken et al. (1999) measured rate of passage of feedstuffs through the gastrointestinal tract in Atlantic salmon and suggested that transit rate could be slightly decreased with SBM in the diet, probably due to viscosity changes attributed to NSP content in different soybean products (Reftsie et al., 1999).

Salmonids appear more sensitive to trypsin inhibitors than other animals (Krogdahl and Holm, 1983). Trypsin inhibitors present in SBM, although inducing some hypersecretion of trypsin, bind with trypsin and chymotrypsin in the digestive tract, thus reducing protein digestibilty in most salmonids (Dabrowski et al., 1989; Olli et al., 1994; Haard et al., 1996). This sensitivity to trypsin inhibitors is also suggested to be size-dependent in rainbow trout, with smaller fish being more sensitive than larger fish (Murai et al., 1989).

Several studies in Norway have shown that the use of SBM in the diets for Atlantic salmon causes an inflammatory response in the distal intestine (Bakke-Mckellep et al., 2000; Refstie et al., 2000, 2001), classified as "a non-infectious subacute enteritis," which occurs within two days of feeding soy-based diets, but can possibly be restored by feeding a fish meal-based diet (Baeverfjord and Krogdahl, 1996). Besides changes in the morphology of epithelial cells, infiltration of inflammatory cells in the distal intestinal mucosa has been observed. These data suggest that toxic or antigenic components of SBM affect the differentiation of the distal intestinal epithelial cells and thus contribute to reduced nutrient digestibilities in Atlantic salmon fed SBM.

Lectins or hemagglutinins present in soybean products affect nutrient absorption through its effects on intestinal function. Buttle et al. (2001) found that fish fed a diet containing pure soybean agglutinin in an amount similar to the level possibly found in diets with high levels of soy products exhibited similar pathological disruptions of the digestive tract that have been observed in fish fed diets containing 60 percent dehulled solvent-extracted SBM. Through immunohistochemical assays, they found binding of soy agglutinin to the enterocytes lining the intestinal villi in both groups of fish fed diets containing either pure soy agglutinin or high levels of SBM.

Alcohol-soluble components in soybeans have been identified as being the cause for morphological changes encountered in the intestine in the Atlantic salmon, leading to associated disorders such as reduced fat digestibility and increased fecal water content (van den Ingh et al., 1996). Krogdahl et al. (2000, 2003) observed enteritis-like changes in the distal intestine in Atlantic salmon fed diets containing either solvent-extracted SBM or soybean molasses (alcohol extract of SBM) but not in fish fed SPC-based diets. Such changes in intestinal mucosa along with changes in some immune parameters led them to suggest that the alcohol extracts would induce decreased defence against pathogens.

Since saponins are alcohol-soluble compounds, Bureau et al. (1998) incorporated alcohol extracts from SBM or soy protein isolate into the diets of rainbow trout and chinook salmon. Based on their results on feed intake and growth of fish fed such purified alcohol extracts, they suggest that soy saponins have a feeding deterrent effect, especially in chinook salmon. This, consequently, also led to changes in intestinal morphology reflecting that from fish under an unfed state. Removal of such alcohol-soluble fractions indeed appears to be beneficial and to improve nutrient utilization in rainbow trout fed such a SPC (Kaushik et al., 1995).

As regards antigenic proteins such as glycinin or beta-conglycinin, leading to allergic reactions, Rumsey et al. (1994) fed rainbow trout with defined diets containing either SBM with high or low levels of antigenic proteins. Fish fed soybean-based diets had poorer growth than those fed fish meal-based diet, and inclusion of glycinin or beta-conglycinin led to significant growth reduction, accompanied by changes in intestinal morphology. Based on a number of serological and immunological parameters, they suggest that antigenic soy protein affects nonspecific defense mechanisms, but sensitivity can be overcome by removal of such antigens in soybean products. In a study with soy protein concentrate, Kaushik et al. (1995) could not detect any of these antigens and consequently no adverse biological effects in rainbow trout fed high levels of SPC.

In a prelilminary study with Siberian sturgeon *(Acipenser baeri)*, isoflavones or phytoestrogens such as daidzein and genistein were incriminated to induce changes in plasma vitellogenin levels (Pelissero et al., 1991). A subsequent study in vitro showed that such flavonoids affect the synthesis of estrogens in rainbow trout (Pelissero et al., 1993). High levels of purified genistein (500 or 1,000 ppm), when included in fish meal-based diets and fed to rainbow trout appeared to induce endocrine disruption, affecting gamete quality, although not in a dose-dependent manner (Bennetau-Pelissero et al., 2001). A later study of the effects of dietary genistein on trout and Siberian sturgeon in vivo showed that sturgeon was sensitive to 20 ppm of genistein, whereas trout was not (Latonnelle et al., 2002). Further knowledge along these lines is definitely needed, given that soybean products appear to have variable levels of such flavonoids (Kaushik et al., 1995; Mambrini et al., 1999; Bennetau-Pelissero et al., 2004).

The hypocholesterolemic effect of soybean as found in terrestrial animals has also been found in salmonids (Kaushik et al., 1995), although this is attributed more to the reduction in fish meal than to a specific effect of SBM, since dietary casein also led to such an effect on plasma cholesterol levels in rainbow trout. Besides, this hypocholeteromic effect of soybean is also linked with other factors such as the iron content of soybean (Mitrenko, 1997).

GROWTH AND NUTRIENT UTILIZATION OF SOY PROTEIN BY SALMONIDS

High-protein soy by-products have been found to hold potential as total fish meal substitutes in the diets for rainbow trout. Available data, however, show that with both Atlantic and Pacific salmons, high levels of substitution of fish meal, even by protein-rich soybean products, have not met with similar success.

Rainbow Trout (Oncorhynchus mykiss)

There is vast amount of literature on the use of soybean products in the diets of rainbow trout. Cho et al. (1974) showed that significant amounts of herring meal could be replaced by SBM in the diets of juvenile rainbow trout. Smith (1977) also indicated the potential of properly heat-treated full-fat SBM in the diets of rainbow trout. Tacon et al. (1983) tested different commercial soybean meals (puffed full-fat, toasted full-fat, hexane extracted, extruded hexane extracted or alcohol extracted) replacing up to 75 percent of the Peruvian fish meal protein with no adverse effects on overall growth performance or feed utilization efficiency. Oliva-Teles et al. (1994) also showed that different processing techniques affect the utilization of soybean products by rainbow trout.

Refstie et al. (2000) found that incorporation of 30 percent of SBM did not affect growth or nutrient utilization in rainbow trout. Comparing soy flour (SF) with SPC, Kaushik et al. (1995) demonstrated that an alcohol-water-extracted SPC, with no detectable levels of any antinutrients, could be used as a total substitute for fish meal, without affecting growth, nitrogen utilization, flesh quality, or vitellogenesis. But, the variability in the quality of such products warrants some caution, since in a subsequent study with another SPC, Mambrini et al. (1998) could not achieve similar success. The SPC used had low antitryptic and antigenic activities, but contained high concentrations of isoflavones. When more than 50 percent of the dietary protein was of soy origin, growth rate was reduced, attributed to a reduction in lipid digestibility. DL-methionine supplementation partially reversed the depressive effects of high levels of dietary SPC and also had beneficial effects in terms of feed intake and decreased N-excretion (Medale et al., 1998).

Atlantic Salmon (Salmo salar)

The early studies of Arnesen et al. (1989) suggested that carbohydrates in soybeans were the main reason behind poor utilization of nutrients by Atlantic salmon. Alcohol-soluble components in soybeans have been identified as the cause for morphological changes often encountered in the intestine of Atlantic salmon (van den Ingh et al., 1996) possibly leading to some associated disorders such as reduced fat digestibility and increased fecal water content. Technological treatments do have beneficial effects in improving the utilization of soybean products by Atlantic salmon. Compared with crude SBM, SBM with a reduced content of different antinutritional factors such as trypsin inhibitors, oligosaccharides, lectins, and soya antigens leads to improved digestibility of protein, fat, and energy, permitting replacement of upto 50 percent of fish meal protein in the diets of Atlantic salmon (Refstie et al., 1998).

Carter and Hauler (2000) did not find any difference in weight gain or feed conversion in Atlantic salmon fed either a fish meal-based commercial diet or test diets containing SBM as a replacement of up to 33 percent of fish meal. In bigger salmon (900 g), a diet in which SBM replaced 15 percent of the fish meal crude protein did not lead to any difference in growth (Storebakken, Shearer, and Roem, 2000), although fish fed SBM-based diets had intestinal mucosal disorders. Refstie et al. (2001) have found that SPC can be incorporated upto a level of 30 percent of fish meal protein replacement.

Pacific Salmons

Early studies with chinook *(Oncorhynchus tshawytscha)* or coho salmon *(Oncorhynchus kisutch)* fed diets with heat-treated full-fat SBM even at levels below 13 percent were not very promising (Fowler, 1980). From the data of Arndt et al. (1999), it appears that reduced feed intake associated with unheated SF was responsible for approximately one-half of the observed reduction in weight gain in coho salmon. They further suggested that the lower nutritional value of the unheated or insufficiently heated SF was responsible for the reduction in growth performance. A combination of proper heat treatment and supplementation with palatability-enhancing dietary ingredients can overcome these problems, even in young Pacific salmon, which are known to be extremely sensitive to SBM in their diets.

Based on an extensive survey of literature data on the utilization of different soybean by-products in the four major species of salmonids where growth performance and nutrient utilization have been comparable to those obtained with diets containing fish meal as the only or at least the major protein source, summary information is provided in Table 10.6.

UTILIZATION OF SOYBEAN OIL BY SALMONIDS

Given the fatty acid composition of soybean oil, rich in n-6 fatty acids, and relatively rich in 18:3 n-3 fatty acid, use of soybean oil in salmonid diets, especially in the diets of rainbow trout has been tested (Cho et al., 1974; Reinitz and Yu, 1981). In this context, the use of full-fat SBM was considered as interesting in supplying digestible energy in an easily utilizable form (Reinitz et al., 1978). Greene and Selivonchick (1990) also did not find any difference in growth or feed efficiency in rainbow trout fed soybean oil-based commercial moist diets, although flesh fatty acid profiles were modified. Hardy et al. (1987) fed Atlantic salmon moist pellets in which herring oil was

TABLE 10.6. Summary information on maximum levels of replacement of fish meal proteins by different soybean products in salmonids (data as % of protein replacement).

	Rainbow trout	Atlantic salmon	Coho salmon	Chinook salmon
Full-fat SBM	25	25		
SBM	40	30	<15	<50
Extruded SBM	50	30	15	50
SF	50	30		
SPC	75-100	>50	?	50
Soy protein isolate	<50	40		50

partially replaced with soybean oil and found that the dietary lipid source did not affect growth or proximate composition of the fillets. They found that although fatty acid composition of the fillets reflected the fatty acid composition of the diets, there was no effect on sensory attributes. Grisdale-Helland et al. (2002) demonstrated that Atlantic salmon can grow well and with high efficiency on high-energy, fish meal-based diets containing up to 100 percent supplementary soybean oil with no detrimental effects on the health of the fish.

Regost (2001) showed that feeding brown trout *(Salmo trutta)* soybean oil instead of fish oil, did not affect growth or feed efficiency, but influenced flesh fatty acid profile. It was also reported that changes in flesh fatty acid composition can be brought to a normal level by feeding the same fish with a fish oil-based "finishing" diet. Although it is generally admitted that the fatty acid content of fish reflects the fatty acid composition of the diet, incorporation of fatty acids into tissues is modulated by various metabolic factors (capacity for elongation/desaturation of 18 carbon fatty acids), the final fatty acid profile being dependent upon the initial total fatty acid content, cumulative intake of dietary fatty acids, growth rate, and duration of feeding (Robin et al., 2003). Such a capacity for tailoring salmonid flesh fatty acid composition is indeed a very interesting means of maintaining the supply of beneficial eicosapentanoic acid (EPA; 20:5 n-3) and docasohexanoic acid (DHA; 22:6 n-3) from farmed fish to human beings.

CONCLUSION AND RECOMMENDATIONS

Salmonids appear to be sensitive to several of the antinutritional factors present in soybean products. There are interspecies differences, but the level of available information is not of equal value in all species to undertake reliable comparisons. High levels of dietary soy products have been related to a number of adverse effects: poor feed intake; decreased nitrogen, phosphorus, or trace element availability; digestive and immune function disorders; antigenic or estrogenic response; and poor nutrient utilization affecting overall performance. At the metabolic level, soybean products appear to induce changes in lipid metabolism, where further insight on the roles of specific antinutritional factors (ANFs) is needed. While the EAA profile can be reasonably well adjusted through adequate supplementation or through judicious mixture of other ingredients, the major problem often encountered is related to one or more ANFs.

There is as yet no clear information as regards which particular ANF specifically exerts its effects on a particular physiological function, affecting overall growth performance and nutrient utilization. In fact, it is often found that specific purified ANFs do not have the same degree of adverse effects as seen with soy products containing the same, showing that a combination of factors induces such effects. This, combined with poorly defined quality criteria for choice of raw materials, often makes it difficult to make progress toward the use of soybean products on a larger scale than is practiced today. Given the urgent need for research on alternatives to fish meals, especially for the salmonids, standardized information based on soybean products, including the technological treatments involved, is warranted. Thus, full advantage of the availability of soybean products in the global market may be taken.

REFERENCES

Anderson, J.S., S.P. Lall, D.M. Anderson, and J. Chandrasoma (1992). Apparent and true availability of amino acids from common feed ingredients for Atlantic salmon *(Salmo salar)* reared in sea water. *Aquaculture* 108(1-2):111-124.

Arndt, R.E., R.W. Hardy, S.H. Sugiura, and F.M. Dong (1999). Effects of heat treatment and substitution level on palatability and nutritional value of soy defatted flour in feeds for coho salmon, *Oncorhynchus kisutch. Aquaculture* 180(1-2): 129-145.

Arnesen, P., L.E. Brattas, J. Olli, and A. Krogdahl (1989). Soybean carbohydrates appear to restrict the utilization of nutrients by Atlantic salmon (*Salmo salar* L.). In *The Current Status of Fish Nutrition in Aquaculture. Proc. III Int. Symp.* Feeding and Nutrition in Fish, M. Takeda and T. Watanabe (eds.). Tokyo University Fisheries, Tokyo, Japan, pp. 273-280.

Baeverfjord, G. and A. Krogdahl (1996). Development and regression of soybean meal induced enteritis in Atlantic salmon, *Salmo salar* L., distal intestine: A comparison with the intestines of fasted fish. *Journal of Fish Diseases* 19(5): 375-387.

Bakke-McKellep, A.M., C.M. Press, G. Baeverfjord, A. Krogdahl, and T. Landsverk (2000). Changes in immune and enzyme histochemical phenotypes of cells in the intestinal mucosa of Atlantic salmon, *Salmo salar* L., with soybean meal-induced enteritis. *Journal of Fish Diseases* 23(2):115-127.

Bennetau-Pelissero, C., B. Breton, B. Bennetau, G. Corraze, F. Le Menn, B. Davail-Cuisset, C. Helou, and S.J. Kaushik (2001). Effect of genistein-enriched diets on the endocrine process of gametogenesis and on reproduction efficiency of the rainbow trout *Oncorhynchus mykiss. General and Comparative Endocrinology* 121(2):173-187.

Bennetau-Pelissero, C., K.G. Latonnelle, V. Lamothe, S. Shinkaruk-Poix, and S.J. Kaushik (2004). Screening for oestrogenic activity of plant and food extracts using *in vitro* trout hepatocyte cultures. *Phytochemical Analysis* 15:40-45.

Bureau, D.P., A.M. Harris, and C. Young Cho (1998). The effects of purified alcohol extracts from soy products on feed intake and growth of chinook salmon *(Oncorhynchus tshawytscha)* and rainbow trout *(Oncorhynchus mykiss). Aquaculture* 161(1-4):27-43.

Buttle, L.G., A.C. Burrells, J.E. Good, P.D. Williams, P.J. Southgate, and C. Burrells (2001). The binding of soybean agglutinin (SBA) to the intestinal epithelium of Atlantic salmon, *Salmo salar* and rainbow trout, *Oncorhynchus mykiss,* fed high levels of soybean meal. *Veterinary Immunology and Immunopathology* 80(3-4): 237-244.

Carter, C.G. and R.C. Hauler (2000). Fish meal replacement by plant meals in extruded feeds for Atlantic salmon, *Salmo salar* L. *Aquaculture* 185(3/4):299-311.

Cheng, Z.J. and R.W. Hardy (2003). Effects of extrusion and expelling processing, and microbial phytase supplementation on apparent digestibility coefficients of nutrients in full-fat soybeans for rainbow trout *(Oncorhynchus mykiss). Aquaculture* 218(1-4):501-514.

Cho, C.Y., H.S. Bayley, and S.J. Slinger (1974). Partial replacement of herring meal with soybean meal and other changes in a diet for rainbow trout *(Salmo gairdneri). Journal of the Fisheries Research Board of Canada* 31(9):1523-1528.

Dabrowski, K., P. Poczyczynski, G. Kock, and B. Berger (1989). Effect of partially or totally replacing fish meal protein by soybean meal protein on growth, food utilization and proteolytic enzyme activities in rainbow trout *(Salmo gairdneri).* New *in vivo* test for exocrine pancreatic secretion. *Aquaculture* 77(1):29-49.

FAO (2002). The State of World Fisheries and Aquaculture 2002. Food and Agriculture Organisation, Rome, Italy, 168 pp.

Fowler, L.G. (1980). Substitution of soybean and cottonseed products for fish meal in diets fed to chinook and coho salmon. *Progressive Fish Culturist* 42(2):87-91.

Greene, D.H.S. and D.P. Selivonchick (1990). Effects of dietary vegetable, animal and marine lipids on muscle lipid and hematology of rainbow trout *(Oncorhynchus mykiss). Aquaculture* 89:165-182.

Grisdale-Helland, B., B. Ruyter, G. Rosenlund, A. Obach, S.J. Helland, M.G. Sandberg, H. Standal et al. (2002). Influence of high contents of dietary soybean oil on growth, feed utilization, tissue fatty acid composition, heart histology and standard oxygen consumption of Atlantic salmon *(Salmo salar)* raised at two temperatures. *Aquaculture* 207:311-329.

Haard, N.F., L.E. Dimes, R.E. Arndt, and F.M. Dong (1996). Estimation of protein digestibility 4. Digestive proteinases from the pyloric caeca of coho salmon *(Oncorhynchus kisutch)* fed diets containing soybean meal. *Comparative Biochemistry and Physiology B-Biochemistry and Molecular Biology* 115(4):533-540.

Hajen, W.E., D.A. Higgs, R.M. Beames, and B.S. Dosanjh (1993). Digestibility of various feedstuffs by post-juvenile chinook salmon *(Oncorhynchus tshawytscha)* in sea water 2. Measurement of digestibility. *Aquaculture* 112(4):333-348.

Hardy, R.W. (2002). Use of soybean meals in diets of salmon and trout. Tech. Review, American Soybean Assn, 14 pp. http://www.asa-europe.org/pdf/salmon.pdf

Hardy, R.W., T.M. Scott, and L.W. Harrell (1987). Replacement of herring oil with menhaden oil, soybean oil, or tallow in the diets of Atlantic salmon raised in marine net-pens. *Aquaculture* 65(3-4):267-277.

Kaushik, S.J., J.P. Cravedi, J.P. Lalles, J. Sumpter, B. Fauconneau, and M. Laroche (1995). Partial or total replacement of fish meal by soybean protein on growth, protein utilization, potential estrogenic or antigenic effects, cholesterolemia and flesh quality in rainbow trout, *Oncorhynchus mykiss. Aquaculture* 133(3): 257-274.

Krogdahl, A., A. Bakke-Mckellep, and G. Baeverfjord (2003). Effects of graded levels of standard soybean meal on intestinal structure, mucosal enzyme activities, and pancreatic response in Atlantic salmon *(Salmo salar* L.). *Aquaculture Nutrition* 9:361-371.

Krogdahl, A., A. Bakke-Mckellep, K. Roed, and G. Baeverfjord (2000). Feeding Atlantic salmon *Salmo salar* L. soybean products: Effects on disease resistance (furunculosis), and lysozyme and IgM levels in the intestinal mucosa. *Aquaculture Nutrition* 6:77-84.

Krogdahl, A. and A. Holm (1983). Pancreatic proteases from man, trout, rat, pig, cow, chicken, mink and fox enzyme activities and inhibition by soybean and lima bean proteinase inhibitors. *Comparative Biochemistry and Physiology* 74B:403-409.

Lall, S. P. and S. Anderson (2005). Amino acid nutrition of salmonids : Dietary requirements and bioavailability. Cahiers Options Méditerranéens, CIHEAM, Zaragoza, Spain. 63: 73-90.

Lall, S.P. and F.J. Bishop (1977). Studies on mineral and protein utilization by Atlantic salmon *(Salmo salar)* grown in seawater. *Technical Report of the Fisheries and Marine Services* N 688, Canada, 21 pp.

Latonnelle, K., F. Le Menn, S.J. Kaushik, and C. Bennetau-Pelissero (2002). Effects of dietary phytoestrogens *in vivo* and *in vitro* in rainbow trout and Siberian sturgeon: Interests and limits of the *in vitro* studies of interspecies differences. *General and Comparative Endocrinology* 126(1):39-51.

Mambrini, M., A.J. Roem, J.P. Carvedi, J.P. Lalles, and S.J. Kaushik (1999). Effects of replacing fish meal with soy protein concentrate and of DL-methionine supplementation in high-energy, extruded diets on the growth and nutrient utilization of rainbow trout, *Oncorhynchus mykiss. Journal of Animal Science* 77: 2990-2999.

Medale, F., T. Boujard, F. Vallee, D. Blanc, M. Mambrini, A. Roem, and S.J. Kaushik (1998). Voluntary feed intake, nitrogen and phosphorus losses in rainbow trout *(Oncorhynchus mykiss)* fed increasing dietary levels of soy protein concentrate. *Aquatic Living Resources* 11(4):239-246.

Mitrenko, A. (1997). Use of a soy protein concentrate as partial or total substitution for fish meal in rainbow trout diets. Masters thesis, ENITA, Bordeaux, France.

Murai, T., H. Ogata, A. Villaneda, and T. Watanabe (1989). Utilization of soy flour by fingerling rainbow trout having different body size. *Bulletin of the Japanese Society of Scientific Fisheries* 55(6):1067-1073.

National Research Council (NRC) (1993). Nutrient requirements of fish. Washington, DC: National Academy Press. 114p.

Oliva-Teles, A., A.J. Gouveia, E. Gomes, and P. Rema (1994). The effect of different processing treatments on soybean meal utilization by rainbow trout, *Oncorhynchus mykiss. Aquaculture* 124:343-349.

Olli, J.J., K. Hjelmeland, and A. Krogdahl (1994). Soybean trypsin inhibitors in diets for Atlantic salmon (*Salmo salar* L.): Effects on nutrient digestibilities and trypsin in pyloric caeca homogenate and intestinal content. *Comparative Biochemistry and Physiology* 109A:923-928.

Pelissero, C., F. Le Menn, and S. Kaushick (1991). Estrogenic effect of dietary soya bean meal on vitellogenesis in cultured Siberian sturgeon *Acipenser baeri. General and Comparative Endocrinology* 83(3):447-57.

Pelissero, C., M. Lenczowski, A. Mones del Pujol, J.P. Sumpter, and A. Fostier (1993). Potential effect of flavonoids from the vegetable component of the diet on the synthesis of estrogens in the rainbow trout *Oncorhynchus mykiss*. An *in vitro* study. Proceedings International Conference Bordeaux Aquaculture, Bordeaux, France, 25-27 March 1992, Special Publication, *European Aquaculture Society* 18:177-184.

Refstie, S., O.J. Korsoen, T. Storebakken, G. Baeverfjord, I. Lein, and A.J. Roem (2000). Differing nutritional responses to dietary soybean meal in rainbow trout *(Oncorhynchus mykiss)* and Atlantic salmon *(Salmo salar). Aquaculture* 190:49-63.

Refstie, S., T. Storebakken, G. Baeverfjord, and A.J. Roem (2001). Long-term protein and lipid growth of Atlantic salmon *(Salmo salar)* fed diets with partial replacement of fish meal by soy protein products at medium or high lipid level. *Aquaculture* 193(1-2):91-106.

Refstie, S., T. Storebakken, and A.J. Roem (1998). Feed consumption and conversion in Atlantic salmon *(Salmo salar)* fed diets with fish meal, extracted soybean meal or soybean meal with reduced content of oligosaccharides, trypsin inhibitors, lectins and soya antigens. *Aquaculture* 162:301-312.

Refstie, S., B. Svihus, K.D. Shearer, and T. Storebakken (1999). Nutrient digestibility in Atlantic salmon and broiler chickens related to viscosity and non-starch polysaccharide content in different soyabean products. *Animal Feed Science and Technology* 79(4):331-345.

Regost, C. (2001). Effects of dietary lipids on the nutritional, physical and organoleptic quality of the flesh in brown trout *(Salmo trutta)* and turbot *(Psetta maxima)*. Ph.D. thesis, University Rennes 1, Rennes (France), 193 pp. (French).

Reinitz, G.L., L.E. Orme, C.A. Lemm, and F.N. Hitzel (1978). Full-fat soybean meal in rainbow trout diets. *Feedstuffs* 50(3):23-24.

Reinitz, G.L. and T.C. Yu (1981). Effects of dietary lipids on growth and fatty acid composition of rainbow trout *(Salmo gairdneri). Aquaculture* 22(4):359-366.

Riche, M. and P.B. Brown (1996). Availability of phosphorus from feedstuffs fed to rainbow trout, *Oncorhynchus mykiss. Aquaculture* 142(3-4):269-282.

Robin, J.H., C. Regost, J. Arzel, and S.J. Kaushik (2003). Fatty acid profile of fish following a change in dietary fatty acid source: Model of fatty acid composition with a dilution hypothesis. *Aquaculture* 225(1-4):283-293.

Rumsey, G.L., A.K. Siwicki, D.P. Anderson, and P.R. Bowser (1994). Effect of soybean protein on serological response, non-specific defense mechanisms, growth, and protein utilization in rainbow trout. *Veterinary Immunology and Immunopathology* 41:323-339.

Sajjadi, M. and C.G. Carter (2004). Effect of phytic acid and phytase on feed intake, growth, digestibility and trypsin activity in Atlantic salmon (*Salmo salar* L.). *Aquaculture Nutrition* 10(2):135-142.

Satoh, S., M. Takanezawa, A. Akimoto, V. Kiron, and T. Watanabe (2002). Changes of phosphorus absorption from several feed ingredients in rainbow trout during growing stages and effect of extrusion of soybean meal. *Fisheries Science* 68(2):325-331.

Smith, R.R. (1977). Recent research involving full-fat soybean meal in salmonid diets. *Salmonid* 1:8-18.

Storebakken, T., I.S. Kvien, K.D. Shearer, B. Grisdale-Helland, and S.J. Helland (1999). Estimation of gastrointestinal evacuation rate in Atlantic salmon (*Salmo salar*) using inert markers and collection of faeces by sieving: Evacuation of diets with fish meal, soybean meal or bacterial meal. *Aquaculture* 172(3/4):291-299.

Storebakken, T., K.D. Shearer, G. Baeverfjord, B.G. Nielsen, T. Asgard, T. Scott, and A. De Laporte (2000). Digestibility of macronutrients, energy and amino acids, absorption of elements and absence of intestinal enteritis in Atlantic salmon, *Salmo salar*, fed diets with wheat gluten. *Aquaculture* 184(1-2):115-132.

Storebakken, T., K.D. Shearer, and A.J. Roem (1998). Availability of protein, phosphorus and other elements in fish meal, soy-protein concentrate and phytase-treated soy-protein-concentrate-based diets to Atlantic salmon, *Salmo salar*. *Aquaculture* 161(1-4):365-379.

Storebakken, T., K. Shearer, and A. Roem (2000). Growth, uptake and retention of nitrogen and phosphorus, and absorption of other minerals in Atlantic salmon *Salmo salar* fed diets with fish meal and soy-protein concentrate as the main sources of protein. *Aquaculture Nutrition* 6(2):103-108.

Sugiura, S. H., F.M. Dong, C.K. Rathbone, and R.W. Hardy (1998). Apparent protein digestibility and mineral availabilities in various feed ingredients for salmonid feeds. *Aquaculture* 159(3/4):177-202.

Tacon, A.G.J., J.V. Haaster, P.B. Featherstone, K. Kerr, and A.J. Jackson (1983). Studies on the utilization of full-fat soybean and solvent extracted soybean meal in a complete diet for rainbow trout. *Bulletin of the Japanese Society of Scientific Fisheries* 49(10):1437-1443.

Van Den Ingh, T., J.J. Olli, and A. Krogdahl (1996). Alcohol-soluble components in soybeans cause morphological changes in the distal intestine of Atlantic salmon, *Salmo salar* L. *Journal of Fish Diseases* 19:47-53.

Vielma, J., T. Makinen, P. Ekholm, and J. Koskela (2000). Influence of dietary soy and phytase levels on performance and body composition of large rainbow trout (*Oncorhynchus mykiss*) and algal availability of phosphorus load. *Aquaculture* 183:349-362.

Vielma, J., K. Ruohonen, J. Gabaudan, and K. Vogel (2004). Top-spraying soybean meal-based diets with phytase improves protein and mineral digestibilities but not lysine utilization in rainbow trout, *Oncorhynchus mykiss* (Walbaum). *Aquaculture Research* 35(10):955-964.

Vielma, J., K. Ruohonen, and M. Peisker (2002). Dephytinization of two soy proteins increases phosphorus and protein utilization by rainbow trout, *Oncorhynchus mykiss*. *Aquaculture* 204(1-2):145-156.

Yamamoto, T., K. Ikeda, and T. Akiyama (1998). Apparent availabilities of amino acids and minerals from several protein sources for fingerling Japanese flounder. *Bulletin of the National Research Institute of Aquaculture* 27:27-35.

Yamamoto, T., K. Ikeda, T. Unuma, and T. Akiyama (1997). Apparent availabilities of amino acids and minerals from several protein sources for fingerling rainbow trout. *Fisheries Science* 63(6):995-1001.

Chapter 11

Utilization of Soybean Products in Diets of Nonsalmonid Marine Finfish

Helena Peres
Chhorn Lim

INTRODUCTION

Intensive aquaculture of marine finfish with high market value, such as yellowtail *(Seriola quinqueradiata)*, European seabass *(Dicentrarchus labrax)*, gilthead seabream *(Sparus aurata)*, and turbot *(Scophthalmus maximux)*, has expanded rapidly during the previous decade and is expected to continue to grow in the foreseeable future as demand for quality seafood continues to rise with world population growth and supply from capture fisheries becomes limited. From 1990 to 2001, aquaculture production of marine fish has grown at an annual rate of 27.8 percent, with a total production of 2.34 million tons in 2001 (FAO, 2004). Given that intensive aquaculture is a diet-based production system, availablity of low-cost, high-quality diets is required to sustain industry growth.

The majority of marine finfish that are currently being farmed are highly carnivorous and generally require high dietary levels of high-quality protein. Fish meal (FM), because of its high nutritional value, has traditionally been used as the major ingredient in diets of carnivorous fish species. Fish meal has high protein and energy content, excellent amino acid profile, and high palatability, and is a good source of

Alternative Protein Sources in Aquaculture Diets
doi:10.1300/5892_11
281

essential fatty acids (n-3 highly unsaturated fatty acids, n-3 HUFA), and minerals. At present, FM is still the primary major source in diets of high value carnivorous fish species. However, since annual global production of FM has been relatively constant, between 6 and 7 million tons since 1989 (FAO, 2004) and the outlook for the next decade has been predicted to be similar to the past, considerable efforts have been devoted by aquaculture nutritionists toward finding alternative protein sources that are readily available and less expensive as substitutes for FM.

Soybean products, because of their low cost and consistent supply and quality, have received the most attention among alternative plant protein sources. According to the American Soybean Association, soybeans rank first among plant oilseeds with a total production of 190 million metric tons in 2003, representing 56 percent of the total world oilseed production. Lim and Akiyama (1991) reported that, from nutritional, economic, and market availability standpoints, soybean products such as full-fat soybean meal (FFSBM), soybean meal (SBM), and soybean protein concentrate (SPC), will likely be the key ingredients in future aquaculture diets.

This chapter will provide an overview on the nutritional values and use of soybean protein products in diets of nonsalmonid marine finfish. The species covered include European seabass, gilthead seabream, yellowtail, red seabream *(Pagrus major),* red drum *(Sciaenops ocellatus),* Japanese flounder *(Paralichthys olivaceus),* turbot *(Scophthalmus maximus),* Asian seabass *(Lates calcarifer),* Atlantic halibut *(Hippoglossus hippoglossus),* and milkfish *(Chanos chanos).*

CHEMICAL COMPOSITION

Proximate Composition and Mineral Content

Proximate composition and mineral contents of soybean protein products are presented in Table 11.1. Soy protein concentrate has crude protein content approximately equal to that of FM and considerably higher than those of dehulled-SBM, SBM with hulls, and FFSBM. Solvent-extracted SBM contain 0.9 to 1.1 percent lipid as compared to 18 percent for FFSBM. Soybean protein products contain higher crude fiber and much lower ash than those of FM. Mineral contents

TABLE 11.1. Proximate composition (% as fed) and mineral content (g/kg as fed) of different soybean protein products and FMs.[a]

	FFSBM	SBM (with hulls)	SBM (dehulled)	SPC	FM (anchovy)	FM (menhaden)
Intl. feed number	5-04-597	5-04-604	5-04-612	—	5-01-985	5-02-009
Proximate composition						
Dry matter	90	90	89	92.5	92	92
Protein	38	44	47.5	66.6	65.5	64.5
Fat	18	1.1	0.9	—	7.6	9.6
Fiber	5	7.3	3.4	3.5	1	0.7
Ash	4.5	6.3	5.8	5.5	14.3	19
Mineral						
Calcium	3	4.2	2.5	2.2	37.3	51.9
Phosphorus	6.5	9.4	5.9	7	24.3	21.88
Potassium	21.1	11.9	17	21	9	7.0
Magnesium	2.9	6.9	2.1	2.5	2.4	1.5
Copper	0.023	0.004	0.016	0.016	0.009	0.01
Iron	0.14	0.031	0.08	0.11	0.22	0.554
Manganese	0.031	0.019	0.03	0.03	0.01	0.037
Zinc	0.052	0.098	0.54	0.061	0.103	0.144

Note: SBM = solvent-extracted soybean meal; FFSBM = full-fat soybean meal, heat-processed; SPC = soy protein concentrate; FM = fish meal.
[a]Data adapted from Perkins (1995) and NRC (1993, 1998).

appear to vary greatly among soybean protein products. Soybean products contain substantially higher potassium, but lower calcium and phosphorus, than FM. Moreover, most phosphorus in soybean products is in the form of phytic acid that is relatively unavailable for fish.

Essential Amino Acid (EAA) Composition

It is generally known that, among plant protein sources, SBM protein has one of the best amino acid profiles for meeting the EAA requirements of most fish species. The EAA composition of FFSBM, SBM with hulls, dehulled-SBM, SPC, and anchovy and menhaden FM are present in Table 11.2. When expressed as percentage of protein, the EAA profile is very consistent among the different soybean products. Despite the higher crude protein level, EAA profile of SPC is similar to that of SBM. Compared to the EAA composition of FM, soybean products contain higher levels of arginine, but are lower in sulfur-containing amino acids (methionine + cystine).

NUTRIENT AVAILABILITY

Gross Nutrient Digestibility

The apparent digestibility coefficients of gross nutrients in soybean products have been determined for a number marine fish species. Table 11.3 summarizes the range of nutrient digestibility coefficients of FFSBM, dehulled-SBM, SPC, and FM for various species considered in this chapter. The digestibility values for various nutrients appear to be similar to those of other finfish species. Protein in soybean products is digested relatively well by most nonsalmonid marine fish. Protein digestibility coefficients of soybean products are similar to that of FM, although greater variations exist among the values for soybean products. These variations may be related to species differences, experimental procedures, and processing conditions of soybean products, which affect protein quality and levels of antinutritional factors, particularly trypsin inhibitors (TIs). Lipid in FFSBM is well digested by marine fish, but the value is slightly lower than that in FM. Energy digestibility of soybean products is much lower than that of FM, possibly due to lower fat content (except FFSBM) and the presence of

TABLE 11.2. Essential amino acid composition (% of crude protein) of different soybean products and FMs. [a]

	FFSBM	SBM (with hulls)	SBM (dehulled)	SPC	FM (anchovy)	FM (menhaden)
Intl. feed number	5-04-597	5-04-604	5-04-612	—	5-01-985	5-02-009
Protein (%)	31.2	44.8	50.0	64.5	65.5	64.5
Arginine	8.11	7.57	7.34	9.05	5.88	5.92
Histidine	2.76	2.66	2.44	2.81	2.46	2.25
Isoleucine	5.13	4.53	4.28	5.16	4.84	4.12
Leucine	8.43	7.79	7.26	8.28	7.71	6.95
Lysine	7.18	6.36	6.16	6.56	7.69	7.32
Methionine +	1.47	1.27	1.36	1.41	3.04	2.71
Cystine	1.09	1.56	1.50	1.56	0.92	0.87
Phenylalanine +	5.51	4.96	4.88	5.31	4.24	3.74
Tyrosine	4.01	1.27	3.52	3.91	3.42	3.01
Threonine	4.52	3.97	3.78	4.38	4.31	3.88
Tryptophan	1.67	1.43	1.38	1.41	1.15	1.01
Valine	6.47	4.51	5.10	5.31	5.34	4.99

Note: SBM = solvent-extracted soybean meal; FFSBM = full-fat soybean meal, heat-processed; SPC = soy protein concentrate; FM = fish meal.

[a] Data adapted from NRC (1993, 1998).

TABLE 11.3. Summary of apparent digestibility coefficient of gross nutrient (%) in dehulled, defatted solvent-extracted soybean meal (SBM), full-fat soybean meal (FFSBM), soy protein concentrate (SPC), and fish meal (FM), for several marine fish species.

Nutrient	SBM (dehulled)	FFSBM	SPC	FM
Protein	80-95	73-92	83-96	85-96
Energy	61-74	62	–	93-97
Lipid	63	85	–	87-97
Carbohydrates	45-58	–	–	–
Phosphorus	36-45	–	–	73-80

carbohydrates (CHOs). Soybean products contain oligosaccharides (mainly stachyose and raffinose) and nonstarch polysaccharides (pectins, galactans, cellulose, and lignin), which are poorly digested by marine carnivorous fish. The digestibility of phosphorus in FM varies depending on the quality of the raw material, but generally is higher than that in soybean products. A significant portion of phosphorus in soybeans is in the form of phytic acid, which is relatively unavailable to fish due to the lack of the enzyme phytase, necessary for phytin hydrolysis (NRC, 1993). Phytic acid also forms complexes with protein and minerals, thereby reducing their bioavailability (Davies and Gatlin, 1991).

The following paragraphs provide more detailed information on apparent gross nutrient digestibility of soybean products by species (Table 11.4).

European Seabass

The apparent protein digestibility of SBM by European seabass is generally high, ranging from 85 to 91 percent, and is inversely related to dietary inclusion levels (Alexis, 1997; Lanari et al., 1998; Tibaldi and Tulli, 1998; Tulli and Tibaldi, 2001). Protein in SPC is well digested and comparable to that of FM (Silva and Oliva-Teles, 1998; Tibaldi and Tulli, 1998; Tulli and Tibaldi, 2001). Energy digestibility of SBM (69 to 82 percent) is lower than that of SPC (88 to 92 percent), due probably to the presence of higher levels of CHO in SBM. These values, however, are lower than that of FM (97 percent). Phosphorus availability of SBM is relatively lower, averaging 40 percent,

TABLE 11.4. Apparent digestibility coefficients of gross nutrients (%) in dehulled, defatted solvent-extracted soybean meal (SBM), full-fat soybean meal (FFSBM), soy protein concentrate (SPC), and fish meal (FM) by several marine fish species.

Species	Ingredient	Apparent digestibility coefficients (%)					Reference
		Protein	Energy	Lipid	CHO[a]	P[b]	
European seabass	SBM	85-91	69-82	–	–	36	Alexis, 1997; Silva and Oliva-Teles, 1998; Lanari et al., 1998; Tibaldi and Tulli, 1998; Tulli and Tibaldi, 2001
	SPC	97	88-92	–	–	–	Tibaldi and Tulli, 1998; Tulli and Tibaldi, 2001
	FM	95	97	–	–	81	Silva and Oliva-Teles, 1998
Gilthead seabream	FFSBM	76	62	85	–	–	Nengas et al., 1995
	SBM	86-93	45-98	–	49	–	Robaina et al., 1995; Lupatsch et al., 1997
	FM	83-96	80-94	94	–	–	Lupatsch et al., 1997
Yellowtail	FFSBM	83	–	–	–	–	Masumoto et al., 1996
	SPC	73-87	–	–	–	–	Masumoto et al., 1996; Ruchimat et al., 1997
	SBM	85-91	68-74	–	45-58	–	Watanabe et al., 1992
	FM	87-94	–	–	–	–	Masumoto et al., 1996; Ruchimat et al., 1997
Red seabream	SBM	92-95	–	–	–	–	Yamamoto et al., 1998
	FM	93	–	–	–	–	
Red drum	SBM	86	93	63	–	47	Gaylord and Gatlin, 1996

TABLE 11.4 (continued)

Species	Ingredient	Apparent digestibility coefficients (%)					Reference
		Protein	Energy	Lipid	CHO[a]	P[b]	
	SBM with hulls	80	38	–	–	–	McGoogan and Reigh, 1996
	FM	88-96	60-95	87	–	50.3	Gaylord and Gatlin, 1996; McGoogan and Reigh, 1996
Japanese flounder	SBM	81-87	61	–	–	8.9	Yamamoto et al., 1998; Masumoto et al., 2001
	FM	91	–	–	–	–	Yamamoto et al., 1998
Turbot	SPC	83-88	–	–	–	–	Day and Gonzalez, 2000
	FM	85	–	–	–	–	
Asian seabass	SBM	91	–	–	–	–	Boonyaratpalin et al., 1998
	FFSBM	93-94	–	–	–	–	
	FM	93	–	–	–	–	
Atlantic halibut	SPC	86	–	97	–	–	Berge et al., 1999
	FM	84	–	97	–	–	

[a]CHO = carbohydrates.
[b]P = phosphorus.

compared to that of FM which is estimated to be 81 percent (Silva and Oliva-Teles, 1998). However, the addition of microbial phytase to a diet containing 66 percent SBM significantly increased the availability of dietary phosphorus from 25 to 72-79 percent (Oliva-Teles et al., 1998). Zinc digestibility of low-temperature FM (averaging 61 percent) is significantly higher than those of SBM and SPC, averaging 41 and 37 percent, respectively (Tulli and Tibaldi, 2001).

Gilthead Seabream

Gilthead seabream digests protein in SBM relatively well, ranging from 86 to 93 percent (Nengas et al., 1995; Lupatsch et al., 1997), depending on the inclusion levels (Robaina et al., 1995). Energy digestibility varies greatly and ranges from 45 to 72 percent. The low energy digestibility values may be attributed to the presence of CHO that has only a 49 percent digestibility coefficient (Lupatsch et al., 1997). Nengas et al. (1995) reported a protein digestibility of 76 percent for FFSBM, which is considerably lower than that of SBM (91 percent), but energy digestibility was higher (62 percent) than that of SBM (45 percent).

Yellowtail

Protein in SBM is well digested by yellowtail (85 to 91 percent). This value is similar to that obtained with protein in brown FM (87 to 94 percent) (Masumoto et al., 1996; Ruchimat et al., 1997). The apparent protein digestibility coefficients of FFSBM and SPC were estimated to be 83 percent and 87 percent, respectively (Masumoto et al., 1996), which are only slightly lower than that of FM. Ruchimat et al. (1997) reported that protein digestibility coefficients for SPC was inversely related to the dietary inclusion rates. Carbohydrate and energy digestibility of SBM also varied depending on the dietary inclusion level. In diets with 30 to 40 percent of SBM, energy digestibility averaged 74 percent and CHO digestibility averaged 58 percent. Increasing the level of SBM in the diet to 50 percent decreased the energy and CHO digestibility to 68 and 45 percent, respectively (Watanabe et al., 1992).

Red Seabream

Red seabream digested the protein in SBM and extruded SBM relatively well, averaging 92 and 95 percent, respectively. The true EAA and non-EAA digestibility were estimated to be 92 to 93 percent for SBM and 94 to 95 percent for extruded SBM (Yamamoto, Akimoto, et al., 1998). Extrusion processing improved the availability of amino acids in SBM.

Red Drum

Protein and energy digestibility of SBM with hulls were measured to be 80 and 38 percent, respectively (McGoogan and Reigh, 1996). Dehulling SBM significantly increased protein digestibility to 86 percent and energy digestibility to 93 percent (Gaylord and Gatlin, 1996; McGoogan and Reigh, 1996). Lipid and phosphorus digestibility of SBM were estimated to be 63 and 47 percent, respectively (Gaylord and Gatlin, 1996).

Japanese Flounder

For Japanese flounder, the digestibility of nutrients in defatted SBM is lower than that of white FM. Apparent digestibility coefficients of protein and amino acids of white FM averaged 91 and 94 percent, respectively, as compared to 80 percent for protein and 81 percent for amino acids in SBM (Yamamoto, Unuma, et al., 1998).

Phytic acid-bound phosphorus is barely utilized by Japanese flounder. In diets containing 67 percent of SBM, the incorporation of phytase-pretreated SBM or the supplementation of phytase to the diet significantly increased the digestibility of phosphorus from 8.9 to 87-95 percent but had no effect on protein or energy digestibility (Masumoto et al., 2001).

Other Marine Fish Species

Turbot: The apparent protein digestibility of FM was similar to that of SPC, ranging from 83 to 88 percent, irrespective of the dietary incorporation level of SPC (Day and Gonzalez, 2000).

Asian Seabass: Boonyaratpalin et al. (1998) evaluated the protein digestibility of solvent-extracted SBM, and extruded, steamed, and soaked raw FFSBM. Protein digestibility values of extruded FFSBM, steamed FFSBM and solvent-extracted SBM were 93, 94, and 91 percent, respectively, which were significantly higher that that of soaked raw FFSBM (21 percent).

Atlantic Halibut: Berge et al. (1999) determined the apparent protein and energy digestibility of a FM diet and a diet containing 28 percent SPC. They reported similar digestibility of protein and energy in both diets, which averaged 85 and 97 percent, respectively.

Amino Acid Availability

The nutritional quality of a protein is primarily based on the EAA content in the protein source and their bioavailability. A protein source having an EAA profile that matches closely the EAA requirements of the fish is of high nutritional value. The true digestibility coefficients of EAAs in FFSBM, SPC, and Chilean brown FM for yellowtail are given in Table 11.5. Yellowtail digested most EAA in SPC relatively

TABLE 11.5. True digestibility coefficient of essential amino acids (EAA) in soybean protein concentrate (SPC), full-fat soybean meal (FFSBM) and Chilean brown fish meal (BFM) for yellowtail.[a]

Amino acid	FFSBM	SPC	FM
Arginine	88.2	92.3	94.9
Histidine	60.0	94.5	94.8
Isoleucine	82.7	90.7	92.8
Leucine	81.4	89.7	93.3
Lysine	86.4	93.7	94.8
Methionine (Cystine)	80.3 (86.7)	90.3 (88.6)	93.7 (92.8)
Phenylalanine	82.7	91.6	91.8
Threonine	81.0	88.1	93.0
Tryptophan	–	–	–
Valine	77.8	86.8	91.6
Average	80.7	90.6	93.3

[a]Data adapted from Masumoto et al. (1996).

well, comparable to that of the brown FM. The average EAA digestibility coefficient for FFSBM is much lower than that of SPC, due possibly to the presence of indigestible CHOs and antinutritional factors. The apparent digestibility values of EAA in SBM, extruded SBM, and white FM for red seabream are presented in Table 11.6. The average apparent digestibility of EAA in SBM, extruded SBM, and white FM is relatively high and are 92.2, 94.0, and 94.5 percent, respectively. Extrusion processing through a twin screw extruder improved the EAA digestibility in SBM, due partly to protein denaturation and inactivation of antinutritional factors, such as TI. The average apparent digestibility of EAA in SBM for Japanese flounder is low (80.5 percent) compared to that of white FM (93.9 percent) (Table 11.7). The digestibility of individual EAA in soybean products for other nonsalmonid marine finfish, however, has not been determined.

Available information on quantitative EAA requirements of various nonsalmonid fish is presented in Table 11.8. Except for the EAA requirements of milkfish, the requirements of only five EAA (arginine, lysine, methionine + cystine, threonine, and tryptophan) have been determined for European seabass, only four EAA (arginine, lysine,

TABLE 11.6. Apparent digestibility coefficient of essential amino acids (EAA) in soybean meal (SBM), extruded soybean meal (ExSBM), and white fish meal (WFM) for red seabream.[a]

Amino acid	SBM	ExSBM	WFM
Arginine	95.6	97.5	96.2
Histidine	93.6	95.4	94.4
Isoleucine	88.7	90.2	93.8
Leucine	90.6	93.4	95.1
Lysine	94.2	96.4	96.3
Methionine (+ Cystine)	93.3 (92.0)	94.1(92.9)	96.0 (90.1)
Phenylalanine (+ Tyrosine)	92.1(93.7)	95.2 (95.4)	94.6 (94.2)
Threonine	92.0	93.4	94.8
Tryptophan	88.8	89.5	92.9
Valine	91.3	94.4	95.1
Average	92.2	94.0	94.5

[a]Data adapted from Yamamoto et al. (1998).

TABLE 11.7. Apparent digestibility coefficient of essential amino acids (EAA) in soybean meal (SBM) and white fish meal (WFM) for Japanese flounder.[a]

Amino acid	SBM	WFM
Arginine	85.8	95.4
Histidine	82.5	93.3
Isoleucine	79.7	94.1
Leucine	78.0	94.3
Lysine	82.7	95.4
Methionine (+ Cystine)	80.2 (79.3)	94.9 (89.3)
Phenylalanine (+ Tyrosine)	79.8 (81.6)	93.7 (94.4)
Threonine	80.2	93.6
Tryptophan	77.5	94.9
Valine	78.6	93.2
Average	80.5	93.9

[a]Data adapted from Yamamoto et al. (1998).

TABLE 11.8. Quantitative essential amino acid requirements (% of dietary protein) of some marine fish species.[a]

Amino acid	Milkfish	Gilthead seabream	European seabass	Asian seabass	Red drum	Yellowtail
Arginine	5.6	<6.0	4.1	3.6	3.7	3.4
Histidine	2.0	–	–	–	–	–
Isoleucine	4.0	–	–	–	–	–
Leucine	5.1	–	–	–	–	–
Lysine	4.0	5.0	4.8	4.5	5.7	4.1
Methionine + cystine	4.8	4.0	4.4	2.4	3.0	3.3
Phenylalanine + tyrosine	5.2	–	–	–	–	–
Threonine	4.9	–	–	–	–	–
Tryptophan	0.6	0.6	0.5	0.5	0.8	–
Valine	3.0	–	–	–	–	–

[a]Data adapted from NRC (1993).

methionine + cystine, and tryptophan) for gilthead seabream, Asian seabass, and red drum, and only three EAA (arginine, lysine, and methionine + cystine) for yellowtail (NRC, 1993; Webster and Lim, 2002) (Table 11.8).

Taking into consideration the true EAA bioavailability in FFSBM and SPC (EAA content, Table 11.3 × true EAA digestibility coefficient, Table 11.5) and the EAA requirements (Table 11.8), these soybean protein products are deficient in sulfur-containing amino acids for yellowtail. Using the same methods of computation, SBM and extruded SBM are also deficient in sulfur-containing amino acids for red seabream. For Japanese flounder, although the apparent digestibility EAA in SBM is available, assessments of its nutritional quality cannot be done due the lack of information on EAA requirements.

Due to the lack of information on EAA requirements and EAA bioavailability, accurate assessment of the nutritional value of various soybean products is difficult. However, it is generally accepted that, among plant protein sources, soybean products have one of the best amino acid profiles for meeting the EAA requirements of fish. The availability of EAA in soybean products for most fish species studied is relatively high compared to other alternative protein sources. The EAA composition of soybean protein products expressed as a percentage of protein (Table 11.3) is relatively consistent. Comparing the EAA composition of soybean products to the available information on EAA requirements (Table 11.8), protein in soy products satisfies the known EAA requirements of nonsalmonid marine fish, except in sulfur-amino acids (methionine + cystine) for milkfish, gilthead seabream, and European seabass.

UTILIZATION OF SOYBEAN PRODUCTS IN DIETS OF MARINE FISH

In recent years, as the cost and demand of FM increased, and its availability inconsistent, considerable efforts have been devoted to evaluate the nutritional values of soybean products as substitutes for FM in diets of aquaculture species. The most commonly studied soy products in nonsalmonid marine finfish diets are FFSBM, SBM with or without hulls (dehulled), and SPC. Most studies were conducted using juvenile fish held under well-controlled laboratory conditions.

Optimum dietary inclusion levels were commonly determined based on growth performance, feed utilization efficiency, biochemical composition of fish tissues, and hematological characteristics.

European Seabass

Alliot et al. (1979) conducted the first study evaluating the nutritional value of SBM as partial replacement of FM in diets of European seabass. They observed that, based on growth performance and diet intake, 40 percent of SBM can be used as a replacement of 20 percent FM in diets of this species. Amerio et al. (1991) evaluated the effect of residual TI in SBM on the growth of European seabass. Defatted SBM was heat-treated at 100°C for zero, three, eight, twelve, and twenty minutes prior to incorporation in diets at a level of 30 percent with methionine supplementation. Fish growth significantly improved as the level of TI gradually decreased to 6.2 mg TI/g of SBM which corresponded to twelve-minutes heating or a destruction of 80 percent of TI activity. Further heat treatment reduced growth due probably to decreased nutritional value of the protein caused by excessive heating. These authors also compared the performance of a commercial diet to those of experimental diets containing 25 percent SBM and 28 percent FFSBM supplemented with methionine. The growth of fish fed the FFSBM diet was similar to that of the commercial diet, but slower growth was obtained in fish fed the SBM diet. Feed conversion ratios of both SBM-containing diets were poorer than that of the commercial diet. Lanari et al. (1998) reported that SBM can replace 25 percent of a mixture of FM + blood meal + yeast without adverse effect on growth performance of young European seabass, while 50 percent replacement significantly decreased growth and feed utilization efficiency.

Tibaldi and Tulli (1998) evaluated the performance of diets in which 0, 20, 40, and 60 percent of low-temperature FM were substituted by SBM and SPC (60 percent substitution only). Fish fed the two highest levels of SBM diets and the SPC diet had higher diet intake than those fed the control-diet. However, growth, diet efficiency, protein and energy retention were significantly lower in fish fed the highest SBM diet. These authors concluded that 40 and 60 percent of low-temperature FM can be replaced by SBM and SPC, respectively, without adverse effects on growth performances and diet utilization efficiency.

A later study at the same laboratory (Tulli et al., 1999) also showed that replacement of 60 percent of dietary low-temperature FM by SBM or SPC did not affect growth performance, but increased voluntary diet intake when compared to the FM control-diet. However, Oliva-Teles et al. (1998) obtained significantly poorer growth rate and feed efficiency in fish fed the diet in which 67 percent of FM was replaced by SBM as compared to the group fed the FM-based diet. Replacement of 67 or 100 percent dietary FM by SPC (D'Agaro and Ravarotto, 1999) has also been shown to significantly reduce growth performance and feed efficiency. The acceptable level of FM that can be replaced by SPC was estimated to be 33 percent.

Soybean meal, at a level of 15 percent, in combination with 20 percent corn gluten meal (CGM), and 5 to 24 percent wheat gluten can be used to replace 95 percent of dietary FM without any adverse effects on voluntary diet intake and somatic growth of juveniles (Kaushik et al., 2004).

A major factor contributing to the poor performance of diets containing high levels of soybean products is poor diet palatability. Dias et al. (1997) compared the performance of a 38 percent FM diets to those in which all FM was replaced by SPC from two different sources with and without addition of an amino acid mixture as an attractant. Addition of attractant improved growth, diet consumption, and feed efficiency, but these values remained inferior to those obtained with the FM control-diet. However, Tibaldi et al. (1999) reported that addition of an attractant (2 percent squid extract) to a 40 percent SBM diet had no effect on diet consumption or growth as compared to those of the unsupplemented diet. It was suggested that, with appropriate dietary supplementation of limiting amino acids, such as lysine and methionine, addition of an attractant was unnecessary (Tibaldi and Tulli, 1998; Tibaldi et al., 1999).

The poor performance of diets containing high levels of soybean protein products (more than 60 percent FM replacement) has also been suggested to be due to the decrease in nitrogen retention (Oliva-Teles et al., 1998; Tibaldi and Tulli, 1998; Dias, 1999; Tulli et al., 2000) which results in the reduction of protein digestibility and increase of ammonia excretion (D'Agaro and Ravarotto, 1999; Dias, 1999). Soybean meal, at a level of 15 percent, in combination with 20 percent corn gluten and 5 to 24 percent wheat gluten, can replace up to 95 percent of dietary FM without negative effect on the nitrogen retention (Kaushik et al., 2004).

Few studies have evaluated the effect of dietary levels of soybean products on immune response of European seabass. Tulli et al. (2000) reported an increase of lymphocytes, serum protein, and immunoglobulin of European seabass when 20, 40, or 60 percent of FM was replaced by SBM, or 60 percent of FM was replaced by SPC, as compared with the FM control-diet. However, a reduction of the number of granulocytes and monocytes were observed in fish fed all the soybean product-containing diets; superoxide anion production was reduced when 20 and 40 percent of FM was replaced by SBM or 60 percent of FM was replaced by SPC. It has also been reported that increasing levels of FM replacement by soybean products decreased plasma cholesterol concentrations of European seabass (D'Agaro and Ravarotto, 1999; Dias, 1999; Tulli et al., 1999). Setchell and Cassidy (1999) attributed the hypocholesterolemic effect of soybean products to the presence of phytoestrogens and flavonoids.

Gilthead Seabream

Robaina et al. (1995) evaluated the performance of gilthead seabream fed diets containing 0, 10, 20, and 30 percent SBM as replacements of FM. Growth and diet intake were not influenced by dietary levels of SBM. Feed and protein efficiency, and protein productive value were not significantly affected by dietary SBM levels, although there appeared to be a general trend of these values decreasing with increasing levels of SBM. Proximate body composition and hepatosomatic index had not significantly affected dietary SBM levels. However, fish fed diets containing high levels of SBM had increased lipid and decreased glycogen deposition in livers, signs reported to be caused by phosphorus deficiency. In a subsequent study, Robaina et al. (1998) supplemented the diet containing 29 percent SBM as a replacement of sardine FM with phosphorus, zinc, phytase or increased level of sardine oil (to increase the n-3 to n-6 ratio to a level similar to that of the FM-control diet). All diets provided similar growth, and feed and protein efficiency ratios as those fed the FM- and SBM-control diets, even though fish fed the FM diet consumed significantly more feed than other groups of fish. They also reported that supplementation of phosphorus increased the protein and decreased the lipid content in muscle, but increasing n-3 to n-6 ratio decreased the protein and increased the lipid content in muscle. However, addition of phosphorus or increasing n-3 to n-6 ratio reduced histopathological abnormalities in livers.

Nengas et al. (1996) evaluated the growth performance of gilthead seabream fed diets containing 0, 12, 24, 36, and 47 percent SBM as replacements of 0, 10, 20, 30, and 40 percent of white FM. The results showed that 24 percent SBM can be included as a replacement of 20 percent white FM in gilthead seabream diets without adversely affecting growth and feed efficiency. The values of these parameters significantly decreased when dietary SBM levels were increased to 36 percent or higher. Similarly, Venou et al. (1997) reported that 22 percent was the acceptable inclusion level of SBM when used as a substitute of high-quality FM. These workers (1996) also examined the nutritional value of various soybean products by replacing of 35 percent of white FM by FFSBM heated for five, twenty, or forty-five minutes at 110°C, solvent-extracted SBM, and SPC, on an equal protein and lipid basis. They observed that the overall performance of diets with FFSBM heated for twenty or forty minutes and FM diets were similar, and were significantly higher than those of SBM- and SPC-diets. It was suggested that at least 85 percent destruction of TI activity is required at the dietary inclusion level of FFSBM used (44.3 percent), since growth and protein efficiency were still improving for the diet containing the forty-minute heated FFSBM. The poor performance of diets containing SBM and SPC was suggested to be due to insufficient inactivation of TI and low available lysine, respectively.

Kissil et al. (2000) assessed the nutritional value of SPC as partial or total replacements (0, 30, 60, and 100 percent) of Chilean FM, on equal apparent digestible protein and energy values basis. Voluntary diet intake decreased with increasing dietary SPC levels and the values were significantly depressed at 60 and 100 percent of FM replacement levels. Growth performances were significantly reduced for all dietary inclusion levels of SPC. The same trend was observed for energy retention. Whole-body lipid and energy contents were significantly reduced in fish fed the 60 and 100 percent SPC-replacement diets, compared to the group fed the FM-based diet. This was attributed to the reduction of energy utilization due to reduced diet intake.

They indicated that SPC may be a promising protein source for use in gilthead seabream diets, but there is a need to identify its possible amino acid deficiencies.

Since palatability is the major limiting factor observed in the study of Kissil et al. (2000), they suggested that there is a need to identify an effective palatability enhancer that can be used with SPC in diets of gilthead seabream. In a study to evaluate the influence of diets containing either FM, SPC, or SPC supplemented with methionine on voluntary diet intake and daily feeding rhythm, Sanchez-Muros et al. (2003) observed that the feeding pattern of juvenile gilthead seabream fed a 100 percent SPC-based diet was the same as that of the FM control-diet, but methionine supplementation to the SPC diet advanced the time of feeding and lengthened ingestion phases.

Yellowtail

Several studies have investigated the nutritional values of soybean protein products as substitutes for FM in diets of yellowtail. Results of earlier studies have demonstrated that approximately 30 percent FFSBM (Shimeno et al., 1997), 20 to 30 percent of commercial SBM or properly heated raw SBM, 14 percent SPC without amino acid (AA) supplementation (Shimeno, Hosokawa, Kunon, et al., 1992; Shimeno, Hosokawa, Yamane, et al., 1992; Shimeno, Kumon, et al., 1993), and 20 percent of SPC with supplemental AA (Takii et al., 1989) can be used as replacements of 30 to 50 percent of FM without affecting the growth performance of yellowtail. At the same level of FM replacement, SPC is a better protein source than SBM (Shimeno, Hosokawa, Kunon, et al., 1992, Shimeno et al., 1995). However, total substitution of FM in basal diets containing 30, 35, 40, and 45 percent protein supplied by 44, 51, 58, and 71 percent brown FM, respectively by 46, 53, 61, and 74 percent SPC resulted in a significant decrease in growth and diet utilization efficiency, but these value were better than those obtained with CGM diets (Ruchimat et al., 1997). The poor nutritional value of SPC and CGM was attributed to imbalanced AA profiles and also poor digestibility for CGM.

Raw SBM was poorly utilized by yellow tail. Fish fed diets containing 20 or 30 percent raw SBM (27.8 mg TI/g SBM) had weight loss, and inferior hematological characteristics and body composition due to the presence of a high level of TI that lead to low nutrient digestibility and absorption (Shimeno, Hosokawa, Yamane, et al., 1992; Shimeno et al., 1994, 1995). The nutritional value of raw SBM heated at 108°C for twenty to thirty minutes (2.9-4.5 mg TI/g SBM) improved considerably, but decreased slightly after being heated for forty minutes (Shimeno, Hosokawa, Yamane, et al., 1992). Fermentation of SBM with *Aspergillus oryzae* or *Eurotium repens* also improved its nutritional value. Fermented SBM at a level of 30 percent in single-moist pellet diets provided better growth and feed efficiency than the unfermented SBM diets, but all diets were still slightly inferior to the FM control-diet (Shimeno, Mima, Yamamoto, and Ando, 1993).

Watanabe et al. (1992) and Viyakarn et al. (1992) evaluated the nutritional value of SBM as substitutes for FM in a newly developed soft-dry pellet that has been reported by Watanabe et al. (2001) to be highly palatable and acceptable to yellowtail, as well as superior to moist pellet and raw fish in nutritional value. Inclusion of dietary SBM at levels up to 50 percent as a replacement of 90 percent FM had no effect on diet palatability and acceptability for either juvenile or adult yellowtail, but the growth rate was significantly reduced when compared to FM-based diets. A similar trend was observed when high proportion of SPC replaced FM in soft-dry pellets for yellowtail (Watanabe et al. 1995). Growth and feed efficiency were slightly reduced at the 30 percent SBM level, but there were no noticeable effect on hematology, body composition, and flesh quality (taste, texture, and pigmentation). They suggested, however, that SBM could be included at levels up to 30 percent as a substitute for 55 percent FM in soft-dry pellets for yellowtail.

Shimeno, Mima, Imanaga, and Tomaru (1993) used various combinations of SBM, meat meal (MM), and CGM in an attempt to further increase the substitution level of dietary FM. They observed that diets containing a combination of 20 percent MM or CGM and 20 percent SBM provided similar growth performance as that of the 20 percent SBM-only diet. The diet containing 10 percent MM and 20 percent SBM provided the best performance, which was superior, not only to that of the 20 percent SBM diet, but also the FM control-diet. Watanabe et al. (1998) evaluated the nutritional value of non-FM diets in which the protein (47 percent) was supplied by 2 percent krill meal, 10 percent SBM, and varied amounts of SPC, CGM, and MM for juvenile and young yellowtail. Diet palatability was poor for juvenile yellowtail, leading to poor growth and diet efficiency. Young fish actively consumed the test diets and grew normally for the first forty-six days, but thereafter had stagnant growth, poor diet efficiency and high mortality. The poor performances of these diets were thought to be associated with the development of green liver and poor blood characteristics. The effects of diet types (soft-dry pellets, extruded pellets, and single-moist pellets) with and without supplementation of an EAA mixture on the nutritional value of a non-FM diet were further assessed by Watanabe et al. (2001). All diets (a commercial FM control-diet and test diets) were readily consumed by fish, but growth, diet efficiency, and physiological conditions of fish receiving the control-diet were superior to those of fish fed the test diets. Among the non-FM diets, the best overall performance was obtained in the group fed the soft-dry pellets with EAA supplementation, and was followed in descending order by the soft-dry pellet without EAA, and extruded pellets and single-moist pellets, with and without EAA addition.

Red Seabream

An early study on the use of SBM as replacement of FM in diets of red seabream showed that inclusion of 25 percent SBM as a replacement of 33 percent of Chilean brown FM provided good growth performance similar to that obtained with the FM control-diet. Increasing the dietary SBM level to 40 percent as a substitute for 50 percent of FM resulted in significant reduction in growth and feed efficiency (Ukawa et al., 1994). A more recent study by Aoki et al. (1998) showed that SBM, at a level of 30 percent in diets processed into dry pellets or soft-dry pellets, promoted growth and feed performances comparable to those of a FM control-diet.

The nutritional value of SPC for juvenile and yearling red seabream was evaluated by Takagi et al. (2001) using diets containing 52 percent SPC and 3 percent krill meal, with and without supplementation of lysine and/or methionine, as substitutes for 90 percent of FM. Superior growth performances were obtained for both sizes of fish fed the FM control-diet. In juveniles, the growth performance of fish fed the SPC diet supplemented with methionine was significantly improved relative to that of the unsupplemented diet. The growth performance was further improved by supplementation of both methionine and lysine. For yearlings, however, the beneficial effect of methionine and/ or lysine supplementation was not observed. It was suggested that the EAA requirements of red seabream may vary with size or age. In adult red seabream, Aoki et al. (1996) observed that fish fed a diet in which protein was supplied by a combination of 40 percent SPC, 10 percent SBM, 3 percent CGM, and 12 percent MM had comparable growth and flesh quality as those fed a commercial diet containing 64 percent FM.

Goto et al. (2001) studied the effect of replacement of 70 and 100 percent FM in a control-diet containing 50 percent FM by a combination of 24 percent SBM and 20 percent CGM, or 24 percent SBM, 20 percent CGM, and 20 percent poultry meal, respectively, on the development of green liver in yearling red seabream. Growth rate and diet utilization efficiency decreased with increasing levels of FM substitution. Green liver symptoms were observed in fish in all treatments, but the incidence of occurrence was much higher in fish fed both test diets. The authors concluded that taurine deficiency was the probable factor responsible for the occurrence of green liver in red seabream fed the substitute protein diets. Unlike freshwater species, marine finfish generally lack the ability to synthesize taurine, which is present in high quantities in FM (Watanabe, 2002).

Red Drum

Reigh and Ellis (1992), in a study to evaluate the nutritional value of SBM as replacements of 0, 25, 50, 75, and 100 percent menhaden FM showed that approximately 50 percent of dietary FM (52.2 percent of diet) can be replaced by 35.5 percent SBM without significantly affecting the growth of juvenile red drum. However, relative to diets containing less SBM, fish fed this diet consumed significantly less feed, but had significantly improved feed efficiency, apparent net protein retention, and apparent net energy retention. Diets in which 75 and 100 percent of FM were replaced by SBM, and 100 percent FM by SBM and supplemental methionine were poorly consumed and utilized by fish.

Davis et al. (1995) conducted a series of four studies to examine the nutritional value of soybean protein products (SBM, SPC, and soy protein isolate) with and without addition of EAA (methionine and lysine) and feed attractants (glycine, fish flavor #2, shrimp flavor, and seafood flavor) as replacements for menhaden FM. They reported that growth, survival, and feed efficiency decreased with increasing substitution levels of FM by SBM or SPC. The poor performance of fish fed diets high in SBM or SPC was not likely attributed to the presence of anti-nutritional factors since the low-FM diets containing SBM or a soy protein isolate provided similar growth, but was due to the deficiency of the sulfur-containing amino acids because addition of methionine to the diets improved growth performances. In subsequent experiments, they observed that addition of fish-soluble, glycine, methionine, or lysine to the 10 percent FM diet had no significant effect on improving the nutritional value or palatability of the diets. However, supplementation of low-FM diets with seafood flavor or fish flavor #2 significantly improved weight gain and feed efficiency to values comparable to those obtained with the high-FM diet. It was suggested that, with suitable restrictions and diet modification, soybean protein products are acceptable for use in diets of red drum. However, protein from soybean appears to be deficient in sulfur-containing amino acids and has reduced palatability. Thus, diets containing increasing levels of soybean protein products as substitutes for marine proteins could require supplementation of amino acids and/or palatability enhancers.

McGoogan and Gatlin (1997) reported that red drum can tolerate relatively high dietary levels of SBM. Fish fed diets in which 90 percent of anchovy FM was replaced by 66.6 percent of SBM gained more weight than fish fed the 100 percent FM (53.4 percent) diet. Supplementation of 2 percent glycine or fish solubles to the diet containing 66.6 percent SBM tended to further improve the growth relative to fish fed the unsupplemented diet. Increasing substitution levels of dietary FM to 95 and 100 percent by SBM resulted in a significant decrease in weight gains as compared to the FM control-diet. However, supplementation of 2 percent glycine in the diet with 95 percent of protein supplied by SBM significantly enhanced weight gain relative to the similar unsupplemented diet. These authors suggested, however, that for red drum a minimum of 10 percent of dietary protein from FM appears necessary in practical diets containing high levels of SBM to prevent impaired growth and feed utilization efficiency.

Japanese Flounder

A preliminary study on the use of soybean protein products in diets of Japanese flounder showed that, with supplementation of deficient

EAA, up to 50 percent of dietary FM could be replaced by SBM without adversely affecting fish growth (Kikuchi et al., 1994). However, feed and protein efficiency ratios decreased with increasing levels of dietary SBM. A subsequent study by Kikuchi (1999) also showed that weight gain of fish fed diets containing 40 percent SBM and 30 percent SBM in combination with 10 percent blood meal or CGM as replacements of 47 percent of FM, was comparable to that of the control group, but feed efficiency and protein efficiency were significantly lower, except the feed efficiency of the diet with 30 percent SBM and 10 percent blood meal. Diets with 40 percent SBM in combination with 10 percent blood meal or 10 percent CGM as substitutes for 60 percent of FM provided significantly lower weight gain and feed efficiency than those obtained with the control-diet. However, fish fed diets containing a combination of 25 percent SBM, 10 percent blood meal, and 5 percent freeze-dried blue mussel meat, or a diet containing 25 percent SBM, 10 percent CGM, and 5 percent freeze-dried blue mussel meat as replacements of 47 percent of FM consumed more diet and outperformed other groups of fish (including the control group) in terms of weight gain and feed efficiency. Masumoto et al. (2001) reported that supplementation of phosphorus or phytase to a diet containing 40 percent SPC as a substitute for 75 percent FM significantly improved growth, phosphorus retention, and plasma phosphorous level in juvenile Japanese flounder.

Other Marine Fish Species

Turbot: Day and Gonzalez (2000) conducted two feeding studies to examine the nutritional value of SPC as replacements of FM in diets of juvenile turbot. The first study which evaluated the effect of gradual substitution of dietary FM by SPC showed that 18.5 percent of SPC can be incorporated in the diet of this species as a replacement of 25 percent of dietary FM without adverse affects on growth performance, feed intake, and feed efficiency. Increasing FM replacement levels to higher than 25 percent significantly decreased growth and feed efficiency. A reduction of diet intake was observed only at FM substitution levels of 75 percent or higher. In the second experiment, the beneficial effect of supplementation with uncoated methionine and lysine, and coated with a protein or an emulsified fish oil-protein of diets containing 44.1 percent SPC as replacements of 60 percent dietary FM was evaluated. Fish fed the FM control-diet exhibited significantly better growth than those fed diets containing SPC. However, supplementation of methionine and lysine coated with protein or emulsified fish oil provided a nonsignificant increase in growth rate over those without EAA supplementation or with uncoated EAA.

Asian Seabass: Different soybean products (solved extracted SBM, and extruded, steamed, and soaked raw FFSBM at levels of 21.0, 27.0, 28.5, and 28.5 percent, respectively) were included as replacements of 37.5 percent of FM in diets of juvenile Asian seabass (Boonyaratpalin et al., 1998). Fish fed the control-diet grew significantly better that those fed extruded, steamed, and soaked raw FFSBM diets, but did not differ from the group fed the SBM diet. Feed efficiency and survival were not affected by dietary treatments, except for fish fed the soaked raw FFSBM, which had significantly low weight gain, diet intake, and feed utilization efficiency. The authors suggested that, since the apparent protein digestibility of the control-diet was similar to that of diets containing SBM, and extruded and steamed FFSBM diets, the slower growth of fish fed extruded and steamed FFSBM diets could be due to diet palatability as lower diet intake occurred as early as during the first two weeks of feeding. Suhaimee et al. (1999) evaluated the growth performance of juvenile seabass fed diets containing graded levels of SBM as replacement of FM. They observed that incorporation of up to 50 percent SBM as a replacement of 60 percent of dietary FM had no adverse effects on growth or feed efficiency.

Atlantic Halibut: Soy protein concentrate, at a level of 28 percent and addition of 5 percent DL-methionine, with and without the addition of attractant (2 percent squid meal), was used as substitutes for 40 percent of FM in grow-out diets of Atlantic halibut (Berge et al., 1999). Growth rate was essentially the same for fish under all treatments, but feed efficiency was significantly lower for the groups fed diets containing SPC. Fish fed the SPC diets consumed more diet than those fed the FM diet. Addition of squid meal had no effect on diet intake. The use of FFSBM in diets for yearling Atlantic halibut was studied by Grisdale-Helland et al. (2002). They reported that inclusion of up to 36 percent FFSBM as a replacement of 28 percent of dietary FM had no significant effect on growth performance, feed consumption, feed efficiency, and protein and energy retention.

Silver Seabream: A study by El-Sayed (1994) to examine the effect of replacing FM by SBM in diets of juvenile silver seabream *Rhabdosargus sarba* showed that replacement of 25 percent of dietary FM by 21 percent SBM and 0.5 percent methionine had no adverse effects on the growth rate and feed efficiency. Increasing dietary SBM beyond this level (21 percent of diet) resulted in significant growth retardation.

Milkfish: The feasibility of using SBM to replace 0, 33, 67, and 100 percent of FM in milkfish diets containing 30 and 40 percent protein was studied by Shiau et al. (1988). All diets having SBM were supplemented with methionine to the level equal to that of the FM control-diet. Results showed that, irrespective of the dietary protein level (30 or 40 percent), SBM supplemented with methionine can be used to replace up to 67 percent of FM in diets of juvenile milkfish with no significant effects on growth and feed utilization efficiency.

CONCLUSION

Soybean protein products are the most commonly studied and used alternative plant protein sources in aquaculture diets. Nonsalmonid marine finfish appear to tolerate higher levels of dietary soybean protein products than salmonids, but lower than freshwater, warmwater omnivores such as channel catfish, tilapia, and carp. The ability of nonsalmonid fish to utilize soybean protein products appears to vary among the type and quality of soy products, fish species, and fish size. Generally, however, high levels of soybean protein products have led to reduced diet intake, decreased growth, poor diet utilization, and the development of a number of physiological, histological, immunological, and biochemical abnormalities. These adverse effects are thought to be related to several factors, such as the presence of antinutritional factors and indigestible nutrients, imbalance of EAA, energy, and minerals, and poor palatability. Various methods have been suggested to improve the nutritional value of soybean products in aquaculture diets. Dehulling of soybean seeds has been shown to considerably increase energy digestibility of SBM. Proper heat treatments, such as cooking or extrusion, of SBM increase not only CHO and energy digestibility but also the protein digestibility due to the disruption of the cell walls, denaturation of proteins, and inactivation of antinutritional factors, particularly TIs. Energy and protein digestibility in soybean products can be increased by removal of oligosaccharides by alcohol or aqueous extraction, or by enzymatic degradation. Pretreatment of soybean products with or addition of fungal enzyme phytase to the diets greatly increases the phosphorus availability. Adequate supplementation of deficient EAA and minerals, and increase of dietary energy levels improve the nutritional value of soybean products. Addition of palatability enhancers or attractants has been reported to stimulate the consumption of diets containing high levels of soybean products.

REFERENCES

Alexis, M.N. (1997). Fish meal and fish oil replacers in Mediterranean marine fish diets. In *Feeding Tomorrow's Fish,* A. Tacon and B. Basurco (eds.). *Cahiers Options Méditerranéennes* 22:183-204.

Alliot, E., A. Pastoureaud, J.P. Hudlet, and R. Métaillar (1979). Utilisation des farines végétales et des levures cultivées sur alcanes pour l'alimentation du bar *(Dicentrarchus labrax).* In *Proceedings of the Symposium on Finfish Nutrition and Fishfeed Technology,* J.E. Halver and K. Tiews (eds.). Berlin, Germany: Heeneman Verlag, pp. 229-238.

Amerio, M., M. Mazzola, C. Caridi, E. Crisafi, and L. Genovese (1991). Soybean products in feeds for sea bass. *Proceedings of the IX National ASPA Congress.* St Louis, MO: American Soybean Association, pp. 1099-1110.

Aoki, H., M. Furuichi, V. Viyakarn, Y. Yamagata, and T. Watanabe (1998). Feed protein ingredients for red sea bream. *Suisanzoshoku* 46:121-127.

Aoki, H., H. Shimazu, T. Fukushige, H. Akano, Y. Yamagata, and T. Watanabe (1996). Flesh quality in red sea bream fed with diet containing a combination of different protein sources as total substitution for fish meal. *Bulletin of the Fisheries Research Institute at Mie* 6:47-54.

Berge, G.M., B. Grisdale-Helland, and S.J. Helland (1999). Soy protein concentrate in diets for Atlantic halibut *(Hippoglossus hippoglossus). Aquaculture* 178:139-148.

Boonyaratpalin M., P. Suraneiranat, and T. Tunpibal (1998). Replacement of fish meal with various types of soybean products in diets for the Asian seabass, *Lates calcarifer. Aquaculture* 161:67-78.

D'Agaro, E. and L. Ravarotto (1999). Partial and total fish meal substitution in sea bass diets using soybean and wheat protein concentrates. In *Proceedings of the ASPA XIII Congress.* Piacenza, Italy, pp. 755-757.

Davies, D.A. and D.M. Gatlin (1991). Dietary mineral requirements of fish and shrimp. In *Proceedings of the Aquaculture Feed Processing and Nutrition Workshop,* D.M. Akiyama and R. Tan (eds.). Singapore: American Soybean Association, pp. 49-67.

Davies, D.A., D. Jirsa, and C.R. Arnold (1995). Evaluation of soybean proteins as replacements for menhaden fish meal in practical diets for the red drum *Sciaenops ocellatus. Journal of the World Aquaculture Society* 26:48-58.

Day, O.J. and H.G.P. Gonzalez (2000). Soybean protein concentrates as a protein source for turbot *Scophthalmus maximus* L. *Aquaculture Nutrition* 6:221-228.

Dias, J., E.F. Gomes, and S.J. Kaushik (1997). Improvement of feed intake through supplementation with an attractant mix in European seabass fed plant-protein rich diets. *Aquatic Living Resources* 10:385-389.

El-Sayed, A.-F.M. (1994). Evaluation of soybean meal, *spiralina* meal and chicken offal meal as protein sources for silver seabream *(Rhabdosargus sarba)* fingerlings. *Aquaculture* 127:169-176.

FAO (Food and Agriculture Organization) (2004). FAO Fisheries Department, Fisheries Information, Data and Statistic Unit. Universal Software for Fishery Statistical Time Series, 1970-2000. Version 2.3.

Gaylord, T.G. and D.M. Gatlin (1996). Determination of digestibility coefficients of various feedstuffs for red drum *(Sciaenops ocellatus). Aquaculture* 139:303-314.

Goto, T., S. Takagi, T. Ichiki, T. Sakai, M. Endo, T. Yoshida, M. Ukawa, and et al. (2001). Studies on the green liver in cultured red sea bream fed low level and non-fish meal diets: Relationship between hepatic taurine and biliverdin levels. *Fisheries Science* 67:58-63.

Grisdale-Helland, B., S.J. Helland, G. Baeverfjord, and G.M. Berge (2002). Full-fat soybean meal in diets for Atlantic halibut: Growth, metabolism and intestinal histology. *Aquaculture Nutrition* 8:265-270.

Kaushik, S.J., D. Coves, G. Dutto, and D. Blanc (2004). Almost total replacement of fish meal by plant protein sources in the diet of a marine teleost, the European seabass, *Dicentrarchus labrax. Aquaculture* 230:391-404.

Kikuchi, K. (1999). Use of defatted soybean meal as a substitute for fish meal in the diet of Japanese flounder. *Aquaculture* 179:3-11.

Kikuchi, K., T. Furuta, and H. Honda (1994). Utilization of soybean meal as a protein source in the diet of juvenile Japanese flounder, *Paralichthys olivaceus. Suisanzoshoku* 42:601-604.

Kissil, G.W., I. Lupatsch, D.A. Higgs, and R.W. Hardy (2000). Dietary substitution of soy and rapeseed protein concentrates for fish meal, and their effects on growth and nutrient utilization in gilthead seabream *Sparus aurata* L. *Aquaculture Research* 31:595-601.

Koven, W. (2002). Gilt-head sea bream, *Sparus aurata.* In *Nutrient Requirements and Feeding of Finfish for Aquaculture,* C.D. Webster and C.E. Lim (eds.). New York, NY: Cabi Publishing, pp 64-78.

Lanari, D., M. Yones, R. Ballestrazzi, and E. D'Agaro, (1998). Alternative dietary protein sources (soybean, rapeseed and potato) in diets for sea bass. In *Recent advances in finfish and crustacean Nutrition. Proceedings of the VIII International Symposium on Nutrition and Feeding in Fish.* Las Palmas De Gran Canaria, Spain, pp. 148.

Lim, C. and D.M. Akiyama (1991). Full-fat soybean meal utilization by fish. In *Proceedings of the Aquaculture Feed Processing and Nutrition Workshop,*

D.M. Akiyama and R. Tan (eds.). Singapore: American Soybean Association, pp. 188-198.

Lupatsch, I., G.W. Kissil, D. Sklan, and E. Pfeffer (1997). Apparent digestibility coefficients of feed ingredients and their predictability in compound diets for gilthead seabream, *Sparus aurata* L. *Aquaculture Nutrition* 3:81-89.

Masumoto, T., T. Ruchimat, Y. Ito, H. Hosokawa, and S. Shimeno (1996). Amino acid availability values for several protein sources for yellowtail *(Seriola quinqueradiata)*. *Aquaculture* 146:109-119.

Masumoto, T., B. Tamura, and S. Shimeno (2001). Effects of phytase on bio-availability of phosphorus in soybean meal-based diets for Japanese flounder *Paralichthys olivaceus*. *Fisheries Science* 67:1075-1080.

McGoogan, B.B. and D.M. Gatlin (1997). Effects of replacing fish meal with soybean meal in diets for red drum *Sciaenops ocellatus* and potential for palatability enhancement. *Journal of the World Aquaculture Society* 28:374-385.

McGoogan, B.B. and R.C. Reigh (1996). Apparent digestibility of selected ingredients in red drum *(Sciaenops ocellatus)* diets. *Aquaculture* 141:233-244.

NRC (National Research Council) (1993). *Nutrient Requirements of Fish.* Washington, DC: National Academy Press.

NRC (National Research Council) (1998). *Nutrient Requirements of Swine.* Washington, DC: National Academy Press.

Nengas, I., M.N. Alexis, and S.J. Davies (1996). Partial substitution of fishmeal with soybean meal products and derivatives in diets for the gilthead sea bream *Sparus aurata* (L.). *Aquaculture Research* 27:147-156.

Nengas, I., M.N. Alexis, S.J. Davies, and G. Petichakis (1995). Investigation to determine digestibility coefficients of various raw materials in diets for gilthead sea bream, *Sparus auratus* L. *Aquaculture Research* 26:185-194.

Oliva-Teles, A., J.P. Pereira, A. Gouveia, and E. Gomes (1998). Utilization of diets supplemented with microbial phytase by seabass *(Dicentrarchus labrax)* juveniles. *Aquatic Living Resources* 11:255-259.

Perkins, E.G. (1995). Composition of soybeans and soybean products. In *Practical Handbook of Soybean Processing and Utilization,* D.R. Erickson (ed.). Champaign, IL: AOAC Press, pp. 9-28.

Reigh, R.C. and S.C. Ellis (1992). Effects of dietary soybean and fish-protein ratios on growth and body composition of red drum *(Sciaenops ocellatus)* fed isonitrogenous diets. *Aquaculture* 104:279-292.

Robaina, L., M.S. Izquierdo, F.J. Moyano, J. Socorro, J.M. Vergara, and D. Montero (1998). Increase of the dietary n-3/n-6 fatty acid ratio and addition of phosphorus improves liver histological alterations induced by feeding diets containing soybean meal to gilthead seabream, *Sparus aurata*. *Aquaculture* 161:281-293.

Robaina, L., M.S. Izquierdo, F.J. Moyano, J. Socorro, J.M. Vergara, D. Montero, and H. Fernandez-Palacios (1995). Soybean and lupin seed meals as protein sources in diets for gilthead seabream *(Sparus aurata)*—Nutritional and histological implications. *Aquaculture* 130:219-233.

Ruchimat, T., T. Masumoto, H. Kosokawa, and S. Shimeno (1997). Nutritional evaluation of several protein sources for yellowtail *(Seriola quinqueradiata)*. *Bulletin of Marine Sciences and Fisheries* 17:69-78.

Sanchez-Muros, M.J., V. Corchete, M.D. Suarez, G. Cardenete, E. Gomez-Milan, and M. de la Higuera (2003). Effect of feeding method and protein source on *Sparus aurata* feeding patterns. *Aquaculture* 224:89-103.

Setchell, K.D.R. and A. Cassidy (1999). Dietary isoflavones: Biological effects and relevance to human health. *The Journal of Nutrition* 129:758S-767S.

Shiau, S.Y., B.S. Pan, S. Chen, H.L. Yu, and S.L. Lin (1988). Successful use of soybean meal with a methionine supplement to replace fish meal in diets fed to milkfish *Chanos chanos* Forskaal. *Journal of the World Aquaculture Society* 19:14-19.

Shimeno, S., H. Hosokawa, M. Kunon, T. Masumoto, and M. Ukawa (1992). Inclusion of defatted soybean meal in diet for fingerling yellowtail. *Nippon Suisangakkaishi* 58:1319-1325.

Shimeno, S., H. Hosokawa, R. Yamane, T. Masumoto, and S.-I. Ueno (1992). Change in nutritive value of defatted soybean meal with duration of heating time for young yellowtail. *Nippon Suisangakkaishi* 58:1351-1359.

Shimeno, S., Y. Kanetaka, T. Ruchimat, and M. Matsumoto (1995). Nutritional evaluation of several soy proteins for fingerling yellowtail. *Nippon Suisangakkaishi* 61:919-926.

Shimeno, S., M. Kumon, H. Ando, and M. Ukawa (1993). The growth performance and body composition of young yellowtail fed with diets containing defatted soybean meal for a long period. *Nippon Suisangakkaishi* 59:821-825.

Shimeno, S., T. Mima, T. Imanaga, and K. Tomaru (1993). Inclusion of combination of defatted soybean meal, meat meal and corn gluten meal to yellowtail diets. *Nippon Suisangakkaishi* 59:1889-1895.

Shimeno, S., T. Mima, O. Yamamoto, and Y. Ando (1993). Effects of fermented defatted soybean meal in diet on the growth, feed conversion, and body composition of juvenile yellowtail. *Nippon Suisangakkaishi* 59:1883-1888.

Shimeno, S., T. Ruchimat, M. Matsumoto, and M. Ukawa (1997). Inclusion of full-fat soybean meal in diet for fingerling yellowtail. *Nippon Suisangakkaishi* 63:70-76.

Shimeno, S., S.-I. Seki, T. Masumoto, and H. Hosokawa (1994). Post feeding changes in digestion and serum constituent in juvenile yellowtail force-fed with raw and heated defatted soybean meals. *Nippon Suisangakkaishi* 60:95-99.

Silva, J.G. and A. Oliva-Teles (1998). Apparent digestibility coefficients of feedstuffs in seabass *(Dicentrarchus labrax)* juveniles. *Aquatic Living Resources* 11:187-191.

Suhaimee, A.M., C.M. Utama, M.B. Khan, and A. Daud (1999). Substitution of fish meal in seabass *(Lates calcarifer)* diet with soybean meal. *Friday Newsletter* 4:16-17.

Takagi, S., S. Shimeno, H. Hosokawa, and M. Ukama (2001). Effects of lysine and methionine supplementation to a soy protein concentrate diet for red sea bream *Pagrus major. Fisheries Science* 67:1088-1096.

Takii, K., S. Shimeno, M. Nakamura, Y. Itoh, H. Obatake, H. Kumai, and M. Takeda (1989). Evaluation of soy protein concentrate as partial substitute for fish meal protein in practical diet for yellowtail. In *Proceeding of the Third International Symposium on Nutrition and Feeding in Fish*. Toba, Japan, pp. 281-288.

Tibaldi, E. and F. Tulli (1998). Partial replacement of fish meal with soybean products in diets for juvenile sea bass (Dicentrarchus labrax). In *Recent Advances in Finfish and Crustacean Nutrition. Proceedings of the VIII International Symposium on Nutrition and Feeding in Fish.* Las Palmas De Gran Canaria, Spain, pp. 149.

Tibaldi, E., F. Tulli, and M. Amerio (1999). Feed intake and growth responses of sea bass *(D. labrax)* fed different plant-protein sources are not affected by supplementation with a feeding stimulant. In *Recent Progress in Animal Production Science,* I.G. Piva, G. Bertoni, F. Masoero, P. Bani, and L. Calamari (eds.). Milano, Italia: Franco Angeli, pp. 752-754.

Tulli, F., M. Ramelii, E. Tibaldi, F. Manetti, D. Volpatti, and M. Galeotti (2000). Feeding seabass *(Dicentrarchus labrax)* juveniles with soybean products: Effects on growth, feed utilization, serological response and non-specific defense. *Bolletino Societa Italiana di Patologia Ittica* 29:3-9.

Tulli, F. and E. Tibaldi (2001). Apparent nutrient digestibility of different protein sources for sea bass *(D. labrax).* In *Recent Progress in Animal Production Science. II. Proceedings of the ASPA XIV Congress.* Florence, Italy, pp. 697-699.

Tulli, F., E. Tibaldi, and A. Comin (1999). Dietary protein sources differently affect plasma lipid levels and body fat deposition in juvenile sea bass. In *Recent Progress in Animal Production Science. Proceedings of the ASPA XIII Congress.* Piacenza, Italy, pp. 782-784.

Ukawa, M., K. Takii, M. Nakamura, and H. Kumai (1994). Utilization of soybean meal for red seabream diet. *Suisanzoshoku* 42:335-358.

Venou, B., M.N. Alexis, E. Fountoulaki, and P. Mattila (1997). Partial substitution of fish meal with extruded and non extruded soybean meal diets for the gilthead seabream *(Sparus aurata* L.). *Proceedings of the Martinique 1997 Island Aquaculture & Tropical Aquaculture.* European Aquaculture Society, pp. 314-315.

Viyakarn, V., T. Watanabe, H. Aoki, H. Tsuda, H. Sakamoto, N. Okamoto, N. Iso, et al. (1992). Use soybean meal as substitute for fish meal in a newly developed soft-dry pellet for yellowtail. *Nippon Suisangakkaishi* 58: 1991-2000.

Watanabe, T. (2002). Strategies for further development of aquatic feeds. *Fisheries Science* 68:242-252.

Watanabe, T., H. Aoki, K. Shimamoto, M. Hadzuma, M. Maita, Y. Yamagata, V. Kiron, et al. (1998). A trial to culture yellowtail with non-fish meal diets. *Fisheries Science* 64:1415-1423.

Watanabe, T., H. Aoki, V. Viyakarn, M. Maita, Y. Yamagata, S. Satoh, and T. Takeichi (1995). Combined use of alternative protein source as a partial replacement for fish meal in a newly developed soft-dry pellet for yellowtail. *Suisanzoshoku* 43:511-520.

Watanabe, T., H. Aoki, K. Watanabe, M. Maita, Y. Yamagata, and S. Satoh (2001). Quality evaluation of different types of non-fish meal diets for yellowtail. *Fisheries Science* 67:461-469.

Watanabe, T., V. Viyakarn, H. Kimura, K. Ogawa, N. Okamoto, and N. Iso (1992). Utilization of soybean meal as a protein source in a newly developed soft-dry pellet for yellowtail. *Nippon Suisangakkaishi* 58:1761-1773.

Webster, C.D. and C. Lim (2002). *Nutrient Requirements and Feeding of Finfish for Aquaculture.* New York, NY: Cabi Publishing.

Yamamoto, T., A. Akimoto, S. Kishi, T. Unuma, and T. Akiyama (1998). Apparent and true availability of amino acids from several protein sources for fingerlings rainbow trout, common carp and red sea bream. *Fisheries Science* 64:448-458.

Yamamoto, T.. T. Unuma, and T. Akiyama (1998). Apparent availability of amino acids from several protein sources for fingerling Japanese flounder. *Bulletin of the National Research Institute of Aquaculture* 27:27-35.

Chapter 12

Cottonseed Meal in Fish Diets

Chhorn Lim
Menghe H. Li
Edwin Robinson
Mediha Yildirim-Aksoy

INTRODUCTION

Aquaculture is growing more rapidly than all other food animal production sectors. Since 1970, aquaculture has increased at an average
compounded rate of 9.2 percent per year, compared to only 1.4 percent for capture fisheries and 2.8 percent for terrestrial farmed animal
production. Global supplies of fish, crustaceans, and mollusks from
aquaculture have grown from 3.9 percent of total production by weight
in 1970 to 27.3 percent in 2000 (FAO, 2002). Paralleling the growth
of the industry has been the intensification of aquaculture production
as well as product diversification. The continued expansion of the industry depends, among other factors, on availability of cost-effective
diets and proper feeding strategies, because feeding is the single most
costly component in intensive aquaculture operations.

Protein is the most expensive component in diets of aquatic species. Fish meal (FM), because of its high protein content, balanced
essential amino acid profile, high digestibility, excellent source of n-3
highly unsaturated fatty acids, and minerals, has traditionally been
used as a major protein source in aquaculture diets, particularly those
of carnivorous species. The shortage of FM supply, coupled with its
increased demand, has lead to increased costs of FM. In recent years,

Alternative Protein Sources in Aquaculture Diets
© 2008 by The Haworth Press, Taylor & Francis Group. All rights reserved.
doi:10.1300/5892_12   *313*

however, extensive research has been conducted to evaluate the nutritional values of alternative protein sources to partially or totally replace FM. Soybean meal (SBM), because of its high nutritional value, availability, and low cost, is the most commonly investigated plant feedstuff in aquaculture diets. Currently, SBM is the major protein source used in omnivorous and herbivorous warmwater fish diets and comprises up to 50 percent of the diet of channel catfish, *Ictalurus punctatus* (NRC, 1993). Replacement of SBM with less expensive plant proteins would be beneficial in reducing diet cost. Cottonseed meal (CSM), which ranks second in the United States (Robinson and Li, 1995) and third in the world (Swick and Tan, 1995) in terms of tonnage among the plant protein feedstuffs produced, is less expensive than SBM on a per unit of protein basis (Robinson and Li, 1995). The use of CSM in aquatic animal diets has recently been reviewed by Li and Robinson (2006). Cottonseed meal contains gossypol, a compound that, when present at high concentrations in the diets, can be toxic to fish. Cottonseed meal is also low in some essential amino acids, particularly lysine and the sulfur-containing amino acids methionine and cystine. Thus, the amount of CSM that can be included in fish diets depends mainly on the levels of free gossypol (FG), available lysine, and sulfur-containing amino acids.

This chapter will provide an overview on toxicity of gossypol, effects of dietary factors on gossypol toxicity, tissue gossypol accumulation and elimination, gossypol and reproductive performance, gossypol and fish health, nutritional value of CSM, and use of CSM in finfish diets.

GOSSYPOL TOXICITY

Gossypol is a naturally occurring phenolic pigment indigenous to the cotton plant genus *Gossypium* and related genera. Gossypol exists in all parts of the cotton plant, but is concentrated in the pigment glands in the seeds (Moore and Rollins, 1961). The gossypol in whole seed is in the free form (FG), to which toxicity to monogastric animals is attributed (Wedegaertner, 1981; Jones, 1987). The amount of FG in raw seed varies considerably (0.39 to 1.75 percent) depending on the environmental factors (Cherry et al., 1978), and cotton species and varieties (Boatner et al., 1949). The processing conditions applied to the production of CSM also affect the concentration of FG in the meal. During processing, varying amounts of FG bind with protein and/or minerals and are converted to bound form that is of little significance to animals because it passes through the gastrointestinal tract unabsorbed (Braham et al., 1967; Wedegaertner, 1981). The total content of gossypol in CSM is not affected by the processing methods used because total gossypol is equal to the free plus bound amounts (Jones, 1991). Free gossypol concentrations in CSM, however, are affected by the processing method used to extract the oil from cottonseed. Heat, moisture, air oxygen, compression, and friction promote a reaction between FG and free epsilon amino group of lysine to form an indigestible complex (Damaty and Hudson, 1979) which lowers lysine availability to about 66 percent for channel catfish (Wilson et al., 1981). The mole ratio of bound gossypol to bound lysine is 1:2 (Reiser and Fu, 1962). Other amino acids that are prone to take part in the formation of bound gossypol are serine, threonine, methionine, and the highly hydrophobic amino acids (Damaty and Hudson, 1979).

Free gossypol also reacts with other constituents (peptides, phospholipids, etc.) to form the corresponding condensation products that are soluble in aqueous acetone and represent the so-called soluble bound gossypol, which is toxic. Also, some gossypol is oxidized and degraded to various products, most of which have not been identified, and differ in their properties and special characteristics. Mean concentrations of FG in CSM produced by three major processes (screw pressed, prepress solvent-extracted, and direct solvent-extracted) are given in Table 12.1. Direct solvent-extracted CSM contains relatively high level of FG, averaging 0.3 percent. The average value for expander solvent-extracted (0.10 percent) is about twice that of the prepress solvent-extracted CSM (0.05 percent). Mechanically extracted (screw pressed) CSM contains the lowest level of FG (0.04 percent). Mean values of total gossypol range from 1.02 to 1.13 percent.

Gossypol exists in the cotton plant as a naturally occurring mixture of two enantiomers (isomers), [(+)-gossypol and (−)-gossypol], with different optical properties (King and DeSilva, 1968). The proportion of isomers varies with the species and variety of cotton (Cass et al., 1991) and the individual isomers exhibit different toxicological and pharmacokinetic effects. Zhou and Lin (1988) and Cass et al. (1991) reported that the (−)-gossypol isomer predominates in *G. barbadense*

TABLE 12.1. Mean free and total gossypol concentrations (as-fed basis) in cottonseed meal produced by three major processes and U.S. process capacity.[a]

Process	Free gossypol (%)	Total gossypol (%)	Percent U.S. process capacity (1991-1992)
Screw pressed	0.04	1.02	9
Direct solvent-extracted	0.3	1.04	8
Expander solvent-extracted	0.10 (0.073-0.131)		83[a]
Prepress solvent-extracted	0.05	1.13	

[a]Process capacity for both expander and prepress solvent-extracted CSM. *Sources:* Jones (1987, 1991).

whereas the (+)-isomer predominates in *G. hirsutum*. Processing of cottonseed into meal does not alter the proportion of isomers. It has been shown that the (−)-isomer is more toxic to rats (Wang et al., 1987) and broilers (Bailey et al., 2000) than the (+)-isomer. In contrast, Chen et al. (1986) reported that effect of the (−)-isomer on the number of DNA-strand breaks in human leukocytes was only about half of that of the (+)-isomer or the racemate.

Free gossypol, when present in sufficient amount in the diet, has been shown to be toxic to monogastric animals, including fish. Toxic concentrations of FG to fish depend on species, differences in strains within the same species, age or size, environmental conditions, dietary levels of nutrients, and feeding management, such as duration of feeding and feeding rate. Poor growth and feed efficiency, anorexia, abnormal hematological parameters, poor reproduction, and histopathological changes have been reported as signs of gossypol toxicity (Berardi and Goldblatt, 1980; Barros et al., 2000; Yildirim et al., 2003).

Juvenile rainbow trout *(Oncorhynchus mykiss)* have been reported to tolerate dietary FG from gossypol-acetic acid (GAA) at levels up to 250 mg/kg diet (Roehm et al., 1967). Herman (1970) observed that rainbow trout fed a diet containing 290 mg FG from CSM/kg had a growth rate similar to that of the control group for about thirty weeks, but a slight reduction was evident thereafter. It was suggested that 300 mg FG/kg diet is near the minimum level affecting growth performance. These values are considerably lower than 1,000 mg/kg reported by Sinnhuber et al. (1968) as the maximum tolerance limit for rainbow trout. Histopathological changes, including thickening of the glomerular basement membrane of the kidney and liver necrosis, were observed in trout fed as low as 95 mg FG from CSM/kg (Herman, 1970). Yellow pigment (ceroid) deposition in liver of rainbow trout receiving CSM diets containing 290 mg FG/kg or higher and in the spleen and posterior kidney of fish fed the 531-mg FG diet was also observed.

For adult rainbow trout, Dabrowski et al. (2000) found no differences in weight gain of fish fed CSM diets containing up to 990 mg FG/kg for 131 days. However, a significant decrease in hematocrit and hemoglobin was noted in fish fed the 990-mg FG diet as compared to those fed diets containing 520 mg/kg or less. Blom et al. (2001) reported that growth and survival of adult rainbow trout were not affected by feeding diet having 619 mg FG/kg from CSM for ten months, but hematocrit and hemoglobin decreased with increasing dietary FG levels. In another study, Dabrowski et al. (2001) obtained lower weight gain and hemoglobin in adult rainbow trout fed CSM diets containing 362 mg FG/kg or more for 258 days.

Published information on the effect of dietary levels of FG on the growth performance of channel catfish is not consistent. Dorsa et al. (1982) reported that juvenile channel catfish could tolerate up to 900 mg FG/diet from either GAA or CSM. Reduction in growth was observed in fish fed the 1,137-mg FG diet. A more recent study by Barros et al. (2000, 2002) showed a reduction in weight gain and feed consumption of juvenile catfish fed the CSM diet containing 671 mg FG/kg. However, total cell count, red and white cell counts, hematocrit, and hemoglobin were not affected. Yildirim et al. (2003) reported a significant growth reduction in juvenile catfish fed purified diets containing FG from GAA as low as 300 mg/kg diet. Hematocrit, hemoglobin, and red blood cell count decreased at dietary FG concentrations of 600, 900, and 1,200 mg/kg or higher, respectively.

As there was variation among the reported concentrations of FG adversely affecting channel catfish, Yildirim-Aksoy, Lim, Wan et al. (2004) conducted a study to evaluate the toxicity of natural FG from CSM and FG from GAA using practical diets containing graded levels of glanded CSM (for natural FG diets), and glandless CSM (for GAA diets), respectively. They observed that neither source of dietary

FG at levels up to 800 mg/kg affected weight gain, feed intake, feed efficiency, survival, red and white blood cell counts, hematocrit, and hemoglobin of juvenile catfish. Lim et al. (2005) evaluated the effect of gossypol from GAA in purified or practical diets. They reported that toxic concentrations of gossypol on channel catfish depend on type of diet. Levels of gossypol up to 1,200 mg/kg in practical diet had no negative effect on weight gain, feed intake, feed and protein efficiencies, and survival. These parameters, however, were significantly lower in the group fed purified diets supplemented with 1,200 mg/kg gossypol.

Compared to rainbow trout and channel catfish, tilapia appear to tolerate relatively high levels of dietary FG. Robinson et al. (1984) reported that weight gain, feed efficiency, and survival of juvenile blue tilapia *(Oreochromis aureus)* were not affected by feeding purified diets supplemented with FG from GAA up to 1,800 mg/kg. For juvenile Nile tilapia *(O. niloticus),* Lim et al. (2003) found that supplementation of purified diets with FG from GAA up to 1,600 mg/kg had no effect on growth, survival, hematocrit, white blood cell count, and hemoglobin. Mbahinzireki et al. (2001), however, found that tilapia *(O.* spp.) fed CSM-based diets can only tolerate FG at a level of up to 520 mg/kg. Free gossypol levels of 700 mg/kg diet or higher resulted in decreased growth and feed efficiency.

EFFECTS OF DIETARY FACTORS ON GOSSYPOL TOXICITY

It has long been postulated that the physiological effects of gossypol may vary considerably depending on the composition of the ration, particularly the quality and quantity of protein, and mineral contents (Berardi and Goldblatt, 1980). As has been discussed previously, FG reacts with the epsilon amino group of lysine to form an indigestible complex, thus, reducing the toxicity of FG to monogastric animals. The addition of crystalline lysine, however, did not provide the protective effect of FG toxicity in swine as efficiently as the addition of high levels of good-quality intact proteins, such as FM and SBM, because free lysine-gossypol complex can be absorbed from the digestive tract (Lyman, 1966). Soybean meal diets provided superior performance of swine to peanut meal and CSM diets at all levels of FG

(Kornegay et al., 1961). Dietary levels of protein have also been reported to affect tolerance levels of dietary FG. Hale and Lyman (1957) showed that pigs fed 30 percent-protein diets containing 100, 200, or 300 mg FG/kg showed no symptoms of FG toxicity compared to those fed 15 percent-protein diets. Herman (1970) suggested that the considerably lower FG tolerance level (300 mg/kg) in juvenile rainbow trout obtained in his study relative to that (1,000 mg/kg) of Sinnhuber et al. (1968) was probably due to the higher protein content (50 percent versus 58 percent) in the diets used by Sinnhuber et al. (1968). The higher dietary protein level (38 percent) used in the study of Dorsa et al. (1982) as compared to that (32 percent) of Barros et al. (2000, 2002) may also have contributed to the differences in the toxic concentrations of FG in juvenile catfish reported in these studies.

The mechanism by which extra protein in the diet protects against FG toxicity is through the formation of insoluble, indigestible gossypol-protein complex (Lyman, 1966). Lim et al. (2005) reported that the type or quality of dietary protein may also influence FG toxicity. They observed that, at equal dietary protein levels, gossypol supplemented to practical diets was less toxic than that supplemented to purified diets. Yildirim-Aksoy, Lim, Wan et al. (2004) attributed the variation among the toxic concentrations of dietary FG in channel catfish obtained by various workers to the differences in dietary nutrient contents, particularly minerals.

The benefit of minerals in counteracting the toxic effects of dietary gossypol has long been recognized. The use of iron salts to reduce the toxicity of FG in diets of monogastric animals was reported in 1913 by W.A. Withers and J.F. Brewster (Berardi and Goldblatt, 1980). Free gossypol reacts with ferrous ion to form the strong ferrous-gossypolate complex, thus preventing its absorption across the intestinal tract of animals (Gallup, 1928; Rands, 1966; Herman and Smith, 1973; Clawson et al., 1975). An in vitro study by Jonassen and Demint (1955) showed that each molecule of FG reacts with one ferrous ion. Bressani et al. (1966) reported that this binding of FG by iron is effective in the ration before it is consumed. Braham et al. (1967), however, indicated that the binding of FG by iron probably takes place at the intestinal level.

The efficacy of iron is influenced by the form of iron and dietary levels of other minerals. The addition of ferric oxide to the diets was of no value in counteracting the effects of FG (probably due to its insolubility), but ferric citrate and ferrous ammonium sulfate at a 3:1 weight ratio of iron to FG were equally effective in promoting normal growth of rats. The addition of smaller amounts of these two salts resulted in less favorable growth (Gallup, 1928). Iron from iron sulfate at a 1:1 weight ratio of iron to FG has been successfully used to counteract the toxicity of FG in diets of swine, broilers, and rats (Bressani et al., 1966; Cabell and Stevenson, 1966; Smith and Clawson, 1970; Jones, 1987). El-Saidy and Gaber (2004) reported that adding 972 mg iron/kg from ferrous sulfate to a CSM-based diet containing 972 mg FG reduced the negative effect of gossypol and improved growth, feed efficiency, and hematological parameters in Nile tilapia. In channel catfish, however, Barros et al. (2000) observed that supplementation of iron from iron sulfate at a 1:1 weight ratio of iron to FG to diets containing 336 or 671 mg FG/kg from CSM had no effect on reducing gossypol toxicity. Other studies have reported that addition of iron sulfate in diets of swine (Smith and Clawson, 1970) and poultry (El Boughy and Raterink, 1989) increased their tolerance to FG toxicity to a varying degree.

The inactivation of FG by iron has also been reported to improve the digestion and absorption of the meal. Supplementation of 600 mg iron/kg from ferrous sulfate resulted in increased metabolizable energy of a glanded, high-gossypol CSM for chicks by as much as 24 percent (Rojas and Scott, 1969). Adding iron from iron sulfate at 1:1 to 2:1 weight ratios of iron to FG to Nile tilapia diets containing 972 mg FG/kg from CSM (67 percent CSM as a total replacement of dietary FM) significantly improved apparent protein and energy digestibility to values comparable to those of the FM-control diet (El-Saidy and Gaber, 2004).

Other minerals, such as calcium, zinc, copper, and manganese, have also been shown to alter the physiological effects of FG in CSM. Calcium has been reported to have synergistic effect with iron in detoxification of FG. Addition of calcium increased the effectiveness of FG-iron complex formation leading to full protection of swine against FG toxicity (Braham et al., 1967). Calcium alone, however, did not protect the swine against FG toxicity (Bressani et al., 1966; Braham et al., 1967). Zinc, manganese, and copper, and their combination, have also been evaluated for their effect in detoxifying CSM-based rat diets containing 750 mg FG/kg (Smith and Clawson, 1970). Zinc and manganese (at 500 and 1,600 mg/kg, respectively) alone did not improve the growth of rats but did significantly lower bound gossypol accumulation in liver. Their combination was less effective in lowering liver bound gossypol. The combination of copper and zinc (at 300 and 500 mg/kg, respectively), and copper and manganese (each at 1,600 mg/kg) led to significant increases in liver bound gossypol relative to the control.

TISSUE GOSSYPOL ACCUMULATION AND ELIMINATION

Following feeding, gossypol is absorbed in the digestive tract and accumulated in the tissues in both the free and bound forms. The concentrations of tissue gossypol are directly related to the level of dietary gossypol (Sharma et al., 1966; Roehm et al., 1967; Dorsa et al., 1982; Lee and Dabrowski, 2002; Yildirim et al., 2003; Yildirim-Aksoy, Lim, Wan et al., 2004), feeding duration (Roehm et al., 1967), and type of diet (Lim et al., 2005), but inversely related to dietary protein content (Sharma et al., 1966). The retention of gossypol is is at its highest in the liver and is followed by spleen, kidney, intestine, blood plasma, stomach, and muscle (Roehm et al., 1967; Smith and Clawson, 1970; Dorsa et al., 1982; Lee and Dabrowski, 2002). Tissue levels of total gossypol in young rainbow trout fed the diet containing 1,000 mg/kg of FG from CSM for eighteen months were 560, 471, 399, 391, 101, and 13.6 µg/g for liver, spleen, kidney, heart, GI tract, and muscle, respectively (Roehm et al., 1967). Free gossypol contents in liver, kidney, and muscle of juvenile channel catfish fed a diet containing 1,400 mg FG from GAA for eight weeks were 65.6, 20.7 and 2.2 µg/g tissue, respectively (Dorsa et al., 1982). Yildirim et al. (2003) and Yildirim-Aksoy, Lim, Wan et al. (2004) found significant linear relationships between liver total gossypol levels of juvenile channel catfish and dietary FG concentrations. However, for diets containing 400 and 800 mg FG/kg, fish fed diets in which FG was supplied by CSM has significantly higher levels of total gossypol than fish fed diets containing FG from GAA.

Since diets used in the study of Yildirim-Aksoy, Lim, Wan et al. (2004) were based on FG content and CSM contains bound gossypol in addition to FG, total dietary gossypol levels in diets containing CSM were higher than those of corresponding diets containing GAA. They suggested that, although bound gossypol has been reported to be inactive physiologically, some of the dietary bound gossypol may be absorbed and deposited in the liver of channel catfish. Digestive enzymes may have hydrolyzed gossypol-protein complexes rendering them available for absorption, or gossypol may have been absorbed while bound to individual amino acids. Herman and Smith (1973) reported gossypol deposition in livers of rats fed diets containing bound gossypol. Lyman and Widmer (1966) found that as much as 50 percent of protein-bound gossypol (gossypol-protein complex preparations containing no detectable levels of FG) was converted to FG after digestion with proteolytic enzymes. Low concentrations of gossypol were also detected in ovaries, semen, sperm, and swim-up fry of rainbow trout. Tissue levels were generally directly related to dietary levels of FG fed to broodstock (Dabrowski et al., 2000, 2001; Blom et al., 2001). Rinchard et al. (2002) reported that, in the same group of tilapia sp., the concentration of total gossypol in testis was lower than that in ovaries.

Fish, similar to other monogastric animals, accumulated the (+)-isomer of gossypol in tissues at higher concentrations than the (−)-isomer although diets contained equal proportions of gossypol isomers (Bailey et al., 2000; Blom et al., 2001; Dabrowski et al., 2001; Lee and Dabrowski, 2002; Yildirim et al., 2003; Yildirim-Aksoy, Lim, Wan et al. (2004)). Regardless of the source of dietary gossypol (from CSM or GAA) the accumulation rate of (+)-gossypol in livers of channel catfish was always higher than the (−)-gossypol.

In rats, Wang et al. (1992) reported that the (−)-gossypol concentration in blood plasma decreased at a faster rate than the (+)-gossypol concentration. They attributed this to more specific binding of (+)-isomer to serum proteins, thus, resulting in a higher excretion rate of (−)-isomer. The higher affinity of (+)-gossypol for proteins in blood and seminal plasma, and spermatozoa of rainbow trout has also been reported (Dabrowski et al., 2001; Lee and Dabrowski, 2002). Sampath and Balaran (1986), however, showed that both gossypol enantiomers had almost equal affinities to serum albumin, but the (−)-isomer disappeared at a faster rate from the body than the (+)-isomer.

The retention rate of (+)- and (−)-isomers in fish tissues is dependant on dietary gossypol concentrations. Fish fed higher levels of dietary gossypol retained higher proportion of (−)-gossypol than fish fed lower gossypol containing diet (Lee and Dabrowski, 2002; Yildirim et al., 2003; Yildirim-Aksoy, Lim, Wan et al., 2004).

Lyman and Widmer (1966) reported on the recovery of radioactive gossypol and/or gossypol metabolites (FG, bound gossypol, and decomposition products of gossypol) in feces and tissues of chicks fed diets containing gossypol labeled in the formyl position with ^{14}C. An average of 89.26 percent of the ^{14}C activity was found in feces and only 8.90 percent in tissues. These authors also reported that a much higher level (25.14 percent) of ^{14}C activity was recovered in the tissues of pigs as compared to that of the chicken and only 61.65 percent in feces. They concluded that the higher retention rate of gossypol in pig tissues probably due to the greater sensitivity of the pig to gossypol toxicity than the chicken. Lee et al. (2000) reported that, following feeding of juvenile rainbow trout and Nile tilapia with diets containing gossypol from CSM, the majority of gossypol is excreted through feces.

Absorbed gossypol is eliminated from the liver via the bile and is slowly eliminated from the tissues when removed from the gossypol-containing diets (Sharma et al., 1966; Roehm et al., 1967; Lee, Dabrowski, et al., 2002). Roehm et al. (1967) reported an apparent shift of accumulated gossypol from muscle, stomach, intestines, and heart to the liver, kidney, and spleen of rainbow trout immediately on withdrawal from feeding the diet containing 1,000 mg gossypol/kg for eighteen months. At the time, there was an increase in bound gossypol levels in all tissues except in muscle and GI tract. They also observed a significant decrease in FG levels in all tissues during the ten-week withdrawal period, but bound gossypol levels increased or showed little change. In pigs, Buitrago et al. (1970) found that gossypol from other body tissues was being released and deposited in the liver and spleen, and that the actual rate of depletion may have been more rapid than indicated by analysis of livers and spleens. They observed that, after feeding the gossypol-free diet for three weeks, FG depleted at a slower rate from the spleen than from the liver. They suggested that FG may form a chelated product with iron in the liver and spleen, and that the differences in iron content of these two organs may account for the difference in the rate of gossypol depletion.

GOSSYPOL AND REPRODUCTIVE PERFORMANCE

The antifertility properties of gossypol were first discovered by Chinese scientists around 1950. Gossypol acts as a male antifertility agent and this property has been proven in several species, including human beings with 99.9 percent efficacy (NCGMAA, 1978). The effect, however, is reversible and males usually recover fertility within three or more months following gossypol withdrawal. Gossypol might also reduce female fertility by disrupting the oocyte or embryo (Randel et al., 1992). It has been shown to inhibit spermatosis in human and certain other mammals (Wu, 1989). Gossypol can cause degeneration of tubular epithelium, alter the ultrastructure of mature spermatozoa, inhibit sperm motility, and increase the spermatozoa mortality rate (Hoffer, 1982). Its anti-androgenic effects in terms of inhibition of testosterone production in rats have been demonstrated in vitro as well as in vivo (Hadley et al., 1981).

Robinson and Tiersch (1995) evaluated the effects of long-term feeding of diets containing 0, 25, 37.5, and 52 percent CSM (with FG contents of 0, 200, 300, and 400 mg/kg of diet, respectively) as replacements of SBM on reproductive performance of male channel catfish. They found that testis weight, gonadosomatic index, and sperm motility was not affected by dietary levels of FG from CSM of up to 400 mg/kg diet. Dabrowski et al. (2000) evaluated the reproductive capacity of rainbow trout fed diets containing 0, 14.70, 29.42, 44.11, and 58.80 percent CSM (with FG levels of 0, 130, 280, 470, and 590 mg/kg diet, respectively) as substitutes for 0, 25, 50, 75, and 100 percent FM (equal proportion of herring and menhaden FM) protein for 131 days. After 112 and 131 days of feeding, testis weights, concentration of testosterone, and 11-ketotestosterone of fish fed diets containing 130 and 280 mg FG/kg were elevated in comparison with fish fed other diets. However, sperm quality measured based on sperm concentration, motility, fertilizing rate sampled on day 71 did not differ among fish fed various dietary levels of total gossypol from CSM. In another study in which rainbow trout were fed the same experimental diets for 256 days, Dabrowski et al. (2001) found that sperm concentrations decreased with increasing dietary FG levels from 0 to 470 mg/ kg, but sperm motility and fertilizing rate were not affected by dietary FG levels up to 590 mg/kg. In yellow perch *(Perca flavescens)* FG from GAA at a concentration of 104 mg/L of water can immobilize sperm in vitro (Ciereszko and Dabrowski, 2000).

Lee, Rinchard, et al. (2002) reported a long-term (thirty-five months) effect of replacing 0, 25, 50, 75, and 100 percent FM protein by CSM (with total gossypol levels of 0, 2,003, 4,887, 7,037, and 9,548 mg/ kg diet, respectively) on reproductive performance of rainbow trout broodstock. They observed that, after one year of feeding, reproductive performance (sperm concentration and egg weight and quality) decreased in broodstock fed the diet in which 100 percent FM protein was replaced by CSM with a total gossypol content of 9,548 mg/kg. In year 2, reproductive performance in both male and female as well as the progeny growth were reduced in fish fed the two highest levels of dietary CSM (total gossypol contents of 7,037 and 9,548 mg/kg). In the third year, however, no significant effects of dietary gossypol from CSM on reproductive performances of both male and female were detected. They suggested that 50 percent of dietary FM protein can be replaced by CSM (total dietary gossypol up to 4,887 mg/kg) without affecting rainbow trout broodstock performance.

GOSSYPOL AND FISH HEALTH

Gossypol, although toxic to monogastric animals when present in large amount in the diet, has been shown to possess antimicrobial (Stipanovic et al., 1975), anti-HIV (Lin et al., 1989), antifungal (Bell, 1967), and antitumor (Joseph et al., 1986; Benz et al., 1990; Wang et al., 2000) activities. The (−)-gossypol enantiomer had the greatest antitumor (Benz et al., 1990) and antiviral (Lin et al., 1989) properties. Yildirim-Aksoy, Lim, Dowd et al. (2004) demonstrated that racemic gossypol from GAA and (+)- and (−)-optical isomers of gossypol have antibacterial effect against *Edwardsiella ictaluri,* the causative agent of enteric septicemia of catfish. Concentrations of racemic gossypol, and (+)- and (−)-gossypol isomers of 1.5 µg/mL or higher significantly reduced the number of bacterial colonies compared to that of the control. The growth of *E. ictaluri* was completely inhibited on agar plates supplemented with 3 μg/mL, regardless of the form of gossypol. The inhibitory effect of (+)-gossypol was higher than the (−)-gossypol or GAA. They suggested, however, that the action of these forms of gossypol on *E. ictaluri* is bacteriostatic rather than bactericidal. Gossypol has also been reported to be effective in the immobilization of three parasites, *Trypanosoma cruzi,* the cause of Chagas' disease (Montamat et al., 1982), *Plasmodium falciparum,* the cause of malaria (Heidrich et al., 1983); and *T. brucei* (Eid et al., 1988).

Since gossypol is a biologically active compound that exhibits antimicrobial activity, its presence in the diet may improve the resistance of aquatic animals to infectious diseases. Wood (1968) reported that replacement of corn gluten meal with CSM (FG level was not reported) in diets fed to pacific salmon (*Oncorhynchus* spp.) produced similar growth, but the incidence of mortality due to bacterial kidney disease was much greater in fish fed the corn gluten meal diet as compared to those fed the CSM diet. Barros et al. (2002) fed juvenile channel catfish diets containing 0, 27.5, and 55.0 percent CSM as replacements of 0, 50, and 100 percent SBM and found that macrophage chemotaxis in the presence of *E. ictaluri* exoantigen was significantly higher for fish fed the diet containing 55.0 percent CSM (671 mg gossypol/kg diet). Cumulative mortality of fish experimentally infected with *E. ictaluri* was lower and antibody titers following challenge were higher for fish fed CSM-containing diets, but the values did not differ for fish fed the 27.5 (336 mg gossypol/kg diet) or 55.0 percent CSM diets. In Nile tilapia, Lim et al. (2002) observed that fish fed diets containing 38 percent CSM (as a replacement of 66.7 percent SBM) or higher, although had adversely affected growth performance, had increased superoxide anion production and macrophage phagocytic activity. Lim et al. (2002) and Barros et al. (2002) suggested that gossypol or other compounds present in CSM may have a beneficial effect by improving the immune response and resistance of catfish to *E. ictaluri* infection.

Yildirim et al. (2003) fed juvenile channel catfish purified diets supplemented with graded levels of gossypol (0, 300, 600, 900, 1,200, and 1,500 mg/kg) from GAA. They observed that superoxide anion production, antibody titer, and the onset of first mortality following *E. ictaluri* challenge were not affected by dietary levels of gossypol, but a reduction in serum protein occurred in fish fed 900 mg/kg or higher gossypol diets. However, improved macrophage chemotaxis ratio was observed at dietary gossypol levels of 300 mg/kg or higher, while lysozyme activity and resistance of fish to *E. ictaluri* challenge were enhanced at dietary gossypol levels of 900 mg/kg or higher. A similar study with Nile tilapia indicated that serum protein, and antibody production were not affected by dietary levels of FG from GAA. Serum lysozyme activity was enhanced in fish fed the 200-mg gossypol/kg diets relative to fish fed 50 mg gossypol/kg or lower, but did not differ from those fed diets with 100 mg/kg or higher gossypol. Mortality fourteen-day post-challenge with *S. iniae* was lowest and highest in fish fed the 100-mg and 800-mg gossypol diets, respectively.

In a more recent study, Yildirim-Aksoy, Lim, Wan et al. (2004) utilized practical diets to evaluate the effect of dietary FG from GAA and natural FG from glanded CSM at levels of 0, 100, 200, 400, and 800 mg/kg on immune response and resistance of channel catfish to *E. ictaluri* infection. They found that macrophage chemotaxis ratio, phagocytic activity of leukocytes, and antibody titer against *E. ictaluri* were not affected by dietary sources or levels of dietary FG. However, lysozyme activity and superoxide anion production were enhanced at dietary levels of gossypol of 200 to 800 mg/kg depending on the sources. Natural FG from glanded CSM appeared to have better enhancement effect than FG from GAA. However, neither the sources nor the levels of dietary FG improve the resistance of fish against *E. ictaluri* infection. Since the levels of dietary gossypol found to improve immune response and disease resistance in juvenile channel catfish also adversely affected their growth performance and physiological conditions, gossypol appears to be of limited or no benefit in improving the disease resistance in channel catfish (Yildirim et al., 2003; Lim et al., 2004).

NUTRITIONAL VALUE OF CSM

Nutrient Composition

Proximate composition of CSMs produced using three different processes and dehulled, solvent-extracted SBM is presented in

TABLE 12.2. Proximate composition of prepress, solvent-extracted cottonseed meal and dehulled, solvent-extracted soybean meal (SBM).

	Composition (% on an as-fed basis)			
	Cottonseed meal			
	Screw pressed	Direct solvent	Prepress solvent	Soybean meal
Dry matter	91	90	90	88
Crude protein	41	41	41	47.8
Crude fat	3.9	2.1	1.5	1
Crude fiber	12.6	11.3	12.7	3
Ash	6.2	6.4	6.4	6

Source: Feedstuffs (2006).

Table 12.2. Except the level of fat, processing methods have little effect on proximate nutrient contents of CSM. Mechanically extracted or screw-pressed CSM has the highest level of fat, whereas prepress or expander solvent-extracted meal contains the lowest fat content. Compared to SBM, CSM is lower in crude protein but higher in fat, crude fiber, and ash.

Essential amino acid composition of prepress, solvent-extracted CSM and dehulled, solvent-extracted SBM is given in Table 12.3. CSM contains higher level of arginine, similar levels of histidine, sulfur-containing amino acids (methionine and cystine), and phenylalanine, but lower levels of other essential amino acids than SBM.

Digestibility

The apparent digestibility coefficients of gross nutrients in CSM appear to vary among nutrients and fish species (Table 12.4). Protein and lipid are digested relatively well by most fish species with digestibility coefficients ranging from 79.3 to 84.5 percent for protein and 71.0 to 83.2 for lipid. These values, however, are somewhat lower than those of SBM which average from 80.0 to 95.0 percent and 62.5 to 94.6 percent for protein and lipid respectively. The digestibility coefficients for dry matter (40.0 to 70.2 percent), carbohydrate (42.0 to 53.3 percent) and energy (48.0 to 73.1 percent) in CSM are generally low.

TABLE 12.3. Essential amino acid composition of prepress, solvent-extracted cottonseed meal and dehulled, solvent-extracted soybean meal.

	Composition (%)	
Essential amino acid	Cottonseed meal	Soybean meal
Arginine	4.59 (11.20)	3.60 (7.53)
Histidine	1.10 (2.68)	1.30 (2.72)
Isoleucine	1.33 (3.24)	2.60 (5.44)
Leucine	2.40 (5.85)	3.80 (7.95)
Lysine	1.65 (4.04)	3.02 (6.32)
Methionine	0.52 (1.27)	0.70 (1.46)
Cystine	0.64 (1.56)	0.71 (1.49)
Phenylalanine	2.22 (5.41)	2.70 (5.65)
Threonine	1.32 (3.22)	2.00 (4.18)
Tryptophan	0.47 (1.15)	0.70 (1.46)
Valine	1.88 (4.59)	2.70 (5.65)

Source: Feedstuffs (2006).

Note: Numbers in parentheses are percent of crude protein.

For channel catfish, however, the apparent digestibility coefficient for energy in CSM is similar to that in SBM, but the value for protein is lower in CSM. Hybrid striped bass *(Morone saxatilis* × *Morone chrysops).* and red drum *(Sciaenops ocellatus)* digest various nutrients in CSM as well as or better than those in SBM. Nile tilapia and Australian silver perch *(Bidyanus bidyanus)* do not digest CSM as well as SBM.

The apparent digestibility coefficients of most essential amino acids for Australian silver perch, channel catfish, and rockfish *(Sebastes schlegeli)* are generally lower in CSM (averaging 78.4, 74.4, and 79.6 percent, respectively) than in SBM with averages of 95.7, 82.2, and 81.8, respectively (Table 12.5). Although the data on the apparent availability of essential amino acids in SBM for rainbow trout are not available, it appears that this species digested most essential amino acids in CSM relatively well with an average of 86.9 percent.

Comparing the calculated essential amino acid availabilities in CSM and SBM, and essential amino acid requirements, SBM can satisfy the requirements of all essential amino acid for channel catfish (Table 12.6). Cottonseed meal is severely deficient in lysine (53 percent of

TABLE 12.4. Apparent digestibility coefficients of gross nutrients in cottonseed meal and soybean meal for various fish species.[a]

Fish species	Apparent digestibility coefficient (%)					Reference
	Dry matter	Protein	Lipid	Carbohydrate	Energy	
Australian silver perch (*Bidyanus bidyanus*)	50.5 (77.3)	83.0 (92.3)	–	–	53.1 (79.8)	Allan et al. (2000)
Channel catfish[b]	–	81.5 (95.0)	–	–	72.5 (72.5)	Wilson and Poe (1985)
European eel (*Anguilla anguilla*)	–	74.3	–	46.5	–	Degani (2004)
Nile tilapia	45.5 (84.1)	79.4 (93.0)	83.2 (94.6)	42.0 (89.4)	57.9 (77.2)	Sintayehu et al. (1996)
Hybrid striped bass[c]	60.9 (40.9)	83.8 (80.0)	–	–	73.1 (55.2)	Sullivan and Reigh (1995)
Rainbow trout[d]	–	79.3 (82.3)	–	–	57.0 (62.9)	Smith et al. (1995)
	58.9 (69.5)	81.2 (86.3)	–	53.3 (61.7)[c]	68.8 (76.9)	Morales et al. (1999)
	64.9	84.3	71	–	–	Cheng and Hardy (2002)

Red drum (*Sciaenops ocellatus*)	70.2 (65.2)	84.5 (86.1)	75.4 (62.7)	70.4 (63.3)	Gaylord and Gatlin (1996)
Rockfish (*Sebastes schlegeli*)[e]	40.0 (56.5)	79.5 (82.0)	—	48.0 (62.5)	Lee (2002)

[a]Values in parentheses are for soybean meal.

[b]Data are average values determined for ingredients processed in either extruded or pelletized forms.

[c]*Morone saxatilis* female × *Morone chrysops* male.

[d]Data from Cheng and Hardy (2002) are average values determined for cottonseed meal from different locations.

[e]Data are average values determined for juveniles and growers.

331

TABLE 12.5. Apparent digestibility coefficient (%) of essential amino acids (EAA) in cottonseed meal (CSM) and soybean meal (SBM) for various fish species.

Amino acid	Australian silver perch[a]		Channel catfish[b]		Rockfish[c]		Rainbow trout[d]
	CSM	SBM	CSM	SBM	CSM	SBM	CSM
Arginine	91.3	97.8	89.6	95.4	86.5	87.0	85.5
Histidine	87.4	96.7	77.2	83.6	74.5	77.5	86.2
Isoleucine	74.3	94.9	68.9	77.6	79.0	79.5	86.1
Leucine	75.1	94.8	73.5	81.0	84.0	83.0	86.4
Lysine	60.4	96.7	66.8	90.9	77.5	85.0	84.9
Methionine	74.4	95.7	72.5	80.4	76.0	80.5	89.9
Cystine	79.3	94.1	–	–	63.0	73.5	–
Phenylalanine	82.5	95.6	81.4	81.3	86.0	81.0	87.0
Tyrosine	82.8	96.2	69.2	78.7	87.0	84.5	–
Threonine	77.7	95.5	71.8	77.5	80.5	83.5	83.1
Tryptophan	–	–	–	–	–	–	93.3
Valine	76.8	94.8	73.2	75.5	82.0	85.0	86.4
Average	78.4	95.7	74.4	82.2	79.6	81.8	86.9

[a]From Allan et al. (2000).

[b]From Wilson et al. (1981).

[c]From Lee (2002). Data are average values determined for juveniles and growers.

[d]From Cheng and Hardy (2002). Data are average values determined for CSM from different locations.

the requirement is met) and marginally deficient in isoleucine and the sulfur-amino acids, methionine and cystine. Based on data obtained from growth studies, Robinson (1991) and Robinson and Li (1994) indicated that the availability of lysine in CSM may be higher than previously reported.

USE OF CSM IN FISH DIETS

Cottonseed meal is an important protein source for domestic animals but its use in fish feed is limited because of the presence of FG and the low available lysine. Results of numerous growth trials on the use of CSM as substitutes for FM and/or more expensive plant and

TABLE 12.6. Essential amino acid (EAA) requirements (% of dietary protein), available essential amino acids (% of protein) from cottonseed meal and soybean meal, and percentage of EAA requirement met by these two feedstuffs for channel catfish.

EAA	Requirement[a]	Cottonseed meal		Soybean meal	
		Available[b]	% of requirement[c]	Available[b]	% of requirement[c]
Arginine	4.3	10.3	240	7.2	167
Histidine	1.5	2.1	140	2.3	153
Isoleucine	2.6	2.2	85	4.2	162
Leucine	3.5	4.3	123	6.4	183
Lysine	5.1	2.7	53	5.7	112
Methionine + cystine[d]	2.3	2.1	91	2.4	104
Phenylalanine + tyrosine[e]	5	5.7	148	7.4	148
Threonine	2	2.3	114	3.2	160
Tryptophan	0.5	–	–	–	–
Valine	3	3.4	113	4.3	143

[a]From NRC (1993).

[b]Available amino acid = amino acid content as percent of protein in Table 12.3 × apparent amino acid digestibility in Table 12.5.

[c]Percent of requirement = amino acid availability/amino acid requirement.

[d]Apparent digestibility coefficient of methionine was used for cystine.

[e]Tyrosine value for cottonseed meal in NRC (1993) was used.

animal protein sources appear to indicate that the levels of CSM that can be incorporated in fish diets depend mainly on the levels of FG and available lysine (Robinson and Li, 1995; Li and Robinson, 2006). Dorsa et al. (1982) suggested that 17.4 percent CSM (0.49 percent FG) can be included in channel catfish diets without adverse effect on growth. Robinson and Daniels (1987) reported reduced weight gain in channel catfish fed a diet containing 23.4 percent CSM (0.43 percent FG). They indicated, however, that this problem did not appear to be related to dietary gossypol level and suggested that, for glanded CSM containing 0.043 percent FG, the level used should not exceed 12 to 15 percent of the diet. For CSM with lower FG content (0.022 percent), a level of 24.8 percent as a replacement of 50 percent SBM in channel catfish diets can be used without requiring lysine supplementation and up to 50.8 percent (100 percent SBM replacement) with lysine supplementation (Robinson, 1991). Barros et al. (2000) obtained better growth and feed utilization efficiency in juvenile channel catfish fed the diet in which 50 percent of SBM was replaced by 27.5 percent CSM (0.122 percent FG) supplemented with lysine, as compared to those fed diets without or with 55.0 percent CSM. Although the diet containing 27.5 percent CSM without lysine supplementation was not included, Barros et al. (2000) suggested that the improved performance of the 27.5 percent CSM diet with added lysine obtained in their study, but not in the study of Robinson (1991) in which lysine was supplemented to the 24.8 percent CSM diet, may be attributed to the much higher FG content in their diets.

Glandless CSM products (defatted glandless, cottonseed flour and meal) which contained less than 0.01 percent FG can be used to replace up to 100 percent SBM in diets of channel catfish without affecting their growth and feed efficiency (Robinson and Rawles, 1983). They also showed that supplementation of lysine had no effect on improving the performance of fish fed diets containing glandless CSM. The authors suggested that glandless CSM may contain sufficient lysine to meet channel catfish requirement.

Fowler (1980) showed that chinook salmon *(Oncorhynchus tshawytscha)* and coho salmon *(O. kisutch)* utilized diets containing CSM (FG level was not reported) at levels of 22 and 34 percent, respectively as a replacement of FM. Lee, Dabrowski et al. (2002) suggested that, although the nutritional values of CSM can be different from different geographical locations, CSM can be incorporated at least 15 percent in juvenile rainbow trout diets as a replacement of 25 percent of dietary FM.

Prepress solvent-extracted CSM containing 300 mg gossypol/kg of diet has been reported to be a good protein source for mossambique tilapia *(O. mossambicus)*. The growth of fish was improved when 50 percent of FM was replaced by CSM. The growth was essentially the same even at 100 percent substitution levels (Jackson et al., 1982). In contrast, Ofojekwu and Ejike (1984) reported that cottonseed cake (FG level was not given) by itself, even at a level of 16 to 20 percent, resulted in an unacceptable growth performance of Nile tilapia. Viola and Zohar (1984) showed that low-gossypol CSM can be included at the same level as SBM in the diet of market-size hybrid tilapia *(O. niloticus × O. aureus)* reared in ponds. For blue tilapia, Robinson et al. (1984) found that inclusion of 26.5 percent solvent-extracted CSM (450 mg FG/kg) with lysine supplementation as a replacement of 25 percent peanut meal and 5 percent FM in a 35 percent crude protein diet depressed growth and feed efficiency of juvenile blue tilapia. Lim et al. (2002) reported that substitution of one-third SBM by 19 percent of CSM and lysine had no effect on the growth performance of juvenile Nile tilapia. Increasing CSM levels to 38 percent or higher adversely affected weight gain and hematological parameters. Since Nile and blue tilapia can tolerate at least up to 1,600 and 1,800 mg gossypol/kg of diet from GAA, respectively, and diets used were supplemented with lysine, the negative effect of CSM was thought to be due to the deficiency of essential amino acids other than lysine, or the toxicity of compounds other than gossypol, such as cyclopropenoid fatty acids, which are present in cottonseed lipid (Robinson et al., 1984; Lim et al., 2002, 2003). Ofojekwu and Ejike (1984) reported reduced weight gain and feed efficiency of Nile tilapia juveniles fed the diet containing 19.4 percent cottonseed cake as compared to the groups fed the FM-control diet. With the same species, El-Sayed (1990) obtained a 24 and 35 percent reduction in weight gain and feed efficiency, respectively, of fish fed the diet containing 65 percent CSM relative to fish fed the FM diet. Mbahinzireki et al. (2001) found that 50 percent of FM in the diet of juvenile *Oreochomis* spp. can be replaced by 29.4 percent CSM without affecting weight gain and feed efficiency.

REFERENCES

Allan, G.L., S. Parkinson, M.A. Booth, D.A.J. Stone, S.J. Rowland, J. Frances, and R. Warner-Smith (2000). Replacement of fish meal in diets for Australian silver perch, *Bidyanus bidyanus:* I. Digestibility of alternative ingredients. *Aquaculture* 186:293-310.

Bailey, C.A., R.D. Stipanovic, M.S. Zeihr, A.U. Haq, M. Sattar, L.F. Kubena, H.I. Kim, and R.M. Vieira (2000). Cottonseed with a high (+)- to (−)-gossypol enantiomer ratio favorable to broiler production. *Journal of Agricultural and Food Chemistry* 48:5602-5605.

Barros, M.M., C. Lim, J.J. Evans, and P.H. Klesius (2000). Effect of iron supplementation to cottonseed meal diets on growth performance of channel catfish, *Ictalurus punctatus. Journal of Applied Aquaculture* 10:65-86.

Barros, M.M., C. Lim, and P.H. Klesius (2002). Effect of soybean meal replacement by cottonseed meal and iron supplementation on growth, immune response and resistance of channel catfish *(Ictalurus punctatus)* to *Edwardsiella ictaluri* challenge. *Aquaculture* 207:263-279.

Bell, A.A. (1967). Formation of gossypol in infected or chemically irritated tissues of *Gossypium* species. *Phytopathology* 57:759-764.

Benz, C.C., M.A. Keniry, J.M. Ford, A.J. Townsen, S.W. Cox, S. Palayoor, S.A. Matlin, W.N. Hait, and K.H. Cowan (1990). Biochemical correlates of the antitumor and antimitochondrial properties of gossypol enantiomaers. *Molecular Pharmacology* 37:840-847.

Berardi, L.C. and L.A. Goldblatt (1980). Gossypol. In *Toxic Constituents of Plant Feedstuffs,* 2nd edn., I.E. Liener (ed.). New York, NY: Academic Press, pp. 183-237.

Blom, J.H., K.J. Lee, J. Rinchard, K. Dabrowski, and J. Ottobre (2001). Reproductive efficiency and maternal offspring transfer of gossypol in rainbow trout *(Oncorhynchus mykiss)* fed diets containing cottonseed meal. *Journal of Animal Science* 79:1533-1539.

Boatner, C.H., L.E. Castillon, and C.M. Hall (1949). Gossypol and gossypurpurin in cottonseed of different varieties of *G. barbadense* and *G. hirsutum,* and variation of the pigments during storage of the seed. *Journal of the American Oil Chemists' Society* 26:19-25.

Braham, J.E., R. Jarquin, R. Bressani, J.M. Gonzales, and L.G. Elias (1967). Effect of gossypol on the iron-binding capacity of serum in swine. *Journal of Nutrition* 93:241-248.

Bressani, R., L.G. Elias, R. Jarquin, and J.E. Brahan (1966). All-vegetable protein mixtures for human feeding. XIII. Effect of cooking mixtures containing cottonseed flour on free gossypol content. *Food Technology* 18:1599-1603.

Buitrago, J.A., A.J. Clawson, L.C. Ulberg, and F.H. Smith (1970). Effect of dietary iron on gossypol accumulation in and elimination from porcine liver. *Journal of Animal Science* 31:554-558.

Cabel, C.A. and J.W. Stevenson (1966). Studies on the effect of orally administrated iron salts upon the performance of swine and rats fed gossypol. In *Proceedings Conference on Inactivation of Gossypol with Mineral Salts.* New

Orleans, LA: The National Cottonseed Products Association, Inc. Memphis, TN, pp. 100-103.

Cass, Q.B., E. Tritan, S.A. Matlin, and E. Freire (1991). Gossypol enantiomer ratios in cottonseeds. *Phytochemistry* 30:2655-2657.

Chen, Y., M. Sten, M. Nordenskjold, B. Lambert, S.A. Matlin, and R.H. Zhou (1986). The effect of gossypol on the frequency of DNA-strand breaks in human leukocytes *in vitro*. *Mutation Research* 164:71-78.

Cheng, Z. and R.W. Hardy (2002). Apparent digestibility coefficients and nutritional values of cottonseed meal for rainbow trout *(Oncorhynchus mykiss)*. *Aquaculture* 212:361-372.

Cherry, J.P., J.G. Simmons, and R.J. Kohel (1978). Cottonseed composition of natural variety test cultivars grown in different Texas locations. In *Proceedings of the Beltwide Cotton Production Research Conference*. Dallas, TX: National Cotton Council, pp. 47-56.

Ciereszko, A. and K. Dabrowski (2000). *In vitro* effect of gossypol acetate on yellow perch *(Perca flavescens)*. *Aquatic Toxicology* 49:181-187.

Clawson, A.J., J.H. Maner, G. Gomez, O. Mejia, Z. Flores, and J. Buitrago (1975). Unextracted cottonseed in diets for monogastric animals. I. The effect of ferrous sulfate and calcium hydroxide in reducing gossypol toxicity. *Journal of Animal Science* 40:640-647.

Dabrowski, K., K.J. Lee, J. Rinchard, A. Ciereszko, J.H. Blom, and J. Ottobre (2001). Gossypol isomers bind specifically to blood plasma proteins and spermatozoa of rainbow trout fed diets containing cottonseed meal. *Biochimica et Biophysica Acta* 1525:37-42.

Dabrowski, K., J. Rinchard, K.J. Lee, J.H. Blom, A. Ciereszko, and J. Ottobre (2000). Effects of diets containing gossypol on reproductive capacity of rainbow trout *(Oncorhynchus mykiss)*. *Biology of Reproduction* 62:227-234.

Damaty, S.M. and B.J.F. Hudson (1979). The interaction between gossypol and cottonseed proteins. *Journal of The Science of Food and Agriculture* 30:1050-1056.

Degani, G. (2004). Digestibility of carbohydrates for European eel. *Journal of Aquaculture in the Tropics* 19:29-36.

Dorsa, W.J., H.R. Robinette, E.H. Robinson, and W.E. Poe (1982). Effects of dietary cottonseed meal and gossypol on growth of young channel catfish. *Transactions of the American Fisheries Society* 111:651-655.

Eid, J.E., H. Ueno, C.C. Wang, and J.E. Donelson (1988). Gossypol-induced death of African *Trypanosomes*. *Experimental Pathology* 66:140-142.

El Boughy, A.R. and R. Raterink (1989). Replacement of soybean meal by cottonseed meal and peanut meal or both in low energy diets for boilers. *Poultry Science* 68:799-804.

El-Saidy, D.M.S.D. and M.M. Gaber (2004). Use of cottonseed meal supplemented with iron for detoxification of gossypol as a total replacement of fish meal in Nile tilapia, *Oreochromis niloticus* (L) diets. *Aquaculture Research* 35:859-865.

El-Sayed, A.F.M. (1990). Long term evaluation of cottonseed meal as a protein source for Nile tilapia, *Oreochromis niloticus* (Linn.) *Aquaculture* 84:315-320.

FAO (2002). *World Review of Fisheries and Aquaculture—Fisheries Resources: Trends in Utilization, Production and Trade.* Rome, Italy: Food and Agriculture Organization.

Feedstuffs (2006). Feedstuff ingredient analysis table. *Reference Issue and Buyer's Guide* 77(38):16-18.

Fowler, L.G. (1980). Substitution of soybean and cottonseed products for fish meal in diets fed to chinook and coho salmon. *Progressive Fish-Culturist* 42:87-91.

Gallup, W.D (1928). The value of iron salts in counteracting the toxic effects of gossypol. *Journal of Biological Chemistry* 43:437-449.

Gaylord, T.G. and D.M. Gatlin, III (1996). Determination of digestibility coefficients of various feedstuffs for red drum *(Sciaenops ocellatus)*. *Aquaculture* 139:303-314.

Hadley, M.A., Y.C. Lin, and Y. Dym (1981). Effect of gossypol on reproductive performance of male rats. *Journal of Androcology* 2:190-199.

Hale, F. and C.M. Lyman (1957). Effect of protein level in the ration on gossypol tolerance in growing-fattening pigs. *Journal of Animal Science* 16:364-369.

Heidrich, J.E., L.A. Hunsaker, and D.L. Vander Jagt (1983). Gossypol, and antifertility agent exhibits antimalarial activity *in vitro*. *IRCS Medical Science* 11:304.

Herman, D.L. and F.S. Smith (1973). Effects of bound gossypol on the absorption of iron by rats. *Journal of Nutrition* 103:882-889.

Herman, R.L. (1970). Effect of gossypol on rainbow trout *Salmo gairdneri* Richardson. *Journal of Fish Biology* 2:293-304.

Hoffer, A.P. (1982). Ultrastructural studies of spermatozoa and the epithelial lining of the epidermis and vas deferens in rats treated with gossypol. *Archives of Andrology* 8:233-246.

Jones, L.A. (1987). Recent advances in using cottonseed products. In *Proceedings of the Florida Nutrition Conference.* Daytona Beach, FL: National Cottonseed Products Association, pp. 119-138.

Jones, L.A. (1991). Definition of gossypol and its prevalence in cottonseed products. In *Cattle Research with Gossypol Containing Feeds—A Collection of Paper Addressing Gossypol Effect in Cattle,* L.A. Jones, D.H. Kinard, and J.S. Mils (eds.). Memphis, TN: National Cottonseed Products Association, pp. 1-18.

Jonassen H.B. and R.J. Demint (1955). Interaction of gossypol with the ferrous ion. *Journal of the American Oil Chemists Society* 32:424-426.

Joseph, A.E.A., S.A. Martin, and P. Knox (1986). Cytotoxicity of enantiomers of gossypol. *British Journal of Cancer* 54:511-513.

King, J.J. and L.B. DeSilva (1968). Optically active gossypol from *Thespesia populnea. Tetrahedron Letters* 3:261-263.

Kornegay, E.T., A.J. Clawson, F.H. Smith, and E.R. Baric (1961). Influence of protein sources on toxicity of gossypol in swine rations. *Journal of Animal Science* 20:597-602.

Lee, K.J. and K. Dabrowski (2002). Gossypol and gossypolone enantiomers in tissue of rainbow trout fed low and high levels of dietary cottonseed meal. *Journal of Agricultural and Food Chemistry* 50:3056-3061.

Lee, K.J., K. Dabrowski, J.H. Blom, S.C. Bai, and P.C. Stromberg (2002) A mixture of cottonseed meal, soybean meal and animal byproduct mixture as a fish meal substitute: Growth and tissue gossypol enantiomer in juvenile rainbow trout *(Oncorhynchus mykiss). Journal of Animal Physiology and Animal Nutrition* 86:201-213.

Lee, K.J., K. Dabrowski, and G.B. Mbahinzireki (2000). Utilization of cottonseed meal in rainbow trout *Oncorhynchus mykiss* and Nile tilapia, *Oreochromis niloticus:* Gossypol enantiomers in fish tissues. *Book of Abstracts, Aquaculture America 2000,* New Orleans, LA, 195 pp.

Lee, K.J., J. Rinchard, K. Dabrowski, J.H. Blom, A. Ciereszko, and J. Ottobre (2002). Long-term effects of dietary cottonseed meal on growth and reproductive performance of rainbow trout broodstock. *Book of Abstracts, Aquaculture America 2002*, San Diego, California, 80 pp.

Lee, S.M. (2002). Apparent digestible coefficients of various feed ingredients for juvenile and grower rockfish *(Sebastes schlegeli). Aquaculture* 207:79-95.

Li, M.H. and E.H Robinson (2006). Use of cottonseed meal in aquatic animal diets: A review. *North American Journal of Aquaculture* 68:14-22.

Lim, C., M. Yildirim-Aksoy, and P.H. Klesius (2002). Dietary cottonseed meal affects growth, hematology, immune response of Nile tilapia. *Global Aquaculture* 5:29-30.

Lim, C., M. Yildirim-Aksoy, and P.H. Klesius (2003). Levels of dietary gossypol affect growth and bacterial resistance of Nile tilapia. *Global Aquaculture* 6:42-43.

Lim, C., M. Yildirim-Aksoy, and P.H. Klesius (2004). Dietary gossypol benefits limited in improving ESC resistance in catfish *Global Aquaculture* 7:47-48.

Lim, C., M. Yildirim, P. Wan, and P.H. Klesius (2005). Gossypol in purified and particle diets: Effect on growth performance, immune responses and resistance of channel catfish, *Ictalurus punctatus* to *Edwardsiella ictaluri* infection. Aquaculture America Conference in New Orleans, LA, 242 pp.

Lin, T.S., R. Schinazi, B.P. Griffith, E.M. August, B.F.H. Eriksson, D.K. Zheng, L. Huang et al. (1989). Selective inhibition of human immunodeficiency virus type I replication by the (−) but not by the (+) enantiomer of gossypol. *Antimicrobial Agents and Chemotherapy* 33:2149-2151.

Lyman, C.M. (1966). The effect of gossypol and gossypol-like compounds upon swine in the presence and absence of iron salt and/or protein of high biological value. *Proceedings, Conference on Inactivation of Gossypol with Mineral Salts.* New Orleans, LA: The National Cottonseed Products Association, Inc. Memphis, TN, pp. 104-117.

Lyman, C.M. and C. Widmer (1966). Mode of action, site of concentration and metabolic fate of gossypol and gossypol-like compounds. *Proceedings, Conference on Inactivation of Gossypol with Mineral Salts.* New Orleans, LA: The National Cottonseed Products Association, Inc. Memphis, TN, pp. 43-53.

Mbahinzireki, G.B., K. Dabrowski, K.J. Lee, D. El-Saidy, and E.R. Wisner (2001). Growth, feed utilization, and body composition of tilapia *(Oreochromis* spp.) fed with cottonseed meal based diets in a recirculating system. *Aquaculture Nutrition* 7:189-200.

Montamat, E.E., C. Burgos, N.M. Gerez de Burgos, L.E. Rovai, and A. Blanco (1982). Inhibitor action of enzymes and growth of *Tryponosoma cruzi. Science* 218:288-289.

Moore, A.T. and M.L. Rollins (1961). New information on the morphology of the gossypol pigment gland of cottonseed. *The Journal of the Oil Chemists' Society* 38:156-160.

Morales, A.E., G. Cardenete, A. Sanz, and M. de la Higuera (1999). Re-evaluation of crude fiber and acid-insoluble ash as inert markers, alternative to chromic oxide, in digestibility studies with rainbow trout *(Oncorhynchus mykiss)*. *Aquaculture* 179:71-79.

NCGMAA (National Coordinating Group on Male Antifertility Agents) (1978). Gossypol—a new antifertility agent for males. *Gynecologic and Obstetric Investigation* 10:163-176.

NRC (National Research Council) (1993). *Nutrient Requirements of Warmwater Fish*. Washington, DC: National Academic Press.

Ofojekwu, P.C. and C. Ejike (1984). Growth response and feed utilization in the tropical cichlid *Oreochromis niloticus* (Linn.) Fed on cottonseed-based artificial diets. *Aquaculture* 42:27-36.

Randel, R.D., C.C. Chase Jr., and S.J. Wyse (1992). Effect of gossypol and cottonseed products on reproduction of mammals. *Journal of American Science* 70: 1628-1638.

Rands, R.D. (1966). Potential sources of mineral salts for use as ration additives to prevent gossypol toxicity. In *Proceedings, Conference on Inactivation of Gossypol with Mineral Salts*. New Orleans, LA: The National Cottonseed Products Association, Inc. Memphis, TN, pp. 54-59.

Reiser, R. and H.C. Fu (1962). The mechanism of gossypol detoxification. *Journal of Nutrition* 76:215-218.

Rinchard, J., G.B. Mbahinzireki, K. Dabrowski, K.J. Lee, M.A. Gacia-Abiabo, and J. Ottobre (2002). Effect of dietary cottonseed meal protein levels on growth, gonad development and plasma sex steroid hormones of tropical fish *Oreochromis* sp. *Aquaculture International* 227:77-87.

Robinson, E.H (1991). Improvement of cottonseed meal protein with supplement lysine in feeds for channel catfish. *Journal of Applied Aquaculture* 1:1-14.

Robinson, E.H. and W.H. Daniels (1987). Substitution of soybean meal with cottonseed meal in pond feeds for channel catfish reared at low densities. *Journal of the World Aquaculture Society* 18:101-106.

Robinson, E.H. and M.H. Li (1994). Use of plant proteins in catfish feeds: Replacement of soybean meal with cottonseed meal and replacement of fish meal with soybean meal and cottonseed meal. *Journal of the World Aquaculture Society* 25:271-276.

Robinson, E.H. and M.H. Li (1995). Use of cottonseed meal in aquaculture feeds. In *Nutrition and Utilization Technology in Aquaculture,* C. Lim and D.J. Sessa (eds.). Champaign, IL: AOCS Press, pp. 157-165.

Robinson, E.H. and S.D. Rawles (1983). Use of defatted, glandless cottonseed flour and meal in channel catfish diets. *Proceedings of the Southeastern Association of Fish and Wildlife Agencies* 37:358-363.

Robinson, E.H., S.D. Rawles, P.W. Oldenburg, and R.R. Stickney (1984). Effects of feeding glandless or glanded cottonseed products and gossypol to *Tilapia aurea*. *Aquaculture* 38:145-154.

Robinson, E.H. and T.R. Tiersch (1995). Effects of long-term feeding of cotton-seed meal on growth, testis development, and sperm motility of male channel catfish *Ictalurus punctatus* broodfish. *Journal of the World Aquaculture Society* 26:426-431.

Roehm, J.M., D.J. Lee, and R.O. Sinnhubuer (1967). Accumulation and elimina-tion of dietary gossypol in the organs of rainbow trout. *Journal of Nutrition* 92:425-428.

Rojas, S.W. and M.L. Scott (1969). Factor affecting the nutritive value of cottonseed meal as a protein source in chick diets. *Poultry Science* 48:819-835.

Sampath, D.S. and P. Balaran (1986). Resolution of racemic gossypol and interaction of individual enantiomers with serum albumine and model peptides. *Biochimica et Biophysica Acta* 882:183-186.

Sharma, M.P., F.H. Smith, and A.J. Clawson (1966). Effects of levels of protein and gossypol, and length of feeding period on the accumulation of gossypol in tis-sues of swine. *Journal of Nutrition* 88:434-438.

Sinnhuber, R.O., D.J. Lee, J.H. Wales, and J.L. Ayres (1968). Dietary factors and hepatoma in rainbow trout *(Salmo gairdneri)*, II. Cocarcinogenesis by cyclo-propenoid fatty acids and the effect of gossypol and altered lipids on alfatoxin-induced liver cancer. *Journal of the National Cancer Institute* 41:1293-1301.

Sintayehu, A., E. Mathies, K.-H. Meyer-Burgdorff, H. Rosenow, and K.-D. Gunther (1996). Apparent digestibilities and growth experiment *(Oreochromis niloticus)* fed soybean meal, cottonseed meal, and sunflower seed meal. *Journal of Applied Ichthyology* 12:125-130.

Smith, R.R. and A.J. Clawson (1970). The effects of dietary gossypol on animals. *Journal of the American Oil Chemists' Society* 47:443-447.

Smith, R.R., R.A. Winfree, G.W. Rumset, A. Allred, and M. Peterson (1995). Ap-parent digestion coefficients and metabolizable energy of feed ingredients for rainbow trout *Oncorhynchus mykiss*. *Journal of the World Aquaculture Society* 26:432-437.

Stipanovic, R.D., A.A. Bell, M.E. Mace, and C.R. Howell (1975). Antimicrobial terpenoids of gossypiun: 6-methoxygossypol and 6,6' dimethoxygossypol. *Phy-tochemistry* 14:1077-108.

Sullivan, A.J. and R.C. Reigh (1995). Apparent digestibility of selected feedstuffs in diets for hybrid striped bass (*Morone saxatilis* female × *Morone chrysops* male). *Aquaculture* 138:313-322.

Swick, R.A. and P.H. Tan (1995). *Considerations in Using Common Asian Protein Meal*. American Soybean Association Technical Bulletin MITA (P) No. 083/3/94, Singapore.

Viola, S. and G. Zohar (1984). Nutritional studies with market size hybrid tilapia, *Oreochromis* in intensive culture, 3. Protein levels and sources. *Israeli Journal of Aquaculture Bamidgeh* 36:3-15.

Wang, J.M., L. Tao, X.L. Wu, L.X. Lin, J. Wu, M. Wang, and J.Y. Zhang (1992). Differential binding of (+) and (−) gossypol to plasma protein and their entry into rat testis. *Journal of Reproduction and Fertility* 95:277-282.

Wang, N.G., L.F. Zhou, M.Z. Guan, and H.P. Lei (1987). Effects of (−) and (+) gossypol on fertility of male rats. *Journal of Ethnopharmacology* 20:21-24.

342 ALTERNATIVE PROTEIN SOURCES IN AQUACULTURE DIETS

Wang, X., J. Wang, S.C.H. Wong, L.S.N. Chow, J.M. Nicolls, Y.C. Wong, Y. Liu et al. (2000). Cytotoxic effect of gossypol on colon carcinoma. *Life Science* 67:2663-2671.

Wedegaertner, T.C. (1981). Making the most of cottonseed meal. *Feed Management Magazine* 32:1-3.

Wilson, R.P. and W.E. Poe (1985). Apparent digestible protein and energy co-efficients of common feed ingredients for channel catfish. *Progressive Fish-Culturist* 47:154-158.

Wilson, R.P., E.H. Robinson, and W.E. Poe (1981). Apparent and true availability of amino acids from common feed ingredients for channel catfish. *Journal of Nutrition* 111:923-929.

Wood, J.W. (1968). *Disease of Pacific Salmon: Their Prevention and Treatment.* Olympia, Washington: State of Washington, Department of Fisheries, Hatchery Division.

Wu, D. (1989). An overview of the clinical pharmacology and therapeutic potential of gossypol as a male contraceptive agent and in gynecological disease. *Drugs* 38:333-341.

Yildirim, M., C. Lim, P.J. Wan, and P.H. Klesius (2003). Growth performance and immune response of channel catfish *(Ictalurus punctatus)* fed diets containing graded levels of gossypol-acetic acid. *Aquaculture* 219:751-768.

Yildirim-Aksoy, M., C. Lim, M.K. Dowd, P.J. Wan, P.H. Klesius, and C. Shoemaker (2004). *In vitro* inhibitory effect of gossypol for gossypol-acetic acid, and (+)- and (−)-isomers of gossypol on the growth of *Edwardsiella ictaluri*. *Journal of Applied Microbiology* 97:87-92.

Yildirim-Aksoy, M., C. Lim, P.J. Wan, and P.H. Klesius (2004). Effect of natural free gossypol and gossypol-acetic acid on growth performance and resistance of channel catfish *(Ictalurus punctatus)* to *Edwardsiella ictaluri* challenge. *Aquaculture Nutrition* 10:153-165.

Zhou, R.H. and X.D. Lin (1988). Isolation of (−)-gossypol from natural plant. *Contraception* 37:239-245.

Chapter 13

Use of Rapeseed/Canola in Diets of Aquaculture Species

C. Burel
Sadasivam J. Kaushik

INTRODUCTION

Rapeseed was grown originally for supplying steam engine lubricant, but it is nowadays an oil source (40 to 45 percent DM) in human and animal diet. It was characterized by a high erucic acid content (25 to 55 percent of total fatty acids), which is well known for its adverse effects on human health (cardiac lipidosis, impaired mitochondrial functions, enlarged pale adrenal), but this content has been reduced by selection of new varieties ("double-low" or "canola").

Rapeseed is now the world's number two oilseed crop, ranking behind soybeans and ahead of cottonseed, peanuts, and sunflower seeds. The reason for this position is that the oil from the double-low or canola variety of rapeseed has one of the best fatty acid compositions of all the vegetal oils. Over the years, this has gained considerable public praise from international nutrition experts as one of the healthiest oils for use in human foods.

While rapeseed is best known for its oil, it is also a good source of protein. The seed meal remaining after the oil has been extracted contains approximately 40 percent protein by weight, and the protein is rich in amino acids, such as methionine and cysteine. Compositionally, it is as good as other plant protein sources, and in fact, it is globally the second largest protein supplement after soybean meal in animal diets, although a distant second.

Alternative Protein Sources in Aquaculture Diets
doi:10.1300/5892_13

The main production areas of rapeseed are Canada, parts of Europe, Australia, and Asia. Global supplies of rapeseed protein exceed those of fish meal protein (Higgs et al., 1996) and presently, the cost of rapeseed meal is less than half that of premium-quality fish meal on a per kilogram protein basis. This price difference means that the scope for processing rapeseed, or its meal, to produce economical products with increased nutritive value is considerable. Some of the rapeseed/canola protein is not, however, directed to animal diets. For instance, in Asia, rapeseed, rather than canola seed, is grown and the meal is used as an organic fertilizer, as well as a livestock diet supplement. In North America and some European countries, canola-type rapeseed is grown and the seed is used as a protein supplement in animal diets.

The genetic selection of rapeseed aims at improving the yield and the agronomic and the technological characteristics of the plants (oil and meal). The first aim of the selection of the technological characteristics was to increase of the oil content of the seeds and to decrease the level of the erucic acid, C22:1, and the second, to develop varieties with reduced content of glucosinolates (GLS).

Indeed, during the 1960s, nutritionists have pointed out the problem of the long chain fatty acids (C22), such as erucic acids that are carcinogenic. Thus, from 1965, plant breeders tried to produce some varieties of rapeseeds containing much lower levels of erucic acid in the oil fraction than the original variety (25 to 50 percent of the oil). Following the success of producing varieties with low levels of erucic acid, attempts were made by plant breeders to develop varieties with low levels of GLS, antinutritional factors (ANFs) with adverse physiological effects (see section "ANFs").

In 1974, a Canadian plant breeder was the first to breed the double-low (00) variety of rapeseed, *Brassica napus* and *Brassica campestris,* that contain reduced levels of erucic acid (<2 percent in oil) and GLS (< 30 µmole/g meal). These varieties are known in their native Canada, Australia, Japan, and the United States as canola. Although these double-low varieties of rapeseed have been exported around the world and form the majority of global rapeseed production, this oilseed is still best known in Asia and Europe as rapeseed or oilseed rape.

From those humble beginnings more than two decades ago, rapeseed has now become an important agricultural source of edible oil and animal diets in many countries. In fact, the global production of rapeseed and canola has expanded to the point where it now ranks second, with 45 million metric tons produced in 2001, or 15 percent of world oilseed production of 286 million metric tons (Morand-Fehr and Tran, 2001).

Rapeseed meal is commonly used as a source of protein for mammals and poultry, due to its good amino acid profile and its lower price compared to fish meal. Since the 1970s, many studies have focused on the incorporation of rapeseed meal into fish diets, such as those of Yurkowski et al. (1978), Teskeredzic et al. (1995), and Webster et al. (1997). Comprehensive data are available from the review of Higgs et al. (1996). Rapeseeds are, nevertheless, known to contain ANFs, such as tannins, phytic acid, and GLS. The incorporation of rapeseed meals containing high levels of GLS into animal diets has caused problems such as reduced diet intake, enlarged thyroids, reduced plasma thyroid hormone levels and, occasionally, organ abnormalities (liver and kidney) and death (Bunting, 1981; VanEtten and Tookey, 1983). In fish, because of the deleterious effects of their high levels of fiber and GLS, the incorporation of rapeseed meal into the diet is generally restricted to approximately 20 percent (Yurkowski et al., 1978; Hardy and Sullivan, 1983; Hilton and Slinger, 1986; Leatherland et al., 1987; Gomes and Kaushik, 1989; Gomes et al., 1993; Higgs et al., 1996). However, in the late 1980s, the content of GLS of rapeseed seeds has been further reduced to below 10 μmole/g by genetic selection of new cultivars (Vermorel et al., 1986) and the improvements made in the processing techniques lead to decrease in the content of fiber and other ANFs such as tannins in RM (Bell, 1993).

PROCESSING OF RAPESEED/CANOLA PRODUCTS AND MEALS

Rapeseed/canola seeds are initially processed (crushed) into two components: oil and meals. Further processing of the meal by various technological treatments results in three products: meals, protein concentrates, and protein isolates (Figure 13.1). Essentially two main processes are currently used to extract oil from the seed and to produce rapeseed/canola meal: solvent extraction and expeller extraction.

Canola seed is traditionally crushed and solvent extracted. The process usually includes seed cleaning, seed preconditioning and flaking,

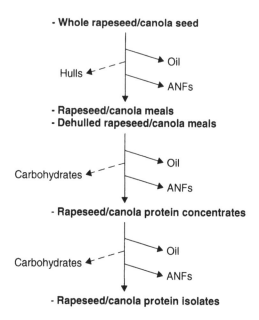

FIGURE 13.1. Schematic presentation of the different rapeseed/canola products obtained after technological treatments.

seed cooking, pressing the flake to mechanically remove a portion of the oil, solvent extraction of the press-cake to remove the remainder of the oil, and desolventizing and toasting of the meal (Hickling, 2001).

Meal quality is influenced by several variables during the process, especially temperature, and there are other operations that can be performed, such as the dehulling of the seed before the oil extraction.

Meals

The main concern in the extraction of rapeseed oil is to achieve a high yield of oil while maintaining high quality of protein (Niewiadomski, 1990). Therefore, operational parameters should be optimized during preliminary operations, processing and separation of oil. Primary processing includes seed cleaning, preconditioning and flaking, seed cooking, pressing, and solvent oil extraction. Following processing, the operations include desolventizing and toasting in order to improve the quality of the by-product: the meal (Hickling, 2001).

In preparing oilseed for the extraction of its oil, energy must be used to rupture or weaken the walls of the oil-containing cells. Both mechanical energy (for breaking, grinding, rolling pressing and pelleting) and thermal energy (to rupture cell walls, reduce oil viscosity, and adjust moisture content) are used (Davie and Vincent, 1980).

Dehulling

To improve the quality of the meal, seed can be dehulled. The purpose of dehulling is to remove the major part of the fiber (cellulose, hemicellulose, and lignin) and a group of pigments which, passing into the meal, would lower its feeding value. The overall dehulling process includes: cooking, dehulling itself, and separation of the hulls (Niewiadomski, 1990). A heating period of 10 to 15 minutes is usually sufficient, since at 90 to 97°C, the seed reaches a moisture content of approximately 6 percent (Niewiadomski, 1990). However, studies on the effect of drying temperature on rapeseed quality showed that elevated drying temperature up to 200°C did not adversely affect oil quality. Also, applying controlled elevated temperature would save up to 80 percent drying time compared to the present practice of drying rapeseed (Pathak et al., 1991). The dehulling itself takes place in a roller mill, in which the roller spacing and the rotational speed can be varied. Grinding wet-heat-treated seeds in a stream of water results in squeezing out the meals intact from the hulls. Correctly performed, dehulling yields completely separated fractions of hulls and meals.

Alternative processes can be used for dehulling. These methods involve either dehulling of rapeseed before further processing, or removal of hulls or fiber by air classification of defatted meal, or milling moisture-adjusted meals followed by sieving.

Removing a significant portion of the hulls from the seed or the meal allows a reduction of fiber (fiber-reduced seeds), as well as tannins and sinapine contents, leading to an increase of the meal's protein content.

Oil Extraction

Extraction of oil from flaked rapeseed can proceed by one of the following processes: direct screw pressing, direct solvent extraction, and prepress solvent extraction (Salunkhe et al., 1992). The prepress solvent extraction process is a classical system of processing rapeseed in which the seed is initially expelled under pressure to release a portion of the oil; the residue is then solvent extracted (Brogan, 1986). This method is still used by many oil producing industries, with some modifications, such as pretreatment of rapeseeds as described by Niewiadomski (1990) and Pathak et al. (1991). Prepress solvent extraction is probably one of the most economical processes (Salunkhe et al., 1992).

The heat treatments applied to the seeds are necessary to inactivate myrosinase, the enzyme responsible for GLS hydrolysis, while the solvent (hexane) and water washes are needed to remove or decrease GLS, fiber, and problem oligosaccharides (e.g., raffinose and stachyose), phenolic compounds such as sinapine. Alternatively, the crushed seeds may be exposed to a two-phase solvent system comprising of methanol, ammonia, and water followed by hexane to almost completely remove GLS and decrease carbohydrate, phenolic compounds, nucleic acids, nonnitrogen compounds, and lipid. Good processing conditions lead to the production of an upgraded rapeseed/canola meal.

Protein Concentrates

Special processing of whole rapeseed/canola seeds or rapeseed/ canola meals can result in products (concentrates or upgraded meals) that have protein content comparable to that of fish meal (Figure 13.2). The further processing of fiber-reduced rapeseed/canola meals entails washing with selected solvents, such as acidified water and ethanol, to decrease the levels of ANFs, for example, GLS, sinapine, and carbohydrate, and concurrently increase protein content. The levels of protein and energy in rapeseed/canola protein concentrates which result from processes as depicted in Figure 13.2 are similar to those found in fishmeals. Hence, based upon their levels of proximate constituents (more than 60 percent protein) and their digestibility (about 96 percent protein digestibility), rapeseed/canola protein concentrates could be considered as potential fish meal replacement in formulated diets for fish. Rapeseed protein concentrates have markedly lower levels of ANFs, except for phytic acid, than the meals.

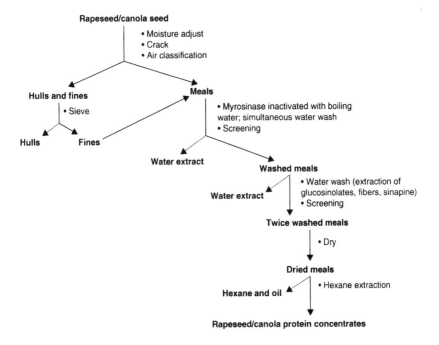

FIGURE 13.2. Schematic presentation of a process to produce protein concentrates from rapeseed/canola seed. *Source:* Adapted from Higgs et al. (1995).

Protein Isolates

The preparation of protein isolates involves the extraction and subsequent recovery of protein by several possible methods using seeds, meal, flour, and protein concentrates that have been subjected to a variety of preprocessing methods as the starting material. Phytic acid is coextracted with the protein as a complex, but is subsequently separated from the protein on the basis of differences in their solubility at dissimilar pH value (Higgs et al., 1996). The protein isolates are more expensive than protein concentrates because of their higher quality (more pure) and their higher biological value. They contain more protein with less fat, carbohydrates, and inorganic matters. The objective of the process is to stay gentle enough in order to produce a high-quality plant protein isolate in terms of nutritional value.

Extrusion Treatment

Since the 1990s, extrusion treatment has been applied to improve the quality of some legume seeds, such as soybeans or rapeseed. In rapeseed/canola, the GLS are hydrolyzed by the myrosinase into toxic compounds when the seed is crushed, which causes metabolic problems in animals, including fish. These hydrolyzates are stronger in wet conditions. Extrusion processing under salt or steam conditions reduced GLS levels in the meal, but no treatment completely inactivates ANFs in the meal, especially GLS breakdown products (Barrett et al., 1997). However, dry extrusion of oilseeds inactivates ANFs and growth inhibitors, improves digestibility, and ruptures the oil cells (removal of up to 75 percent of the available oil from most oilseeds). No steam is used in a dry extrusion system, which guarantees that no external moisture is added to the meal or to the oil and prevents hydrolysis of GLS.

Full-fat rapeseed meal is rarely used due to its high oil content (up to 40 percent). Extrusion is the most efficient processing method to extract oil and yield low-fat meal. Alternatively, rapeseed can be extruded in combination with a low-oil legume, such as field beans or peas (Gomes et al., 1993), which also need heat processing. The resulting product would have about half the oil content of raw rapeseed, low levels of ANFs, and high available energy.

Enzymatic Treatments

Despite the improvement of the quality of the rapeseed/canola products that can be obtained by a variety of selected technological treatments, there are limitations to their use in animal diets, including fish diets. The presence of ANFs, such as GLS, phytic acids, and insoluble and soluble fiber, prevents the full expression of the high-quality rapeseed/canola proteins. Numerous methods have been employed to reduce the levels of ANFs in these products, such as the addition of enzymes, but have produced variable results. The potential enzymes are phytate-degrading enzymes (phytases), oligosaccharide-degrading enzymes, and plant-cell- or fiber-degrading enzymes (Mwachireya et al., 1999).

CHEMICAL COMPOSITION

The Canadian Oilseed Processors Association (COPA) has defined some trading rules for canola meal (COPA, 1999). To be sold in Canada and United States, Canola meal has to contain as minimum 38.6 percent DM of protein, 2.3 percent DM of fat, and as maximum 12 percent of moisture, 13.6 percent DM of fiber, and 34.1 μmole/g DM of GLS. For export, canola meal has to contain as minimum 39.8 percent DM of protein plus fat, and as maximum 12 percent of moisture, 3.4 percent DM of fat, 13.6 percent DM of fiber; 34.1 percent μmole/g DM of GLS, and 1.14 percent DM of sand and/or silica.

Proximate Composition

Most feed ingredient databases around the world have listings of nutrient values for canola and/or rapeseed products as shown in Table 13.1, and it is obvious that the different products vary in their proximate composition. From whole seeds to protein isolates, there is a great variation in protein concentrations: 21 to 23 percent DM in whole seeds, 34 to 48 percent DM in rapeseed/canola meals, 65 percent DM in rapeseed/canola protein concentrates, and 92 percent DM in the protein isolates. The protein isolates are almost pure protein products.

There are some variations in nutrient values of the rapeseed/canola meals between different references, as apparent from Table 13.1. Some of these differences are caused by variations in seed nutrient composition in different countries (due to different environmental conditions), while other differences are due to processing. Canola meal produced in Canada generally has a higher level of oil and lower level of protein than European, Australian, and Asian canola/rapeseed meal. This is because canola crushing companies in Canada usually add some of the gums from crushing and some of the soapstocks from oil refining back into the meal, leading to 1.5 to 2.0 percent of extra oil content. Furthermore, canola/rapeseed crusher influences the protein composition of the meal by adjusting the level of oil and carbohydrate to counteract yearly variations in canola rapeseed composition due to growing conditions (influence of weather and soil conditions) (Hickling, 2001).

Also, it is clear that the quality of the process applied on the seeds to produce meals, and the quality of the subsequent treatments (mainly

TABLE 13.1. Proximate composition of rapeseed/canola products (expressed as % DM).

Protein products	Dry matter %	Crude protein %/DM	Crude fat %/DM	Gross energy kJ/g/DM	Ash %/DM	Reference
Canola seed, Canadian, whole	91.7	22.5	45.2	28.3	4.3	French Feed Database (2005) (n = 111)
Rapeseed 00, European, whole	92.3	20.7	45.9	28.4	4.3	French Feed Database (2005) (n = 5,157)
Rapeseed 00, European, dehulled (kernel)	94.5	20.7	54.3	31.3	3.8	French Feed Database (2005) (n = 7)
Canola meal, Canadian, solvent-extracted	90.4	39.1	4.0	19.8	7.8	Hickling (2001); French Feed Database (2005) (n = 100)
Canola meal, Australian, solvent-extracted	96.2	43.1	2.2	19.7	8.6	Glencross et al. (2004) (n = 4)
Rapeseed meal 00, European, solvent-extracted	88.7	38.4	2.8	19.3	7.9	French Feed Database (2005) (n = 9,906)
Rapeseed meal, Chinese, solvent-extracted	89.7	41.2	2.6	20.7	9.6	French Feed Database (2005) (n = 406)
Rapeseed meal, Indian, solvent-extracted	91.8	41.6	1.4		8.8	French Feed Database (2005) (n = 11)
Rapeseed meal 00, European, dehulled, solvent-extracted	89.4	40.9	2.4	19.3	8.2	French Feed Database (2005) (n = 57)

Ingredient						Reference
Canola meal, Canadian, dehulled, solvent-extracted, solvent-washed	90	43.5	4.0	20.1	7.7	French Feed Database (2005) (n = 2)
Rapeseed meal 00, European, dehulled, solvent-extracted, solvent-washed	92.9	47.9	1.2		8.3	French Feed Database (2005) (n = 6)
Canola meal, Canadian, expeller		34.1	21.2	23.9		French Feed Database (2005) (n = 1)
Rapeseed meal 00, European, expeller	90.6	35.2	10.1	20.8	7.7	French Feed Database (2005) (n = 75)
Rapeseed meal 00, European, dehulled, expeller	89.7	38.9	16.1		7.4	French Feed Database (2005) (n = 2)
Rapeseed meal 00, European, solvent-extracted, extruded	89.0	38.0	3.9		7.8	French Feed Database (2005) (n = 2)
Rapeseed meal 00, European, solvent-extracted, tanned	87.8	37.4	2.9		7.8	French Feed Database (2005) (n = 52)
Rapeseed/canola protein concentrate (dehulled, solvent- and water-extracted)	94.7	65.1	8.9	22.5	6.7	Hajen et al. (1993) (n = 1)
Canola protein isolate	97	92	3.3	24.5	5.6	Higgs et al. (1996) Mwachireya et al. (1999) (n = 2)

the washes), have a direct impact on the protein level of the rapeseed/canola meals. Indeed, the more the oil and undesirable compounds that are removed, the higher the remaining proportion of protein in the final product.

Essential Amino Acid Composition

Rapeseed/canola products have one of the best essential amino acid profiles among vegetal (Table 13.2). The quality of protein in both rapeseed/canola meals and concentrates is similar to that of fish meal and higher than that of soybean meal. As their methionine plus cystine are relatively high compared to other vegetal protein sources, rapeseed/canola protein products (meals and concentrates) are recognized as a good source of sulfur amino acids. In contrast, due to their suboptimal levels of lysine and the sulfur amino acids, rapeseed/canola protein isolates have similar protein quality to soybean meal (Table 13.2) (Higgs et al., 1996).

The technological treatments applied on the rapeseed/canola seeds have an effect on the amino acid composition of the products. Of course, the treatment applied to increase the protein content results in an increase in the amino acid content (Table 13.3). This relationship between amino acid content and protein content has been studied, and there are useful equations to predict amino acids from crude protein (Hickling, 2001).

The lysine level in Chinese rapeseed meal is lower than the meal from other countries as shown in Table 13.3, despite its having a high level of crude protein. It is likely that the high temperatures used in processing rapeseed in China results in the lower lysine values (Hickling, 2001).

Lipids and Fatty Acids

As shown in Table 13.1, the oil content of Canadian canola meals tends to be relatively high, at about 4 percent DM compared to 1 to 3 percent DM oil in rapeseed/canola meals produced in most other countries. The canola gums added back to Canadian canola meal at 1 to 2 percent are obtained during the refining of canola oil and consist of glycolipids, phospholipids, and variable amounts of triglycerides,

TABLE 13.2. Essential and semi-essential amino acid composition of rapeseed/canola products (% DM or % of protein in parentheses).

Protein products	Protein	Lys	Met	Cys	Thr	Arg	Gly	His	Ile	Leu	Phe	Tyr	Val	Ala	Trp	Reference
Canola, Canadian, whole		1.9 (6.9)			1.3 (4.8)	2.5 (9.1)	2.2 (8.3)	0.8 (3.1)	1.0 (3.8)	2.0 (7.4)	1.0 (3.8)	0.9 (3.4)	1.6 (5.8)	1.4 (5.1)		Wang et al. (1999) (n = 1)
Rapeseed 00, European, whole	20.7	1.3 (6.2)	0.4 (2.0)	0.5 (2.6)	0.9 (4.2)	1.2 (6.0)	1.0 (4.6)	0.6 (2.7)	0.8 (4.0)	1.3 (6.4)	0.8 (3.8)	0.6 (2.9)	1.1 (5.2)	0.9 (4.5)	0.3 (1.3)	Sauvant et al. (2004), French Feed Database (2005) (n = 5,157)
Canola, Canadian, whole, micronized		1.7 (6.2)			1.4 (5.1)	2.4 (8.7)	2.2 (8.2)	0.8 (3.1)	0.9 (3.3)	1.9 (7.0)	1.0 (3.5)	0.9 (3.3)	1.5 (5.6)	1.3 (4.7)		Wang et al. (1999) (n = 1)
Canola meal, Canadian, solvent-extracted	35.0	2.2 (5.8)	0.9 (2.2)	1.0 (2.7)	1.7 (4.3)	2.4 (6.1)	1.9 (5.0)	1.3 (3.2)	1.6 (4.0)	2.7 (6.8)	1.7 (4.4)	1.2 (3.0)	1.9 (4.9)	1.7 (4.4)	0.5 (1.3)	Hickling (2001) (n = 9); French Feed Database (2005) (n = 100)
Canola meal, Australian, solvent-extracted	43.1	3.2 (7.4)	3.1 (0.7)		2.6 (6.1)	7.8 (18.1)		1.6 (3.8)	2.5 (5.8)	4.2 (9.6)	2.7 (6.3)		3.0 (6.8)			Glencross et al. (2004) (n = 1)
Rapeseed meal 00, European, solvent-extracted	38.3	2.1 (5.4)	0.8 (2.0)	0.9 (2.4)	1.7 (4.3)	2.3 (6.0)	1.9 (5.0)	1.0 (2.6)	1.5 (4.0)	2.5 (6.7)	1.5 (3.9)	1.1 (2.9)	1.9 (5.1)	1.7 (4.4)	0.5 (1.2)	Sauvant et al. (2004), French Feed Database (2005) (n = 9,906)

TABLE 13. 2 (continued)

Protein products	Protein	Lys	Met	Cys	Thr	Arg	Gly	His	Ile	Leu	Phe	Tyr	Val	Ala	Trp	Reference
Rapeseed meal 00, Chinese, solvent-extracted	41.2	1.8 (4.4)	0.8 (1.9)	1.0 (2.5)	1.7 (4.0)										0.5 (1.2)	French Feed Database (2005) (n = 406)
Rapeseed meal 00, Indian, solvent-extracted	41.6	2.0 (4.7)	0.8 (1.9)	1.1 (2.6)	1.7 (4.1)										0.5 (1.3)	French Feed Database (2005) (n = 11)
Canola meal, Canadian, dehulled, solvent-extracted	41.5	2.3 (5.6)	1.3 (3.2)		1.8 (4.4)	2.7 (6.6)		1.2 (2.9)	1.7 (4.2)	2.8 (6.8)	3.0 (7.2)		2.2 (5.2)		0.5 (1.3)	Mwachireya et al. (1999)
Rapeseed meal 00, European, dehulled, solvent-extracted	40.9	2.4 (5.3)	0.9 (2.1)	1.1 (2.5)	1.9 (4.2)											French Feed Database (2005) (n = 57)
Canola meal, Canadian, expeller	34.1	2.0 (6.0)	0.7 (2.0)	1.0 (2.9)	1.5 (4.5)										0.4 (1.2)	French Feed Database (2005) (n = 1)
Canola meal, Australian, expeller	38.1	3.9 (10.2)	0.3 (0.8)		2.8 (7.3)	6.6 (17.3)		1.9 (5.0)	2.8 (7.3)	4.6 (12.1)	2.9 (7.6)		3.7 (9.7)			Glencross et al. (2004) (n = 1)
Rapeseed meal 00, European, expeller	35.2	2.0 (5.7)	0.7 (2.1)	0.8 (2.4)	1.7 (4.7)										0.5 (1.3)	French Feed Database (2005) (n = 75)

Feed													Reference
Rapeseed meal 00, European, heated, solvent-extracted	39.0	2.2 (5.5)	0.9 (2.2)	1.0 (2.4)	1.8 (4.6)		1.5 (3.3)	1.9 (4.3)	2.7 (6.1)	1.7 (3.8)	1.9 (4.2)	1.9 (4.3)	French Feed Database (2005) (n = 6)
Rapeseed meal 00, European, dehulled, heated, solvent-extracted	43.3	2.9 (6.6)	1.0 (2.3)	1.1 (2.5)	2.5 (5.6)	3.5 (7.9)						0.6 (1.3)	Burel et al. (2000a) (n = 1)
Rapeseed meal 00, European, extruded	38.0	2.0 (5.2)	0.7 (1.9)	0.7 (1.8)	1.5 (4.0)							0.5 (1.2)	French Feed Database (2005) (n = 2)
Rapeseed meal 00, European, Tanned	37.4	1.8 (4.8)	0.6 (1.7)	0.4 (1.2)	1.4 (3.7)								French Feed Database (2005) (n = 52)
Canola protein concentrate	48.3	4.9 (10.1)	0.4 (0.8)		3.5 (7.2)	8.3 (17.2)	2.3 (4.8)	3.5 (7.3)	5.8 (12.0)	3.6 (7.5)		4.7 (9.8)	Glencross et al. (2004) (n = 1)
Canola protein isolate	93.5	3.4 (3.6)	2.6 (2.7)		3.0 (3.2)	6.4 (6.9)	2.3 (2.5)	3.5 (3.7)	6.0 (6.4)	5.5 (5.9)		4.2 (4.5)	1.4 (1.6) Mwachireya et al. (1999) (n = 1)

TABLE 13.3. Essential amino acids index (EAAI) for different protein sources including rapeseed/canola products.

Protein sources	EAAI	Major limiting amino acids for carp and rainbow trout
Fish whole body protein	97	Threonine
Fish muscle	97	Threonine
Fish meal (whole herring meal)	94	Threonine
Soybean meal	91	Methionine + Cystine, Threonine, Lysine
Canola/rapeseed meal	95	Lysine
Canola/rapeseed protein concentrate	94	Lysine

Note: Table adapted from Higgs et al. (1996).

sterols, fatty acids, etc. The objective of this addition of gums is to increase the energy value of canola meal.

The fatty acid composition of the different rapeseed/canola products is similar. Rapeseed/canola oil is a rich source of n-6 fatty acids (particularly linoleic acid; 18:2 n-6) and of monounsaturated fatty acids (especially oleic acid; 18:1 n-9) compared to marine lipids. Linolenic acid (18:3 n-3) is the main (n-3) fatty acid in rapeseed/canola oil (Table 13.4).

Carbohydrates

Rapeseed/canola meal contains comparable nitrogen-free extract (37 percent DM) but considerably more fiber (13 percent DM) than soybean meal (Table 13.5) because, unlike soybean meal, the canola hull stays with the meal and the hull is a relatively high proportion of the canola seed. However, these levels can be reduced in dehulled meals: 27 percent DM of nitrogen-free extract and 7 to 10 percent DM of fiber. The percentage of crude fiber within rapeseed/canola meal does not truly reflect the total level of indigestible carbohydrates, since other indigestible carbohydrates, such as raffinose and stachyose, are not measured as part of the crude fiber (Higgs et al., 1995). The few data available indicate that the proportion of total oligosaccharides is between 6 and 10.5 percent DM in common rapeseed/canola

TABLE 13.4. Lipid and fatty acid composition of rapeseed/canola products.

	Rapeseed meal 00, European, solvent-extracted (00)	Rapeseed, whole (00)	Rapeseed oil 00, European	Canola oil, Canadian
Crude fat (% DM)	2.6	45.6	100	100
Fatty acids (% FA)				
C14:0	0.1	0.1	0.0	0.0
C15:0	0.0	0.0	0.0	0.0
C16:0	4.2	4.2	6.5	3.7
C16:1 (n-7)	0.4	0.4	0.0	0.2
C18:0	1.8	1.8	1.6	1.6
C18:1 (n-9)	58.0	58.0	51.0	58.6
C18:1 (n-7)	0.0	0.0	3.4	0.0
C18:2 (n-6)	20.5	20.5	29.0	22.9
C18:3 (n-6)	0.0	0.0	0.0	0.0
C18:3 (n-3)	9.8	9.8	8.6	12.4
Total saturated (% FA)	6.1	6.1	8.1	5.2
Total unsaturated (% FA)	93.9	93.9	91.9	94.8
Total (n-6) (% FA)	20.5	20.5	29.0	22.9
Total (n-3) (% FA)	9.8	9.8	8.6	12.4
Total n-3 HUFA (% FA)	0.0	0.0	0.0	0.0
References	Sauvant et al. (2004) (n = 2,820)	Sauvant et al. (2004) (n = 4)	Weber et al. (2002) (n = 1)	Dosanjh et al. (1998) (n = 1)

TABLE 13.5. Carbohydrates in rapeseed/canola products (% DM).

Protein products	Crude fiber	Crude cellulose	Hemi-cellulose	NDF[a]	ADF[b]	ADL[c]	Total sugars	Starch	N-free extracts + fiber	Reference
Canola seed, Canadian, whole		7.4		12.8	9.0	4.5		5.7		French Feed Database (2005) (n = 111)
Rapeseeds 00, European, whole	7.8	10.3		20.1	14.5	6.3	5.5	3.7		French Feed Database (2005) (n = 5,157)
Canola meal, Canadian, solvent-extracted	13.3	12.7	5-8	27.0	18.7	7.3	8.9	3.7	36.7	Mwachireya et al. (1999), Hickling (2001), French Feed Database (2005) (n = 100)
Rapeseed meal 00, European, solvent-extracted		13.9		31.7	20.6	9.9	10.5	6.5		French Feed Database (2005) (n = 9,906)
Rapeseed meal 00, Chinese, solvent-extracted		13.5		39.1	24.5	12.4	6.2	2.7		French Feed Database (2005) (n = 406)
Rapeseed meal 00, Indian, solvent-extracted		11.0		29.0	17.4	6.3				French Feed Database (2005) (n = 11)

Ingredient									Reference
Rapeseed meal 00, European, expeller		12.9		26.4	19.9	7.8	7.2	7.1	French Feed Database (2005) (n = 85)
Canola meal, Canadian, dehulled, solvent-extracted	9.7	12.1	2.2	20.1	15.1	12.5			26.8 — McCurdy and March (1992), Higgs et al. (1995), Mwachireya et al. (1999) (n = 12)
Rapeseed meal 00, European, dehulled, solvent-extracted	7.1	10.2		20.3	12.4	6.1	1.5		French Feed Database (2005) (n = 57)
Rapeseed meal 00, European, heat treated, solvent-extracted		13.3		29.5	21.3	10.2			French Feed Database (2005) (n = 6)
Rapeseed protein concentrate	4.9								15.4 — Higgs et al. (1996)
Canola protein isolate	2.1	1.3	0.3	2.6	2.4	0.6			Mwachireya et al. (1999)

[a] Neutral detergent fiber.

[b] Acid detergent fiber.

[c] Acid detergent lignin.

meals and 1.5 percent DM in dehulled meals (Table 13.5). In general, cellulose, hemicellulose, neutral detergent fiber (NDF = lignin + cellulose + hemicellulose), acid detergent fiber (ADF = lignin + cellulose), and acid detergent lignin (ADL = lignin) comprise, on DM basis, 11 to 14, 5 to 8, 26 to 39, 17 to 25, and 6 to 13 percent, respectively in common rapeseed/canola meals and 10 to 12, 2.2, 20, 12 to 15, and 6 to 13 percent in dehulled meals (Table 13.5). The proportion of starch generally comprises 3 to 7 percent DM. The inherently high levels of hulls and indigestible carbohydrates in common rapeseed/canola meals partly contribute to their reduced available energy content relative to fish meal in some fish, especially salmonids (Higgs et al., 1995).

The high variation in the fiber proportion in the rapeseed/canola meals is related to the origin of the seeds and to processing techniques applied. For instance, Canadian values reported for levels of NDF are generally lower (Table 13.5). It is unclear why there is a discrepancy here although the Canadian values are consistent between different laboratories and samples (Hickling, 2001).

Further processing of meals can also result in rapeseed/canola concentrates or isolates with very low levels of fiber (about 5 percent DM in concentrates and 2 percent DM in isolates) (Table 13.5).

Vitamins

Information on the vitamin content of rapeseed/canola meal is very limited. It appears to be rich in choline, biotin, folic acid, niacin, riboflavin, and thiamine. However, compared to fish meal, rapeseed/canola meal is distinguished by a lack of vitamin B_{12} (Table 13.6). As the oil is almost totally removed in the rapeseed/canola meals, the fat-soluble vitamins are also removed.

Minerals

Rapeseed/canola meal is a relatively good source of essential minerals (Tables 13.7 and 13.8) compared to other vegetable-origin oilseed meals. It is an especially good source of selenium and phosphorus. When compared to fish meals, it composition is lower in calcium and higher in magnesium, but it contains either lower (meals) or similar (protein concentrates) levels of phosphorus (Table 13.7). The sodium content of rapeseed/canola meal may vary somewhat depending on

TABLE 13.6. Vitamin content of solvent-extracted rapeseed meal.

Vitamins (mg/kg DM)	Rapeseed meal 00, European, solvent-extracted	Canola meal, Canadian, solvent-extracted
Vit E	15.8	14.4
Vit B_1	3.4	5.8
Vit B_2	4.5	6.5
Vit B_6	12.4	8.0
Vit B_{12}	0.0	0.0
Niacin	187.1	177.8
Pantothenic acid	10.1	10.6
Folic acid	0.92	0.89
Biotin	1.01	1.09
Choline	7,372	7,444
References	Sauvant et al. (2004) (n = 2,820)	Hickling (2001)

whether soap stocks from refining (usually sodium salt of fatty acids) are added to the meal.

Approximately 60 to 90 percent of total oilseed phosphorus is in the form of phytic acid, the hexaphosphate of myoinositol, which occurs as a mixture of calcium, magnesium, and potassium salt (Higgs et al., 1995). The intestinal enzyme phytase, which is necessary for phytin hydrolysis is lacking in fish. Hence, the phytin phosphorus is unavailable for fish. Thus, when rapeseed/canola protein products are used at high dietary levels, it is crucial to ensure that there is sufficient inorganic phosphorus to meet the fishes' dietary needs for growth or to use exogenous phytase as a feed additive.

ANFs

In addition to high levels of crude fiber, other indigestible carbohydrates, and phytic acid, rapeseed/canola meals also contain other ANFs such as tannins, sinapine, and goitrogenic compounds: GLS (Table 13.9). The consequences of the utilization of rapeseed/canola meals at high levels in diets are decreased diet intake, low feed

TABLE 13.7. Macro-mineral contents of rapeseed/canola products (g/kg DM).

Protein products	Ca	P	Phytin P	Phytase activity UI/kg DM	Mg	K	Na	Cl	S	Reference
Canola, Canadian, whole	4.3	6.5								French Feed Database (2005) (n = 111)
Rapeseed 00, European, whole	5.0	7.3	5.5		2.6	8.3	0.1	1.0	3.6	Sauvant et al. (2004) (n = 5,157)
Canola meal, Canadian, solvent-extracted	7.0	12.0	7.7		6.0	13.6	1.1	1.1	9.4	Hickling (2001) (n = 2)
Rapeseed meal 00, European, solvent-extracted	9.4	13.0	9.0	11.3	5.5	13.9	0.5	0.8	6.7	Sauvant et al. (2004) (n = 2,820)
Rapeseed protein concentrate	7.9	2.0			7.6	2.5				Higgs et al. (1995) (n = 1)

TABLE 13.8. Trace mineral contents (mg/kg DM).

Protein products	Mn	Zn	Cu	Fe	Se	Co	Mb	I	Reference
Rapeseed 00, European, whole	36.9	43.4	3.3	234.3	0.84				Sauvant et al. (2004) (n = 4)
Canola meal, Canadian, solvent-extracted	57.8	64.4	6.4	184.4	1.2		1.6		Hickling (2001) (n = 2)
Rapeseed meal 00, European, solvent-extracted	58.6	73.3	7.9	193.9	1.2	0.10	1.8	0.10	Sauvant et al. (2004) (n = 2,820)
Rapeseed protein concentrate	57.0	126.0		95.0					Higgs et al. (1995) (n = 1)

efficiency, and metabolic disturbances (thyroid dysfunction), leading to a decrease of the growth performance in all animal species.

Fiber

The hulls of the rapeseed/canola seeds are rich in parietal carbohydrates and in fiber (Tables 13.5 and 13.9). Fiber in monogastric animal diets may generally be divided into two physicochemical groups. These are the soluble fibers (gar gum, pectins, etc.) which create viscous conditions within the small intestine and can affect digestion and absorption, and the usually insoluble lignified fibers (cellulose, hemicellulose, lignin, etc.), which mainly increase fecal output. Consequently, many analytical methods have been adopted to determine a soluble and insoluble fraction (Graham and Aman, 1991).

Nutrient absorption depends on the rate at which nutrients are in contact with the absorptive epithelium. The influence of dietary fiber on the movement of nutrients along the gastrointestinal tract will likely influence nutrient absorption. The soluble fibers have been reported to delay stomach emptying, attributed to the increased viscosity of the meal, which might influence the absorption rate of nutrients. They give rise to a viscous solution when dissolved, which should slow transport phenomena, such as diffusion and mixing. However, the presence of soluble fiber does not always lead to slower stomach emptying. Other factors, such as pH of the stomach may also be important in this regard (Shiau, 1989).

TABLE 13.9. Antinutritional factors in rapeseed/canola products.

Rapeseed/canola protein products	Glucosinolates (μmoL/g DM)	Phytic acid (% DM)	Sinapine (% DM)	Tannins (% DM)	Reference
Rapeseeds 00, whole, ground, heated (00)	36				Bourdon and Aumaître (1990)
Canola meal, Canadian, solvent-extracted	8.8-19.7	2.6-4.5	1.1-2.6	0.3-1.7	McCurdy and March (1992), Higgs et al. (1996), Satoh et al. (1998), Mwachireya et al. (1999), Hickling (2001) (n = 9)
Rapeseed meal 00, European, solvent-extracted	30.0-40.7	4.2			Bourdon and Aumaître (1990), Burel et al. (2000b) (n = 4)
Rapeseed meal 00, European, dehulled, heated, solvent-washed[a]	5.1-26.3	4.4			Burel et al. (2000b) (n = 2)
Rapeseed meal 00, European, expeller	36.0	4.2			Bourdon and Aumaître (1990) (n = 1)
Canola meal, Canadian, solvent extracted, acid-washed[a]	5.2-19.7	1.5-4.5	0.9-2.5	0.24	McCurdy and March (1992), Higgs et al. (1996), Mwachireya et al. (1999) (n = 7)

Canola meal, Canadian, solvent extracted, ethanol-washed[a]	3.5-4.1	5.5-5.5	0.2-0.3		McCurdy and March (1992), Higgs et al. (1996) (n = 3)
Canola meal, Canadian, solvent extracted, ethanol- and acid washed[a]	0.6-1.3	2.4-3.1	0.2		McCurdy and March (1992) (n = 4)
Canola meal, Canadian, solvent-extracted, ammonia/ethanol-washed[a]	1.1-1.4	4.6-5.8	0.7-0.8	0.11	McCurdy and March (1992), Mwachireya et al. (1999) (n = 4)
Canola meal, Canadian, extruded 90°C[a]		3.8			Satoh et al. (1998) (n = 1)
Canola meal, Canadian, extruded 150°C[a]		3.0			Satoh et al. (1998) (n = 1)
Rapeseed protein concentrate	1.0	5.7			Higgs et al. (1996) (n = 1)
Canola protein isolate	2.0	0.3-0.4		0.10	Higgs et al. (1996), Mwachireya et al. (1999) (n = 2)

[a]Experimental products.

Insoluble fibers increase fecal output, decrease the contact time between the dietary nutrients, the digestive enzymes, and the cells of the intestinal mucosa, site of the absorption process. Fibers in moderate amounts in the diet can nevertheless improve the growth of fish, because they serve as bulk in the alimentary bolus, which regulate the speed of the intestinal transit. But fiber in high amounts (>8 percent) can lead to the opposite effect. It is generally accepted that these compounds have no nutritional value in fish, as in most monogastric animals, because they are not digested in the intestinal tract due to the absence of competent hydrolytic enzymes (Shiau, 1989). However, it is worth noting that a cellulase activity has been reported in the digestive content of some fishes (Stickney and Shumway, 1974), including rainbow trout (Smith, 1971). According to these authors, this activity could be due to the intestinal microflora, which is strongly influenced by the temperature and the water microflora. Another hypothesis is that this activity would come from the prey consumed by the fishes.

Tannins

Tannins in legume seeds have received intermittent attention as ANFs over the past thirty years. Tannins are water-soluble polyphenolic compounds, which are widely distributed in plants, including rape (Table 13.9). Tannins may be regarded as biological pesticides from an agricultural point of view. It has been suggested that tannins play a major role in the plant's defence against fungi and insects. The synthesis of tannins has been attributed to stress conditions (Jansman, 1993).

The two major types of tannins are chemically quite different. Condensed tannins (proanthocyanidins) are flavonoid polymers, with carbon-carbon bonds joining the individual flavonoid monomers. Condensed tannins are not susceptible to hydrolysis, but can be oxidatively degraded in strong acid to yield anthocyanidins. Condensed tannins are labile polyphenolic substances, which are easily oxidized and may deteriorate during extraction and chromatographic procedures. When seeds are stored for some years, the color often changes (darker). These changes in the color are probably related to oxidation of the tannins. Hydrolyzable tannins are gallic or hexahydroxydiphenic acid esters of glucose or other polyols. The ester bonds are acid-, base-, and enzyme-labile, and the hydrolyzable tannins are easily broken down to gallic acid or hexahydroxydiphenic acid subunits and the core polyol (Butler and Bos, 1993).

Tannins in legume seeds are categorized as condensed tannins. They are most commonly found in dicotyledon plants and are found in particular in the hulls. Tannins are encountered in various foods and feedstuffs such as cereal grains (sorghum and barley), in a large variety of legume seeds, and in rapeseed (Jansman and Longstaff, 1993).

Tannins can affect the diet intake in terrestrial animals by affecting the lubricating properties of saliva. It seems, however, that tannins reduce diet intake only when the dietary content is relatively high (>200 g/kg of diet in pigs and >300 g/kg in poultry) (Jansman and Longstaff, 1993).

Tannins have harmful nutritional effects, resulting in a lower feed efficiency in monogastric farm animals, including fish. Tannins complex with proteins and the tannins-protein interactions usually result in precipitation of the complex. Activities of digestive enzymes, such as trypsin and chymotrypsin, may also be inhibited by tannins. It may be suggested that the digestion processes in monogastric animals, including fish, are disturbed when the tannins complex with digestive enzymes. Tannins also complex with mucous cell membranes, which results in increased endogenous losses of protein (Butler and Bos, 1993).

The effects on the digestibility of protein and amino acids seem to be most predominant. This may indicate that tannins, in vivo, have a higher affinity to bind to protein than to other compounds like carbohydrates and fat. Indeed, on average, protein digestibility is reduced to a greater extent than digestibility of starch and lipid in monogastric animals (Jansman and Longstaff, 1993).

Sinapine

Sinapine is a phenolic compound that is undesirable in animal diets. They are associated with poor palatability due to bitterness or astringency, thus affecting diet intake. They also interfere with nutrient uptake in the digestive system. Phenolic compounds exist as small molecules or in polymeric form. Rapeseed/canola sinapine is comprised mostly of small molecules of phenolic compounds (Table 13.9).

Sinapine gained its notoriety when hens started to give eggs that smelled "fishy" or "crabby," and the problem was traced to rapeseed/canola meal that contained sinapine. These hens could not metabolize sinapine fully and trimethylamine, an intermediate, was leached into their eggs, thus giving them the objectionable fishy odor.

The bitterness of sinapine may also decrease the palatability of rapeseed/canola meals for fish, and it is noteworthy that treatment of rapeseed/canola products with various solvents, such as water and ethanol, or a blend of methanol, ammonia, and water, can dramatically reduce the levels of sinapine and other phenolic compounds. Consequently, rapeseed/canola protein concentrates and upgraded meals are expected to have improved nutritive value for finfish.

Recognizing sinapine as a problem, canola breeders have for a long while screened germplasm collections for low-sinapine lines and have concluded that suitable lines do not exist. As a consequence, biotechnological approaches have been applied to solving the long-standing problem of sinapine content in canola. This has involved studying the metabolic routes to sinapine synthesis. Sinapine is a product of the phenylpropanoid metabolism. These pathways occur in all land plants in one form or another. However, in the seeds of rapeseed/canola and related species, one of the metabolic pathways is extended and produces sinapine. A transgenic derivative of canola has been obtained in which the sinapine content was reduced to approximately 60 percent of the original level. The levels in some transgenics were ~5 mg/g of the whole seed weight (seed coat included) as opposed to ~9 mg in the control seeds. Extending such metabolic engineering, the sinapine content has been reduced even further and these new strains of canola contain as little as 20 percent of the original content.

Phytic Acid

Phytic acid and GLS have the greatest influence on reducing the nutritive value of rapeseed/canola protein products for finfish. Although phytic acid may comprise between 1 percent and 5 percent by weight of cereals, nuts, legumes, oilseeds, spores, needles, and pollen, normally it constitutes 1 to 2 percent by weight of most cereals and oilseeds (review of Higgs et al., 1995). The phytic acid content in rapeseed/canola meals is about 3.0 to 4.4 percent DM (Table 13.9), which is higher than that of soybean meal (<2 percent DM). In upgraded rapeseed/canola meals (produced by solvent washing of fiber-reduced meal) and rapeseed protein concentrate, the phytate or phytic acid levels are higher than in commercial rapeseed/canola meal (Table 13.9).

Phytic acid, the hexaphosphate of myoinositol, is strongly negatively charged at all pH values normally encountered in the content of the gastrointestinal tract. Indeed, this compound has strong affinity for proteins at low-to-neutral pH and for cations, such as zinc, at intestinal pH. Consequently, high dietary levels of phytic acid may depress the bioavailability of polyvalent cations to homeotherms and fish, including salmonids (Higgs et al., 1995) and channel catfish (Satoh et al., 1989). High dietary calcium levels may potentiate the complexion of zinc with phytic acid. Hence, there may be a further reduction in zinc bioavailability in both mammals and fish.

The consequences of high dietary phytic acid levels in fish include depressed growth, diet and protein utilization, survival, and thyroid function. High dietary phytate levels also may result in cataract formation (owing to lowered zinc bioavailability) and structural anomalies in the pyloric caecal region of the fish intestine. The latter effect may be caused by reduced magnesium bioavailability, and/or by a toxic effect of phytic acid on the epithelial layer of the pyloric caeca. Impairment of apparent magnesium absorption has been noted in rats ingesting soybean protein or sodium phytate. The effect was attributed to an increased fecal excretion of endogenous magnesium, possibly because of damage to the intestinal epithelial cells. Many of the adverse effects of high dietary levels of phytate may be ameliorated by dietary zinc supplementation. However, complete counteraction of phytate effects may not be possible in cases of pronounced intestinal epithelial cell damage (Higgs et al., 1995).

Phytic acid concentrations in oilseed protein sources can be decreased by limiting the amount of phosphorus available to the oilseed during plant growth, removing the phytic acid from the oilseed during food processing (e.g., by selective precipitation), using the enzyme phytase from microbial (i.e., fungi, bacteria, and yeast) sources to dephosphorylate the phytic acid, and allowing the seed to germinate.

With regard to the previous strategies, the pretreatment of rapeseed/ canola protein sources with microbial phytase, or the direct addition of phytase to the diet may prove to be the most efficacious and economical means of eliminating the antinutritional effects of phytate in finfish. Certainly, this approach has proven to be worthwhile in poultry, swine, and fish fed phytate-rich diets. Alternatively, rapeseed/ canola protein products could be blended with fish silage and wheat bran (naturally rich in phytase) to form acid-stabilized, paste-like, phytase-free products that subsequently can be dried and incorporated into fish diets (Higgs et al., 1995).

Teskeredzic et al. (1995) have shown that dephytinized rapeseed protein concentrates (phytic acid removed) can comprise 39 percent of the dietary protein (fish meal comprises only 11 percent of diet) for rainbow trout without adversely affecting performance. Furthermore, using an improved dephytinization protocol for rapeseed protein concentrate, Prendergast et al. (1994) reported that dephytinized rapeseed protein concentrate could comprise 59 percent of the dietary protein for rainbow trout by complete removal of premium-quality fish meal from the diet. The phytic acid content can also be reduced by extrusion processing. In commercial canola meal, the level of phytate has been reduced approximately by 10 and 30 percent from the original level by extrusion cooking at low (90°C) and high (150°C) temperatures, respectively. It was found that the high-temperature-treated canola meal could comprise 24 percent of the dietary protein (30 percent of incorporation) without adversely affecting performance of chinook salmon (Satoh et al., 1998).

GLS

Nature and Effects of GLS

Glucosinolates are thioethers. They generally consist of a sugar entity, b-D-thioglucose, with an ester bond to an organic aglycone that is an alkyl group yielding derivatives, and a specific side group structure distinguishes one glucosinolate from another (Figure 13.3). They are found exclusively in dicotyledenous plants, with highest concentrations in the Brassicaceae families.

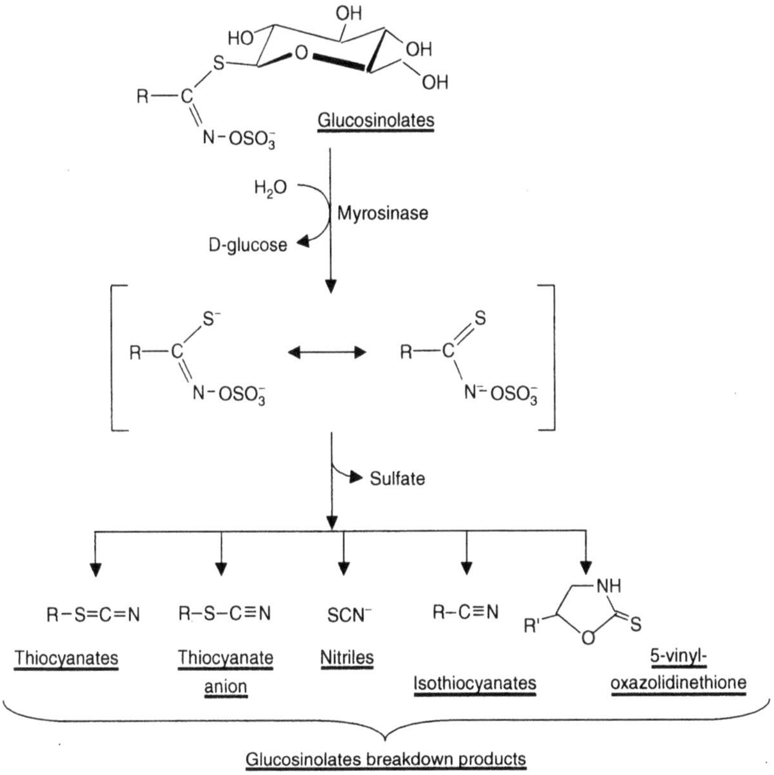

FIGURE 13.3. Enzymatic breakdown of glucosinolates. *Source:* Adapted from Quinsac, 1993.

The levels of GLS in rapeseed/canola meals, severely limited their use as a protein supplement in animal diets. The hydrolysis of GLS is known to result in an array of possible products. These include isothiocyanates, 5-vinyloxazolidine-2-thiones (goitrin), thiocyanates, and nitriles, depending upon the hydrolysis conditions (Figure 13.3). The enzymes for hydrolysis are produced by plants and by rumen organisms. They react with the GLS when plant tissue is crushed, for example by mastication, or when the plant is consumed into the rumen of a ruminant animal. Hydrolysis of GLS can also occur during oil extraction in rapeseed/canola processes (high temperatures in wet conditions) or during diet pelleting (Higgs et al., 1995).

Glucosinolates appear to have little biological impact themselves, but some of their products are biologically active in animals:

- *Isothiocyanates* are irritating to mucous membranes and not readily consumed in sufficient quantities to be toxic. However, if GLS are consumed and then hydrolyzed to isothiocyanates in the gut, they can have powerful antithyroid effects and interfere with the synthesis of necessary thyroid hormones.
- *Oxazolidine-2-thiones* are closely related to isothiocyanates. One way they are produced is by the conversion of the glucosinolate progoitrin in rapeseed meal to goitrin, which in turn is hydrolyzed to oxazolidine-2-thiones. These compounds depress growth and increase the incidence of goiters. They inhibit thyroid function by blocking the incorporation of iodine into thyroxine precursors and by suppressing thyroxine secretion from the thyroid.
- *Nitriles* depress growth, cause liver and kidney lesions, and in severe cases liver necrosis, bile duct hyperplasia, and megalocytosis of tubular epithelium in the kidney.
- *Thiocyanates* inhibit iodine uptake by the thyroid, leading to reduced iodination of tyrosine and resulting in decreased production of the important thyroid hormone thyroxine (Higgs et al., 1995).

Impairment of Thyroid Function in Fish

The effect of GLS on thyroid function in fish has been mainly studied in salmonids, including rainbow trout (Higgs et al., 1995, 1996). Recent studies by Burel, Boujard, Escaffre, et al. (2000), Burel, Boujard, Kaushik, et al. (2000), Burel, Boujard, Tulli et al. (2000), and Burel et al. (2001) showed that incorporation of rapeseed meal in the diet of trout induces thyroidal disturbances. Some changes in plasma concentrations of thyroid hormones can be observed after fourteen days of feeding, and the effects were more pronounced with increasing levels of incorporation of rapeseed meal (Figure 13.4). Plasma concentrations of T_4 decreased at incorporation of 10 percent of rapeseed meal into the diet (1.5 µmole of GLS breakdown products/g DM of diet) and the same observation was made for T_3 at 30 percent level (3.7 µmole/g DM). The thyroidal disturbances were more pronounced

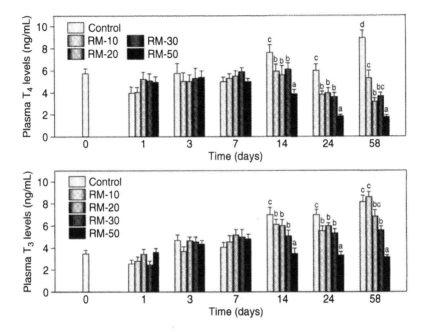

FIGURE 13.4. Changes of plasma concentrations of thyroxine (T_4) and triiodo-thyronine (T_3) over time in trout fed diets with increasing dietary levels of rape-seed meals (0, 10, 20, 30, and 50%) which correspond to dietary levels of GLS toxic derivatives of 0, 1.5, 2.4, 3.7, and 17.4 µmole/g DM, respectively. Means (± standard deviation, n = 15) sharing no common superscript letter are signifi-cantly different ($P < 0.05$). *Source:* Adapted from Burel et al. (2001a).

when the myrosinase has not been inactivated by a thermal treatment due to the production of higher levels of GLS toxic derivatives in the diets during rapeseed processing and diet preparation (Burel et al., 2001).

The mode of action of the GLS toxic derivatives contained in the rapeseed meals seems to be similar to that described in terrestrial ani-mals (Burel et al., 2001). Studies in mammals have shown that thio-cyanate anions are competitors for iodine with respect to the active transport of iodine across the cell membrane and its binding to the tyro-sine residues of thyroglobulin (Fenwick et al., 1983). Vinyloxazoli-dinethiones block the normal coupling reaction whereby two molecules of diiodotyrosine (DIT) are combined to form T_4, or in smaller quan-tities, DIT with monoiodotyrosine (MIT) to form T_3 (Mawson et al.,

1994). The effect of isothiocyanates depends on their conversion. These compounds may be recycled and converted into vinyloxazolidinethione or hydrolyzed into thiocyanate anions (Langer and Stolc, 1965). Data from mammals suggest that the resulting lower plasma T_3 and T_4 levels stimulate the secretion of hypothalamic thyroid-stimulating hormone (TSH), which in turn induces an increased thyroid gland activity, resulting in the hypertrophy of the thyroid tissue (Mawson et al., 1994). In studies conducted in trout (Burel, Boujard, Escaffre et al., 2000; Burel et al., 2001), a hyperactivity of the thyroid follicles is also observed in fish fed a diet containing a high proportion of rapeseed meal (Figure 13.5, Table 13.10) (Higgs, McBride, et al., 1982; Hardy and Sullivan, 1983; Leatherland et al., 1987; Hossain and Jauncey, 1988; Teskeredzic et al., 1995; Burel, Boujard, Escaffre et al., 2000).

Dietary supplementation of iodine, as well as rearing fish in an aquatic environment rich in iodine can limit the thyroid disturbances in trout (Figure 13.6) (Burel et al., 2001), as has been previously shown in growing pigs (Anke et al., 1980; Ludke et al., 1985). These results indicate the effect of thiocyanate anions as competitor of iodine in thyroid metabolism in fish.

Other symptoms of the hypothyroid conditions were also observed. Indeed, the low production of T_4 induces a reaction at the level of the peripheral deiodinations in trout and on the basis of these results, it seems that the deiodinase activities can play an important role in the adjustment of the plasma T_3 levels. The response in terms of adjustments of the deiodinase activities as compensation for the shortage in bioactive T_3 is sensible, reacting readily at the lowest incorporation (10 percent) of rapeseed meal (Figure 13.7). The deiodinase response in liver and brain was observed after seven days of feeding, even though the effect of the toxic compound was not yet visible on plasma thyroid hormone levels. This suggests a fast change in T_3 homeostasis in these tissues. After two months of feeding, the compensatory activities of deiodinases are very high, but efficient only up to a level of 20 percent of dietary rapeseed meal, corresponding to about 4 μmole GLS toxic derivatives (isothiocyanates, oxazolidine-2-thiones, nitriles) per gram of diet. The stimulation of the in vitro deiodinase D2 activity (production of T_3 from T_4) in liver, in relation to the decrease of plasma thyroid hormone levels, suggests the possibility that in vivo,

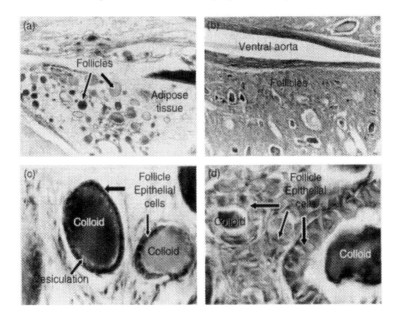

FIGURE 13.5. Effect of rapeseed meal intake on thyroid activity in rainbow trout as shown by sagittal sections of the thyroid tissue situated in the lower jaw, adjacent to the ventral aorta. *Source:* Adapted from Burel et al. (2000a). *Note:* (a). Thyroid follicles of a rainbow trout fed a rapeseed meal free diet (control) for 73 days. The thyroidal tissue is not very dense, the follicles are spread into adipose tissue (×60). (b). Thyroid follicles of a rainbow trout fed a diet containing 50% of rapeseed meal for 73 days. Thyroidal tissue is very dense (×60). (c). Unstimulated thyroidal tissue of a rainbow trout fed a rapeseed meal free diet (control) for 73 days. The follicle epithelial cells are cuboidal in appearance, and the colloid within the lumen is homogeneous. The ratio of nucleus/cytoplasm of the epithelial cells is high. Vesiculation of the colloid is evident in this follicle (×500). (d). Stimulated thyroid tissue of a rainbow trout fed a diet containing 50% of rapeseed meal for 73 days. The follicle epithelial cells are very large and are columnar. The colloid is partly or wholly depleted. The ratio of nucleus/cytoplasm of the epithelial cells is low (×500).

T_3 production from the available T_4 is increased. In addition, the decrease of in vitro D_3 activity observed in brain suggests that the in vivo degradation of T_3 is reduced, protecting the brain against too low T_3 levels (Burel, Boujard, Escaffre et al., 2000; Burel et al., 2001).

A treatment with TSH does not stimulate the thyroid hormone release in trout fed a rapeseed meal-based diet, opposed to that of fish fed a control diet (Figure 13.8). These results showed that, as in ter-

TABLE 13.10. Effects of glucosinolates on plasma concentrations of thyroid hormones and thyroid follicular activity in rainbow trout fed a control diet or a diet containing 50% of rapeseed meal.

| | Glucosinolates (μmole/g DM) in diet | | |
	Control (0)	Rapeseed meal-based (19.3)	Number of fish used
Thyroxin (T_4; ng/mL)	9.8[b]	1.0[a]	10
Triiodothyronin (T_3; ng/mL)	7.2[b]	4.5[a]	10
Thyroid tissue volume (mm^3)	2.8[a]	10.3[b]	7
Thyroid epithelial cell height (μm)	4.3[a]	13.0[b]	10

Note: Table adapted from Burel et al. (2000a).

Mean values (standard error) within a row with different superscript letters were significantly different ($P < 0.05$).

restrial animals (Mawson et al., 1994), the follicular activity of the thyroid tissue was inhibited in fish following the ingestion of GLS derivatives, and that this inhibition was not caused by a lack of pituitary stimulation. Moreover, the hyperactivity of the thyroidal follicles is really a response to an excessive stimulation of the synthesis of the pituitary TSH in trout (1999, personal communication) and in turbot (Pradet-Balade et al., 1999). An increase of the pituitary concentration of RNAm coding for this hormone, in response to a low production of T_4 is observed in fish fed a rapeseed meal-based diet.

Comparing the results obtained in turbot (Burel, Boujard, Kaushik et al., 2000) with those obtained in trout (Burel, Boujard, Escaffre et al., 2000; Burel et al., 2001), the thyroidal metabolism of turbot is less sensitive than that of trout to the ingestion of rapeseed meal (Figure 13.9, Table 13.11), likely because of its marine environment, which is richer in iodine. Thus, the deleterious action of one type of the GLS breakdown products, the thiocyanate anions, could be limited. A decrease of the plasma levels of T_4 was observed at 30 percent of incorporation of heat-treated rapeseed meal (3.6 μmole of GLS derivatives /g DM of diet), but T_3 remained at a normal level, even with a 46 percent dietary incorporation (4.4 μmole/g DM), owing to the compensatory activities of the deiodinases D1 and D2. In contrast to trout, untreated rapeseed meal was less toxic than a heat-treated one

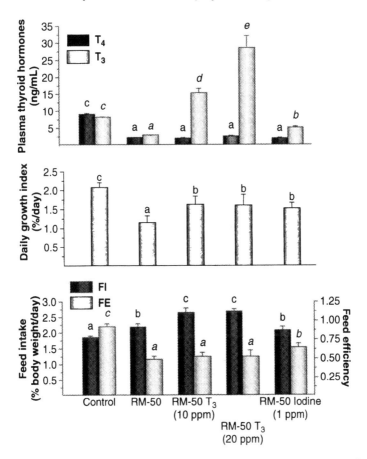

FIGURE 13.6. Effect of rapeseed meal intake on plasma concentrations of T_3 and T_4, daily growth index, voluntary feed intake, and feed efficiency in rainbow trout (final body weight 54 g). Trout were fed for 2 months a control diet or diets containing 50% of rapeseed meal without and with supplementation of 10 or 20 mg/kg of T_3 or 1 mg/kg of iodine. Means ($\pm$ standard deviation, n = 15) with no common letter are significantly different ($P < 0.05$). *Source:* Adapted from Burel et al. (2001).

for thyroid metabolism of turbot. The two rapeseed meals have different goitrogenic activities according to the fish species. This could be due to some differences in the profile of GLS derivatives, and maybe a higher proportion of thiocyanate anions in the untreated rapeseed meal.

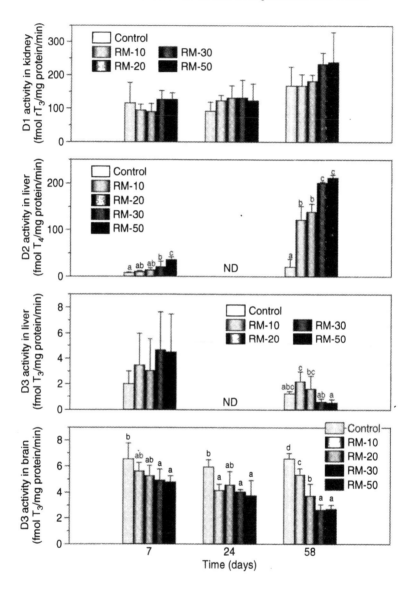

FIGURE 13.7. Effect over time of rapeseed meal intake on the in vitro deiodinase activities in rainbow trout : D1 activity in kidney (deiodination of rT_3), D2 activity in liver (deiodination of T_4), and D3 activity in liver and brain (deiodination of T_3). Means (± standard deviation; n = 3, 5 pooled samples) with no common superscript letter are significantly different ($P < 0.05$). Diets contain 0, 10, 20, 30, or 50% of rapeseed meal. *Source:* Adapted from Burel et al. (2001).

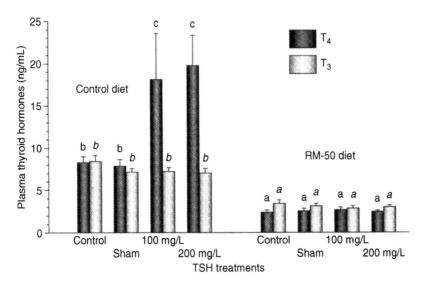

FIGURE 13.8. Effect of thyroid-stimulating hormone (TSH) treatments (sham saline solution, 100 mg/L of TSH, or 200 mg/L of TSH) on plasma levels of T_3 and T_4, measured 24 hours after injection, in rainbow trout (mean weight: 55g) fed a control diet or a diet containing 50% of rapeseed meal for 64 days. Means ($\pm$ standard deviation; n = 9) with no common letter are significantly different ($P < 0.05$). *Source:* Adapted from Burel et al. (2001).

Relation Between Thyroidal Status and Growth

In trout fed rapeseed-meal-based diets, a dietary supplementation of T_3 or iodine, as well as an environment rich in iodine, induced an increase of plasma levels of T_3, accompanied by higher growth performance (Figure 13.6) (Leatherland et al., 1987; Burel et al., 2001). These results confirm that the thyroidal disturbances are, at least in part, responsible for the reduction of the growth in this fish species. A dietary content of approximately 1.5 μmole/g of GLS derivatives was enough to induce some changes in the plasma levels of thyroid hormones, and a content of about 3.7 μmole/g was enough to cause a decrease of the growth performance. It seems that the capacity to growth in trout was affected when their plasma concentration of T_3 was below 4 to 5 ng/ml.

Considering the effect of a dietary supplementation of T_3 on the activity of the intestinal digestive enzymes shown in red seabream

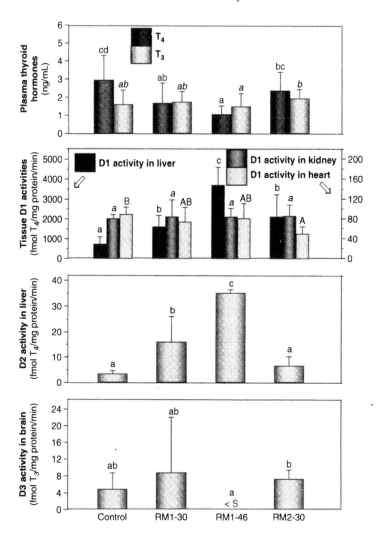

FIGURE 13.9. Effect of rapeseed meal intake on plasma T_3 and T_4, and on deiodinase activities, that is, D1 activity degrading the rT_3 in liver (left axis), in kidney and heart (right axis), D2 activity converting the T_4 into T_3 in liver, and D3 activity degrading the T_3 in brain, measured in blood and tissues samples of turbot after nine weeks of feeding diets containing 30% and 46% of heat-treated rapeseed meal (RM1-30 and RM1-46) and a diet containing 30% of unheated rapeseed meal (RM2-30). Means are given with standard deviation (n = 24 for plasma hormone levels and n = 6 for deiodinase activities of 4 pooled tissues. Means with no common letter are significantly different ($P < 0.05$). *Source:* Adapted from Burel et al. (2000b).

TABLE 13.11. Effect of rapeseed meal intake on plasma concentrations of thyroid hormones, deiodinases (D1, D2, D3) activities, and growth performance of turbot and rainbow trout.

	Fish species	
	Turbot (sea water)	**Rainbow trout (fresh water)**
Effect of the ingestion of GLS breakdown products (μmole/g of diet) on the thyroid status		
Plasma T_4 levels	Affected at 3.6 μmole/g	Affected at 1.5 μmole/g
Plasma T_3 levels	Not affected at least up to 4.4 μmole/g	Affected already at 3.7 μmole/g
Deiodinase compensatory effect	Efficient up to at least 4.4 μmole/g	Affected under 3.7 μmole/g
D1 activity (degradation rT_3)	Affected at a concentration nil[a]	not affected up to at least 17.4 μmole/g
D2 activity (production of T_3 from T_4)	Affected at 3.6 μmole/g	Affected at 1.5 μmole/g
D3 activity (degradation T_3)	Affected at 4.4 μmole/g	Affected at 1.5 μmole/g
Effect of the level of incorporation of rapeseed meal (% of diet) on fish performance		
Growth performance	Affected at 46% of heat-treated rapeseed meal and at 30% of untreated rapeseed meal	Affected at 30% of heat-treated rapeseed meal or untreated Rapeseed meal
Diet intake	Affected at 30% of rapeseed meal	Not affected at a level up to 50% of rapeseed meal
Feed efficiency	Not affected at a level up to 46% of heat treated rapeseed meal and 30% of untreated rapeseed meal	Affected at 30% of rapeseed meal

Note: Table adapted from Burel et al. (2001).

[a] In the case of the untreated rapeseed meal, with the myrosinase not inactivated, a hydrolysis of the dietary GLS, and then a production of goitrogenic derivatives, is possible in the intestinal tract of fish.

(Woo et al., 1991), the shortage of bioactive thyroid hormones may be responsible for the decrease of the capacity of trout to digest energy and phosphorus (Table 13.12). However, in the studies of Burel, Boujard, Escaffre et al. (2000) and Burel et al. (2001), a reduction in the nutrient utilization in trout was only observed when 50 percent of rapeseed meal was incorporated into the diet, even though the plasma levels of T_3 were already low in fish with 30 percent of rapeseed meal. It is possible that the plasma concentration of total T_3 is not sufficient as an indicator of the level of physiological activity of this hormone. Indeed, the fractions of free and bound T_3 should be taken into consideration as the affinity of plasma albumin for thyroxin may be altered by vinyloxazolidinethione in mammals (Mawson et al., 1994). Moreover, the sensibility or the number of the thyroidal nuclear receptors could be also affected. Nevertheless, the decrease of the digestibility of the components of the diet containing 50 percent of rapeseed meal could be also attributed to its higher content of GLS and GLS derivatives, as suggested by Bille et al. (1983).

However, it is very likely that the low tissue deposition of proteins, energy, and phosphorus observed in trout fed a diet containing 50 percent of rapeseed meal in the studies of Burel, Boujard, Escaffre, et al. (2000) and Burel et al. (2001) is the direct consequence of a poor hormonal stimulation. Several studies have shown in fish that thyroid hormones regulate the conversion of the metabolizable energy into net energy (MacKenzie et al., 1998) and have an anabolic action on the protein metabolism, stimulating the incorporation of the amino acids into the tissues (Donaldson et al., 1978; Eales, 1979; Higgs, Fagerlund, et al., 1982; Leatherland, 1982; Plisetskaya et al., 1983). The reduction of the phosphorus deposition observed in trout having no dietary deficiency in digestible phosphorus could be due to the lower protein synthesis essential for the development of the bone structure (Laroche et al., 1966).

A dietary supplementation of iodine induced an increase of the plasma T_3 levels (Figure 13.6) and improved the retention of proteins, energy, and phosphorus (Burel, 1999) in trout. However, there is not sufficient data to determine if this is due to an increase of the digestibility of these dietary components or of their net tissue deposition. In the case of the dietary T_3 supplementation (Figure 13.6) the growth was also increased, but the benefit of this treatment was principally

TABLE 13.12. Apparent digestion coefficient of gross nutrients of rapeseed/canola products in different fish species.

Species	Protein product	Dry matter	Crude protein	Fat	Gross energy	Phosphorus	Reference
Chinook salmon (Oncorhynchus tshawytscha)	Canola meal, Canadian, solvent-extracted	54	85		65		Hajen et al. (1993)
	Canola meal, GLS- freecultivar, Canadian, solvent-extracted	59	88		71		Hajen et al. (1993)
	Canola meal, GLS extracted, Canadian, solvent-extracted	38	79		51		Hajen et al. (1993)
	Rapeseed protein concentrate	70	96		81		Hajen et al. (1993)
Atlantic salmon (Salmo salar)	Canola meal, Canadian, solvent-extracted		74-87		62-73		Higgs et al. (1995, 1996), Anderson et al. (1992)
	Canola meal, Canadian, GLS-free cultivar, solvent-extracted		86		71		Higgs et al. (1996)
	Canola meal, Canadian, upgraded		87		68		Higgs et al. (1996)
	Canola protein concentrate		98		79		Higgs et al. (1996)
Rainbow trout (Oncorhynchus mykiss)	Rapeseed 00, European, whole, full-fat		81		77		Abdou Dade et al. (1990)
	Canola meal, Canadian, solvent-extracted	56-60	64-87	92-93	21-75		Cho and Slinger (1979), Cho et al. (1985), Hilton and Slinger (1986)

TABLE 13.12 *(continued)*

Species	Protein product	Dry matter	Crude protein	Fat	Gross energy	Phosphorus	Reference
	Canola meal, Canadian, solvent-extracted	50	83-88		56-75	5	Higgs et al. (1996), Riche and Brown (1996), Mwachireya et al. (1999)
	Rapeseed meal 00, European, solvent-extracted		76				Kaushik (1989)
	Canola meal, Canadian, dehulled, solvent-extracted	46	86		53		Mwachireya et al. (1999)
	Rapeseed meal 00, European, dehulled, solvent-extracted		93		83		Abdou Dade et al. (1990)
	Rapeseed meal 00, European, dehulled, solvent-extracted	71	91		76	26	Burel et al. (2000a)
	Rapeseed meal 00, European, dehulled, heated, solvent-extracted	67	89		70	42	Burel et al. (2000a)
Rainbow trout (*Oncorhynchus mykiss*)	Canola meal, Canadian, phytase					46	Riche and Brown (1996)
	Canola meal, Canadian, methanol-ammonia-washed	30	84		41		Mwachireya et al. (1999)
	Canola meal, Canadian, phytase-treated	41	84		52		Mwachireya et al. (1999)
	Canola meal, Canadian, SP-249-treated	44	77		53		Mwachireya et al. (1999)

Species	Ingredient					Reference
	Canola meal, Canadian, Alpha-Gal-treated	49	84	56		Mwachireya et al. (1999)
	Canola meal, Canadian, Alpha-Gal/SP-249-treated	31	80	39		Mwachireya et al. (1999)
	Rapeseed protein concentrate		89-96			Higgs et al. (1996)
	Canola protein isolate	77	98	85		Mwachireya et al. (1999)
Turbot (*Psetta maxima*)	Rapeseed meal 00, European, dehulled, solvent-extracted	57	83	69	49	Burel et al. (2000a)
	Rapeseed meal 00, European, dehulled, heated, solvent-extracted	65	92	81	65	Burel et al. (2000a)
Gilthead seabream (*Sparus aurata*)	Rapeseed protein concentrate		92	79		Kissil et al. (2000)
Red seabream (*Pagrus auratus*)	Canola meal, Australian, solvent-extracted		83	44		Glencross et al. (2004)
	Canola meal, Australian, expeller		94	62		Glencross et al. (2004)
	Canola meal, Australian, expeller, heat-treated (120°C)		51	33		Glencross et al. (2004)
	Canola meal, Australian, expeller, heat-treated (150°C)		23	30		Glencross et al. (2004)
	Canola protein concentrate		53	74		Glencross et al. (2004)
Silver perch (*Bidyanus bidyanus*)	Canola seed, Canadian, whole	47	80	59		Allan et al. (2000)
	Canola meal, Canadian, solvent-extracted	52	83	58		Allan et al. (2000)

due to a rise of the appetite, as previously shown by Higgs, Fagerlund, et al. (1982). In fact, in this case, the resultant plasma T_3 levels were supra physiological.

In turbot, a content of approximately 3.6 μmole/g of GLS derivatives in the diet is enough to induce some thyroidal disturbances, and a lower growth is observed for a content of about 4.4 μmole/g. However, it is unlikely that these thyroidal disturbances were responsible for the decrease of the growth performance in turbot, because the plasma levels of T_3, a physiologically active hormone, were not affected. Moreover, the retention of protein, energy, and phosphorus remained unchanged. The low acceptability of the RM-based diets for turbot could be because of their content of sinapin and GLS, as well as their low nutritional value due to high contents of fiber, tannins, and phytic acid. Moreover, as has been suggested, GLS can have a direct action on turbot, decreasing their digestive capacities. The principal hypothesis on the mode of action of GLS and GLS derivatives on the thyroidal metabolism and the growth of trout and turbot is illustrated in Figure 13.10.

NUTRIENT AVAILABILITY

Energy

Digestibility of energy of rapeseed/canola meals reported in the literature for different fish species differs depending on the products and/or the study (from 21 percent up to 83 percent), but generally, the highest values are obtained for upgraded rapeseed/canola meals (Table 13.12), likely due to the decrease in their fiber content. Moreover, the lowest values of energy digestibility have been observed in earlier studies (21 to 45 percent reported in Cho and Slinger [1979] and Cho et al. [1985], respectively), with the exception of the more recent study of Mwachireya et al. (1999), which shows low digestibility values. There is great variation in the observed digestibility of energy from rapeseed/canola meals among different fish species (51 to 71 percent in chinook salmon; 62 to 73 percent in Atlantic salmon; 39 to 83 percent in rainbow trout; 69 to 81 percent in turbot; 79 percent in gilthead seabream; 30 to 62 percent in red seabream; 58 percent in silver perch) and it is impossible to determine if these variations are

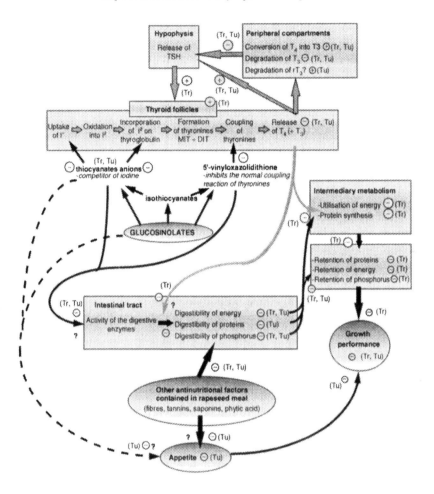

FIGURE 13.10. Principal hypothesis of the mode of action of glucosinolates and glucosinolate derivatives on the thyroidal metabolism and the growth of trout and turbot. The notation "Tr" means that the concerned fact is observed or suspected (?) in rainbow trout and the notation "Tu" concerns turbot. *Source:* Adapted from Burel et al. (2001).

the results of different research methodology used or the quality of the product tested.

The energy of lipid-extracted rapeseed/canola meal is contributed principally by its high protein content. Its utilization in diets of nonruminants was limited because of its high content of GLS, fiber, and tannins, leading to a low energy digestibility (Fenwick, 1982; Higgs et al., 1995). The dehulling of rapeseed meal improves its energy value in nonruminant animals such as fish. In rats, some GLS affect the digestibility (progoitrin, epiprogoitrin, gluconapin, glucoïberin) and the net utilization (progoitrin, sinigrin, sinalbin, gluconapin, epiprogoitrin, glucoïberin, glucorapharin, glucocheirolin, and glucotropaeolin) of dietary protein (Bille et al., 1983; Bjerg et al., 1989). In common carp *(Cyprinus carpio)* the isothiocyanates affect the digestibility of dry matter, protein, and minerals (Hossain and Jauncey, 1988). Bille et al. (1983) suggested that some GLS and GLS breakdown products can perturb the activity of the digestive enzymes and/or decrease the capacity of absorption of the cells of the intestinal brush. As has been earlier indicated, heat treatment of the rapeseed/canola meals, which eliminates a part of the GLS and inactivates the myrosinase, increases the energy value of these ingredients for fish.

In the study of Burel et al. (2001) conducted with trout fed a diet containing a high proportion of rapeseed meal (50 percent), a simultaneous decrease of the digestibility of the energy and the conversion of the metabolizable energy into net energy was observed. This might suggest that GLS interferes with energy metabolism of the fish, probably via disturbance of thyroid metabolism. It is possible that the low plasma concentration of thyroid hormones is also responsible for the decrease of the digestive capacities of trout. It has been shown in mammals that T_3 regulates the synthesis of the digestive enzymes in the cells of the intestinal brush (Hodin et al., 1992), and in seabream *(Chrysophrys major)* a dietary supplementation of T_3 leads to an increase of the activity of the enzymes involved in the digestion of carbohydrates and proteins (Woo et al., 1991). When the level of the ANFs was substantially decreased, as in the case of rapeseed/canola protein concentrates and of rapeseed/canola protein isolates, the digestibility of energy was greatly improved in fish (74 to 81 percent and 85 percent, respectively) (Table 13.12).

Protein and Amino Acids

The digestibility of protein contained in rapeseed/canola meals reported in literature for different fish species shows extreme values depending on the product and/or the study (from 23 percent up to 94 percent) (Table 13.12). However, the digestibility of protein is quite high in most rapeseed meals (>80 percent), upgraded or not. The lowest values (51 and 23 percent) are obtained in red seabream with canola meal expeller-extracted and heat-treated at 120°C and 150°C, respectively (Glencross et al., 2004). Thus, it seems that heat treatments can negatively affect the biological value of protein from rapeseed/canola meals. The digestibility of protein from rapeseed/canola protein concentrates (89 to 98 percent) and rapeseed/canola protein isolates (98 percent) is very high and similar to or even higher than that of fish meal, except in the study of Glencross et al. (2004), where the protein concentrate used seems of low quality. Although the level of phytic acid is high in rapeseed/canola concentrates, and this constituent can depress protein availability in fish, it is likely that this dietary factor probably exerted minimal influence on protein availability.

Digestibility of amino acids is generally high in fish (Table 13.13). Higgs et al. (1995) examined the quality of rapeseed/canola protein by expressing each of the essential amino acids in a protein source, including cystine and tyrosine, as a percent of the total weight of the essential amino acids (g/100 g protein). Subsequently, these values were compared with the corresponding essential amino acid requirements of rainbow trout and carp, for calculation of essential amino acid index (EAAI) (see Table 13.3). Fish whole-body protein and fish muscle protein showed the best correspondence to the pattern of amino acid needs of carp and rainbow trout. They concluded that protein furnished by rapeseed/canola meal and rapeseed/canola protein concentrate was of equal quality to that of herring meal protein and superior to that of soybean meal protein.

However, research by Newkirk et al. (1999) and Newkirk and Classen (2000) has shown that processing temperatures are the main reason for the lower amino acid bioavailability of some plant products. Although processing temperatures are relatively constant at rapeseed/canola crushing companies, it is prudent for rapeseed/canola meal users to monitor amino acid bioavailability as part of their quality control programs.

Phosphorus

Major portion of phosphorus in rapeseed/canola meals is in the phytate form, and it can be liberated only by the hydrolysis catalyzed

TABLE 13.13. Apparent digestibility and availability of amino acids of rapeseed/canola products in different fish species.

Species and Protein products	Lys	Met	Cys	Thr	Arg	Gly	Ser	His	Ile	Leu	Phe	Tyr	Val	Pro	Ala	Glu	Asp	Reference
Apparent digestion coefficient (%)																		
Rainbow trout (Oncorhynchus mykiss)																		
Canola meal, Canadian, solvent-extracted	87.2	88.2		90.5	86.4			87.0	80.0	79.2	85.4		78.6					Hilton and Slinger (1986)
Atlantic salmon (Salmo salar)																		
Canola meal, Canadian, solvent-extracted	77.1	89.0	78.6	74.8	85.1			85.3	74.8	77.4	77.9	75.7	73.4					Anderson et al. (1992)
Silver perch (Bidyanus bidyanus)																		
Canola seed, Canadian whole	85.5	85.2	79.8	85.5	91.2	82.0	81.9	90.3	84.5	86.1	87.3	86.6	84.9	82.0	83.7	90.6	82.0	Allan et al. (2000)
Canola meal, Canadian, solvent-extracted	86.2	89.2	79.8	87.9	91.7	85.0	85.9	91.7	84.6	87.8	88.0	89.8	84.6	84.8	86.0	91.4	82.5	Allan et al. (2000)
Apparent availability (% of protein)																		
Silver perch (Bidyanus bidyanus)																		
Canola seed, Canadian, whole	4.6	2.4	2.8	4.0	6.0	4.4	4.4	2.6	4.0	6.2	3.6	2.7	4.8	5.4	3.9	17.7	6.2	Allan et al. (2000)
Canola meal Canadian, solvent-extracted	4.7	2.2	2.4	4.1	5.5	4.4	4.2	2.5	3.9	6.5	3.6	2.7	4.9	5.6	3.8	16.7	5.6	Allan et al. (2000)

by phytase. Ruminant animals hydrolyze the phytic phosphorus due to the microflora of their rumen. In nonruminants, the capacity to digest phytic phosphorus is much lower (Ogino et al., 1979; Ketola, 1985; Pointillart, 1994).

The phytic acid content of rapeseed/canola meals and concentrates is high (1.5 to 5.8 percent DM, Table 13.9), and the proportion of available phosphorus for fish is low (0.3 to 0.4 percent DM, Table 13.8). Moreover, the activity of the endogenous phytase is generally low (0-280 IU/kg) in these feedstuffs compared to that of wheat (200-1000 IU/kg) or barley (150-1000 UI/kg) (Pointillart, 1994), and is likely totally suppressed by the thermal treatments applied on some rapeseed/canola protein products. Digestibility of phosphorus of rapeseed/canola meals in fish ranges from 5 percent to 65 percent (Table 13.12). It is not known if the large variability observed for the digestibility of phosphorus is due to study effects (methodology used to assess feedstuff digestibility) or product effects. However, it has been suggested that heat treatment (Burel, Boujard, Escaffre, et al., 2000), as well as the addition of phytase into the diet (Riche and Brown, 1996), can improve the digestibility of phosphorus in fish.

USE OF RAPESEED/CANOLA PROTEIN PRODUCTS IN FISH DIETS

Rapeseed/canola meals can be incorporated at higher levels in the diets of warm-water species than in those of cold-water (salmonid) species (Higgs et al., 1995). Among salmonids, trout are generally more sensitive than salmon. As shown in Table 13.14, growth performance is reduced in rainbow trout fed rapeseed/canola-meal-based diets at levels of 30 to 40 percent. However, growth of rainbow trout fed upgraded rapeseed/canola meals that have their fiber content reduced and acid- or solvent-washed is similar to those of trout fed a fish-meal-based diet (McCurdy and March, 1992). An earlier study by Teskeredzic et al. (1995) showed that rapeseed/canola protein concentrates, even dephytinized, can not be incorporated in the diet of rainbow trout at higher levels than the meals. A more recent study, however, showed that they can be utilized up to a level of 42 percent (Forster et al., 1999).

TABLE 13.14. Growth, feeding parameters, nutrient and energy retention, and plasma concentrations of thyroid hormones in fish fed the rapeseed products-based diets.

	% in diet	SGR[1]	VFI[2]	FE[3]	FC[4]	Survival	Ret. Prot.[5]	Ret. Ener.[5]	Ret. Phosp.[5]	T$_4$	T$_3$	Reference
Rapeseed meals 00, European (RM)												
Rainbow trout												
Control	0	2.4^d	2.0^a	0.99^d			36.0^b	38.6^b	28.8^b	10.6^b	6.5^b	Burel et al. (2000a, 2001)
RM, solvent-extracted	30	2.2^c	2.1^{bc}	0.86^c			31.7^{ab}	36.6^b	22.0^a	1.7^a	3.8^a	
RM, dehulled, solvent-extracted	50	2.0^b	2.1^{ab}	0.80^b			32.5^b	34.9^b	21.2^a	1.6^a	4.0^a	
RM, dehulled, heated, solvent-extracted	30	2.0^b	2.0^{ab}	0.81^{bc}			32.3^b	35.6^b	21.8^a	1.5^a	3.6^a	
RM, dehulled, heated, solvent-extracted	50	1.8^a	2.2^c	0.68^a			27.1^a	27.6^a	19.1^a	1.4^a	4.0^a	
Turbot												
Control	0	1.5^b	1.4^b	1.02			30.4	25.4	32.9	2.9^c	1.6	Burel et al. (2000b)
RM, solvent-extracted	30	1.3^a	1.3^a	1.01			30.9	26.7	31.9	1.7^{ab}	1.7	
RM, dehulled, solvent-extracted	46	1.2^a	1.2^a	0.93			28.3	23.9	35.7	1.1^a	1.5	
RM, dehulled, heated, solvent-extracted	30	1.3^a	1.2^a	1.01			29.4	25.1	33.0	2.4^{bc}	1.9	
Canola meals, Canadian (CM)												
Chinook salmon												
Control	0	0.6^b			2.7^b							McCurdy and March (1992)

Ingredient	Inclusion (%)								Reference
CM	25	0.4^{a}			4.5^{c}				
CM, fiber-reduced	25	0.4^{a}			5.1^{c}				
CM, fiber-reduced, ethanol/acid-washed	25	0.9^{d}			2.0^{a}				
CM, fiber-reduced, ethanol/hot acid washed	25	0.7^{c}			2.0^{a}				
CM, fiber-reduced, ethanol/acid-washed, enzyme-treated	25	0.6^{bc}			2.4^{ab}				
Canola meals, Canadian (CM)									
Chinook salmon									Hajen et al. (1993)
Control	0	1.1^{b}	1.7			95.0			
CM, solvent-extracted	30	0.4^{a}	1.5			88.9			
Chinook salmon									Satoh et al. (1998)
Control	0	1.4^{b}	1.7^{b}	0.90					
CM, solvent-extracted	15	1.2^{ab}	1.5^{b}	0.82					
CM, solvent-extracted	30	1.0^{a}	1.3^{a}	0.80					
CM, extruded 90°C	15	1.3^{b}	1.7^{b}	0.86					
CM, extruded 90°C	30	1.2^{ab}	1.7^{b}	0.77					
CM, extruded 150°C	15	1.3^{b}	1.7^{b}	0.79					
CM, extruded 150°C	30	1.3^{b}	1.6^{b}	0.86					
Rainbow trout									Hilton and Slinger (1986)
Control	0	2.7	1.0			99.0	0.9	0.9	
CM, solvent-extracted	14	2.6	1.0			99.7	0.7	0.7	
CM, solvent-extracted	27	2.5	1.1			99.4	0.5	0.6	
CM, solvent-extracted	40	2.4	1.1			99.0	0.4	0.5	

TABLE 13.14 *(continued)*

	% in diet	SGR[1]	VFI[2]	FE[3]	FC[4]	Survival	Ret. Prot.[5]	Ret. Ener.[5]	Ret. Phosp.[5]	T_4	T_3	Reference
Canola meals, Canadian (CM)												
Rainbow trout												McCurdy and
Control	0	3.7^d			1.20^{de}							March (1992)
CM, solvent-extracted	40	3.5^{bc}			1.26^f							
CM, heat-treated, solvent-extracted	40	3.2^a			1.24^{ef}							
CM, dehulled, solvent-extracted	40	3.4^b			1.25^f							
CM, dehulled, heat-treated, solvent-extracted	40	3.4^b			1.20^{de}							
CM, dehulled, solvent-extracted, acid-washed	40	4.0^e			1.13^{bc}							
CM, dehulled, heat-treated, solvent-extracted, acid-washed	40	3.7^d			1.15^{cd}							
CM, dehulled, solvent-extracted, ethanol-washed	40	3.7^d			1.07^a							
CM, dehulled, heat-treated, solvent-extracted, ethanol-washed	40	3.5^{bc}			1.08^{ab}							

396

CM, dehulled, solvent-extracted, ammonia/methanol-washed	40	3.9[e]			1.14[c]						
CM, dehulled, heat-treated, solvent-extracted, ammonia/methanol-washed	40	3.8[de]			1.13[c]						
CM, dehulled, solvent-extracted, ammonia/methanol/hexane-washed	40	3.7[d]			1.19[de]						Lim et al. (1998)
Channel Catfish											
Control	0	2.4[b]	0.7[b]		1.53[a]	96.7	34.6				
CM, solvent-extracted	15	2.4[b]	0.7[b]		1.64[ab]	96.7	32.0				
CM, solvent-extracted	31	2.3[ab]	0.7[b]		1.56[ab]	95.0	32.8				
CM, solvent-extracted	46	2.3[ab]	0.6[a]		1.53[a]	91.6	34.9				
CM, solvent-extracted	62	2.1[a]	0.6[a]		1.68[b]	93.3	31.4				
Canola meals, Australian (CM)											
Red seabream											
Control	0	1.6	0.8	0.71	1.43	97	33.2	28.4[a]	3.6	1.2	
CM, solvent-extracted	20	1.6	0.8	0.71	1.36	100	34.8	31.4[a]			
CM, solvent-extracted	30								2.1	1.2	
CM, solvent-extracted	40	1.9	0.8	0.86	1.16	97	33.3	38.9[b]			
CM, solvent-extracted	60	1.6	0.8	0.71	1.43	98	32.1	34.1[ab]			
Control	0	1.6	0.8	0.71	1.43	97	33.2	28.4[a]	3.6	1.2	Glencross et al. (2004)

TABLE 13.14 (continued)

	% in diet	SGR[1]	VFI[2]	FE[3]	FC[4]	Survival	Ret. Prot.[5]	Ret. Ener.[5]	Ret. Phosp.[5]	T$_4$	T$_3$	Reference
CM, expeller-extracted	20	1.7	0.8	0.76	1.27	100	37.0	33.9[ab]				Lim et al. (1998)
CM, expeller-extracted	30									2.7	1.3	
CM, expeller-extracted	40	1.7	0.8	0.77	1.30	95	37.5	33.6[ab]				
CM, expeller-extracted	60	1.5	0.8	0.65	1.51	98	30.5	28.4[a]				
Channel catfish												
Control	0	2.4[b]	0.7[b]		1.53[a]	96.7	34.6					
CM, solvent-extracted	15	2.4[b]	0.7[b]		1.64[ab]	96.7	32.0					
CM, solvent-extracted	31	2.3[ab]	0.7[b]		1.56[ab]	95.0	32.8					
CM, solvent-extracted	46	2.3[ab]	0.6[a]		1.53[a]	91.6	34.9					
CM, solvent-extracted	62	2.1[a]	0.6[a]		1.68[b]	93.3	31.4					
Rapeseed protein concentrate (RPC)												
Chinook salmon												
Control	0	1.1[b]	1.7			95.0						Hajen et al. (1993)
RPC	30	1.0[b]	1.5			96.7						
Rainbow trout												
Control	0	2.4[de]	4.1[a]	0.61[de]			19.9[d]	22.0[c]				Teskeredzic et al. (1995)
RPC, undephytinized	12	2.5[e]	3.8[a]	0.69[f]			23.5[f]	27.5[d]				
RPC, undephytinized	24	2.2[cd]	3.9[a]	0.56[cd]			19.1[d]	22.5[c]				
RPC, undephytinized	36	1.8[ab]	4.5[b]	0.41[b]			13.5[b]	16.0[b]				
RPC, dephytinized	13	2.3[de]	3.9[a]	0.59[de]			20.5[de]	21.5[c]				
RPC, dephytinized	26	2.2[cd]	4.0[a]	0.54[cd]			19.2[d]	22.0[c]				

RPC, dephytinized	39	1.9^{bc}	4.4^{b}	0.44^{b}		14.5^{b}	16.5^{b}	
RPC, undephytinized, solvent-washed	13	2.4^{de}	3.8^{a}	0.65^{ef}		23.5^{ef}	25.5^{d}	
RPC, undephytinized, solvent-washed	25	2.1^{cd}	4.3^{ab}	0.49^{c}		16.9^{c}	20.0^{c}	
RPC, undephytinized, solvent-washed	38	1.6^{a}	4.8^{b}	0.33^{a}		11.5^{a}	12.5^{a}	
Gilthead seabream								Kissil et al. (2000)
Control	0	2.5^{c}	3.0^{c}		0.97^{a}	38.7	53.2^{b}	
RPC	20	2.4^{c}	2.9^{c}		0.98^{a}	39.2	53.1^{b}	
RPC	43	2.2^{b}	2.6^{b}		0.95^{a}	37.8	49.2^{ab}	
RPC	75	1.6^{a}	2.0^{a}		1.04^{b}	36.9	45.0^{a}	
Canola protein concentrate (CPC)								
Rainbow trout								Forster et al. (1999)
Control	0	1.9	0.7^{b}	1.22		40.4		
CPC	42	1.8	0.6^{a}	1.14		38.8		
CPC + 0.45% P	42	1.9	0.7^{b}	1.17		39.6		
CPC + 0.45% P + phytase (500 FTU)	42	1.9	0.7^{b}	1.18		39.9		
CPC + 0.45% P + phytase (1500 FTU)	42	1.9	0.7^{b}	1.20		40.3		
CPC + 0.45% P + phytase (4500 FTU)	42	1.8	0.7^{b}	1.14		38.4		
CPC + 0.45% P + phytase (4500 FTU)	42	1.7	0.6^{a}	1.22		40.0		

TABLE 13.14 (continued)

	% in diet	SGR[1]	VFI[2]	FE[3]	FC[4]	Survival	Ret. Prot.[5]	Ret. Ener.[5]	Ret. Phosp.[5]	T_4	T_3	Reference
CPC + 0% P + phytase (1500 FTU)	42	1.8	0.7[b]	1.17			40.1					
CPC + 0.22% P + phytase (1500 FTU)	42	1.9	0.7[b]	1.21			40.8					

[1]Daily growth index (DGI) = 100 × (Final body weight)$^{1/3}$ − (Initial body weight$^{1/3}$) / duration in day.

[2]Voluntary feed intake (VFI) = 100 × dry feed intake (g)/((Initial body weight + Final body weight)/2) duration in day.

[3]Feed efficiency (FE) = Wet weight gain (g)/Dry feed intake (g).

[4]Feed conversion (FC) = Dry feed intake (g)/Wet weight gain (g).

[5]Retention of protein, energy, or phosphorus (% of intake) = 100 × (Final body weight = final carcass nutrient content × Initial body weight × initial carcass nutrient content)/nutrient intake.

In chinook salmon a reduction of growth performance has been observed at a level of incorporation of rapeseed/canola meal of 25 percent (Table 13.14), but higher growth performance was obtained when the rapeseed/canola meal had been subjected to a fiber reduction and ethanol/acid washings (McCurdy and March, 1992). A favorable effect of the extrusion process (90 or 150°C) has also been shown in this fish species. Moreover, rapeseed/canola protein concentrates can be utilized without any deleterious effect at least up to a 30 percent incorporation (Hajen et al., 1993).

In red seabream (Table 13.14), an Australian canola meal which had been expeller-extracted can be incorporated up to 60 percent in the diet without any reduction of the growth performance (Glencross et al., 2004), while the gilthead seabream exhibited a growth reduction even at a 43 percent incorporation of a rapeseed protein concentrate (Kissil et al., 2000). In turbot, a reduction of the growth performance is observed at the incorporation of 30 percent of an upgraded rapeseed meal (Table 13.14). In channel catfish *(Ictalurus punctatus),* rapeseed/canola meal can be incorporated in the diet at high levels without any deleterious effect on growth (Table 13.14). Indeed, a slight reduction of the growth rate has been observed only from a level of 62 percent (Lim et al., 1998).

Data given in the Table 13.14 show that the reduction in growth performance depends on the rapeseed/canola protein products and/or the fish species. This is due to a reduction of the diet intake, feed efficiency, or both. It seems that a reduction of the feed efficiency was observed in most cases due to a reduction of the retention of protein, energy, and mainly phosphorus. Burel, Boujard, Escaffre, et al. (2000) and Burel, Boujard, Kaushik, et al. (2000) evaluated the effects of same rapeseed meal diets in trout and turbot. They observed that a diet with high incorporation of rapeseed meal affected the growth of trout through a reduction of feed efficiency, while the growth of turbot was mainly affected through a reduction of diet intake.

The rapeseed/canola protein concentrates can be incorporated into the diets of fish at higher levels (about 40 percent) than the meals without inducing a reduction of the growth performance (Table 13.14). When lower growth is observed at high levels of incorporation, it is mainly related to a decrease of the feed efficiency, but diet intake can also be affected, as has been shown in gilthead seabream at the level of incorporation of 75 percent (Kissil et al., 2000).

CONCLUSION AND RECOMMENDATIONS

The incorporation of rapeseed/canola meal in the diet of most fish species is still problematical despite the reduction of its GLS content by genetic selection and of fiber and tannins by dehulling. From a practical point of view, the use of heat-treated meals is suggested, but the level of incorporation should not exceed 20 to 30 percent, depending on the fish species. Moreover, a dietary supplementation of iodine at a level beyond that recommended by NRC (1993) could limit the risk of thyroidal disturbances induced by the GLS derivatives. Available information from earlier studies shows, however, that GLS as well as fiber, tannins, and phytic acid are likely still the major factors adversely affecting growth performances of fish fed high dietary levels of rapeseed/canola meals. It is recommended that further research into methods to reduce the ANFs and to improve the nutritional quality of canola/rapeseed meals, and into the effects of these ANFs on nutrient absorption, metabolism, physiological function, and fish health should be conducted.

In contrast to rapeseed/canola meals, rapeseed/canola protein concentrates and isolates are well utilized by various fish species. This is mainly due to improved palatability and nutrient utilization as a result of a reduction of the ANFs. Thus, these products seem to be much more promising as fish meal substitutes from a nutritive point of view than the meals.

REFERENCES

Abdou Dade, B., P. Aguirre, D. Blanc, and S.J. Kaushik (1990). Incorporation du colza 00 sous forme de tourteau ou d'amande dans les aliments de la truite arc-en-ciel *(Oncorhynchus mykiss):* Performance zootechnique et digestibilité. *Bulletin Français de la Pêche et de la Pisciculture* 317:50-57.

Allan, G.L., S. Parkinson, M.A. Booth, D.A.J. Stone, S.J. Rowland, J. Frances, and R. Warner-Smith (2000). Replacement of fish meal in diets for Australian silver perch, *Bidyanus bidyanus:* I. Digestibility of alternative ingredient. *Aquaculture* 186:293-310.

Anderson, J.S., S.P. Lall, D.M. Anderson, and J. Chandrasoma (1992). Apparent and true availability of amino acids from common feed ingredients for Atlantic salmon *(Salmo salar)* reared in sea water. *Aquaculture* 108:111-124.
Anke, M., S. Schwarz, A. Hennig, B. Groppel, M. Grun, G. Zenker, and S. Glos (1980). Der einfluss zusächtlicher zink- und jodgaben auf rapsextraktions-schrotbedingte schäden beim schwein. *Monatsh. Veterinärmed* 35:90-94.
Barrett, J.E., C.F. Klopfenstein, and H.W. Leipold (1997). Detoxification of rape-seed meal by extrusion with an added basic salt. *Cereal Chemistry* 74:168-170.
Bell, J.M. (1993). Factors affecting the nutritional value of canola meal: A review. *Canadian Journal of Animal Science* 73:679-697.
Bille, N., B.O. Eggum, I. Jacobsen, O. Olsen, and H. Sorensen (1983). Anti-nutritional and toxic effects in rats of individual glucosinolates ($\pm$myrosinases) added to a standard diet. *Z. Tierphysiol., Tierernährg. und Futtermittelkde* 49:195-210.
Bjerg, B., B.O. Eggum, I. Jacobsen, J. Otte, and H. Sorensen (1989). Antinutritional and toxic effects in rats of individual glucosinolates ($\pm$ myrosinases) added to a standard diet (2). *Journal of Animal Physiology and Animal Nutrition* 61:227-244.
Bourdon, D. and A. Aumaître (1990). Low-glucosinolate rapeseeds and rapeseed meals: Effect of technological treatments on chemical composition, digestible energy content and feeding value for growing pigs. *Animal Feed Science and Technology* 30:175-191.
Brogan, I.W. (1986). Control of rapeseed quality for crushing. In *Oilseed Rape,* D.H. Scarisbrick and R.W. Daniels (Eds). London, England: Collins, pp. 282-300.
Bunting, E.S. (Ed.) (1981). *Production and Utilization of Protein in Oilseed Crops.* The Hague, The Netherlands: Martinus Nijhoff.
Burel, C. (1999). Utilisation de protéines d'origine végétale dans l'alimentation de la truite arc-en-ciel *(Oncorhynchus mykiss)* et du turbot *(Psetta maxima)*: Valeur nutritionnelle et effets sur l'axe thyroïdien (Utilisation of vegetal proteins in the diet of rainbow trout *(Oncorhynchus mykiss)* and turbot *(Psetta maxima)*: Nutritional value and effect on thyroid axis). Doctoral dissertation, Rennes, France: University of Rennes 1.
Burel, C., T. Boujard, A.M. Escaffre, S.J. Kaushik, G. Boeuf, K. Mol, S. Van Der Geyten et al. (2000). Dietary low-glucosinolate rapeseed meal affects thyroid status and nutrient utilization in rainbow trout *(Oncorhynchus mykiss). British Journal of Nutrition* 83:653-664.
Burel, C., T. Boujard, S.J. Kaushik, G. Boeuf, K. Mol, S. Van Der Geyten, V.M. Darras et al. (2001). Effects of rapeseed meal-glucosinolates on thyroid metabo-lism and feed utilization in rainbow trout. *General and Comparative Endocri-nology* 124:343-358.
Burel, C., T. Boujard, S.J. Kaushik, G. Boeuf, S. Van Der Geyten, K. Mol, E.R. Kühn et al. (2000). Potential of plant-protein sources as fish meal substitutes in diets for turbot *(Psetta maxima):* Growth, nutrient utilisation and thyroid status. *Aquaculture* 188:363-382.
Burel, C., T. Boujard, F. Tulli, and S.J. Kaushik (2000). Digestibility of extruded peas, extruded lupin, and rapeseed meal in rainbow trout *(Oncorhynchus mykiss)* and turbot *(Psetta maxima). Aquaculture* 188:285-298.

Butler, L.G. and K.D. Bos (1993). Analysis and characterization of tannins in faba beans, cereals and other seeds. A literature review. In *Second International Workshop on "Antinutritional Factors (ANFs) in Legume Seeds", Recent Advances of Research in Antinutritional Factors in Legume Seeds.* A.F.B van der Poel, J. Huisman, and H.S. Saini (eds.). Wageningen, The Netherlands: EAAP, pp. 81-89.

Cho, C.Y., C.B. Cowey, and T. Watanabe (Eds.) (1985). Finfish Nutrition in Asia. Methodological Approaches to Research and Development. International Development Research Centre, Ottawa, Ont. Publ. No. IDRC-233a.

Cho, C.Y. and S.J. Slinger (1979). Apparent digestibility measurement in feedstuffs for rainbow trout. In J.E. Halver and K. Tiews (eds.), *Finfish Nutrition and Fishfeed Technology, II, Hamburg, 1978, 20-23 June.* Berlin, Germany: Proceeding World Symposium FAO-EIFAC, ICES and IUNS, pp. 239-247.

COPA (Canadian Oilseed Processors Association) (1999). *Trading rules. 1998-1999.* Winnipeg, Manitoba, Canada.

Davie, J. and L. Vincent (1980). Extraction of vegetable oils and fats. In *Fats and Oils: Chemistry and Technology.* R.J. Hamilton and A. Bhati (eds.). London, England: Applied Science Publishers, pp. 123-124.

Donaldson, E.M., U.H.M. Fagerlund, D.A. Higgs, and J.R. McBride (1978). Hormonal enhancement of growth. In *Fish Physiology, Volume 8,* W.S. Hoar, D.J. Randall, and J.R. Brett (eds.). Academic Press, New York, NY: Academic Press, pp. 455-597.

Dosanjh, B.S., D.A. Higgs, D.J. McKenzie, D.J. Randall, J.G. Eales, N. Rowshandeli, M. Rowshandeli et al. (1998). Influence of dietary blends of menhaden oil and canola oil on growth, muscle lipid composition, and thyroidal status of Atlantic salmon *(Salmo salar)* in sea water. *Fish Physiology and Biochemistry* 19:123-134.

Eales, J.G. (1979). Thyroid functions in cyclostomes and fishes. In *Hormones and Evolution, Volume 1,* E.J.W. Barrington (ed.). London, England: Academic Press, pp. 341-436.

Fenwick, G.R. (1982). The assessment of a new protein source—rapeseed. *Proceeding of Nutrition Society* 41:277-288.

Fenwick, G.R., R.K. Heaney, and W.J. Mullin (1983). Glucosinolates and their breakdown products in food and food plants. *Critical Reviews of Food Science and Nutrition* 18:123-201.

Forster, I., D.A. Higgs, B.S. Dosanjh, M. Rowshandeli, and J. Parr (1999). Potential for dietary phytase to improve the nutritive value of canola protein concentrate and decrease phosphorus output in rainbow trout *(Oncorhynchus mykiss)* held in 11C fresh water. *Aquaculture* 179: 109-125.

French Feed Database (2005). *Technical Data.* The French Association of Animal Production (AFZ): http://www.feedbase.com

Glencross, B., W. Hawkins, and J. Curnow (2004). Nutritional assessment of Australian canola meals. I. Evaluation of canola oil extraction method and meal processing conditions on the digestible value of canola meals fed to the red seabream *(Pagrus auratus,* Paulin). *Aquaculture Research* 35:15-24.

Gomes, E.F., G. Corraze, and S.J. Kaushik (1993). Effects of dietary incorporation of a co-extruded plant protein (rapeseed and peas) on growth, nutrient utilization and muscle fatty acid composition of rainbow trout *(Oncorhynchus mykiss)*. *Aquaculture* 113:339-353.

Gomes, E.F. and S.J. Kaushik (1989). Incorporation of lupin seed meal, colzapro or triticale as protein/energy substitutes in rainbow trout diets. *Proc. Third Int. Symp. on Feeding and Nutr. in Fish.* Toba, Japan, September, pp. 315-324.

Graham, H. and P. Aman (1991). Nutritional aspects of dietary fibers. *Animal Feed Science and Technology* 32:143-158.

Hajen, W.E., D.A. Higgs, R.M. Beames, and B.S. Dosanjh (1993). Digestibility of various feedstuffs by post-juvenile chinook salmon *(Oncorhynchus tshawytscha)* in sea water. 2. Measurement of digestibility. *Aquaculture* 112:333-348.

Hardy, R.W. and C.V. Sullivan (1983). Canola meal in rainbow trout *(Salmo gairdneri)* production diets. *Canadian Journal of Fisheries and Aquatic Sciences* 40:281-286.

Hickling, D. (2001). *Canola Meal: Feed Industry Guide,* 3rd ed. Canola Council of Canada: Winnipeg, Manitoba, Canada, 39 pp.

Higgs, D.A., B.S. Dosanjh, R.M. Beames, A.F. Prendergast, S.A. Mwachireya, and G. Deacon (1996). Nutritive value of rapeseed/canola protein products for salmonids. In *Eastern Nutrition Conference,* N. Kent and D. Anderson (eds.). Dartmouth / Halifax, Canada, Canadian Feed Industry Association, pp. 187-196.

Higgs, D.A., B.S. Dosanjh, A.F. Prendergast, R.M. Beames, R.W. Hardy, W. Riley, and G. Deacon (1995). Use of rapeseed/canola protein products in finfish diets. In *Nutrition and Utilization Technology in Aquaculture,* C.E. Lim and D.J. Sess (eds.). Italy: AOAC Press, pp. 187-196.

Higgs, D.A., U.H.M. Fagerlund, J.G. Eales, and J.R. McBride (1982). Application of thyroid and steroid hormones as anabolic agents in fish culture. *Comparative Biochemistry and Physiology* 73B:143-176.

Higgs, D.A., J.R. McBride, J.R. Markert, B.S. Dosanjh, M.D. Plotnikoff, and W.C. Clarke (1982). Evaluation of Tower and Candle rapeseed (Canola) meal and Bronowski rapeseed protein concentrate as protein supplements in practical dry diets for juvenile chinook salmon *(Oncorhynchus tshawytscha)*. *Aquaculture* 29:1-31.

Hilton, J.W. and S.J. Slinger (1986). Digestibility and utilization of canola meal in practical-type diets for rainbow trout *(Salmo gairdneri)*. *Canadian Journal of Fisheries and Aquatic Sciences* 43:1149-1155.

Hodin, R.A., S.M. Chamberlain, and M.P. Upton (1992). Thyroid hormone differentially regulates rat intestinal brush border enzyme gene expression. *Gastroenterology* 103:1529-1536.

Hossain, M.A. and K. Jauncey (1988). Toxic effects of glucosinolate (allyl isothiocyanate) (synthetic and from mustard oilcake) on growth and food utilization in common carp. *Indian Journal of Fisheries* 35:186-196.

Jansman, A.J.M. (1993). Antinutritional effects of tannins in monogastric animal. *Grains Legume* 3:11-12.

Jansman, A.J.M. and M. Longstaff (1993). Nutritional effects of tannins and vicine/convicine in legume seeds. In *Second International Workshop on*

"Antinutritional Factors (ANFs) in Legume Seeds", Recent Advances of Research in Antinutritional Factors in Legume Seeds, A.F.B. van der Poel, J. Huisman, and H.S. Saini (eds.). Wageningen, The Netherlands: EAAP, pp. 301-316.

Kaushik, S.J. (1989). Use of alternative protein sources for intensive rearing of carnivorous fishes. In *Mediterranean Aquaculture,* R. Flos, L. Tort, and P. Torres (eds.). Chichester, United Kingdom: Ellis Horwood Ltd., pp. 125-138.

Ketola, H.G. (1985). Mineral nutrition: Effect of phosphorus in trout and salmon feeds on water pollution. In *Nutrition and Feeding of Fish,* C.B. Cowey, A.M. Mackie, and J.G. Bell (eds.). New York, NY: Academic Press, pp. 465-473.

Kissil, G.W., I. Lupatsch, D.A. Higgs, and R.W. Hardy (2000). Dietary substitution of soy and rapeseed protein concentrates for fish meal, and their effects on growth and nutrient utilization in gilthead seabream *Sparus aurata* L. *Aquaculture Research* 31:595-601.

Langer, P. and V. Stolc (1965). Goitrogenic activity of allyl isothiocyanate—a widespread natural mustard oil. *Endocrinology* 76:151-155.

Laroche, G., A.N. Woodhall, C.L. Johnson, and J.E. Halver (1966). Thyroid function in the rainbow trout *Salmo gairdneri* R. II. Effects of thyroidectomy on the development of young fish. *General and Comparative Endocrinology* 6:249-266.

Leatherland, J.F. (1982). Environmental physiology of the teleostean thyroid gland: A review. *Environmental Biology of Fishes* 7:83-110.

Leatherland, J.F., J.W. Hilton, and S.J. Slinger (1987). Effects of thyroid hormone supplementation of canola meal-based diets on growth, and interrenal and thyroid gland physiology of rainbow trout *(Salmo gairdneri). Fish Physiology and Biochemistry* 3:73-82.

Lim, C., P.H. Klesius, and D.A. Higgs (1998). Substitution of canola meal for soybean meal in diets for channel catfish *Ictalurus punctatus. Journal of the World Aquaculture Society* 29:161-168.

Ludke, H., F. Schone, and A. Hennig (1985). Der einfluss von jodkupfer- und zink-zulagen zu rationen mit hohem rapsextraktionsschrotanteil auf wachstum und schildrüssenfunktion dés mastschweines. 1. Einfluss auf die mastleistung. *Arch. Tierernähr., Berlin* 35:835-845.

MacKenzie, D.S., C. Moore VanPutte, and K.A. Leiner (1998). Nutrient regulation of endocrine function in fish. *Aquaculture* 161:3-25.

Mawson, R., R.K. Heaney, Z. Zdunczyk, and H. Kozlowska (1994). Rapeseed meal-glucosinolates and their antinutritional effects. Part 4. Goitrogenicity and internal organs abnormalities in animals. *Die Nahrung* 38:178-191.

McCurdy, S.M. and B.E. March (1992). Processing of canola meal for incorporation in trout and salmon diets. *Journal of the Association of Official Analytical Chemist* 69:213-220.

Morand-Fehr, P. and G. Tran (2001). La fraction lipidique des aliments et les corps gras utilisés en alimentation animale. *INRA Productions Animales* 14:285-302.

Mwachireya, S.A., R.M. Beames, D.A. Higgs, and B.S. Dosanjh (1999). Digestibility of canola protein products derived from the physical, enzymatic and chemical processing of commercial canola meal in rainbow trout *Oncorhynchus mykiss* (Walbaum) held in fresh water. *Aquaculture Nutrition* 5:73-82.

Newkirk, R.W. and H.L. Classen (2000). The effect of standard oil extraction and processing on the nutritional value of canola meal for broiler chicken. *Poultry Science* 79 (Suppl. 1):10.

Newkirk, R.W., H.L. Classen, T.A. Scott, and M.J. Edney, (1999). Commercial desolventization-toasting conditions reduce the content and availability of amino-acids in canola meal. *Poultry Science* 78 (Suppl. 1):16.

Niederholzer, R. and R. Hofer (1979). The adaptation of digestive enzymes to temperature, season and diet in roach *Rutilus rutilus L.* and rudd *Scardinius erythrophthalmus L.* Cellulase. *Journal of Fish Biology* 15:411-416.

Niewiadomski, H. (1990). *Rapeseed: Chemistry and Technology.* Amsterdam, The Netherlands: Elsevier Press.

NRC (National Research Council) (1993). *Nutrient Requirements of Fish.* Washington, DC: National Academy Press, 114 pp.

Ogino, C., L. Takeuchi, H. Takeda, and T. Watanabe (1979). Availability of dietary phosphorus in carp and rainbow trout. *Bulletin of the Japanese Society of Scientific Fisheries* 45:1527-1532.

Pathak, P.K., Y.C. Agrawal, and B.P.N. Singh (1991). Effect of elevated drying temperature on rapeseed oil quality. *Journal of the American Oil Chemists' Society* 68(8):275-280.

Plisetskaya, E., N.Y.S. Woo, and J.-C. Murat (1983). Thyroid hormones in cyclostomes and fish and their role in regulation of intermediary metabolism. *Comparative Biochemistry and Physiology* 74A:179-187.

Pointillart, A. (1994). Phytates, phytases: Leur importance dans l'alimentation des monogastriques. *Productions Animales* 7(1):29-39.

Pradet-Balade, B., C. Burel, S. Dufour, T. Boujard, S.J. Kaushik, B. Quérat, and G. Boeuf (1999). Thyroid hormones down-regulate thyrotropin béta mRNA level *in vivo* in the turbot (*Psetta maxima*). *Fish Physiology and Biochemistry* 20:193-199.

Prendergast, A.F., D.A. Higgs, D.M. Beames, B.S. Dosanjh, and G. Deacon (1994). Searching for substitutes: Canola. *Northern Aquaculture* 10(3):15-20.

Quinsac, A. (1993). Les glucosinolates et leurs dérivés dans les crucifères. Analyses par Chromatographie en Phase Liquide et perspectives d'utilisation de l'Electrophorèse Capillaire. Doctoral dissertation, Orléans, France: University of Orléans, 143 pp.

Riche, M. and P.B. Brown (1996). Availability of phosphorus from feedstuffs fed to rainbow trout, *Oncorhynchus mykiss. Aquaculture* 142:269-282.

Salunkhe, D.K., J.K. Chavan, R.N. Adsule, and S.S. Kadam (1992). *World Oilseeds: Chemistry, Technology, and Utilization.* New York, NY: Van Nostrand Reinhold, 554 pp.

Satoh, S., D.A. Higgs, B.S. Dosanjh, R.W. Hardy, J.G. Eales, and G. Deacon (1998). Effect of extrusion processing on the nutritive value of canola meal for chinook salmon *(Oncorhynchus tshawytscha)* in seawater. *Aquaculture Nutrition* 4:115-122.

Satoh, S., W.E. Poe, and R.P. Wilson (1989). Effect of supplemental phytate and/or tricalcium phosphate on weight gain, feed efficiency and zinc content in vertebrae of channel catfish. *Aquaculture* 80:155-161.

Sauvant, D., J.M. Perez, and G.E. Tran (Eds.) 2004. *Tables de composition et de valeur nutritive des matières premières destinées aux animaux d'élevage: Porcs, volailles, bovins, ovins, caprins, lapins, chevaux, poissons.* Paris, France, 301 pp.

Shiau, S.-Y. (1989). Role of fiber in fish feed. In *Proceedings of the Fish Nutrition Symposium, Progress in Fish Nutrition, 6-7 September 1989, Keelung, Taiwan,* S.-Y. Shiau (ed.). China: pp. 93-119.

Smith, R.R. (1971). A method for measuring digestibility and metabolizable energy of fish feeds. *Progressive Fish-Culturist* 33:132-134.

Stickney, R.R. and S.E. Shumway (1974). Occurrence of cellulase activity in the stomach of fishes. *Journal of Fish Biology* 6:779-790.

Teskeredzic, Z., D.A. Higgs, B.S. Dosanjh, J.R. McBride, R.W. Hardy, R.M. Beames, J.D. Jones et al. (1995). Assessment of undephytinized and dephytinized rapeseed protein concentrate as sources of dietary protein concentrate as sources of dietary protein for juvenile rainbow trout *(Oncorhynchus mykiss)*. *Aquaculture* 131:261-277.

VanEtten, C.H. and H.L. Tookey (1983). Glucosinolates. In *Handbook of Naturally Occuring Food Toxicants,* M. Rechcigl (ed.). Boca Raton, FL: CRC Press, 15 pp.

Vermorel, M., R.K. Heaney, and G.R. Fenwick (1986). Nutritive value of rapeseed meal: effects of individual glucosinolates. *Journal of Sciences of Food and Agriculture* 37:1197-1202.

Wang, Y., T.A. McAllister, M.D. Pickard, Z. Xu, L.M. Rode, and K.J. Cheng (1999). Effect of micronizing full-fat canola seed on amino acid disappearance in the gastrointestinal tract of dairy cows. *Journal of Dairy Science* 82:537-544.

Weber, N., E. Klein, and K.D. Mukherjee (2002). The composition of the major molecular species of adipose tissue triacylglycerols of rats reflects those of dietary rapeseed, olive and sunflower oils. *Journal of Nutrition* 132:726-732.

Webster, C.D., L.G. Tiu, J.H. Tidwell, and J.M. Grizzle (1997). Growth and body composition of channel catfish *(Ictalurus punctatus)* fed diets containing various percentages of canola meal. *Aquaculture* 150:103-112.

Woo, N.Y.S., A.S.B. Chung, and T.B. Ng (1991). Influence of oral administration of 3,5,3'-triiodo-L-thyronine on growth, digestion, food conversion and metabolism in the underyearling red seabream, *Chrysophrys major* (Temminck & Schlegel). *Journal of Fish Biology* 39:459-468.

Yurkowski, M., J.K. Bailey, R.E. Evans, J.-A.L. Tabachek, and G. Burton Ayles (1978). Acceptability of rapeseed proteins in diets of rainbow trout *(Salmo gairdneri)*. *Journal of Fisheries Research of Board Canada* 35:951-962.

Chapter 14

Lupins in Fish and Shrimp Diets

D.M. Smith
B.D. Glencross
G.L. Allan

INTRODUCTION

Lupins are the harvested seeds of leguminous plants within the genus *Lupinus,* a group within the bean and pea family, the Fabaceae. Plants within this group are also known as grain legumes or pulses. Apart from their value as a cash crop, lupins are of great value in crop rotation programs with wheat and other grains, with the additional benefit that they fix and return nitrogen to the soil. With the global interest in the development of protein-rich diet ingredients derived from agriculture to replace proteins of marine origin in aquaculture diets, lupins have been found to have considerable potential in diets for various species of fish and shrimp. Global production of lupins is relatively modest and variable, with a maximum production of about two million metric tons (mmt) in 1999. In that year, about 90 percent of the global production was in Australia. Very low rainfall in the growing areas of Australia in recent years has resulted in less than 1 mmt being produced (FAO, 2004). However, as lupins are generally not considered a culinary grain, there is little competition for its use as human food.

The nutritional value of a number of species and cultivars of lupins has been assessed for a wide variety of fish and shrimp species (reviewed by Glencross, 2001). Rainbow trout *(Oncorhynchus mykiss)* have been used extensively as a test species, though there is an increasing body of work with Atlantic salmon *(Salmo salar)*. In addition,

Alternative Protein Sources in Aquaculture Diets
© 2008 by The Haworth Press, Taylor & Francis Group. All rights reserved.
doi:10.1300/5892_14

there have been studies with seabreams, silver perch, Asian seabass, carps, tilapia, milkfish, and turbot, and with marine shrimp and freshwater crayfish. In this chapter we will summarize this research and discuss the results of more recent research.

PRODUCT NAMES

There are three main commercial species of lupins: *L. angustifolius* (Narrow-leafed lupin or Australian sweet lupin), *L. albus* (White lupin), and *L. luteus* (Yellow lupin). *L. angustifolius* is the most widely grown species in Australia, whereas *L. albus* is more widely grown in Chile, Egypt, South Africa, and Eastern Europe. There is increasing interest in a fourth species from South America, *L. mutabilis* (Pearl or Andean lupin) that appears to have commercial potential. With all species, there are a number of different varieties and cultivars that are better suited to particular climatic and soil conditions. Hence, they tend to differ in their yield, chemical composition, and nutritional value. In Australia, lupins are generally marketed in bulk with little segregation of seed according to species or cultivar. This practice is changing, with premium prices being paid to growers for seed with high protein content.

The lupin products that have been evaluated in aquaculture diets are meals prepared from both whole seed and dehulled seed (lupin kernel meal). In addition, a lupin protein concentrate, prepared by air classification of kernel meal from *L. angustifolius,* cv. Gungurru, has been evaluated in a number of studies (Sarac et al., 1999; Sudaryono et al., 1999; Carter and Hauler, 2000; Booth et al., 2001). More recently, aqueous ethanol extraction of lupin kernel meals has been carried out to remove a proportion of the nonstarch polysaccharide and hence increase the protein content of the meal (Glencross, Boujard, et al., 2003).

Lupin kernel meal is produced through the removal of the largely indigestible seed coat (dehulling). This product appears more suitable for use in diets for carnivorous fish species and for marine shrimp than whole-seed meal. The removal of the seed coat involves additional expense and produces a low-value by-product (the seed coat), but at the same time increases the protein and lipid content of the lupin meal, and hence its value. *L. albus* is more easily dehulled than the other species as it has a thinner seed coat and as a consequence, the yield of kernel is about 80 percent of the dry weight of seed (Table 14.1; Sipsas, 2003). The seed coat of *L. angustifolius* comprises 26 percent of the dry weight of the seed, while that of *L. luteus* is 30 percent, with the consequence that the maximum yield of *L. luteus* kernels is about 70 percent of whole-seed dry weight. The dry matter (DM) content of lupin seed and kernel meals is typically about 90 percent. All data on the composition of lupin products are reported here on a DM basis, and in the absence of a reported DM content in the source document, an assumed DM content of 90 percent has been used.

CHEMICAL COMPOSITION

Nutrient Content

Crude Protein

Although the nutrient composition of whole lupins has been widely reported (Petterson et al., 1997) and that of lupin kernel meals much

TABLE 14.1. Gross chemical composition (% DM) of whole seed and kernel of four species of lupins.

Species	L. angustifolius		L. albus		L. luteus		L. mutabilis	
	Seed	Kernel	Seed	Kernel	Seed	Kernel	Seed	Kernel
Moisture[a]	9	12	9	11	9	12	9[a,b]	11[a,b]
Seed coat	26	0	20	0	30	0	18	0
Protein	35	43	39	44	42	57	48	58
Fat	7	8	10	12	6	8	15	19
Ash	3	3	3	5	3	5	3	5
Lignin	1	1	1	1	1	1	1	1
Polysaccharides	25	32	19	24	9	11	10	10
Oligosaccharides	5	7	8	9	10	14	?	?

Source: Adapted from Sipsas, 2003.

[a]% as "received."

[b]Assumed data for moisture content of *L. mutabilis* as actual data are not available.

less, so it is the latter that will have greater interest to the aquaculture diet industry. Lupin seeds typically have a higher crude protein content (31 to 42 percent DM) than most other grain legumes. There is considerable variation in protein content among species and cultivars and even within cultivars, as a result of the climatic conditions during the growing season and the soil type (Petterson et al., 1997). Of the commercial species, *L. luteus* is generally considered to have the highest protein content of whole seed (40 to 45 percent DM). The kernel typically contains about 53 percent DM protein though content of 57 percent DM has been reported (Petterson et al., 1997) (Table 14.1; Sipsas, 2003). The whole seed of *L. albus* is also high in protein, having a protein content of between 32 and 44 percent DM, and the kernel having approximately 44 percent DM. The protein content of *L. angustifolius* seed typically ranges from 30 to 41 percent DM, with the kernel having about 44 percent DM. Though *L. mutabilis* is not grown in commercial quantities, it has high protein content, 42 to 55 percent DM (Buirchell, 2003; Molina-Poveda, C, CENAIN-ESPOL, Ecuador, personal communication). The seed coat of *L. mutabilis* is relatively thin, hence dehulling yields about 82 percent kernel (Sipsas, 2003) with an estimated protein content of 50 to 65 percent DM.

Amino Acid Composition

The amino acid composition of lupin protein is similar to that of soybean but is characterized by relatively high levels of arginine, 11.3 to 12.2 g/16 g N, which is about twice the level in soybean protein (Table 14.2). However, it has relatively low levels of methionine, 0.65 g/16 g N, or about half that of soybean protein. Hence, the total sulfur amino acid content (methionine+cysteine) is also low, 2.1 g/16 g N. A transgenic variety of *L. angustifolius* has been developed with 94 percent more methionine than in the parent cultivar, 0.59 percent DM (1.24 g/100 g protein) (Molvig et al., 1997) but this cultivar is not currently in commercial production.

Lipids and Fatty Acids

The crude fat content of lupins also varies considerably among species and cultivars. Of the commercial species, *L. luteus* generally has the lowest fat content, from 5.7 to 6.8 percent DM, and *L. albus* the highest, 8.3 to 14.5 percent DM (Petterson et al., 1997; Petterson,

TABLE 14.2. Amino acid composition (g amino acid/16 g N) of protein from three species of lupins and soybean.

Amino acid	*L. angustifolius*	*L. albus*	*L. luteus*	Soybean
Arginine	11.62	12.20	11.30	5.42
Cysteine	1.36	1.34	2.28	1.41
Histidine	2.57	1.86	3.30	2.46
Isoleucine	3.91	3.80	2.70	4.51
Leucine	6.61	6.90	7.89	6.81
Lysine	4.66	4.75	5.35	5.66
Methionine	0.72	0.66	0.70	1.28
Phenylalanine	3.65	3.85	4.04	3.60
Threonine	3.54	3.29	3.51	3.56
Tryptophan	1.00	0.97	n/a	1.35
Tyrosine	3.66	4.26	3.10	1.67

Source: Glencross, 2001.

2000; Table 14.1). *L. mutabilis* is reported to have a relatively high fat content (13 to 24 percent DM) (Buirchell, 2003). The fat in lupins comprises predominantly triacylglycerides (71.1 percent) and phospholipids (14.9 percent), with lesser amounts of free sterols (5.2 percent), glycolipids (3.5 percent), and other lipid material (5.3 percent) (van Barneveld, 1999; Petterson, 2000).

The fatty acid profile of the lipid in lupins is typical of that of most legumes, being high in monounsaturated fatty acids (MUFA) and polyunsaturated fatty acids (PUFA), mainly linoleic acid and linolenic acid, but with none of the long-chain highly unsaturated fatty acids found in marine oils (Table 14.3). Erucic acid is rarely detected in most lupin varieties but is reported at very low concentrations in *L. mutabilis* (Buirchell, 2003).

Carbohydrates

The carbohydrate composition of lupins is quite different from that of most legumes, as it contains high levels (>50 percent whole seed) of nonstarch polysaccharides (NSPs) and very low levels of starch (van Barneveld, 1999; Petterson, 2000). The content and chemical

TABLE 14.3. Fatty acid composition (% of total fatty acids) of three species of lupins and soybean (Glencross, 2001). Fatty acids not listed had levels less than 0.5% in all species.

Fatty acid	*L. angustifolius*	*L. albus*	*L. luteus*	Soybean
16:0	11.0	7.8	4.8	10.3
18:0	3.8	1.6	2.5	4.5
18:1n-9	38.2	53.0	21.0	23.9
18:2n-6	37.1	17.2	47.3	51.8
18:3n-3	5.3	9.5	7.5	6.5
20:0	0.9	1.2	2.7	–
20:1n-9	0.3	4.3	1.8	–
22:0	1.9	3.9	7.1	–
22:1n-11	–	1.9	0.8	–
Total SFA	17.6	14.5	17.1	14.8
Total MUFA	38.5	59.2	23.6	23.9
Total PUFA	42.4	26.7	54.8	58.3
n-3 (omega-3)	5.3	9.5	7.5	6.5
n-6 (omega-6)	37.1	17.2	47.3	51.8

Source: Adapted from Tacon, 1990; Petterson et al., 1998; and van Barneveld, 1999).

composition of lupin NSPs varies among species and cultivars, but their structures appear to be quite similar (Cheung, 1990). In whole-seed and in kernel meal, insoluble NSPs (or dietary fiber) predomi-nates with lesser proportions of soluble NSPs and free sugars. The seed coat is comprised mainly of insoluble NSPs, and hence, the ker-nel meal has markedly less NSPs than whole-seed meal. Neverthe-less, the kernel meal typically contains about 40 percent NSPs on a DM basis, but can contain up to 50 percent. The soluble NSPs consist predominantly of oligosaccharides, mainly raffinose, stachyose, and verbascose, with a total content in the seed typically <10 percent DM (Table 14.4).

Vitamins and Minerals

There is little information available on the vitamin and mineral composition of lupins, and it is highly likely to be variable according to species, cultivar, and agronomic conditions. Petterson (2000) reported

TABLE 14.4. Antinutritional factors in the whole seed (g/kg DM) of various lupin species and defatted soybean meal.

Antinutrient	L. angustifolius	L. luteus	L. albus	Soybean meal
Trypsin inhibitor	0.13	0.18	0.09	3.46
Alkaloids	<0.20	<0.50	<0.10	nd
Oligosaccharides	46	99	73	58
Phytate	5.6	10.3	6.3	17.7
Saponins	0.63	nr	nd	nr
Tannins	0.10	0.30	0.20	nr

Source: Adapted from Petterson et al., 1997.

Note: nd = not detected; nr = not reported.

the vitamin content (DM, mg/kg) of *L. angustifolius* whole-seed meal to be: thiamin, 5.9; riboflavin, 3.1; biotin, 0.04; folate, 0.4; choline, 3,035; niacin, 40; pantothenate, 1.8; and α-tocopherol, 2.4. In addition, β-carotene was detected at 3.9 mg/kg.

The variability of most of the major mineral elements in whole-lupin seed samples throughout Australia is low, but there is more variability in the concentration of the trace elements (Table 14.5) (Petterson et al., 1997). Sodium content is variable (30 to 110 mg/kg DM) but is of little relevance in aquaculture diets. Calcium concentrations range from an average of 220 mg/kg DM in *L. albus* to 240 mg/kg DM in both *L. angustifolius* and *L. luteus.* Phosphorous varied from an average of 330 mg/kg DM in *L. angustifolius* to 570 mg/kg in *L. luteus.* Potassium concentration varied from an average of 890 mg/kg DM in *L. angustifolius* to 1,080 mg/kg DM in *L. luteus.* Of over 300 samples of *L. angustifolius* that were tested from a wide range of growing conditions, none contained concentrations of heavy metals that exceeded the proposed Codex Alimentarius (World Health Organization and the Food and Agriculture Organization of the United Nations) limit of 0.1 mg/kg for cadmium or 2.0 mg/kg for lead (Petterson et al., 1997).

Antinutritional Factors

Current commercial varieties of lupins contain low levels of antinutritional factors (Table 14.4). Selective breeding programs over the

TABLE 14.5. Mineral content of whole seeds of narrow-leafed lupins, *L. angustifolius* and white lupins, *L. albus*.

Mineral	L. angustifolius		L. albus	
	Mean	Range	Mean	Range
	mg/kg DM			
Copper	5.2	2.8-7.6	5.7	3.4-9.0
Iron	76.1	34.4-166.7	30.1	23.3-48.9
Manganese	21.1	7.4-84.4	995.5	25.6-4191.1
Molybdenum	1.8	0.8-3.2	2.3	0.9-3.4
Zinc	37.9	27.4-50.0	33.4	24.4-42.2
	μg/kg DM			
Cobalt	87	11-289	229	11-478
Selenium	99	20-267	94	22-400

Source: Adapted from Petterson et al., 1997.

last forty years have resulted in the development of varieties that have low levels of alkaloids, such that *L. angustifolius* seed contains <0.02 percent, while *L. luteus* contains <0.05 percent. The exception is *L. mutabilis*. Cultivars currently being grown in South America contain relatively high levels of alkaloids (3.18 percent as lupanin), which are generally reduced to about 0.05 percent using aqueous extraction processes (Molina-Poveda, C., CENAIN-ESPOL, Ecuador, personal communication). Other potential antinutritional factors present in lupins include oligosaccharides, phytate, saponins, and protease inhibitors. However, none of these are considered to have a significant effect on the nutritional value of lupins (Pettersen et al., 1997).

NUTRIENT AVAILABILITY

Carnivorous Finfish

The apparent digestibility of dry matter (ADMD), energy (ADE), and crude protein (ACPD) of a number of species and cultivars of lupins, as whole-seed meal or as kernel meal, has been determined for a number of carnivorous fish species. The ADE of lupins has been found to be less than that of fish meal or other pulse or oilseed meals, and appears closely related to the protein and lipid content of the meal; the carbohydrate fraction appears to be poorly digested and contributes little to the digestible energy of the ingredient (Burel et al., 2000; Glencross, Curnow, et al., 2003). However, lupin kernel meals of *L. albus* and *L. luteus* have relatively high fat contents (Table 14.1), and hence higher gross energy contents. As a consequence, the ADE of these kernel meals is greater than that of solvent-extracted soybean meal (Glencross and Hawkins, 2004).

The ADMD of both lupin seed and kernel has been reported to be less that that of fish meal, soybean meal, and various meals derived from legumes and oilseeds, with the ADMD of lupin seed meal being less than that of kernel meal (Glencross, 2001), reflecting the low digestibility of the fibrous seed coat. In another study, Glencross, Boujard et al. (2003) reported that endogenous lupin oligosaccharides affected the apparent digestible value of the ingredient. In that study, cross-referencing of both chemical-extraction and enzymatic methods provided direct evidence of the negative influence of the lupin oligosaccharides when fed to rainbow trout.

In all cases where the digestibility of lupin protein has been measured, the protein has been found to be highly digestible. In rainbow trout diets, the ACPD of lupin seed *(L. albus)* was found to be greater than that of full-fat soybean meal (85.2 percent compared to 79.5 percent) (Hughes, 1988), similar to that of fish meal (85.2 percent compared to 83.6 percent) (Morales et al., 1994) and similar to peas and faba beans (85.5 percent compared to 80.4 percent and 80.2 percent, respectively) (Gomes et al., 1995).

The ACPD of kernel meals was found to be higher than that of whole-seed meals, reflecting the interaction of the high levels of NSPs present in the whole-seed meal. The ACPD of *L. angustifolius* kernel meal was found to be greater than that of fishmeal with red seabream *(Pagrus auratus),* 99 percent and 81 percent, respectively (Glencross, Curnow, Hawkins, et al., 2003), and was greater than that of defatted soybean meal with Asian seabass (or barramundi; *Lates calcarifer*), 98 percent and 86 percent, respectively (McMeniman, 1998). Apparent crude protein digestibility of *L. luteus* in diets for rainbow trout was also greater than that of solvent-extracted soybean meal, 93 percent and 87 percent, respectively (Glencross and Hawkins, 2004). In a study with turbot *(Pselta maxima),* Burel et al. (2000) found that the ACPD of extruded *L. albus* kernels was 98 percent, which was similar to that of extruded peas (93 percent) and rapeseed that had been dehulled, solvent-extracted and heat-treated, (92 percent), but significantly greater than that of rapeseed that had been dehulled and solvent-extracted, (83 percent). Glencross, Curnow, et al. (2003), using five different lupin kernel meals in a study with rainbow trout, found that the total-tract apparent digestibility of nitrogen (or crude protein) increased with increasing protein content of the kernel meal. The nitrogen digestibility varied from 78 percent with the kernel meal with the lowest crude protein content (35.1 percent DM) to 90 percent for the kernel meal with the highest crude protein content (48.2 percent). It seems reasonable to attribute this variability in digestibility to the NSP content of the meal, which would be less in the high-protein kernel meals and greater in the low-protein meals. Across the carnivorous fish species that have been studied, the ACPD of lupin seed meal has been consistently found to be less (approximately 86 percent) than that of kernel meals (approximately 96 percent), which would suggest that the kernel meal would be a better ingredient in diets for these species.

Carter and Hauler (2000) studied the effect on digestibility of replacing 25 percent and 33 percent of the fish meal protein with a lupin protein concentrate (52 percent CP, from *L. angustifolius,* Gungurru) in diets for Atlantic salmon *(Salmo salar).* The increase in protein concentration in this ingredient was brought about by a reduction in the proportion of NSPs in the meal using an air classification process (Evans, 1998). The ACPD of the two diets containing the lupin protein concentrate were significantly greater than that of the basal diet containing the fish meal (95.7 percent and 95.9 percent, respectively, compared to 92.7 percent. In addition, they found that the ACPD of diets containing the solvent-extracted soybean meal, at the 25 percent and 33 percent protein replacement levels, was not significantly different from that of the diets containing lupin protein, 95.3 percent and 95.9 percent, respectively. However, Carter and Hauler (2000) did not compare lupin protein concentrate with lupin kernel meal which would have enabled evaluation of the benefit of the additional processing of the lupin meal on diet digestibility.

In a study with red seabream, Glencross, Curnow, Hawkins, et al. (2003) compared the nutritional and biological value of a transgenic, high-methionine strain of *L. angustifolius* with a nontransgenic, commercial variety, *L. angustifolius*, cv. Warrah. The ACPD of both species was similar, but the transgenic variety had higher ADE (56 percent compared to 64 percent) reflecting the higher lipid and lower NSP contents of the transgenic variety.

Omnivorous and Herbivorous Finfish

There is little information on the digestibility of lupin meal for omnivorous and herbivorous finfish other than for omnivorous freshwater fish, silver perch *(Bidyanus bidyanus)*. The nutritional value of seed and kernel meals of both *L. angustifolius* and *L. albus* for silver perch has been studied extensively by Allan and co-workers (Allan, Frances et al., 1998; Allan, Gleeson, et al., 1998; Allan et al., 2000; Booth et al., 2001). The whole-seed meals of both *L. angustifolius* and *L. albus* had substantially lower ADMD and ADE than that of solvent-extracted soybean meal, but the ACPD were similar (Allan, Gleeson, et al., 1998; Allan et al., 2000). Dehulling of the lupin seed *(L. angustifolius)* resulted in a significant increase in the ACPD (100 percent compared to 97 percent for whole seed) though this was less marked with *L. albus* (Booth et al., 2001). Similarly, the ADMD and ADE of lupin kernels were significantly greater than those of whole seed (Table 14.6) (Allan, Gleeson, et al., 1998).

There has been considerable variability reported in the availability of the amino acids from lupin protein when fed to pigs (van Barneveld, 1999). However, Allan et al. (2000), from a study with silver perch, reported relatively low variability in the apparent digestibility coefficients of amino acids in *L. angustifolius* and *L. albus*. The apparent digestibility for most of the amino acids was >95 percent with the exception for cysteine, which was about 80 percent for both species, and methionine, which was about 87 percent for *L. angustifolius*.

It is possible that other omnivorous or herbivorous fish such as carps, tilapias, and milkfish *(Chanos chanos)*, would be able to digest and utilize the NSPs in lupins. In these species, the use of lupin seed meal might be more cost-effective than kernel meal. However, this remains to be tested rigorously.

TABLE 14.6. Apparent digestibility (%) of dry matter, crude protein, and energy of whole-seed and kernel meals from *L. angustifolius* and *L. albus* when included at two levels in diets for silver perch *(Bidyanus bidyanus)*.

	Whole-seed meal		Kernel meal	
L. angustifolius				
Inclusion	30%	50%	30%	50%
Dry matter	50.3	50.8	67.6	68.9
Crude protein	96.6	95.8	100	99.9
Energy	59.4	58.4	74.0	75.0
L. albus				
Inclusion	30%	50%	30%	50%
Dry matter	64.7	59.4	77.8	68.2
Crude protein	96.1	97.0	100	97.3
Energy	72.7	67.1	85.2	74.7

Source: Adapted from Allan, Gleeson et al., 1998.

Penaeid Shrimp

The apparent digestibility of lupin seed meal and lupin kernel meal (*L. angustifolius* cv. Gungurru) was determined by Smith et al. (1998) in a study with 10 to 15 g shrimp *(Penaeus monodon)*. The lupin kernel meal had higher apparent digestibility than whole lupins: ADMD 73 percent and 39 percent; ACPD 95 percent and 88 percent; and ADE 74 percent and 45 percent, respectively. From the digestibility data, the kernel meal appeared to be a much better ingredient than the whole-seed meal for use in shrimp diets.

In a study to identify the factors that limit the utilization of *L. angustifolius* (cv. Gungurru) kernel meal in *P. monodon* diets, Smith (2002) confirmed the relatively high digestibility of CP in lupin kernel. However, the ADMD and ACPD of diets containing equivalent inclusion levels of air-classified lupin protein concentrate were significantly greater than that of the diets containing kernel meal at inclusion levels of 20 percent, 40 percent, and 60 percent (Table 14.7). This highlights the need to assess whether a diet containing lupin protein concentrate would be more cost-effective than one containing lupin kernel meal

TABLE 14.7. Apparent dry matter digestibility (% ADMD) and apparent crude protein digestibility (% ACPD) of shrimp diets containing incrementally increasing levels of lupin kernel meal and lupin protein concentrate from *L. angustifolius*, included in the basal diet (0%) at the expense of fish meal and wheat flour.

	Kernel meal		Protein concentrate	
Inclusion level (%)	ADMD	ACPD	ADMD	ACPD
0	68	82	68	82
20	69	84	68	82
40	62	85	70	86
60	60	86	74	90

Source: Adapted from Smith, 2002.

and whether the enhanced digestibility would offset the additional costs involved in making the lupin protein concentrate.

UTILIZATION OF LUPINS IN AQUAFEEDS

Carnivorous Finfish

Burel et al. (1998) conducted a series of growth trials examining the effect of a range of inclusion levels of *L. albus* kernel meal in diets for rainbow trout. The lupin meal was added to a basal diet formulation (CP = 40 percent DM; GE = 22 MJ/kg DM) at the expense of fish meal with adjustments made to the flaked corn and fish oil contents to maintain the diets iso-nitrogenous and iso-energetic. Addition of lupin kernel meal at 50 percent of the diet did not have a significant effect on growth rate, feed conversion ratio (FCR), or nitrogen retention. However, at an inclusion level of 70 percent, diet intake, and hence growth rate, decreased, as did feed efficiency and nitrogen retention. The results of a subsequent experiment suggested that the decrease in diet intake was not associated with a decrease in attractants or feeding stimulants in the diet.

Glencross, Curnow, Hawkins, et al. (2003) compared the response to dietary inclusions of a transgenic, high-methionine strain of *L. angustifolius* with a nontransgenic, commercial variety, *L. angustifolius*, cv. Warrah in diets for red seabream (or pink snapper, *Pagrus auratus*). The experiment was designed to ensure that dietary protein intake was limited (dietary CP approximately 35 percent DM). Under these conditions, a significant benefit of the enhanced methionine was demonstrated. However, economic modeling suggested that there was greater benefit in the marginally higher protein content of the transgenic lupin kernel meal than the additional methionine content. The authors suggested that a lupin meal with greater digestible protein and digestible energy content would be of greater value to the aquaculture diet industry than alterations to the amino acid profile of the ingredient.

In experiments with gilthead seabream *(Sparus aurata)* using *L. angustifolius* "seed meal" (Robaina et al., 1995) and turbot using *L. albus* kernel meal (Burel et al., 2000), there was a trend of decreasing daily growth coefficient of fish with increasing inclusion level. When individual treatments were compared with those using other vegetable protein sources studied in the same experiment (soybean meal; Robaina et al., 1995) and rapeseed meal (Burel et al., 2000) there were no significant differences in growth response of fish among the matched inclusion levels of lupin and the other protein sources, or with the fish fed the basal diet, suggesting a similar biological value of these ingredients.

The results of these studies (Robaina et al., 1995; Burel et al., 1998, 2000; Glencross, Curnow, Hawkins, et al., 2003) suggest that carnivorous finfish are able to utilize lupin kernel meals well at dietary inclusion levels of <50 percent. Growth rate of fish was not adversely affected by the inclusion of lupins at levels generally considered to be practical for diets for carnivorous species. However, increasing inclusion of lupin kernel meal above 30 percent appears to result in a decrease in growth rate, though this decrease does not become statistically significant until the inclusion level is above 50 percent.

Omnivorous and Herbivorous Finfish

Whole-seed meal *(L. angustifolius)* was used in diets for carp and Nile tilapia *(Oreochromis niloticus)* with a CP content of 30 percent (Viola et al., 1988). In the first experiment, the lupin meal was added to the diets at the expense of soybean meal and sorghum at an inclusion rate of 30 percent of the diet. This diet resulted in significantly better growth of both tilapia and carp than obtained with the basal diet. The reason for the better performance with the lupin diet is not clear. However, from the diet formulation, it appears possible that the digestible protein content of the basal diet was limiting performance. Hence, the difference in growth rates may be due to the lupin diet having a higher digestible protein content (Glencross, 2001). In a second experiment with carp only, lupin meal was added to the diet, at the expense of fish meal and sorghum, at an inclusion rate of 45 percent of the diet. In this experiment there were no significant differences in the performance of fish fed the two diets.

Lupins have been tested in several other studies with Nile tilapia that have been reported in various conference proceedings. However, these reports do not provide sufficient detail of the methods and results to enable critical appraisal of the conclusions. Nevertheless, in a study by Petterson et al. (1998), using a basal diet containing 15 percent fish meal and 15 percent soybean meal as the main protein sources, *L. angustifolius* kernel meal was used to replace half of the soybean meal (inclusion level reduced to 7.5 percent). A third treatment consisted of using the kernel meal at 17 percent of the diet, to partially replace both the soybean meal (reduced to 8.5 percent of the diet) and the fish meal (reduced to 8.5 percent of the diet). The diets were formulated on an iso-nitrogenous (CP = 30 percent) and iso-energetic basis. There were no significant differences in the growth rates of tilapia fed the diets containing the lupin kernel meal and the basal diet. However, the protein efficiency ratio of fish fed the diet with 17 percent lupin kernel meal was less than with the other two diets. In addition, fish fed the 17 percent kernel meal diet had a lower overall sensory evaluation score than those fed the other two diets. Whether this was due to the inclusion of lupin meal at this level in the diet, or due to the reduction of fish meal to 8.5 percent of the diet, with the consequential reduction of its endogenous highly unsaturated fatty acids, cannot be ascertained from the study.

In a summit dilution growth assay, Allan, Frances, et al. (1998) found that the growth rate of silver perch fed diets (CP = 38.5 percent) containing 60 percent of *L. angustifolius* kernel meal, was not significantly different from those fed the summit diet (CP = 36.9 percent) that was based on fish meal (26.6 percent of diet) and soybean meal (20.2 percent of diet). However, at 70 and 80 percent inclusion of lupin meal, there was a decline in nitrogen retention efficiency that was attributed to a methionine deficiency.

The work of Allan, Frances, et al. (1998) provides the most detailed and interpretable results on the efficacy of lupin kernel meal in diets for herbivorous and omnivorous finfish. Lupin kernel meal appears to be readily interchangeable with fish meal in these diets up to a maximum inclusion level of 60 percent. However, whole-seed meal may be a more cost-effective ingredient for lower value species. There is a need for further, well-designed studies to evaluate the benefit of lupin seed meal in diets for carp, tilapia, and milkfish, particularly to determine the effect of the lupin NSPs on growth rates and feed conversion ratios.

Penaeid Shrimp

Lupin kernel meal (*L. angustifolius,* cv. Warrah) and a transgenic, high-methionine variety, were evaluated in a feeding trial in cages in a raceway pond with *P. monodon* (initial weight of 3 g) (Smith and Tabrett, 2003). The lupin meals were included at 25 percent of the diet, replacing fish meal and wheat flour, in a diet containing 40 percent CP. The results demonstrated that *L. angustifolius* kernel meal could be included at up to 25 percent in a practical shrimp diet formulation without adversely affecting performance. In addition, the results suggested that at this inclusion level, the dietary concentration of the essential amino acid, methionine, was not limiting the performance of the shrimp. Smith (2002) investigated the effect of including up to 60 percent lupin kernel meal (*L. angustifolius,* cv. Gungurru) in *P. monodon* diets. The study demonstrated that the growth rate of shrimp was adversely affected by the kernel meal of this cultivar when used at dietary inclusion levels greater than 25 percent. From subsequent studies there is evidence that the kernel meal of more recently developed cultivars of *L. angustifolius,* such as Myallie, can be included at 50 percent of a diet containing 40 percent CP, without having a negative effect on growth rate (Smith and Tabrett, unpublished data).

Sudaryono et al. (1999a) evaluated *L. albus* kernel meal in diets for *P. monodon* containing 40 percent CP; replacing 0, 25, 50, 75, and 100 percent of the fish meal protein with an equivalent amount of lupin protein. Reanalysis of the data demonstrated a progressive decline in weight gain when more than 25 percent of the dietary fish meal protein was replaced with *L. albus* kernel meal. In another study, Sudaryono et al. (1999) compared the performance of whole-seed meal and kernel meal of *L. albus,* kernel meal of air-classified lupin protein concentrate *(L. angustifolius)* and defatted soybean meal. In that study, they concluded that dehulling seed or concentrating lupin protein did not improve the nutritive value of lupin meal.

This result is in contrast to findings by Smith (2002) who found that diets containing lupin protein concentrate were more highly digestible than those containing a protein level equivalent to that of lupin kernel meal. Smith (2002) also found that the insoluble NSPs in the kernel meal, which was proportionally reduced in the lupin protein concentrate, was a major factor in reducing digestibility of diets containing lupin kernel meal. In addition, lupin kernel fiber had no significant effect on the growth of shrimp even at high (30 percent) inclusion levels. These results demonstrate that a reduction in the proportion of insoluble NSPs in a lupin product will have a beneficial effect on the ingredient digestibility, but that the insoluble NSPs do not have an antinutritional effect.

Sudaryono et al. (1999) also concluded that *L. angustifolius* meal generally performed better than *L. albus* meal and that *L. angustifolius* meal was comparable to soybean meal. In a third study with *P. monodon,* Sudaryono et al. (1999b) serially replaced defatted soybean meal, included at 30 percent, with *L. albus* kernel meal in a fish meal–based basal diet containing 40.4 percent CP. The results show a progressive decline in growth rate of the shrimp with increasing replacement of soybean meal with lupin meal. Both of these studies demonstrate that the *L. albus* kernel meal that was used in the diets had a lower nutritional value for shrimp than defatted soybean meal.

From the various studies with shrimp, it appears that there are differences in the growth response of shrimp to the inclusion of different species or cultivars of lupins in the diet. However, the reason for the reduction in growth response with increasing inclusion levels of lupin meal has not been identified. For lupin kernel meal to be used with confidence by diet manufacturers in shrimp diets, it is important that the factor or factors limiting the utilization of the kernel meal in the diets be identified. The two factors most likely responsible for the decline in performance of shrimp associated with the feeding of high dietary inclusions of lupin kernel meal are: (1) the relatively low methionine content of lupin protein and (2) the oligosaccharide content of the kernel meal. A satisfactory method to minimize the leaching loss of supplementary methionine from diets will need to be developed to address unequivocally the low methionine issue.

EFFECT OF LUPINS ON FEED PROCESSING

In formulating commercial aquaculture diets, the functional properties of an ingredient, as well as its nutritional content, is of interest to the manufacturer. The functional value is a measure of the effect the ingredient has on the processing characteristics of the diet mash and on the physical characteristics of the finished product. At the laboratory scale, diets are typically prepared by forcing the moist diet mash through a die plate of a heavy duty meat grinder to form spaghetti-like strands. For shrimp feeds, these strands are placed in a food steamer for a period to activate natural or added binding agents before being air dried, whereas with fish feeds, the strands are simply air dried. The processing involved at the laboratory scale is quite different from the commercial manufacture of diets with either steam pellet presses or extruders. In the commercial processing, typically greater sheer forces, pressures, and temperatures are involved. Hence, the functional value of an ingredient as assessed at the laboratory scale is likely to have little bearing on its value in a commercial manufacturing process. Gleeson et al. (1998) carried out a study to optimize the physical characteristics of Atlantic salmon diets by modeling a twin-screw extrusion process using response surface methodology. The process variables were product temperature at the die and screw speed. They examined the effect of replacing one third of the fishmeal protein in the diet (41 percent CP, 22 percent lipid) with a plant protein. The diets were prepared on an iso-nitrogenous and iso-lipidic basis, with the replacement of fish meal (66.5 percent CP) with soybean meal (49.1 percent CP,), lupin protein concentrate (45.9 percent CP), and a field pea protein concentrate (48.6 percent CP). The characteristics of the diet that were measured were: radial expansion, bulk density, oil absorption, sinking rate, durability, shear strength, hardness, work to shear, and degree of starch gelatinization. The results of this study indicated that at least 33 percent of the fish meal protein in

salmon diets can be replaced with plant protein from the three ingredients without having a deleterious effect on pellet quality. However,
the pellets produced using the lupin product did not absorb oil readily
during the post-extrusion coating step. Hence, for salmon diets that
have >20 percent lipid, the authors recommended that the lupin protein concentrate be used at an inclusion level less than the 29 percent
used in the study. However, the effects of alterations to the starch content of the diet to counter this issue were not examined.

CONCLUSIONS AND RECOMMENDATIONS

In this review, generally only papers that have clear and readily
interpretable results have been cited. A number of other papers have
not been cited because factors, such as the type of lupins (species or
cultivar) or starting product (whole-seed meal, kernel meal, or protein
concentrate), have not been reported. In other studies that have not
been cited, a number of ingredients have been changed with each dietary treatment in the growth response studies, resulting in difficulty in
interpreting the results. However, from the literature a number of conclusions can be made about the use of lupin meal in aquaculture diets:

- The nutritional value of lupin kernel meal is substantially greater
  than that of the whole-seed meal.
- Lupin kernel meal appears to be similar to soybean meal for a
  number of species of fish and for shrimp. In one particular study
  with shrimp, *L. albus* meal resulted in a reduction in growth rate
  when used to replace defatted soybean meal.
- The digestible energy derived from lupins is primarily obtained
  from digestion of protein and lipid with little being derived from
  the carbohydrate. However, there appears to be no data available for the more herbivorous finfish such as carp, tilapia, and
  milkfish.
- The protein in lupin meal is highly digestible (generally >90 percent). However, the digestibility of protein in whole-seed meal
  is adversely affected by the relatively high levels of insoluble
  NSPs from the seed coat.
- Lupins have low levels of antinutritional factors, particularly
  trypsin inhibitors. Current cultivars of *L. mutabilis* grown in

South America appear to contain relatively high levels of alkaloids, which can be reduced to acceptable levels through aqueous extraction processes.

- Maximum inclusion levels of lupin meal in fish and shrimp diets vary with lupin species or cultivar, and whether the meal contains the seed coat or not. As a guide, including lupin kernel at levels of up to 30 percent in nutritionally balanced diets is unlikely to adversely affect growth.

- Recently developed cultivars of lupins, particularly of *L. angustifolius* appear to be better utilized than older cultivars. The kernel meal of some of these cultivars may be used at inclusion levels of up to 50 percent.

REFERENCES

Allan, G.L., J. Frances, and M. Booth (1998). Replacement of fishmeal in diets for silver perch: IV. Effects of increasing dehulled lupins *Lupinus angustifolius* (var. Gungurru) on growth and body composition. In *Fishmeal Replacement in Aquaculture Diets for Silver Perch*. G.L. Allan and S.J. Rowland (eds.). Final Report of Project 93/120-01 to the Fisheries Research and Development Corporation, Canberra, Australia, pp. 165-180.

Allan, G.L., V.P. Gleeson, A.J. Evans, and D.A.J. Stone (1998). Replacement of fishmeal in diets for silver perch: II. Digestibility of Australian lupins. In *Fishmeal Replacement in Aquaculture Diets for Silver Perch*. G.L. Allan and S.J. Rowland (eds.). Final Report of Project 93/120-01 to the Fisheries Research and Development Corporation, Canberra, Australia, pp. 53-71.

Allan, G.L., S. Parkinson, M.J. Booth, D.A.J. Stone, and W. Jantrarotai (2000). Replacement of fishmeal in diets for Australian silver perch *Bidyanus bidyanus:* Digestibility of alternative ingredients. *Aquaculture* 186:293-310.

Booth, M., G.L. Allan, J. Frances, and S. Parkinson (2001). Replacement of fish meal in diets for Australian silver perch, *Bidyanus bidyanus:* IV. Effects of dehulling and protein concentration on the digestibility of grain legumes in diets for silver perch. *Aquaculture* 196:67-85.

Buirchell, B. (2003). Grain varieties and species potential. In *Seeding a Future for Grains in Aquaculture Feeds*. Proceedings of a Workshop held in Fremantle, Western Australia, 28 May, 2003. Grains Research and Development Corporation, Canberra, Australia, pp. 11-13.

Burel, C., T. Boujard, G. Corraze, S.J. Kaushik, G. Boeuf, K.A. Moi, S. van der Geyten, et al. (1998). Incorporation of high levels of extruded lupin in diets for rainbow trout *(Oncorhynchus mykiss):* Nutritional value and effect on thyroid status. *Aquaculture* 163:325-345.

Burel, C., T. Boujard, F. Tulli, and S. Kaushik (2000). Digestibility of extruded peas, extruded lupin, and rapeseed meal in rainbow trout *(Oncorhynchus mykiss)* and *(Psetta maxima)*. *Aquaculture* 188:285-298.

Carter, C.G. and R.C. Hauler (2000). Fish meal replacement by plant meals in extruded feeds for Atlantic salmon, *Salmo salar* L. *Aquaculture* 185:299-311.

Cheung, P.C.K. (1990). Carbohydrates of *Lupinus angustifolius*. A composite study of the seeds and structural elucidation of thje kernel cell wall polysaccharides of *Lupinus angustifolius*. PhD Thesis, Australia: University of New South Wales.

Evans, A.J. (1998). Preparation of protein enriched fractions from grain legumes. In *Fishmeal Replacement in Aquaculture Diets—Feed Processing*. A. Evans (ed.). Final Report of Project 93/120-06 to the Fisheries Research and Development Corporation, Canberra, Australia, pp. 107-109.

FAO (2004). Food and Agriculture Organization of the United Nations Statistics. http://www.fao.org/es/ess/top/commodity.jsp?commodity=210&lang=EN&year=2002

Gleeson, V., M. O'Sullivan, and A. Evans (1998). Optimisation of the physical quality of aquaculture feeds processed by twin-screw extrusion. An empirical modelling approach using response surface methodology. I. Partial replacement of fish meal by grain legume protein concentrates in feeds for Atlantic salmon *(Salmo salar)*. In *Fishmeal Replacement in Aquaculture Diets—Feed Processing*. A. Evans (ed.). Final Report of Project 93/120-06 to the Fisheries Research and Development Corporation, Canberra, Australia, pp. 46-80.

Glencross, G.D. (2001). Feeding lupins to fish: Understanding the nutritional and biological value of lupins in aquaculture feeds. North Beach, WA, Australia: Fisheries Western Australia, pp. 119.

Glencross, B.D., T. Boujard, and S.J. Kaushik (2003). Influence of oligosaccharides on the digestibility of lupin meals when fed to rainbow trout, *Oncorhynchus mykiss*. *Aquaculture* 219:703-713.

Glencross, B.D., J.G. Curnow, and W.E. Hawkins (2003). Evaluation of the variability in chemical composition and digestibility of different lupin *(Lupinus angustifolius)* kernel meals when fed to rainbow trout *(Oncorhynchus mykiss)*. *Animal Feed Science and Technology* 107:117-128.

Glencross, B.D., J.G. Curnow, W.E. Hawkins, G. Wm. Kissil, and D.S. Petterson (2003). Evaluation of the feed value of a transgenic strain of the narrow-leaf lupin *(Lupinus angustifolius)* in the diet of the marine fish *Pagrus auratus*. *Aquacult. Nutr.* 9:197-206.

Glencross, B.D. and W.E. Hawkins (2004). A comparison of the digestibility of lupin (*Lupinus* spp.) kernel meals as dietary protein resources when fed to either, rainbow trout, *Oncorhynchus mykiss* or red seabream, *Pagrus auratus*. *Aquacult. Nutr.* 10:65-73.

Gomes, E.F., P. Rema, and S.J. Kaushik (1995). Replacement of fish meal by plant proteins in the diet of Rainbow Trout *(Oncorhynchus mykiss):* Digestibility and growth performance. *Aquaculture* 130:177-186.

Hughes, S.G. (1988). Assessment of lupin flour as a diet ingredient for rainbow trout *(Salmo gairdneri)*. *Aquaculture* 7:379-385.

McMeniman, K.C. (1998). The apparent digestibility of feed ingredients based on stripping methods. In *Fishmeal Replacement in Aquaculture Feeds for Barramundi.* K.C. Williams (ed.). Final Report of Project 93/120-03 to the Fisheries Research and Development Corporation, Canberra, Australia, pp. 34-39.

Molvig, L., L.M. Tabe, B.O. Eggum, A.E. Moore, S. Craig, D. Spencer, and T.J.V. Higgins (1997). Enhanced methionine levels and increased nutritive value of seeds of transgenic lupins (*Lupinus angustifolius* L.) expressing a sunflower seed albumen gene. *Proceedings of the National Academy of Science USA* 94:8393-8398.

Morales, A.E., G. Cardenete, M. De la Higuera, and A. Sanz (1994). Effects of dietary protein source on growth, feed conversion and energy utilisation in rainbow trout, *Oncorhynchus mykiss. Aquaculture* 124:117-126.

Petterson, D.S. (2000). The use of lupins in feeding systems—Review. *Asian Australian Journal of Animal Science* 13:861-882.

Petterson, D.S., G.I. Jenkins, and J. Hutabarat (1998). The use of lupins in aquaculture. *Proc. 3rd AEP Conference*, Valladolid, Spain, pp. 456-457

Petterson, D.S., S. Sipsas, and J.B. Mackintosh (1997). *The Chemical Composition and Nutritive Value of Australian Pulses.* Canberra, Australia: Grains Research Corporation (GRDC), 65 pp.

Robaina, L., M.S. Izquierdo, F.J. Moyano, J. Socorro, J.M. Vergara, D. Montero, and H. Fernandez-Palacios (1995). Soybean and lupin seed meals as protein sources in diets for gilthead seabream *(Sparus aurata):* Nutritional and histological implications. *Aquaculture* 130:219-233.

Saraç, H.Z., H.B. Thaggard, M.E. Rose, B.J. Kelly, and M.R. Gravel, (1998). Utilisation of plant protein sources. In *Fishmeal Replacement in Aquaculture Feeds for Prawns.* D.M. Smith (ed.). Final Report to Fisheries Research and Development Corporation, (FRDC), Canberra, Australia, Project 93/120-02, pp. 58-80.

Sipsas, S. (2003). Protein concentrates and isolates. In *Seeding a Future for Grains in Aquaculture Feeds.* B.D. Glencross (ed.). Proceedings of a Workshop held in Fremantle, Western Australia, 28 May, 2003. Grains Research and Development Corporation, Canberra, Australia, pp. 8-10.

Smith, D.M. (2002). Improving the efficacy of lupins as fishmeal replacements in aquaculture diets for prawns. Final Report to the Grains Research and Development Corporation (GRDC), Project CSM1, Canberra, Australia, pp. 23.

Smith, D.M. and S.J. Tabrett (2003). Efficacy of dehulled lupin meal in diets for the black tiger prawn, *Penaeus monodon.* In *Aquaculture Diet Development Subprogram: Diet Validation and Feeding Strategies,* C.G. Barlow, G.L. Allan, K.C. Williams, S.J. Rowland, and D.M. Smith (eds.). Final Report to the Fisheries Research and Development Corporation (FRDC), Canberra, Australia, Project 96/363, pp. 70-75.

Smith, D.M., S.J. Tabrett, and L.E. Moore (1998). Digestibility of plant protein sources. In *Fishmeal Replacement in Aquaculture Feeds for Prawns.* D.M. Smith (ed.). Final Report to Fisheries Research and Development Corporation, (FRDC), Canberra, Australia, Project 93/120-02, pp. 51-57.

Sudaryono, A., E. Tsvetnenko, and L.H. Evans (1999a). Evaluation of potential of lupin meal as an alternative to fish meal in juvenile *Penaeus monodon* diets. *Aquaculture Nutrition* 5:277-285.

Sudaryono, A., E. Tsvetnenko, and L.H. Evans (1999b). Replacement of soybean meal by lupin meal in practical diets for juvenile *Penaeus monodon. J. World Aquaculture Society* 30:46-57.

Sudaryono, A., E. Tsvetnenko, J. Hutabarat, Supriharyono, and L.H. Evans (1999). Lupin ingredients in shrimp *(Penaeus monodon)* diets: Influence of lupin species and types of meals. *Aquaculture* 171:121-133.

Tacon, A.G.J. (1990). *Standard methods: For the Nutrition and Feeding of Farmed Fish and Shrimp.* Redmond, Washington, DC: Argent laboratories press.

van Barneveld, R.J. (1999). Understanding the nutritional chemistry of lupin (*Lupinus* spp.) seed to improve livestock production efficiency. *Nutrition Research Reviews* 12:203-230.

Viola, S., Y. Arieli, and G. Zohar (1988). Unusual feedstuffs (tapioca and lupin) as ingredients for carp and tilapia in intensive culture. *Israeli Journal of Aquaculture-Bamidgeh* 40:29-34.

Chapter 15

Underutilized and Unconventional Plant Protein Supplements

Robert C. Reigh

INTRODUCTION

Fish meal and high-protein soybean meal are staple ingredients of many of today's manufactured aquafeeds. Alone or in combination, fish meal and soybean meal often comprise 50 percent or more of the formulations of many aquatic animal diets, where they typically serve as the primary sources of dietary protein. The costs of fish meal and soybean meal are likely to increase in the future as new markets that are more profitable than animal diets are developed for these materials and their high-value derivatives. The realization that future price increases, or reductions in the availability, of fish meal and soybean meal could someday preclude the use of these ingredients as major components of aquafeeds has stimulated research to identify alternative protein supplements and to develop recommendations for the use of these alternative protein sources in prepared diets.

According to the U.S. Department of Agriculture, approximately 185.61 million metric tons (mmt) of major protein meals were produced worldwide in 2002-2003 (USDA, 2004). Soybean meal (130.89 mmt) constituted 70.5 percent of the total global supply of major protein meals. Smaller quantities of other oilseed products and fish meal comprised the remainder: rapeseed meal, 10 percent (18.49 mmt); cottonseed meal, 6.1 percent (11.23 mmt); sunflower seed meal, 4.9 percent (9.14 mmt); peanut meal, 2.9 percent (5.34 mmt); fish meal,

Alternative Protein Sources in Aquaculture Diets
© 2008 by The Haworth Press, Taylor & Francis Group. All rights reserved.
doi:10.1300/5892_15

2.7 percent (4.95 mmt); palm kernel meal, 2.1 percent (3.93 mmt); and copra (coconut) meal, 0.9 percent (1.64 mmt).

The United States, Brazil, and Argentina produced a combined total of 80 percent of the world's supply of soybeans and 58 percent of the world's supply of soybean meal in 2003 (FAO, 2004). The United States is the world's single largest producer of soybeans and soybean meal. In 2003, the United States generated 35 percent of the total global production of soybeans and 26 percent of the global production of soybean meal (FAO, 2004).

Despite the relative abundance of soybean meal and its status as the world's primary protein supplement, its availability varies greatly in different parts of the world. It is probable that soybean meal will remain a cost-effective protein source for aquatic animal diets produced in the Western Hemisphere, where the majority of soybeans are grown, but the high cost of shipping soybean meal to markets far from the primary sites of soybean production in North America and South America is likely to raise the cost of soybean protein in many parts of the world, perhaps to levels that could be prohibitive to aquafeed producers.

Alternative protein supplements are needed to reduce the current dependence on fish meal and soybean meal as the primary protein sources for aquatic animal diets. A number of other plant protein products available in different parts of the world could serve as partial replacements for fish meal and soybean meal in aquafeeds. Many of these products are underutilized or unconventional ingredients whose usefulness in prepared diets could be increased as research provides more information on the nutritional value of alternative feedstuffs for aquatic animals.

USE OF ALTERNATIVE PROTEIN SUPPLEMENTS IN AQUAFEEDS

Sources, World Production, Applications

The amount of research that has been conducted with unconventional protein supplements is quite limited. Products that are not consistently available in adequate quantities in ingredient markets will not be used as regular ingredients in compounded diets. Feedstuffs that are perceived to be unpromising, because availability is erratic or supplies are insufficient, will be of less interest to researchers than materials that are abundant and widely available. Thus, research with unconventional feedstuffs often has been restricted to regions of the world where the ingredient of interest was locally available and where it could be used in prepared diets in a cost-effective manner. An emphasis on replacing expensive, imported protein supplements with ingredients grown closer to the site of diet production will undoubtedly continue as the cost of shipping agricultural products increases. Continued research is needed to maximize the value of locally produced protein supplements in aquafeeds for economically important species in different parts of the world.

Peanut Meal (from **Arachis hypogaea***)*

Peanut (or groundnut) meal is a plant protein supplement used in animal diets and human foods. Peanut meal is made from the residue (cake) remaining after the oil has been removed from the peanut seed by either mechanical extraction (pressing) or solvent extraction (hexane solubilization). Meal made from mechanically extracted peanut seeds contains approximately 45 percent crude protein (CP) and 5 percent ether extract (EE), or lipid (Dale and Batal, 2003). Meal made from solvent-extracted seeds contains 48 to 49 percent CP, 1 to 2 percent EE, 7 to 10 percent crude fiber (CF), and 26 to 29 percent nitrogen-free extract (NFE) comprised primarily of starches from the endosperm (Dale and Batal, 2003; National Research Council, 1993).

In 2003, approximately 35.66 mmt of peanuts (in shell) and 7.45 mmt of peanut meal were produced worldwide (FAO, 2004). The People's Republic of China and India, the world's largest peanut producers, accounted for 59 percent (38 and 21 percent, respectively) of total global peanut production and 65 percent (36 and 29 percent, respectively) of peanut meal production. The United States produced 5 percent of the world's supply of peanuts and 2 percent of the global supply of peanut meal in 2003 (FAO, 2004). In the United States, the demand for peanut meal in human foods has made it less available for use in animal diets in recent years. This trend is likely to continue if the production of peanuts remains near current levels.

Solvent-extracted peanut meal has high protein content and a favorable amino acid composition for aquatic animal diets (Table 15.1). Although it contains approximately 41 percent less lysine and 19 percent less sulfur amino acids by weight than high-protein (48 percent CP) soybean meal, it has been used successfully as a partial substitute for soybean meal and fish meal in diets for marine fishes, freshwater fishes, crustaceans, and mollusks.

Small et al. (1999) determined that in striped bass, *Morone saxatilis,* the true amino acid availability (TAAA) of histidine (65 percent), isoleucine (88 percent), leucine (88 percent), and lysine (85 percent) in peanut meal were significantly lower than those in herring fish meal (89 to 93 percent) or soybean meal (92 to 95 percent) (Table 15.2). Apparent dry matter digestibility (ADMD; 60 to 72 percent) did not differ among the ingredients, but the apparent crude protein digestibility (ACPD) of peanut meal (85 percent) was significantly lower than the ACPD of soybean meal (93 percent). Amino-nitrogen digestibility of peanut meal (83 percent), corn gluten meal (91 percent), and herring fish meal (88 percent) did not differ significantly. It appeared from the results of this study that amino acid unsupplemented peanut meal was not an adequate replacement for fish meal in diets for striped bass.

In research with the freshwater channel catfish, *Ictalurus punctatus,* Brown et al. (1985) reported that the ACPD of peanut meal (86 percent; Table 15.2) was not significantly different from that of corn gluten meal (92 percent), menhaden fish meal (86 percent), soybean meal (85 percent), or meat and bonemeal (82 percent), but was significantly higher than the ACPD of blood meal (74 percent) and poultry by-product meal (65 percent). The protein sources in that study were fed in semipurified, isonitrogenous (20 percent CP) diets comprised primarily of corn starch, dextrose, dextrin, cellulose, and cod liver oil, with mineral and vitamin premixes (92.8 percent of diet). Wilson and Poe (1985) determined that the ACPD of extruded peanut meal in a mixed diet for channel catfish, in which the fecal samples were collected by dissection, was 76 percent (Table 15.2) and not different from the ACPD of menhaden fish meal (76 percent). The ACPD of pelleted peanut meal (74 percent) did not differ significantly from the ACPD of extruded peanut meal.

Apparent energy digestibility (AED) coefficients of extruded and pelleted peanut meal (GE, 18.5 kJ/g as fed) were 68 percent and

TABLE 15.1. Essential amino acid (+ cystine and tyrosine) composition (%)[a,b] of selected plant protein products and fish meals.

Ingredient	DM[c]	CP	Arg	His	Iso	Leu	Lys	Met/Cys	Phe/Tyr	Thr	Try	Val
Coconut meal	92	22	2.4	0.4	0.7	1.4	0.6	0.3/0.3	0.8/0.6	0.7	0.2	1.1
Fababean seed meal	87	25	2.3	0.7	1.0	1.9	1.6	0.2/0.3	1.0/0.9	0.9	0.2	1.1
Fish meal, anchovy	92	65	3.7	1.6	3.1	5.0	5.1	1.9/0.6	2.7/2.1	2.8	0.8	3.5
Fish meal, herring	93	68	4.0	1.5	2.9	5.2	5.5	2.0/0.7	2.7/2.2	3.0	0.7	3.5
Fish meal, menhaden	92	63	3.7	1.8	2.6	4.5	4.8	1.8/0.6	2.5/2.0	2.6	0.7	3.0
Lentil seed meal	89	24	2.1	0.8	1.0	1.8	1.7	0.2/0.3	1.3/0.7	0.8	0.2	1.3
Leucaena leaf meal[d]	92	26	1.2	0.5	2.3	1.9	1.3	0.4/0.4	1.2/1.1	0.9	–	1.4
Linseed meal	90	34	3.0	0.7	1.6	2.1	1.2	0.6/0.6	1.6/1.0	1.3	0.5	1.7
Mustard seed meal	93	25	1.6	0.5	1.1	1.6	1.6	0.5/0.5	1.1/0.5	1.1	0.5	1.1
Palm kernel meal	91	16	2.2	0.3	0.6	1.1	0.6	0.3/0.2	0.7/0.4	0.5	0.2	0.9
Pea seed meal	89	23	1.9	0.5	0.9	1.5	1.5	0.2/0.3	1.0/0.7	0.8	0.2	1.0
Peanut meal	92	49	5.1	1.1	1.8	2.8	1.7	0.5/0.7	2.3/1.8	1.3	0.5	2.0
Potato protein concentrate	91	74	3.8	1.7	4.1	7.6	5.8	1.7/1.2	4.9/4.3	4.3	1.0	4.9
Safflower seed meal	92	43	3.6	1.1	1.7	2.6	1.2	0.7/0.7	2.0/1.1	1.3	0.5	2.3
Sesame seed meal	93	43	4.9	1.0	1.5	2.7	1.0	1.1/0.8	1.8/1.5	1.4	0.5	1.9
Soybean meal	90	48	3.5	1.3	2.2	3.7	3.0	0.7/0.7	2.4/1.8	1.9	0.7	2.3
Sunflower seed meal	93	42	2.9	0.9	1.4	2.3	1.2	0.8/0.7	1.7/1.0	1.3	0.4	1.7

[a]Percent of ingredient weight.

[b]Data from National Research Council (1993), Dale and Batal (2003), and other sources cited in the text.

[c]DM, dry matter; CP, crude protein; Arg, arginine; His, histidine; Iso, isoleucine; Leu, leucine; Lys, lysine; Met/Cys, methionine/cystine; Phe/Tyr, phenylalanine/tyrosine; Thr, threonine; Try, tryptophan; Val, valine.

[d]Dash indicates data not available.

TABLE 15.2. Digestibility/availability (%) of protein, energy, and amino acids in selected plant protein products.

Ingredient	Species[a]	ACPD[b]	AED[c]	ADMD[d]	AAAA[e]	TAAA[f]	Reference
Fababean seed meal	Bidyanus bidyanus	91	59	58	78-89		Allan et al. (2000)
	Haliotis midae	93			89-97	94-99	Sales and Britz (2003)
Fish meal, herring	Morone saxatilis					89-93	Small et al. (1999)
Fish meal, menhaden	Clarias gariepinus	93	85				Fagbenro (1998)
	Ictalurus punctatus	86					Brown et al. (1985)
		76-85	92-95				Wilson and Poe (1985)
Fish meal, other[g]	Bidyanus bidyanus	89-94	89-98	77-94			Allan et al. (2000)
	Haliotis midae	83			77-85	78-86	Sales and Britz (2003)
	Oncorhynchus mykiss	84-89	85-86				Sanz et al. (1994)
Linseed meal	Bidyanus bidyanus	78	44	30	72-92		Allan et al. (2000)
	Cyprinus carpio	86	72		86[h]		Hossain and Jauncey (1989)
Mustard seed meal	Cyprinus carpio	85	71		86[h]		Hossain and Jauncey (1989)
Pea seed meal	Litopenaeus vannamei	90-100[i]	72-94[i]				Davis et al. (2002)
	Oncorhynchus mykiss	88	69				Burel et al. (2000)

Feedstuff	Species						Reference
	Oreochromis niloticus	93					Fontanhas-Fernandes et al. (1999)
	Penaeus monodon	93					Bautista-Teruel et al. (2003)
Peanut meal	Psetta maxima	92	78				Burel et al. (2000)
	Bidyanus bidyanus	98	77	74	86-98		Allan et al. (2000)
	Clarias gariepinus	87	76				Fagbenro (1998)
	Haliotis midae	79-87	68	77-94	80-95		Sales and Britz (2002, 2003)
	Ictalurus punctatus	86					Brown et al. (1985)
	Ictalurus punctatus	74-76	68-76				Wilson and Poe (1985)
Sesame seed meal	Morone saxatilis	85		60		65-93	Small et al. (1999)
	Procambarus clarkii		94	89			Brown et al. (1989)
	Clarias gariepinus	80	65				Fagbenro (1998)
	Cyprinus carpio	79	70		82[h]		Hossain and Jauncey (1989)
Soybean meal	Clarias gariepinus	87	77				Fagbenro (1998)
	Haliotis midae	96			95-98	97-99	Sales and Britz (2003)
	Ictalurus punctatus	85					Brown et al. (1985)
	Ictalurus punctatus	93-97	72-73				Wilson and Poe (1985)
	Morone saxatilis	93				92-95	Small et al. (1999)
	Oncorhynchus mykiss	90	78				Sanz et al. (1994)

TABLE 15.2 (continued)

Ingredient	Species[a]	ACPD[b]	AED[c]	ADMD[d]	AAAA[e]	TAAA[f]	Reference
Sunflower seed meal	Clarias gariepinus	87	79				Fagbenro (1998)
	Haliotis midae	92	87	71	89-96	91-98	Sales and Britz (2002, 2003)
	Oncorhynchus mykiss	93	68				Sanz et al. (1994)

[a] Bidyanus bidyanus, Australian silver perch; Clarias gariepinus, African catfish; Cyprinus carpio, common carp; Haliotis midae, South African abalone; Ictalurus punctatus, channel catfish; Morone saxatilis, striped bass; Oncorhynchus mykiss, rainbow trout; Oreochromis niloticus, Nile tilapia; Penaeus monodon, tiger shrimp; Procambarus clarkii, red swamp crayfish; Psetta maxima, turbot.

[b] Apparent crude protein digestibility.

[c] Apparent energy digestibility.

[d] Apparent dry matter digestibility.

[e] Apparent amino acid availability.

[f] True amino acid availability.

[g] Sanz et al. (1994), country of origin and species not identified; Allan et al. (2000), Australian, Danish, and Peruvian origin, species not identified; Sales and Britz (2003), Danish origin, species not identified.

[h] Average of all amino acids.

[i] Depending on whether the product was whole or dehulled, raw or extruded, or micronized.

440

76 percent, respectively, and also did not differ significantly. By comparison, AED coefficients of extruded and pelleted menhaden fish meal (GE, 18.5 kJ/g as fed) were higher (95 and 92 percent, respectively) than AEDs of peanut meal, while AEDs of extruded and pelleted soybean meal (GE, 17.5 kJ/g as fed) were comparable to those of peanut meal.

Despite the relatively high apparent digestibility of peanut meal in catfish diets, substitution of peanut meal for soybean meal has been problematic. Replacement of soybean meal (50.5 percent of diet) with peanut meal (22.5 or 38 percent, respectively) and cottonseed meal (10 or 15 percent, respectively) in a diet with 5 percent menhaden fish meal reduced the growth of restriction-fed (65 kg/ha/day) and satiation-fed channel catfish (Munsiri and Lovell, 1993). However, the growth reduction of satiation-fed fish (mean weights 556 g, 507 g, and 472 g among treatments) was not as large as that of restriction-fed fish (mean weights 468 g, 440 g, and 417 g among treatments). Thus catfish appear to be sensitive to changes in dietary protein quality caused by the substitution of peanut meal for soybean meal, at both high and low feeding rates.

Similarly, Zarate et al. (1999) reported that weight gain of catfish fed a diet with 30 percent peanut meal and 7.5 percent corn gluten meal was depressed relative to fish fed a 39 percent soybean meal diet, even when both diets contained 10 percent fish meal and 0.8 percent free lysine was added to the peanut meal diet. The soybean meal diet produced higher weight gain, higher body protein percentage, and lower body fat percentage in catfish than the lysine-supplemented peanut meal and corn gluten meal diet. Zarate and Lovell (1997) observed that a diet containing 31 percent peanut meal and 8 percent corn gluten meal could produce weight gain of channel catfish equivalent to a diet containing 39 percent soybean meal, if the peanut meal was fortified with free lysine at twice the concentration present in soybean meal.

Rainbow trout, *Oncorhynchus mykiss,* fed an all-plant diet containing 20 percent peanut meal and 64 percent corn products (corn gluten meal, ground yellow corn, and corn gluten feed), exhibited significantly lower weight gain than fish fed similar diets containing 39 percent soy flour or 43 percent soybean meal (Adelizi et al., 1998). However, the phosphorus availability of the diet was increased by partial replacement of fish meal with peanut meal. Riche and Brown (1999) reported that the apparent phosphorus availability (APA) and true phosphorus availability (TPA) of a diet with 20 percent peanut meal and 41 percent fish meal was significantly higher (39 percent APA, 41 percent TPA) for rainbow trout than a diet with 52 percent fish meal as the sole protein source (19 percent APA, 21 percent TPA). Apparent phosphorus availability and TPA of diets with soybean meal and canola meal substituted at the same 20 percent level with 41 percent and 45 percent fish meal (39 percent APA and 43 percent TPA for soy, and 40 percent APA and 42 percent TPA for canola) did not differ significantly from the peanut meal diet, but were significantly higher than the APA and TPA of fish meal alone. In addition, it was established in an earlier study that the APA and TPA of peanut meal (0.85 percent P) were 22 and 42 percent, respectively, for rainbow trout and that treatment of peanut meal with phytase (3,750 U/kg dry diet) raised its APA to 76 percent (Riche and Brown, 1996). The TPA of fish meals (menhaden, anchovy, sardine, and herring in ascending order) ranged from 21 to 55 percent compared with 10 to 48 percent TPA for plant feedstuffs (three types of soybean meal, corn gluten meal, and canola meal). Phytase treatment raised the TPA of plant feedstuffs to 46 to 77 percent.

Studies with other freshwater fishes have shown that peanut meal can serve as a suitable partial replacement for fish meal. Regression analysis of the growth performance of Australian silver perch, *Bidyanus bidyanus*, fed peanut meal at levels of 15, 30, 45, 60, or 75 percent of a mixed diet (containing 85, 70, 55, 40, and 25 percent, respectively, of a premixed combination of other practical ingredients) indicated that inclusion of peanut meal at or above levels of 45 percent of the diet produced outliers (outside 95 percent confidence limits) when protein deposition was plotted against digestible protein intake (Booth and Allan, 2003). The same results were observed for the diet containing 75 percent peanut meal when energy deposition was plotted against digestible energy intake. Results suggested that the Australian silver perch can effectively utilize peanut meal as a protein source at levels less than 45 percent of the diet. Allan et al. (2000) reported that the ACPD of peanut meal for Australian silver perch was 98 percent, with ADMD and AED coefficients of 74 and 77 percent, respectively (Table 15.2). Apparent amino acid availability (AAAA) coefficients ranged from 86 percent (glycine) to 98 percent (methionine). Apparent crude protein digestibilities of the plant products tested ranged from 71 percent (vetch) to 100 percent (wheat), while ACPDs of fish meals ranged from 89 to 94 percent. Legumes tended to have lower apparent availability coefficients for sulfur amino acids than for other amino acids.

The specific growth rate and feed conversion ratio (FCR) of common carp, *Cyprinus carpio,* fry were not significantly different when peanut meal comprised 25 percent of protein (17 percent peanut, with 40 percent fish meal) in a 41 percent-CP diet, compared with the same parameters among fry fed a 43 percent-CP control diet with 100 percent of protein from fish meal (53 percent fish meal) (Hasan et al., 1997). However, protein efficiency ratio and apparent net protein utilization were significantly poorer among carp fry fed 25 percent peanut meal protein (PMP) than among fry fed the control diet with 100 percent of protein from fish meal, and all production parameters were significantly poorer among carp fry fed 75 percent PMP (51 percent peanut, with 13 percent fish meal) than among fry fed 25 percent PMP or fry fed the control diet (75 percent PMP < 25 percent PMP ≤ control). The high level of PMP caused dietary deficiencies in lysine, methionine, and threonine, which may have been responsible for the growth depression that occurred at the highest level of peanut meal inclusion. Peanut meal protein was a suitable replacement for 25 percent of fish meal protein in diets for common carp fry, but linseed protein was reported to produce better results than PMP. Sesame protein produced results equivalent to PMP at the same level of inclusion.

Khan et al. (2003) reported that peanut meal was inferior to (tuna) fish meal and soybean meal as a dietary protein source for rohu (or Indian major carp), *Labeo rohita.* Complete replacement of fish meal (50 percent of diet) with peanut meal (68 percent of diet) produced deficiencies of sulfur amino acids and lysine, and a reduction in consumption of the fish meal-free diet, which led to significantly poorer feed conversion and significantly reduced growth of rohu. A combination of peanut meal, soybean meal, and canola meal (70 percent of diet) also did not significantly improve rohu growth performance or the efficiency of dietary protein utilization. A diet containing equal amounts (16 percent) of fish meal, soybean meal, peanut meal, and canola meal was as good as a diet containing soybean meal (62 percent) as the sole protein source, but was still inferior to a diet containing fish meal (50 percent) as the sole protein source. Peanut meal was deemed inferior to soybean meal as a replacement for fish meal in rohu diets.

Jackson et al. (1982) reported that PMP could replace up to 25 percent of the fish meal protein in a 30 percent-CP diet for Mozambique tilapia, *Oreochromis mossambicus,* without reducing growth. Growth of *O. mossambicus* was significantly reduced when PMP replaced 50 percent or more of the fish meal protein in a 30 percent-CP diet. Ogunji and Wirth (2001) reported that 10 percent peanut meal (with 26 percent fish meal) in a mixed diet did not reduce growth of Nile tilapia, *Oreochromis niloticus,* relative to fish fed a diet with 43 percent fish meal, but 15 percent peanut meal (with 5 percent fish meal) significantly reduced growth of Nile tilapia compared with fish fed the control diet. However, the amino acid profile of the peanut meal used in the study was considered poor, and aflatoxin effects also were a possibility, so additional research may indicate that a restriction of less than 15 percent peanut meal in Nile tilapia diets is not necessary in all cases.

In a study designed to evaluate the effects of dietary protein concentration on the weight gain and specific growth rate of snakehead, *Channa striata,* Mohanty and Samantaray (1996) demonstrated that *C. striata* fry could use up to 43 percent peanut meal (in combination with 43 percent brown fish meal) as a dietary protein source. Weight gain of snakehead fry was best when fed a diet containing 41 percent peanut meal and 41 percent fish meal as the primary protein sources.

The addition of 30 percent peanut meal to a mixed composition reference diet fed to red swamp crayfish, *Procambarus clarkii,* as part of a digestibility trial reduced diet consumption (Brown et al., 1989). However, the ADMD of peanut meal (when present as 98 percent of the diet pellet) was high for red swamp crayfish (89 percent; Table 15.2) even when peanut meal comprised 98 percent of the feed pellet. The AED of peanut meal (GE, 20.7 kJ/g dry matter) also was high at 94 percent.

Lim (1997) reported that peanut meal (12 percent of diet) could replace 20 percent of a mixture of animal proteins—comprised of 53 percent menhaden fish meal, 34 percent shrimp waste meal, and 13 percent squid meal—in diets for juvenile Pacific white shrimp, *Litopenaeus vannamei,* with no effect on diet intake or weight gain. Levels of peanut meal at or above approximately 23 percent of the diet significantly reduced diet consumption and weight gain. Levels higher than 47 percent of the diet also reduced FCR, protein efficiency ratio, and apparent protein utilization, perhaps because of poor amino acid balance. Results of the study indicated that, if the palatability of a high peanut meal diet could be improved, 35 percent peanut meal could replace 60 percent of the animal protein in a diet for white shrimp.

Sales and Britz (2003) reported that protein and amino acids in oilseeds were highly digestible by the South African abalone, *Haliotis midae*. The ACPD of peanut meal for *H. midae* was 87 percent (Table 15.2)—higher than the ACPD of corn gluten meal (76 percent), comparable with that of cottonseed meal (86 percent) and fish meal (83 percent), and lower than the ACPD of soybean meal (96 percent) and lupin seed meal (97 percent). Apparent crude protein digestibilities of sunflower seed meal and canola meal were above 90 percent (92 percent and 94 percent, respectively). Apparent amino acid availabilty coefficients and TAAA coefficients (range of all IAA and DAA) for peanut meal were 77 to 94 percent and 80 to 95 percent, respectively, in *H. midae*. In an earlier study, Sales and Britz (2002) determined that the digestibility of compound diets by South African abalone can be determined additively from coefficients calculated for single protein diets, and they recommended that single ingredient diets be used to determine ACPD and amino acid availability coefficients for *H. midae*. The calculated digestibility coefficient determined with a single protein diet predicted the digestibility of that ingredient in a compound diet for *H. midae* with an error of less than 3 percent, compared with a deviation of 13 to 16 percent when apparent digestibility coefficients were measured with ingredient substitution trials.

Sunflower Seed Meal (from **Helianthus annuus***)*

Sunflower seed meal is produced from the residue remaining after most of the oil has been removed from sunflower seeds by mechanical extraction or solvent extraction. Meal made from mechanically extracted seeds contains approximately 41 percent CP and 8 percent EE (Dale and Batal, 2003). Meal made from solvent-extracted seeds contains 42 to 46 percent CP, 2 to 3 percent EE, 12 to 21 percent CF, and 21 to 25 percent NFE (Dale and Batal, 2003; National Research Council, 1993).

Approximately 27.74 mmt of sunflower seeds were produced worldwide in 2003 (FAO, 2004). The Russian Federation, Ukraine, Argentina, and China produced more than half (53 percent) of the world's supply of sunflower seeds. In 2003, approximately 9.53 mmt of sunflower seed meal were produced worldwide (FAO, 2004). Argentina, the Russian Federation, Ukraine, and China generated a combined 46 percent (13, 12, 11, and 10 percent, respectively) of the total global production of sunflower seed meal. The United States generated 2 percent of the world's supply of sunflower seeds and less than 2 percent of the world's supply of sunflower seed meal in 2003 (FAO, 2004).

Solvent-extracted sunflower seed meal has high protein content and a favorable amino acid composition for aquafeeds (Table 15.1). It contains approximately 44 percent less lysine, but 56 percent more sulfur amino acids by weight, than high-protein soybean meal. The average price of sunflower seed meal is less than half that of high-protein soybean meal and sunflower seed meal is readily available in the feed ingredient market. Sunflower seed meal has been used in a few cases as a partial substitute for fish meal and soybean meal in diets for marine and freshwater fishes, and mollusks, with mixed results. Additional research is needed to determine the efficacy of sunflower seed meal as an ingredient in aquatic animal diets.

European eels, *Anguilla anguilla,* fed a diet containing sunflower seed meal exhibited reduced diet utilization and lower growth indices than eels fed a control diet with fish meal as the only protein source (de la Higuera et al., 1999). Supplementation of the sunflower seed meal diet with selected dietary essential amino acids (i.e., lysine, methionine, histidine, and threonine) doubled growth performance, which increased growth almost to the level of fish fed the control diet. Liver protein turnover rates increased in eels fed the amino-acid-unsupplemented sunflower seed meal, but muscle protein turnover decreased. The rate of protein degradation in the liver of eels fed the unsupplemented sunflower seed meal diet exceeded that of protein synthesis.

Supplementation of sunflower seed meal with dietary essential amino acids increased the rate of synthesis of muscle protein, relative to eels fed unsupplemented sunflower seed meal, and reduced the rate of protein synthesis in the liver to the level in fish fed the control diet. The regulatory function of the liver resulted in increased protein synthesis and protein degradation during periods of nutritional imbalance. Rates of protein synthesis and the efficiency of protein deposition in muscle, however, were more closely related to fish growth than rates of protein synthesis in the liver.

Rainbow trout fed diets in which sunflower seed meal replaced up to 40 percent of fish meal protein showed no reduction in weight gain (Sanz et al., 1994). The ACPD of sunflower seed meal (93 percent; Table 15.2) was significantly higher for rainbow trout than the ACPD of soybean meal (90 percent) or fish meal (84 to 89 percent), while the digestibility of carbohydrate (40 percent) in sunflower seed meal was significantly lower for rainbow trout than the carbohydrate in soybean meal (78 percent).

In a study with the Mozambique tilapia, Jackson et al. (1982) reported that protein from sunflower seed meal could replace 25 to 50 percent of the fish meal protein in a 30 percent-CP diet.

The protein and amino acids in sunflower seed meal, and other oilseed meals, have been shown to be highly digestible by the South African abalone. Although fish meal was determined to be the best protein source for formulated abalone diets, plant proteins were found to produce good growth and good protein digestibility (Sales and Britz, 2001). The ACPD of sunflower seed meal (92 percent; Table 15.2) was higher for South African abalone than the ACPD of corn gluten meal (76 percent), but lower than the ACPD of soybean meal (96 percent) and lupin seed meal (97 percent) (Sales and Britz, 2003). The protein digestibility of canola meal (94 percent), cottonseed meal (86 percent), peanut meal (87 percent), and fish meal (83 percent) also were high. Apparent amino acid availability coefficients and TAAA coefficients for sunflower seed meal (all amino acids) ranged 89 to 96 percent and 91 to 98 percent, respectively, in South African abalone. The ADMD of sunflower seed meal was 71 percent; organic matter digestibility, 83 percent; and AED, 87 percent (Sales and Britz, 2002).

Potato Protein Concentrate (from **Solanum tuberosum***)*

Potato protein concentrate (PPC) is a by-product of the production of potato starch. The Netherlands is the world's largest producer of potato starch and PPC. Potato protein concentrate contains approximately 74 percent CP. It provides higher levels of all dietary essential amino acids than high-protein soybean meal (Table 15.1). It also contains an antinutritional factor, solanine (a steroid glycoalkaloid found in plants belonging to the nightshade family; woody nightshade, potato, and tomato), which can reduce fish growth when present in sufficient quantity in the diet.

Solanine functions as a natural insect repellent in potato plants, and it has been used in some parts of the world as an agricultural insecticide. The amount of solanine in PPC appears to affect its efficacy as a protein source for aquatic animals. Standard PPC contains 1.5 to 2.5 g of solanine per kilogram dry weight. However, solanine can be removed from PPC during the manufacturing process to produce low-solanine products. Recommended inclusion levels of PPC in aquafeeds have ranged up to 20 percent of the diet, depending on the type of PPC used and the species being fed.

PPC without solanine can be a suitable protein source for fishes. Refstie and Tiekstra (2003) reported that PPC with a very low level of solanine (< 0.1 g/kg) could be fed to Atlantic salmon, *Salmo salar,* at levels up to 21 percent of the diet without adverse effects. In their study, low-solanine PPC replaced up to 40 percent of the dietary protein provided by low-temperature processed fish meal with no reduction in salmon growth. Appetite, weight gain, feed conversion efficiency, and retention of nitrogen and energy were unaffected by low-solanine PPC at levels up to 21 percent of the diet. The protein and essential amino acids in PPC were highly digestible by salmon, and no significant differences in the apparent utilization of macronutrients or the availability of amino acids were observed among treatment groups fed 0 to 21 percent PPC in prepared diets. The authors noted that the high protein content of low-solanine PPC makes it particularly suited for nutrient-dense diets.

Results of experiments with rainbow trout fed a standard PPC product differed from those obtained with salmon fed low-solanine PPC. Moyano et al. (1992) reported that growth and dietary protein utilization of rainbow trout were negatively correlated with the amount of PPC in the diet, although measurable amounts of solanine were not detected. Weight gain of rainbow trout fed a 45 percent-CP diet with 20, 30, or 40 percent of dietary protein provided by PPC was significantly reduced at all levels of potato protein inclusion, relative to a control diet that contained only fish meal protein. Weight gain of fish fed a 35 percent-CP diet with 20, 30, 40, or 60 percent of dietary protein provided by PPC showed a significant trend toward reduced growth at higher levels of potato protein inclusion. Weight gain of all fish fed PPC was lower than that of fish fed the control diet, but weight gain was not different between fish fed 20 or 30 percent potato protein, those fed 30 or 40 percent potato protein, or those fed 40 or 60 percent potato protein. Diets with PPC were poorly accepted. Fish growth, protein efficiency ratio, productive protein value (net protein retention), and energy utilization were negatively correlated with the level of PPC in the diet. Fish fed more than 20 percent of dietary protein from PPC did not double their body weight in eight weeks. The authors concluded that diet intake was reduced by the presence of PPC in the diet, perhaps because of the presence of alkaloids or the high levels of free amino acids in PPC. Dietary energy utilization also was poor among fish fed PPC. Potato protein concentrate was deemed to be an unacceptable protein source for rainbow trout unless the product was free of solanine.

Xie and Jokumsen (1997b) observed that rainbow trout fed 0 to 51 percent dietary PPC exhibited decreased growth, reduced diet utilization and condition factor, and increased mortality when potato protein was the sole dietary protein source. Dietary PPC also significantly decreased the dry matter, protein, and fat content of the body, and increased body ash content. Specific growth rate, feed efficiency, protein efficiency ratio, and net protein retention decreased significantly with each incremental increase in the level of potato protein in the diet (51 <30 <20 <9 <0 percent). Final body weight and final condition factor followed the same trend, but decreased more gradually as graded levels of PPC (maximum 51 percent) completely replaced low-temperature processed fish meal (maximum 54 percent) in the diet. The PPC used in the study contained heat-stable solanine at a concentration of 1 g/kg, which was much higher than the recommended maximum of 0.06 g/kg. In another study involving PPC with heat-stable solanine at a concentration of 1 g/kg, Xie and Jokumsen (1997a) observed that feed efficiency and net protein retention of rainbow trout fed 2 to 5.6 percent PPC were not different from fish fed a control diet without potato protein, but were lower among fish fed 9 to 11 percent potato protein than among those fed 2 to 5.6 percent potato protein or the control diet. Specific growth rate did not differ between fish fed 2 percent potato protein and fish fed the control diet, but was lower in fish fed 5.6 percent potato protein, lower still in fish fed 9 percent of the concentrate, and lowest in fish fed 11 percent PPC. Feed consumption was lowest among fish fed 9 to 11 percent potato protein, higher among fish fed 2 to 5.6 percent potato protein, and highest among control-fed fish. Increased levels of dietary PPC led to decreased body dry matter and fat content and increased body protein and ash content. A maximum of 2 percent PPC was recommended for rainbow trout diets.

Subsequently, Xie and Jokumsen (1998) reported that the specific growth rate of rainbow trout fed PPC at a level of 2 percent of the diet (3.65 percent/day) was not significantly different from that of fish fed a control diet with no potato protein (3.55 percent/day). Fish fed 5.6 percent potato protein had significantly lower specific growth rate (3.28 percent/day) than the fish fed 2 percent potato protein, but did not differ in specific growth rate from control fish. The specific growth rate of fish fed 11 percent PPC (2.24 percent/day) was lower than that of fish fed the other potato protein diets or the control diet. Feed efficiency, net protein retention, and final condition factor did not differ significantly among fish fed 0 to 5.6 percent PPC, while 11 percent dietary potato protein significantly reduced all of these parameters. Body dry matter and fat content were inversely related to the amount of PPC in the diet, but no other differences in body composition were observed. A level of 5.6 percent PPC appeared to be suitable for rainbow trout diets based on the results of this study. As the same source of high-solanine PPC was used in each of the feeding trials, Xie and Jokumsen attributed the favorable results obtained in the later study— where 5.6 percent dietary PPC was deemed acceptable for rainbow trout rather than the 2 percent PPC recommended in an earlier study— to the use of automatic feeders and thus increased access to diet.

Safflower Seed Meal (from Carthamus tinctorius)

Safflower seed meal is produced from the cake remaining after oil is extracted from the safflower seed. Safflower seed meal made from mechanically extracted seeds contains approximately 20 percent CP and 7 percent EE (Dale and Batal, 2003). Solvent-extracted safflower meal contains approximately 42 percent CP, 1 percent EE, 15 percent CF, and 25 percent NFE (Dale and Batal, 2003).

In 2003, the total world production of safflower seed was approximately 648,000 metric tons (mt) (FAO, 2004). India, Mexico, and the United States produced a combined total of approximately 75 percent of the world's supply of safflower seed and safflower seed meal in 2003. India, the largest safflower producer, generated 44 percent of the world's supply of safflower seed meal in 2003. The United States produced 19 percent of the global supply of safflower seed meal that year.

The research conducted with safflower seed products in aquatic animal diets has concentrated on the use of safflower oil as a source of energy and essential fatty acids in diets, rather than safflower seed meal as a source of dietary protein. To date, virtually no research on the nutritional value of safflower seed meal as a protein source for fishes or other cultured aquatic animals has been published.

Sesame Seed Meal (from Sesamum indicum)

Sesame seed meal is produced from the material remaining after the oil has been removed from sesame seeds, usually by a mechanical extraction process. Meal made from mechanically extracted sesame seed contains approximately 42 percent CP, 7 percent EE, 7 percent CF, and 27 percent NFE (Dale and Batal, 2003).

Total world production of sesame seed was approximately 2.94 mmt in 2003 (FAO, 2004). China, India, and Myanmar produced 62 percent of the total world's supply of sesame seed in 2003. Global production of sesame seed meal was 804,641 mt in 2003, of which China, Myanmar, and India produced 66 percent of the total (FAO, 2004).

Hossain and Jauncey (1989) reported that the ACPD of sesame seed meal for common carp was 79 percent (Table 15.2); TCPD, 83 percent, and AED, 70 percent. The mean AAAA coefficient was 82 percent (average of all amino acids). True amino acid availability coefficients were approximately 3 percent higher than apparent values.

In a study with channel catfish, Zarate and Lovell (1997) observed that weight gain of fish fed a diet with 43 percent sesame seed meal and 1 percent fish meal was lower than weight gain of fish fed 39 percent soybean meal and 1 percent fish meal, even after the sesame seed meal diet was supplemented with free lysine to the same level of lysine that was present in the soybean-based diet. Similarly, the specific growth rate, protein efficiency ratio, and apparent net protein utilization of common carp fry were significantly reduced when sesame seed meal protein (SMP) comprised 25 percent, 50 percent, or 75 percent of protein (20, 39, or 59 percent sesame, with 40, 27, or 13 percent fish meal) in 40 to 41 percent CP diets, compared with the same parameters among fry fed a 43 percent-CP diet with 100 percent of protein from fish meal (53 percent fish meal) (75 percent SMP < 50 percent SMP < 25 percent SMP < control) (Hasan et al., 1997). A significant reduction in these performance parameters occurred at each stepwise increase in sesame seed meal inclusion level. The FCR of fry fed 25 percent SMP was not different from the FCR of fry fed the control diet, and the FCRs of fry fed 25 percent SMP and 50 percent SMP were not different. The FCR of fry fed 75 percent SMP was significantly poorer than the FCRs of fry fed the other sesame seed meal diets or the control diet. Essential amino acid deficiencies (i.e., lysine and threonine) and reduced diet intake caused by the relatively poor palatability of sesame seed meal could have contributed to the negative effects obtained at the higher levels of sesame seed meal inclusion (50 and 75 percent of protein). Despite these drawbacks, 25 percent SMP was a better protein source in diets for common carp fry than protein from mustard seed meal, *Leucaena* leaf meal, or coconut meal at the same level of inclusion.

In an experiment with rohu, fermented SMP supplemented with deficient amino acids (lysine and methionine-cystine) was used to replace up to 50 percent of the fish meal protein in the diet without adverse effect (Mukhopadhyay and Ray, 1999). Fermentation was used to remove possible antinutritional factors (primarily phytic acid) from the sesame seed meal prior to feeding. Weight gain of rohu was significantly reduced when amino-acid-unsupplemented SMP replaced 30 to 50 percent of fish meal protein in the diet. Supplementation of the diet with lysine significantly increased rohu weight gain; supplementation with methionine-cystine alone increased weight gain more than supplementation with lysine alone; and supplementation with lysine plus methionine-cystine produced the greatest weight gain increase. Results indicated that amino-acid-supplemented sesame seed meal can be used effectively in prepared diets for rohu, and that rohu can use free amino acids in prepared diets.

Mustard Seed Meal (from **Brassica juncea***)*

Mustard seed meal is produced from the cake remaining after the oil has been removed from mustard seed by mechanical extraction or solvent extraction. Meal produced from solvent-extracted mustard seed contains approximately 26 percent CP, 2.3 percent EE, 18 percent CF, and 46 percent NFE (AFRIS, 2004).

Approximately 632,354 mt of mustard seed and 111,730 mt of mustard seed meal were produced globally in 2003 (FAO, 2004). Canada and Nepal produced 57 percent of the world's supply of mustard seed in 2003. Nepal and Myanmar produced 91 percent of the world's supply of mustard seed meal that year. India, Pakistan, and Bangladesh are also important mustard seed producers, but data on mustard seed production in these countries are combined with rapeseed production for reporting purposes, so statistics on production of mustard seed alone are unavailable.

Mustard seed contains glucosinolates that can be converted to highly toxic isothiocyanates, which have been shown to depress growth and cause organ abnormalities in rodents and poultry. The presence of these antinutritional compounds in mustard seed meal could limit its potential as a major protein source in aquatic animal diets.

The availability of protein in mustard seed meal appears to be adequate for common carp, the only species in which it has been measured. Hossain and Jauncey (1989) reported the ACPD (Table 15.2), true protein digestibility, and AED of mustard seed meal for common carp to be 85, 89, and 71 percent, respectively. The mean AAAA coefficient is approximately 86 percent (average of all amino acids). Apparent amino acid availability coefficients for lysine and arginine were lower than those for other amino acids, and TAAA coefficients were approximately 3 percent higher than apparent values.

Hasan et al. (1997) reported that the specific growth rate, protein efficiency ratio, and apparent net protein utilization of common carp fry were significantly reduced, and FCR was significantly increased, when mustard seed meal protein (MMP) comprised 25 or 50 percent of dietary protein (29 or 58 percent mustard seed meal, with 40 or

27 percent fish meal) in a 41 percent-CP diet, when compared with the same parameters among fry fed a 43 percent-CP diet with 100 percent of protein from fish meal (53 percent fish meal). These production parameters also were significantly poorer among fry fed 50 percent MMP than among fry fed 25 percent MMP (50 percent MMP <25 percent MMP < control). The MMP diets did not have amino acid deficiencies, so antinutritional factors could have been responsible for the growth depression that occurred at the higher level of mustard seed meal inclusion. Mustard seed meal was deemed an unsuitable protein source for carp fry at a high (50 percent of protein) dietary inclusion level. Although acceptable at a relatively low (25 percent of protein) level of inclusion, mustard seed meal still was inferior to linseed meal, peanut meal, and sesame seed meal in diets for common carp fry.

Linseed Meal (from **Linum usitatissimum***)*

Linseed (flaxseed) meal is a by-product of linseed oil production. Linseed meal made from mechanically extracted seed contains approximately 32 percent CP and 3 percent EE (Dale and Batal, 2003). Meal made from solvent-extracted linseed contains 33 percent CP, 1 percent EE, 10 percent CF, and 39 percent NFE (Dale and Batal, 2003).

The total world production of linseed and linseed meal was 2.09 mmt and 1.12 mmt, respectively, in 2003 (FAO, 2004). Canada, China, the United States, and India produced 80 percent of the world's supply of linseed in 2003 and 48 percent of the world's supply of linseed meal. The United States generated 13 percent of global linseed production and 13 percent of the global production of linseed meal in 2003.

Linseed meal has a favorable amino acid composition (Table 15.1), but it contains a glucoside, linamarin, that is converted by linase enzyme to hydrogen cyanide, a highly toxic compound. High-temperature processed linseed meal is safe, but meal processed at low temperature may contain enough linamarin and linase to be toxic.

Hossain and Jauncey (1989) determined that the ACPD of linseed meal for common carp was 86 percent (Table 15.2); true protein digestibility, 90 percent; and AED, 72 percent. The mean AAAA coefficient was 86 percent (average of all amino acids). The TAAA coefficients for lysine and arginine were lower than those for other amino acids.

True amino acid availability coefficients were approximately 3 percent higher than apparent values.

In common carp fry, the specific growth rate, protein efficiency ratio, apparent net protein utilization, and FCR were not significantly different when linseed meal protein (LMP) comprised 25 percent of protein (26 percent linseed, with 40 percent fish meal) in a 41 percent-CP diet, compared with the same parameters among fry fed a 43 percent-CP control diet in which 100 percent of dietary protein came from fish meal (53 percent fish meal) (Hasan et al., 1997). These production parameters were significantly poorer among fry fed 50 percent LMP (53 percent linseed, with 27 percent fish meal) than among fry fed 25 percent LMP or fry fed the control diet with 100 percent of protein from fish meal (50 percent LMP < 25 percent LMP = control). The linseed meal diets did not have amino acid deficiencies, so antinutritional factors were thought to have been responsible for the growth depression that occurred at the higher level of linseed meal inclusion. Linseed meal effectively replaced 25 percent of fish meal protein in diets for common carp fry with no reduction in growth, and linseed performed as well or better than peanut meal at the same level of inclusion.

The ACPD of linseed meal in Australian silver perch was reported to be 78 percent (Table 15.2); ADMD, 30 percent; and energy digestibility, 44 percent (Allan et al., 2000). Apparent availability of amino acids ranged from 72 percent (proline) to 92 percent (arginine).

In rohu, protein from fermented linseed meal can replace up to 50 percent of the fish meal protein in the diet, if the LMP is supplemented with deficient amino acids (i.e., lysine and methionine plus cystine) (Mukhopadhyay and Ray, 2001). The linseed meal used in the study was fermented with *Lactobacillus acidophilus* to remove tannins and phytic acid, which are antinutritional factors. Weight gain, specific growth rate, FCR, and protein efficiency ratio were significantly reduced when unsupplemented LMP replaced 30 to 50 percent of dietary fish meal protein. Supplementation of the diet with lysine or methionine-cystine did not significantly increase these parameters, but supplementation with lysine and methionine-cystine significantly increased weight gain, specific growth rate, protein efficiency ratio, and apparent net protein utilization of fish fed 25 to 50 percent LMP, relative to the response obtained with the control diet. Results indicated that "unprocessed" linseed meal cannot be used in diets for rohu at a level of more than 20 to 30 percent of the diet without reducing growth. Inclusion rates can be increased to 30 to 40 percent of the diet by eliminating or reducing tannins and phytic acid with fermentation. Protein from fermented linseed meal can replace 50 percent of the fish meal protein in diets for rohu fingerlings, if the linseed meal is supplemented with deficient amino acids.

Leucaena Leaf Meal (from Leucaena leucocephala)

Leucaena leaf meal is made from the dried leaves of a leguminous tree, *Leucaena leucocephala,* grown in many tropical areas of the world as forage for livestock. Native to southern Mexico and the northern part of Central America, *L. leucocephala* has been introduced throughout the tropics since its discovery in the New World. The Philippines is the world's largest producer of *Leucaena* leaves.

Leucaena leaf meal contains 26 to 27 percent CP and approximately 6 percent EE. Although the protein in the leaves can be extracted and concentrated relatively easily to produce a material with 30 to 38 percent CP (Telek, 1983), development of *Leucaena* leaf protein concentrate as a commercial product has not occurred.

Leucaena leaf contains a potent antinutritional compound, called mimosine [β-N(3-hydroxy-4-oxo-pyridine) α–amino propionic acid], which is a toxic nonprotein amino acid that serves the tree as a natural insecticide. Mimosine comprises 3 to 5 percent of the dry weight of the CP in the leaf. Unprocessed *Leucaena* leaves are fed to cattle and other ruminants, usually without adverse effect, because microbes in the rumen break down mimosine into nontoxic compounds. Nonruminants cannot detoxify mimosine, and unprocessed *Leucaena* should not be fed to monogastric animals.

Moist-heat treatment of *Leucaena* leaves can greatly reduce mimosine concentrations. Soaking the leaves in water for forty-eight hours at 30°C eliminates virtually all of the mimosine. Other heating, drying, soaking, and pH treatments also have been shown to be effective in reducing the mimosine content of *Leucaena* leaves (Sethi and Kulkarni, 1995).

Leucaena leaf meal has been used in prepared diets for a few species of fish with mixed results. Cage-cultured Nile tilapia fed 24 percent CP and 22 percent CP diets with 10 and 5 percent *Leucaena* leaf meal (plus 20 percent fish meal; and 70 percent rice bran, or 65 percent rice bran + 10 percent coconut meal, respectively) had significantly lower weight gains and significantly higher FCRs than tilapia fed similar 24 percent CP diets without *Leucaena* but with, and without, coconut meal (Guerrero, 1980). Addition of 10 percent coconut meal to the diet had no effect on tilapia growth, but addition of 5 percent *Leucaena* leaf meal with 10 percent coconut meal, or 10 percent *Leucaena* with no coconut meal, significantly reduced the growth of Nile tilapia. Jackson et al. (1982) also reported that *Leucaena* leaf protein was not a suitable replacement for 25 percent of the fish meal protein in a 30 percent-CP diet for Mozambique tilapia.

Santiago et al. (1983) reported that milkfish, *Chanos chanos,* fry fed 40 percent-CP diets with 13 percent and 19 percent *Leucaena* leaf meal (plus 57 percent fish meal, 9 percent shrimp meal, and 5 percent soybean meal + 3 percent rice bran, or 0 percent soybean meal + 0.7 percent rice bran) had weight gains and FCRs equivalent to milkfish fry fed diets in which soybean meal and/or fish meal replaced *Leucaena* leaf meal. Substitution of up to 5 percent CP from soybean meal and/or *Leucaena* leaf meal did not affect growth. However, survival of milkfish fry fed diets with *Leucaena* leaf meal was lower than that of fry fed a similar diet without *Leucaena* but containing soybean meal, which suggested that antinutritional factors, such as mimosine, in *Leucaena* leaf meal could have had toxic effects on milkfish fry.

Conversely, the specific growth rate, protein efficiency ratio, and apparent net protein utilization of common carp fry were significantly reduced, and FCR was significantly increased, when *Leucaena* leaf meal comprised 25 percent of protein (33 percent *Leucaena,* with 40 percent fish meal) in a 41 percent-CP diet, when compared with the same parameters among fry fed a 43 percent-CP diet with 100 percent of dietary protein from fish meal (53 percent fish meal) (Hasan et al., 1997). *Leucaena* leaf meal was judged to be an unsuitable protein source in diets for common carp fry, even at a relatively low level of inclusion.

Fababean Seed Meal (from Vicia faba)

The total world production of fababean (broad bean) seed was approximately 4.02 mmt in 2003 (FAO, 2004). China, the world's largest producer, accounted for 45 percent of global production. Other significant producers of fababeans are Ethiopia (11 percent), Egypt (11 percent), Australia (7 percent), and France (7 percent). Production of fababeans in the United States is negligible. The seed is eaten by humans in some areas of Europe, North Africa, and the Middle East. In North America it is used as animal feed.

Fababean seed (dry) contains approximately 25 percent CP, 3 percent EE, 20 percent CF, and 35 percent NFE (ESCHA Research, 2003). It provides less than half the lysine and less than a third of the methionine-cystine available in soybean meal, but is otherwise a reasonably good protein source, similar in its amino acid composition to lentil seed and pea seed (Table 15.1), which also are derived from leguminous plants. As it contains no significant antinutritional compounds, fababean seed can be fed to animals or humans with no additional processing other than crushing or grinding to a suitable particle size.

Although fababean seed meal appears to have potential as a dietary protein source in some areas of the world, little research on the use of this material in aquatic animal diets has been published. Gouveia et al. (1993) reported that fababean seed meal could provide 20 percent of the dietary protein for rainbow trout without negative effects on the protein or energy digestibility of the diet, and with no effect on fish growth or carcass composition. Both cooked and uncooked fababean seed meal was tested with rainbow trout, and either product could replace part of the fish meal in a rainbow trout diet effectively, although fababean seed meal that was cooked (145°C) during the diet manufacturing process had a slightly higher nutritional value for rainbow trout than uncooked meal. Among the three legume seeds (lupin, pea, and fababean) tested in rainbow trout diets, lupin seed meal produced the best dietary protein utilization. Fontaínhas-Fernandes et al. (1999) reported that the protein and energy digestibility of fababean seed meal by Nile tilapia also was similar to that of lupin seed meal. However, both fababean seed meal and lupin seed meal were determined to be inferior to pea seed meal, defatted soybean meal, full-fat toasted soybean, and micronized wheat as protein sources for Nile tilapia.

The Australian silver perch can digest the protein in fababean seed effectively (Table 15.2), although the availability of different amino acids in the protein can vary (Allan et al., 2000). Booth et al. (2001) observed that dehulled fababean seed had significantly improved protein digestibility for Australian silver perch compared with whole seed.

Dry matter and energy digestibility was not improved by dehulling. The protein digestibility of dehulled fababean seed and a fababean seed protein concentrate was determined to be similar in Australian silver perch, indicating that extraction and concentration of the fababean seed protein did not improve its bioavailability for this species.

Sales and Britz (2003) reported that the protein and amino acids in fababean seed meal were well digested by the South African abalone. The ACPD of fababean seed meal for *H. midae* was 93 percent (Table 15.2)—comparable with that of sunflower seed meal (92 percent) and canola meal (94 percent). Apparent amino acid availability coefficients and TAAA coefficients (range of all IAA and DAA) for fababean seed meal were 89 to 97 percent and 94 to 99 percent, respectively, in *H. midae*.

Lentil Seed Meal (from Lens culinaris)

Approximately 3.09 mmt of lentil seed was produced worldwide in 2003 (FAO, 2004). India, Turkey, and Canada generated 61 percent of the world's production of lentils. The United States produced less than 4 percent, primarily in the states of Washington and Idaho, where it is a specialty crop. American consumers use very little lentil seed, and nearly three-quarters of the total production of lentils in the United States is exported.

The protein content of lentil seed ranges from 22 to 35 percent. On average, lentil seed (dry) contains approximately 24 percent CP, 1 percent EE, 31 percent CF, and 26 percent NFE (ESCHA Research, 2003). It is deficient in sulfur amino acids, containing approximately one-third of the methionine-cystine present in high-protein soybean meal (Table 15.1).

Lentil seed appears to have potential as a protein supplement for aquatic animal diets, similar to that of other legume seeds. However, virtually no research on the use of lentil seed meal as a protein source in aquafeeds has been published to date.

Pea Seed Meal (from Pisum sativum)

The total world production of pea seed was approximately 10.25 mmt in 2003 (FAO, 2004). Canada, France, the Russian Federation, and

China produced 59 percent of that total. The United States produced less than 3 percent of the world's supply of pea seed in 2003.

Pea seed (dry) contains approximately 22 percent CP, 1 percent EE, 6 percent CF, and 59 percent NFE (Dale and Batal, 2003). It has an amino acid profile similar to that of lentils and fababeans (Table 15.1). Gouveia et al. (1993) determined that pea seed meal (PSM) could replace a portion of the fish meal in rainbow trout diets without adverse effect. Rainbow trout fed diets in which PSM provided 20 percent of dietary protein exhibited growth and body composition not significantly different from trout fed a control diet without plant proteins. Cooking during the diet manufacturing process (Gouveia et al., 1993) or prior to diet mixing (Pfeffer et al., 1995) increased the nutritional value of pea seed for rainbow trout. Burel et al. (2000) reported that the protein digestibility of extruded peas was 88 percent (Table 15.2) for rainbow trout and 92 percent for turbot, *Psetta maxima*. The digestibility of energy in extruded peas, which comes primarily from starch, was 69 and 78 percent for rainbow trout and turbot, respectively.

Gomes et al. (1993, 1995) observed that a coextruded mixture of rapeseed and peas could replace 15 percent of the fish meal protein in rainbow trout diets effectively. A pea protein concentrate was used by Carter and Hauler (2000) to replace 33 percent of the fish meal protein in a diet for Atlantic salmon with no reduction in weight gain. Among the ingredients tested, pea protein concentrate produced one of the highest feed efficiency ratios and productive protein values in Atlantic salmon, indicating the potential of pea protein as a fish meal replacement in salmon diets.

Pea seed appears to be an acceptable source of protein for the Australian silver perch, especially when it is dehulled (Allan et al., 2000; Booth et al., 2001). Amerio et al. (1998) determined, based on evaluation of several indices of amino acid quality, that methionine was likely to be the most deficient essential amino acid in PSM used as a protein source in diets for gilthead seabream, *Sparus aurata*. Recently, a mixture of extruded peas and other plant protein sources supplemented with amino acids was used successfully to replace 50 to 75 percent of the fish meal in diets for gilthead seabream (Gómez-Requeni et al., 2004).

Gouveia and Davies (1998) reported that up to 40 percent PSM could be used in diets for juvenile European sea bass, *Dicentrarchus labrax,* without negative effects on growth or feed utilization. This level of PSM inclusion allowed a 12 percent reduction in the fish meal content of the diet without adverse effect. However, because of the apparently low digestibility of starches in PSM, the digestible energy content of the diet was significantly reduced at a high level of PSM inclusion. Subsequent research indicated that up to 30 percent PSM could be used in European sea bass diets as a replacement for both fish meal and nonprotein energy without negative effects on growth, diet utilization, protein efficiency ratio, nitrogen deposition, or digestibility of protein, lipid, and carbohydrates (Gouveia and Davies, 2000). Indeed, protein assimilation improved as the amount of PSM in the diet increased, indicating that PSM may actually provide beneficial effects in diets for European sea bass in addition to its value as a fish meal replacement.

Fontaínhas-Fernandes et al. (1999) reported that the protein digestibility of PSM was high (93 percent) for Nile tilapia (Table 15.2). It was one of the better plant protein supplements tested in that study and was deemed a suitable protein source for use in Nile tilapia diets as a partial replacement for fish meal. Similarly, PSM has been shown to be suitable for providing up to 20 percent of the protein in milkfish diets (Borlongan et al., 2003).

Whole raw peas were determined to be an acceptable ingredient for blue shrimp, *Litopenaeus stylirostris,* diets (Cruz-Suarez et al., 2001). However, extrusion-cooked peas produced better feed conversion and improved the dietary protein utilization of blue shrimp. Davis et al. (2002) observed that extrusion cooking and infrared cooking significantly increased the apparent protein digestibility and AED of PSM for Pacific white shrimp (Table 15.2). Based on the results of growth trials with several differently processed PSMs, they recommended that PSM be considered as an alternative protein source in shrimp diets. Bautista-Teruel et al. (2003) reported that the apparent protein digestibility of PSM for tiger shrimp, *Penaeus monodon,* was approximately 93 percent (Table 15.2), and PSM could effectively replace all of the soybean meal protein (25 percent of total dietary protein in this case) in a tiger shrimp diet without affecting growth. Diets containing up to 42 percent PSM were fed to juvenile *P. monodon* without adverse effect.

Coconut Meal (from Cocos nucifera*)*

World production of coconuts and coconut meal was 52.94 mmt and 1.93 mmt, respectively, in 2003 (FAO, 2004). The Philippines, Indonesia, and India accounted for 73 percent of the total world production of coconuts and 80 percent of the world production of coconut meal.

Coconut meal is made from the residue remaining after the oil is extracted from dried coconut meat, called copra. Meal made from mechanically extracted copra contains approximately 22 percent CP and 6 percent EE. Solvent-extracted coconut meal contains approximately the same amount of protein, with 3 percent EE, 14 percent CF, and 46 percent NFE (National Research Council, 1983). Coconut meal contains no known antinutritional factors and its protein digestibility is good, but, in comparison with fish meal and soybean meal, it is deficient in all of the dietary essential amino acids required by fishes. It also is a poorer source of essential amino acids than most of the other plant protein supplements derived from grains and legumes (Table 15.1). Nonetheless, coconut meal is a useful diet ingredient in those areas of the world where it is locally available in quantity.

Guerrero (1980) reported no significant difference in weight gains and FCRs of cage-cultured Nile tilapia fed diets containing 10, 20, or 30 percent coconut meal (plus 20 percent fish meal and 70, 60, or 50 percent rice bran, respectively), and concluded that up to 30 percent coconut meal could be used in diets for Nile tilapia. Jackson et al. (1982) reported that coconut protein could replace 25 percent of the fish meal protein in a 30 percent CP diet for Mozambique tilapia without negative effects. However, grass carp, *Ctenopharyngodon idella,* fry fed coconut meal exhibited a significantly lower growth rate than fry fed a mixed diet of coconut meal and zooplankton (De Silva and Weerakoon, 1981). The rate of weight increase (i.e., incremental gains) of fry fed coconut meal decreased with time, while the rate of weight increase of fry fed coconut meal plus zooplankton increased with time.

Hasan et al. (1997) observed that the specific growth rate, protein efficiency ratio, and apparent net protein utilization of common carp fry also were significantly reduced, and FCR was significantly increased, when coconut meal comprised 25 percent of protein (41 percent coconut, 40 percent fish meal) in a 40 percent CP diet, when compared with the same parameters among fry fed a 43 percent CP diet with 100 percent of dietary protein from fish meal (53 percent fish meal). Coconut meal protein was determined to be unsuitable as a replacement for fish meal protein in carp fry diets at a level of 25 percent of dietary protein. The authors noted that poor pelleting characteristics, apparently caused by the presence of coconut meal in the formulation, prevented them from maintaining the desired diet particle size during the feeding trial, which could have affected the performance of common carp fry fed a diet containing coconut meal.

Palm Kernel Meal (from Elaeis guineensis*)*

Palm kernel meal is produced from the seeds of several species of oil palm trees, which grow at low elevations in tropical areas of the world. Oil palms produce compact bunches of fruits, each of which has a single seed, called the palm kernel, from which edible palm kernel oil is extracted. Palm kernel meal is made from the residue remaining after palm kernel oil has been extracted from the seed.

The global production of palm kernels in 2003 was approximately 7.50 mmt, from which 4.19 mmt of palm kernel meal were produced (FAO, 2004). Malaysia and Indonesia are the primary producers of palm kernels (76 percent of world production) and palm kernel meal (80 percent of world production). There is virtually no production of palm kernels or palm kernel meal in the United States.

Palm kernel meal contains 14 to 16 percent CP and 2 to 10 percent EE (Tang, no date; Chin, 2001), depending on the oil content of the cake used to make the meal. Meal produced from mechanically extracted palm kernel cake contains 5 to 12 percent residual oil. Meal from solvent-extracted cake contains 0.5 to 3 percent oil. Malaysia, the world's largest producer of palm kernel products, does not use the solvent-extraction process to produce palm kernel oil because of its high cost, so much of the world's supply of palm kernel meal has been mechanically extracted (often called palm kernel expeller because of the pressing process used), and it has a relatively high lipid content.

Although the amino acid composition of palm kernel meal is not as favorable as many other protein supplements (Table 15.1), it is inexpensive and has been used as a diet for livestock for many years. It contains a relatively large amount of fiber (15 to 23 percent) and NFE

(56 to 64 percent), which does not affect its value as a ruminant diet but may limit its usefulness in aquafeeds. There are no known antinutritional compounds in palm kernel meal, but aflatoxin contamination can be a problem with meal that has been dried or stored improperly, especially in the warm, humid climates where palm kernel meal is typically produced. Despite these drawbacks, efforts to increase the value of palm kernel meal as a primary protein supplement in aquatic animal diets are in progress in southeastern Asia where palm kernel meal is abundant and aquaculture is growing rapidly (Ng, 2003).

Few reports have been published on the use of palm kernel meal in aquatic animal diets. Omoregie and Ogbemudia (1993) reported that the best growth of Nile tilapia and the best feed efficiency were obtained with a 29 percent-CP diet containing 15 percent palm kernel meal (PKM) and 25 percent fish meal, with 13 percent fiber. Nile tilapia fed a 28 percent-CP diet containing 30 percent PKM and 10 percent fish meal, with 19 percent fiber, exhibited the poorest growth and the poorest feed efficiency. Lower and higher levels of palm kernel meal (10, 20, 25, and 30 percent PKM at 28 to 29 percent dietary CP) decreased growth and feed efficiency. The high fiber content of palm kernel meal was thought to have reduced diet digestibility at higher inclusion levels. Replacement of fish meal with palm kernel meal at a level of 15 percent of the diet appeared to be optimum for Nile tilapia.

African carp, *Labeo senegalensis,* fed 10 percent palm kernel meal with 29 percent dietary fish meal, exhibited weight gain, specific growth rate, FCR, and protein efficiency ratio that were not different from fish fed a control diet without PKM (Omoregie, 2001). All of these parameters were significantly reduced in fish fed 20 or 30 percent PKM, in combination with 28 and 22 percent fish meal, respectively, relative to fish fed the control diet. The ACPD (89 percent) and carbohydrate digestibility (47 percent) of diets containing 10 percent PKM were significantly higher than diets containing 30 percent PKM (86 and 46 percent, respectively). Palm kernel meal protein appeared to be a suitable replacement for up to 10 percent of the fish meal protein in a diet for African carp.

Palm kernel meal, with or without naturally produced aflatoxin from *Aspergillus flavus,* was used to replace 20 or 50 percent of soybean meal protein (up to 30 percent of the diet) in a fish meal protein

(30 percent) plus soybean meal protein (10 percent) based diet for Mozambique tilapia (Lim et al., 2001). Aflatoxin levels in the tainted diets were 75-100 µg/kg diet. Performance of fish fed the PKM replacement diets without aflatoxin was not significantly different from fish fed the control diet; however, the presence of PKM reduced the ADMD and ACPD of the diet at the higher level of PKM inclusion. Fish fed aflatoxin-contaminated PKM had significantly reduced growth performance relative to fish fed the control diet, but growth was not different in most cases from fish fed a diet with uncontaminated PKM. The addition of a commercially available aflatoxin adsorbent (sodium hydrated aluminosilicate) at the level recommended for live-stock diets (0.5 percent of the diet) had no effect on tilapia fed contaminated or uncontaminated palm kernel meal. Results indicated that palm kernel meal can replace up to 50 percent of the soybean meal protein in diets for Mozambique tilapia, thereby reducing the need to use imported soybean meal in tilapia diets.

CONCLUSION

Biological problems associated with the use of alternative plant protein sources must be overcome to effectively increase the use of these products in aquafeeds. Among these problems are: (1) lower CP levels in the replacement ingredient than in the ingredient(s) being replaced; (2) possible amino acid deficiencies caused by the replacement of high-quality ingredients with substitutes possessing less favorable amino acid profiles; and (3) the presence of antinutritional (growth-inhibiting) factors in many of the plant products that have potential as alternative protein supplements. How favorably a potential feedstuff is viewed as an alternative protein source will be determined by how many negative attributes it possesses and how easily the problems these attributes create can be overcome.

While we often see references to the "protein requirements" of different species of aquatic animals, the requirements are actually for minimum quantities of specific amino acids that are provided in differing amounts in the naturally occurring proteins in feedstuffs. To use practical ingredients most effectively to satisfy the amino acid requirements of a particular aquatic animal, we should know three things: (1) the amino acid composition of each ingredient available for use in the diet formulation; (2) the biological availability to the target species of each dietary essential amino acid in each of the feedstuffs available for use; and (3) the minimum dietary requirement of the target species for each of its dietary essential amino acids. The first of these, the dietary amino acid composition of an ingredient of interest, can often be obtained from published databases or, if necessary, can be determined relatively easily by a commercial analytical laboratory.

The second item of information, the biological availability to the target species of amino acids in the ingredient of interest, is often not available, especially for the less commonly used feedstuffs. This information is usually obtained from laboratory feeding trials in which diets comprised of carefully measured amounts of feedstuffs are fed to a target species and the digestibility of each nutrient of interest is measured by comparing the composition of the food consumed with that of the feces produced. Such in vivo measurements of nutrient availability are laborious to obtain and expensive to produce, and thus much digestibility information is currently lacking for a number of ingredients in diets for aquatic species of aquacultural interest. In vitro methods that estimate the digestibility of an ingredient by subjecting it to mixtures of enzymes similar to those found in the digestive tract of the target species have potential to reduce the time and cost involved in measuring nutrient availability in a wide variety of feedstuffs, but in vitro methods that are as reliable and accurate as the in vivo methods now in use have yet to be developed.

The third item of information, the minimum dietary requirements of the target species for dietary essential amino acids, is often unavailable for species of aquacultural interest. Minimum dietary requirements for nutrients are typically determined in laboratory growth trials in which diets with graded levels of the nutrient of interest are fed to the target species under controlled conditions. While it is unlikely that this biological approach will be replaced by nonbiological methods of determining amino acid requirements, mathematical methods that rely on modeling to predict the amino acid requirements of a species might come to be used more often in the future.

Much additional information on the digestibility of unconventional feedstuffs by aquacultured species and the bioavailability of the nutrients in these materials will be needed in the future to allow these ingredients to become viable alternatives to the primary protein supplements currently in use. Research is needed to maximize the use of alternative plant proteins in areas of the world where each of these alternative materials is locally abundant. This research should be conducted with aquatic animals of local or regional value to maximize the benefits to producers in different geographical areas.

An additional problem associated with the use of underutilized plant protein supplements, which often contain lower levels of CP than the ingredients they are intended to replace, is the potential for creation of dietary amino acid deficiencies caused by ingredient substitutions. Most alternative plant protein supplements possess poorer amino acid profiles than the ingredients they replace (i.e., fish meal and soybean meal). Table 15.3 provides an index of the relative value (i.e., replacement score) of selected plant protein supplements that could be used as substitutes for fish meal and soybean meal. Replacement scores (RS) are based on a comparison of the amino acid composition and protein content of each ingredient in relation to that of fish meal or soybean meal, as follows:

$$RS = (\Sigma EAA_{alt}/\Sigma EAA_{fish/soy}) \times (CP_{alt}/CP_{fish/soy}),$$

where
ΣEAA_{alt} = sum (%) of essential amino acids (EAA) in the alternative ingredient;
$\Sigma EAA_{fish/soy}$ = sum (%) of EAA in fish meal (average of anchovy, herring, and menhaden) or soybean meal (dehulled, solvent extracted);
CP_{alt} = amount (%) of crude protein (CP) in the alternative ingredient; and
$CP_{fish/soy}$ = amount (%) of CP in fish meal or soybean meal.

Replacement score values may range above or below 1.0 depending on the similarity between the amino acid composition and protein content of the alternative ingredient and that of fish meal or soybean meal. A score of 1.0 or higher suggests that the ingredient could be a suitable replacement for fish meal or soybean meal with little or no modification. Scores below 1.0 indicate the presence of amino acid deficiencies that should be corrected (e.g., by amino acid supplementation) before the ingredient is used as a primary source of dietary protein.

TABLE 15.3. Replacement scores of selected plant protein products.[a]

Ingredient	Fish meal[b]	Soybean meal[b]
Potato protein concentrate	1.58	2.89
Peanut meal	0.50	0.91
Sesame seed meal	0.41	0.74
Safflower seed meal	0.38	0.70
Sunflower seed meal	0.32	0.59
Linseed meal	0.25	0.47
Leucaena leaf meal	0.15	0.28
Fababean seed meal	0.14	0.26
Mustard seed meal	0.14	0.25
Lentil seed meal	0.14	0.25
Pea seed meal	0.11	0.21
Coconut meal	0.10	0.18
Palm kernel meal	0.06	0.11

[a]Replacement score = $(\Sigma AA_{alt}/\Sigma AA_{fish/soy}) \times (CP_{alt}/CP_{fish/soy})$, where ΣAA_{alt} = sum (%) of essential amino acids (ΣAA) in the alternative ingredient; $\Sigma AA_{fish/soy}$ = sum (%) of ΣAA in fish meal (average of anchovy, herring, and menhaden) or soybean meal (dehulled, solvent-extracted); CP_{alt} = amount (%) of CP in the alternative ingredient; and $CP_{fish/soy}$ = amount (%) of CP in fish meal or soybean meal.

[b]Scores in each column represent the relative value of an ingredient as a 1:1 replacement for fish meal or soybean meal, based on the ingredient's amino acid composition and protein content. A score of 1.0 or higher suggests that the ingredient could be a suitable replacement for fish meal or soybean meal with little or no modification. Scores below 1.0 indicate the presence of amino acid deficiencies that should be corrected (e.g., by amino acid supplementation) before the ingredient is used as a primary protein source.

Amino acid deficiencies can occur when ingredients with desirable amino acid profiles are replaced by feedstuffs possessing less favorable amino acid compositions. The best way to avoid this problem is to provide a mixture of dietary protein sources that, in combination, provide the quantities of essential amino acids needed to meet minimum dietary requirements. A less desirable approach is to supplement the diet with purified, free amino acids in the quantities needed to eliminate specific amino acid deficiencies. Additional research is needed to determine the amino acid requirements of many aquacultured species, as well as the bioavailability of amino acids in alternative protein supplements fed to these species, so high-quality diets can be formulated effectively from a wider range of ingredients at least cost.

Antinutritional factors are often present in plant protein products in quantities sufficient to inhibit growth when the material is used in a diet at the level required to provide benefit as an alternative protein source. Antinutritional factors are compounds in diet ingredients that produce physiological effects that result in reduced fish growth. A comprehensive discussion of the types of antinutritional factors present in plant protein products, their modes of action, and effects on fish physiology are beyond the scope of this chapter. The National Research Council (1993) provides a brief overview of some antinutritional factors of importance in fish diets, and a review of the literature generated in the decade since that report was published will yield considerable additional information on the subject. Nonetheless, additional research is needed to develop effective methods for reducing concentrations of antinutritional compounds in plant products to levels that are safe for aquatic animals.

A number of underutilized and unconventional plant protein products are currently available for use in aquafeeds, and new products will be developed as agricultural processing technologies continue to improve. Research aimed at maximizing the use of these materials in aquatic animal diets is already under way, but much work remains to be done before the use of unconventional products is optimized in prepared diets. As the biological and economic constraints limiting the use of these feedstuffs are overcome, some of today's unconventional protein sources could become staple ingredients in the aquafeeds of tomorrow.

REFERENCES

Adelizi, P.D., R.R. Rosati, K. Warner, Y.V. Wu, T.R. Muench, M.R. White, and P.B. Brown (1998). Evaluation of fish-meal free diets for rainbow trout, *Oncorhynchus mykiss. Aquaculture Nutrition* 4:255-262.

AFRIS (2004). Animal feed resources information system, *Brassica* spp., mustard, http://www.fao.org/ag/AGA/AGAP/FRG/afris/Data/491.HTM.

Allan, G.L., S. Parkinson, M.A. Booth, D.A.J. Stone, S.J. Rowland, J. Frances, and R. Warner-Smith (2000). Replacement of fish meal in diets for Australian silver perch, *Bidyanus bidyanus:* I. Digestibility of alternative ingredients. *Aquaculture* 186:293-310.

Amerio, M., C. Vignali, L. Castelli, L. Fiorentini, and E. Tibaldi (1998). Vegetable protein sources, protein evaluation indices and "ideal protein" of sea bream *(Sparus aurata). Rivista Italiana di Acquacoltura* 33:135-145.

Bautista-Teruel, M.N., P.S. Eusebio, and T.P. Welsh (2003). Utilization of feed pea, *Pisum sativum,* meal as a protein source in practical diets for juvenile tiger shrimp, *Penaeus monodon. Aquaculture* 225:121-131.

Booth, M.A. and G.L. Allan (2003). Utilization of digestible nitrogen and energy from four agricultural ingredients by juvenile silver perch *Bidyanus bidyanus. Aquaculture Nutrition* 9:317-326.

Booth, M.A., G.L. Allan, J. Frances, and S. Parkinson (2001). Replacement of fish meal in diets for Australian silver perch, *Bidyanus bidyanus:* IV. Effects of dehulling and protein concentration on digestibility of grain legumes. *Aquaculture* 196:67-85.

Borlongan, I.G., P.S. Eusebio, and T. Welsh (2003). Potential of feed pea *(Pisum sativum)* meal as a protein source in practical diets for milkfish *(Chanos chanos Forsskal). Aquaculture* 225:89-98

Brown, P.B., E.H. Robinson, A.E. Clark, and A.L. Lawrence (1989). Apparent digestible energy coefficients and associative effects in practical diets for red swamp crayfish. *Journal of the World Aquaculture Society* 20:122-126.

Brown, P.B., R.J. Strange, and K.R. Robbins (1985). Protein digestibility coefficients for yearling channel catfish fed high protein feedstuffs. *The Progressive Fish-Culturist* 47:94-97.

Burel, C., T. Boujard, F. Tulli, and S.J. Kaushik (2000). Digestibility of extruded peas, extruded lupin, and rapeseed meal in rainbow trout *(Oncorhynchus mykiss)* and turbot *(Psetta maxima). Aquaculture* 188:285-298.

Carter, C.G. and R.C. Hauler (2000). Fish meal replacement by plant meals in extruded feeds for Atlantic salmon, *Salmo salar* L. *Aquaculture* 185:299-311.

Chin, F.Y. (2001). Palm kernel cake (PKC) as a supplement for fattening and dairy cattle in Malaysia. In *Proceedings of the 7th Meeting of the Regional Working Group on Grazing and Feed Resources for Southeast Asia.* Manado, Indonesia, July 2001, http://www.fao.org/ag/AGP/AGPC/doc/Proceedings/manado/chap25.htm

Cruz-Suarez, L.E., D. Ricque-Marie, M. Tapia-Salazar, I.M. McCallum, and D. Hickling (2001). Assessment of differently processed feed pea *(Pisum sativum)* meals and canola meal *(Brassica* sp.) in diets for blue shrimp *(Litopenaeus stylirostris). Aquaculture* 196:87-104.

Dale, N. and A. Batal (2003). Ingredient analysis table, 2003 edition. *Feedstuffs 2003-04 Reference Issue and Buyers Guide* 75:16-17.

Davis, D.A., C.R. Arnold, and I. McCallum (2002). Nutritional value of feed peas *(Pisum sativum)* in practical diet formulations for *Litopenaeus vannamei. Aquaculture Nutrition* 8:87-94.

de la Higuera, M., H. Akharbach, M.C. Hidalgo, J. Peragón, J.A. Lupiáñez, and M. García-Gallego (1999). Liver and white muscle protein turnover rates in the

European eel *(Anguilla anguilla):* Effects of dietary protein quality. *Aquaculture* 179:203-216.

De Silva, S.S. and D.E.M. Weerakoon (1981). Growth, food intake and evacuation rates of grass carp, *Ctenopharyngodon idella* fry. *Aquaculture* 25:67-76.

ESCHA Research (2003). *Professional Nutrition Analysis Software and Databases.* Salem, Oregon.

Fagbenro, O.A. (1998). Apparent digestibility of various oilseed cakes/meals in African catfish diets. *Aquaculture International* 6:317-322.

FAO (2004). FAOSTAT, http://faostat.fao.org/default.jsp?language=EN.

Fontaínhas-Fernandes, A., E. Gomes, M.A. Reis-Henriques, and J. Coimbra (1999). Replacement of fish meal by plant proteins in the diet of Nile tilapia: Digestibility and growth performance. *Aquaculture International* 7:57-67.

Gomes, E.F., G. Corraze, and S. Kaushik (1993). Effects of dietary incorporation of a co-extruded plant protein (rapeseed and peas) on growth, nutrient utilization and muscle fatty acid composition of rainbow trout *(Oncorhynchus mykiss). Aquaculture* 113:339-353.

Gomes, E.F., P. Rema, and S.J. Kaushik (1995). Replacement of fish meal by plant proteins in the diet of rainbow trout *(Oncorhynchus mykiss):* Digestibility and growth performance. *Aquaculture* 130:177-186.

Gómez-Requeni, P., M. Mingarro, J.A. Calduch-Giner, F. Médale, S.A.M. Martin, D.F. Houlihan, S. Kaushik et al. (2004). Protein growth performance, amino acid utilization and somatotropic axis responsiveness to fish meal replacement by plant protein sources in gilthead sea bream *(Sparus aurata). Aquaculture* 232: 493-510.

Gouveia, A. and S.J. Davies (1998). Preliminary nutritional evaluation of pea seed meal *(Pisum sativum)* for juvenile European sea bass *(Dicentrarchus labrax). Aquaculture* 166:311-320.

Gouveia, A. and S.J. Davies (2000). Inclusion of an extruded dehulled pea seed meal in diets for juvenile European sea bass *(Dicentrarchus labrax). Aquaculture* 182:183-193.

Gouveia, A., A. Olivia Teles, E. Gomes, and P. Rema (1993). Effect of cooking/expansion of three legume seeds on growth and food utilization by rainbow trout. In *Fish Nutrition in Practice,* S.J. Kaushik and P. Luquet (eds.). Paris, France: Institut National de la Recherche Agronomique, pp. 933-938.

Guerrero, R.D. (1980). Studies on the feeding of *Tilapia nilotica* in floating cages. *Aquaculture* 20:169-175.

Hasan, M.R., D.J. Macintosh, and K. Jauncey (1997). Evaluation of some plant ingredients as dietary protein sources for common carp (*Cyprinus carpio* L.) fry. *Aquaculture* 151:55-70.

Hossain, M.A. and K. Jauncey (1989). Studies on the protein, energy and amino acid digestibility of fish meal, mustard oilcake, linseed and sesame meal for common carp (*Cyprinus carpio* L.). *Aquaculture* 83:59-72.

Jackson, A.J., B.S. Capper, and A.J. Matty (1982). Evaluation of some plant proteins in complete diets for the tilapia *Sarotherodon mossambicus. Aquaculture* 27:97-109.

Khan, M.A., A.K. Jafri, N.K. Chadha, and N. Usmani (2003). Growth and body composition of rohu *(Labeo rohita)* fed diets containing oilseed meals: Partial or total replacement of fish meal with soybean meal. *Aquaculture Nutrition* 9:391-396.

Lim, C. (1997). Replacement of marine animal protein with peanut meal in diets for juvenile white shrimp, *Penaeus vannamei. Journal of Applied Aquaculture* 7:67-78.

Lim, H.-A., W.-K. Ng, S.-L. Lim, and C.O. Ibrahim (2001). Contamination of palm kernel meal with *Aspergillus flavus* affects its nutritive value in pelleted feed for tilapia *Oreochromis mossambicus. Aquaculture Research* 32:895-905.

Mohanty, S.S. and K. Samantaray (1996). Effect of varying levels of dietary protein on the growth performance and feed conversion efficiency of snakehead *Channa striata* fry. *Aquaculture Nutrition* 2:89-94.

Moyano, F.J., G. Cardenete, and M. de la Higuera (1992). Use of two vegetable by-products as protein sources in rainbow trout feeding. *Animal Production* 55:277-284.

Mukhopadhyay, N. and A.K. Ray (1999). Improvement of quality sesame *Sesamum indicum* seed meal protein with supplemental amino acids in feeds for rohu *Labeo rohita* (Hamilton) fingerlings. *Aquaculture Research* 30:549-557.

Mukhopadhyay, N. and A.K. Ray (2001). Effects of amino acid supplementation on the nutritive quality of fermented linseed meal protein in the diets for rohu, *Labeo rohita*, fingerlings. *Journal of Applied Ichthyology* 17:220-226.

Munsiri, P. and R.T. Lovell (1993). Comparison of satiate and restricted feeding of channel catfish with diets of varying protein quality in production ponds. *Journal of the World Aquaculture Society* 24:459-465.

National Research Council (1983). Composition of feeds. In *Nutrient Requirements of Warmwater Fishes and Shellfishes, revised edition.* Washington, DC: National Academy Press, pp. 71-94.

National Research Council (1993). Composition of feed ingredients. In *Nutrient Requirements of Fish.* Washington, DC: National Academy Press, pp. 64-71.

Ng, W.K. (2003). The potential use of palm kernel meal in aquaculture feeds. *Aquaculture Asia* 8(1):38-39.

Omoregie, E. (2001). Utilization and nutrient digestibility of mango seeds and palm kernel meal by juvenile *Labeo senegalensis (Antheriniformes: Cyprinidae). Aquaculture Research* 32:681-687.

Omoregie, E. and F.I. Ogbemudia (1993). Effect of substituting fish meal with palm kernel meal on growth and food utilization of the Nile tilapia, *Oreochromis niloticus. The Israeli Journal of Aquaculture—Bamidgeh* 45:113-119.

Ogunji, J.O. and M. Wirth (2001). Alternative protein sources as substitutes for fish meal in the diet of young tilapia *Oreochromis niloticus* (Linn.). *The Israeli Journal of Aquaculture—Bamidgeh* 53:34-43.

Pfeffer, E., S. Kinzinger, and M. Rodehutscord (1995). Influence of the proportion of poultry slaughter by-products and of untreated or hydrothermically treated legume seeds in diets for rainbow trout, *Oncorhynchus mykiss* (Walbaum), on apparent digestibilities of their energy and organic compounds. *Aquaculture Nutrition* 1:111-117.

Refstie, S. and H.A.J. Tiekstra (2003). Potato protein concentrate with low content of solanidine glycoalkaloids in diets for Atlantic salmon *(Salmo salar)*. *Aquaculture* 216:283-298.

Riche, M. and P.B. Brown (1996). Availability of phosphorus from feedstuffs fed to rainbow trout, *Oncorhynchus mykiss. Aquaculture* 142:269-282.

Riche, M. and P.B. Brown (1999). Incorporation of plant protein feedstuffs into fish meal diets for rainbow trout increases phosphorus availability. *Aquaculture Nutrition* 5:101-105.

Sales, J. and P.J. Britz (2001). Review: Research on abalone (*Haliotis midae* L.) cultivation in South Africa. *Aquaculture Research* 32:863-874.

Sales, J. and P.J. Britz (2002). Evaluation of the reference diet substitution method for determination of apparent nutrient digestibility coefficients of feed ingredients for South African abalone (*Haliotis midae* L.). *Aquaculture* 207:113-123.

Sales, J. and P.J. Britz (2003). Apparent and true availability of amino acids from common feed ingredients for South African abalone (*Haliotis midae* L.). *Aquaculture Nutrition* 9:55-64.

Santiago, C.B., M. Bañes-Aldaba, and E.T. Songalia (1983). Effect of artificial diets on growth and survival of milkfish fry in fresh water. *Aquaculture* 34:247-252.

Sanz, A., A.E. Morales, M. de la Higuera, and G. Cardenete (1994). Sunflower meal compared with soybean meal as partial substitutes for fish meal in rainbow trout *(Oncorhynchus mykiss)* diets: Protein and energy utilization. *Aquaculture* 128: 287-300.

Sethi, P. and P.R. Kulkarni (1995). Food-science: *Leucaena leucocephala* A nutrition profile. *Food and Nutrition Bulletin* 16(3). The United Nations University Press, Tokyo, Japan, http://www.unu.edu/unupress/food/8F163e/8F163E08.htm.

Small, B.C., R.E. Austic, and J.H. Soares, Jr. (1999). Amino acid availability of four practical feed ingredients fed to striped bass *Morone saxatilis. Journal of the World Aquaculture Society* 30:58-64.

Tang, T.S. (no date). Quality and characteristics of Malaysian palm kernel cakes/expellers. Palm Oil Information Online Service, http://palmoilis.mpob.gov.my/webbased/pod34-tang.pdf.

Telek, L. (1983). Leaf protein extraction from tropical plants. In *Plants: The Potentials for Extracting Protein, Medicines, and Other Useful Chemicals—Workshop Proceedings*. U.S. Congress, Office of Technology Assessment, OTA-BP-F-23, September 1983, Washington, DC, pp. 78-110.

USDA (2004). Production, supply and distribution (PS&D) online database, oilseeds pre-defined tables, http://www.fas.usda.gov/psd/intro.asp?circ_id=2.

Wilson, R.P. and W.E. Poe (1985). Apparent digestible protein and energy coefficients of common feed ingredients for channel catfish. *The Progressive Fish-Culturist* 47:154-158.

Xie, S. and A. Jokumsen (1997a). Incorporation of potato protein concentrate in diets for rainbow trout: Effect on feed intake, growth and feed utilization. *Aquaculture Nutrition* 3:223-226.

Xie, S. and A. Jokumsen (1997b). Replacement of fish meal by potato protein concentrate in diets for rainbow trout, *Oncorhynchus mykiss* (Walbaum): Growth, feed utilization and body composition. *Aquaculture Nutrition* 3:65-69.

Xie, S. and A. Jokumsen (1998). Effects of dietary incorporation of potato protein concentrate and supplementation of methionine on growth and feed utilization of rainbow trout. *Aquaculture Nutrition* 4:183-186.

Zarate, D.D. and R.T. Lovell (1997). Free lysine (L-lysine.HCl) is utilized for growth less efficiently than protein-bound lysine (soybean meal) in practical diets by young channel catfish *(Ictalurus punctatus)*. *Aquaculture* 159:87-100.

Zarate, D.D., R.T. Lovell, and M. Payne (1999). Effects of feeding frequency and rate of stomach evacuation on utilization of dietary free and protein-bound lysine for growth by channel catfish *Ictalurus punctatus*. *Aquaculture Nutrition* 5:17-22.

Chapter 16

Use of Distillers Grains with Solubles and Brewery By-Products in Fish and Crustacean Diets

Carl D. Webster
Kenneth R. Thompson
Linda S. Metts
Laura A. Muzinic

INTRODUCTION

Intensively cultured fish and crustaceans rely on prepared diets to be fed to the cultured organism. These diets need to supply all the nutrients required by the organism for growth, reproduction, health, and other physiological functions. The nutrients required by fish and crustaceans are similar to those of land animals, for example, protein, essential amino acids (EAAs), minerals, vitamins, energy, and essential lipids. Proteins are the principal organic constituent of animal tissue and are the most expensive component in fish diets. Since body protein is constantly undergoing two major processes, protein synthesis (anabolism) and protein breakdown (catabolism), animals need a continuous supply of protein throughout life. Inadequate intake of protein will result in retardation, cessation of growth, or loss of weight due to the withdrawal of protein from less vital tissues to maintain the function of more vital ones. If too much protein is supplied, however, only part will be used to synthesize new tissues and the remainder will be converted to energy or excreted (NRC, 1993).

Alternative Protein Sources in Aquaculture Diets
© 2008 by The Haworth Press, Taylor & Francis Group. All rights reserved.
doi:10.1300/5892_16

Fish do not have an absolute protein requirement per se but have a requirement for EAAs. Furthermore, the diet should have a well-balanced mixture of essential and nonessential amino acids. Numerous investigators have utilized semipurified, purified, and practical diets to determine the protein requirements of numerous fish and crustacean species. These values have been obtained mostly by measuring the growth response of fish fed diets containing graded levels of good quality protein sources.

It has been reported that the dietary protein requirements of fish is affected by a number of factors such as size or age of the fish, protein quality, nonprotein energy level, water temperature and salinity, presence of natural foods in the culture environment, and diet allowance (NRC, 1993), but limited studies have been conducted to determine the effect of these parameters on protein requirement of fish. Younger or smaller fish have a higher protein requirement than older or larger fish. Higher dietary levels of protein are required if poor quality protein sources are used. Insufficient nonprotein energy in the diet will lead to higher dietary protein requirements because fish will utilize part of the protein as energy to meet their metabolic energy needs.

Energy is not a nutrient but is a property of nutrients that are released during metabolic oxidation of proteins, carbohydrates, and lipids. Generally, protein is given the first priority in formulating fish diets because it is the most expensive component of the prepared feeds. However, energy is important since fish eat to satisfy their energy needs. If insufficient nonprotein energy is available, part of the protein will be broken down into energy. An excess of energy in the diet can limit consumption, thus reducing the intake of protein and other nutrients.

Aquaculture has become the fastest growing agriculture industry over the past three decades in the United States and is presently worth approximately US$1 billion annually, placing it in the top-thirty agricultural industries in the United States. Globally, aquaculture has been expanding by 9 percent per year over the past ten years, with production in Asia increasing by 85 percent during that time. Worldwide, aquaculture is valued at US$70 billion and this value is increasing. However, there are numerous challenges to the global aquaculture industry including low-cost imports, high diet and energy costs, and lower prices paid to producers for their products. Diet costs represent a significant expenditure in most aquaculture operations. One way of reducing diet costs is to decrease the use of marine fish meal (FM) in the diet. Marine FM is considered the most desirable animal protein ingredient in diets for fish and crustaceans due to its high protein content, balanced amino acid profile, high digestibility and palatability, and as a source of essential n-3 polyenoic fatty acids. However, FM is one of the most expensive macroingredients (used in high percentages) in an aquaculture diet, currently costing approximately $1,000/ton in the United States. With the static or declining fish populations that are used to produce FM, any negative disturbance, supply disruption, or availability problems lead to dramatic increases in the commodity price. Furthermore, the capture of wild fish to be used to feed farm-raised fish is thought to be unsustainable by many critics of aquaculture.

The International Fish Meal and Fish Oil Organization (IFFO) has reported that the annual global production of FM is in the range of 6-7 million tons and a little below 1 million tons for fish oil. As a result, this demands an annual catch of 25 to 30 million tons of feed-grade fish and unwanted fish processing waste. In 2004, the estimated global FM production was 6.33 million tons with a firsthand sales turnover exceeding $US3.0 billion, while global production of fish oil is estimated at approximately 930,000 tons, representing a firsthand sales turnover of around $US560 million, as reported by IFFO. Aquaculture's share of FM and marine fish oil usage is forecast to expand from 34 percent and 56 percent, respectively, to 48 percent and 79 percent of global production by 2010. While various organizations may or may not present distortions with the effects of aquaculture on world fishery supplies, one fact is inescapable; as the global aquaculture industry expands, it must dramatically reduce or eliminate its reliance on FM as a diet ingredient. There has been substantial progress made in this area over the past ten to fifteen years.

One approach to partially, or totally, eliminate FM use from aquaculture diets is to replace it with alternative, less expensive, animal- and/or plant-protein ingredients that will help reduce diet costs, potentially increase profits for producers, and allow for continued expansion of the global aquaculture industry by utilizing renewable ingredients.

The nutritional value of protein sources, commonly referred to as protein quality, is determined based on the EAA content and their digestibility (or bioavailability). A protein source having an EAA profile that closely matches the EAA requirements of fish is likely to have high nutritional value. Protein and amino acid digestibility of good quality FM is relatively high for most fish species. Fish meal is also a good source of essential fatty acids and minerals. Currently FM still constitutes a substantial part of diet formulations of many aquaculture species. However, the increased demand for FM coupled with the rising cost and uncertain availability have initiated much interest in evaluating alternative protein ingredients to partially or totally replace FM in diets of various aquaculture species.

Animal by-product meals such as poultry by-product meal (PBM), hydrolyzed feather meal (HFM), meat and bonemeal (MBM), and blood meal (BM) have been evaluated as substitutes for FM with varying results, generally due to differences in ingredient quality or deficiencies/imbalances in EAA. Among plant-protein sources, soybean meal (SBM), because of its availability, consistent quality, high protein content with good amino acid profile, and low cost, is the most studied plant feedstuff in aquaculture diets. Cottonseed meal (CSM) is an important protein source for domestic animals, but its use in fish diets is limited due to the presence of high fiber and other indigestible carbohydrates, antinutritional factors (protease inhibitors), and phenolic compounds (such as glucosinolates, sinapine, tannins, and phytic acid).

The reasons for low nutritional value of plant proteins as compared to FM have not been thoroughly studied, but several hypotheses have been suggested: (1) presence of antinutritional factors or toxic substances; (2) improper balance of essential nutrients such as amino acid, energy, and minerals; (3) presence of high amounts of fiber and indigestible carbohydrates; (4) decreased palatability of the diet; and (5) reduction of pellet quality especially its water stability.

DISTILLERS DRIED GRAINS WITH SOLUBLES

In the early 1900s, some distillers also owned feedlots and fed their cattle the mash by-product from alcohol fermentation. It was not until the 1940s, however, that distillers' feeds were recognized for use in dairy cows. Research at Cornell University indicated that distiller feeds were highly palatable and had a low starch content. This allows for up to 40 percent distillery by-products to be added in dairy cattle diet formulations. There are actually four distillery by-products: (1) distiller's dried grains with solubles (DDGS), which is the product after removal of ethyl alcohol by distillation from yeast fermentation of grain or grain mixture by condensing and drying at least 75 percent of the solids from the whole stillage; (2) distiller's dried grains (DDG) is obtained after the removal of ethyl alcohol by distillation from yeast fermentation of grain by separating the resultant course grain fraction of the whole stillage and drying it; (3) distiller's dried solubles (DDS) is obtained after the removal of ethyl alcohol by distillation from yeast fermentation of grain by condensing the thin stillage fraction and drying it; and (4) condensed distiller's solubles (CDS) which is obtained after the removal of ethyl alcohol by distillation from yeast fermentation of grain by condensing the thin stillage fraction to a semisolid. For this chapter, discussion will be generally confined to the use of DDGS in aquaculture diets as this is the most readily available product.

For many years, the majority of DDGS was produced from beverage alcohol distillation. However, ethanol production (either using corn or milo) has tremendously increased the amount of DDGS produced in the United States. In 2002, the United States produced more than 1.6 billion gallons of ethanol. Most ethanol is corn based. For every bushel of corn (25.5 kg), 2.5 gallons of ethanol and 8.1 kg of DDGS are produced. According to the American Coalition for Ethanol, more than 1 billion gallons of ethanol was blended with gasoline in 2002 and is used in more than 11 percent of all gasoline in the United States. This percentage will likely rise as more ethanol will be added to gasoline to be more environmentally friendly, and to conserve oil. One gallon of ethanol produced domestically displaces seven gallons of imported oil. In 2005, ethanol production increased to 2.7 billion gallons; however, under the new energy bill, ethanol production is proposed to increase to 5.0 billion gallons by 2015.

Currently, DDGS is a stable, free-flowing granular product that is yellow/tan to brown/black in color with a bulk density of 14 to 19 kg per cubic foot. Product color varies with variations in the distillation process. While it used to be stated that there was no correlation with the color of DDGS and nutrient quality, Cromwell et al.

(1993) reported that swine fed a diet containing a dark-colored DDGS had reduced growth compared to animals fed a diet containing a light-colored DDGS. The dark (brown/black) color was probably a result of overheating the DDGS during the drying process. Excessive heating binds lysine and partially destroys amino acids. It was also reported that fuel alcohol-derived DDGS had nutritional values that were within the range of beverage alcohol-produced DDGS, indicating that there appears to be no difference in how DDGS are produced. However, there are subtle differences between corn DDGS and milo DDGS in terms of nutrient composition.

Proximate composition and energy values for corn DDGS are reported in Table 16.1; amino acid composition of corn DDGS are reported in Table 16.2; vitamin and mineral composition of corn DDGS are reported in Table 16.3; proximate composition and energy values for livestock for milo-derived DDGS are reported in Table 16.4; amino acid composition of milo DDGS are reported in Table 16.5; and vitamin and mineral composition of milo DDGS are reported in Table 16.6.

Distiller's dried grains with solubles has been shown to be a suitable ingredient in diets for turkeys at inclusion rates of 3 to 5 percent (Atkinson et al., 1955; Manley et al., 1978); dairy cows (Hawkins and

TABLE 16.1. Percentage dry matter, protein, lipid, fiber, ash, and energy values of distillers dried grains with solubles (corn DDGS).

Dry matter	91.0
Protein[a]	27.0
Lipid[a]	9.0
Fiber[a]	9.1
Ash[a]	4.9
ME (cattle; Mcal/lb)	3.04
ME (swine; Mcal/lb)	3.39
ME (poultry; Mcal/lb)	2.75
DE (horses; Mcal/lb)	3.49
DE (rainbow trout; kcal/kg)	2,265

Note: ME = metabolizable energy; DE = digestible energy.

[a]Dry-matter basis.

TABLE 16.2. Amino acid composition of distillers dried grains with solubles (corn DDGS).

Amino acid	%
Alanine	1.90
Arginine	1.12-1.31
Aspartic acid	1.70
Cystine	0.46-0.76
Glutamic acid	4.20
Glycine	1.00
Histidine	0.64-0.90
Isoleucine	1.04-1.52
Leucine	2.89-3.77
Lysine	0.55-0.99
Methionine	0.50-0.76
Phenylalanine	1.40-1.66
Proline	2.80
Serine	1.00
Threonine	0.93-1.15
Tryptophan	0.10-0.25
Tyrosine	0.80-1.00
Valine	1.45-1.69

Note: Ranges are provided due to differences in distilleries processes used to make DDGS.

Little, 1967); beef cattle (Horn and Beeson, 1968); chicken at inclusion levels of 5 to 10 percent (Marvel et al., 1946; Harms et al., 1970); as well as use in diets for ducks, swine, rabbits, and dogs. There are no antinutritional factors in DDGS other than the possibility of mycotoxin contamination. If contaminated corn is used to produce ethanol or beverage alcohol, the mycotoxin content of the DDGS will be two to three times higher than the concentration found in the contaminated corn because of the removal of starch. Mycotoxins resist degradation during the fermentation process; however, aflatoxins can be reduced by 60 percent by heating when a floating diet is made. If longer extruder times are allowed in making the diet, a destruction of 90 percent of aflatoxins was recorded (personal communication; Bruce Manning, Mississippi State University).

TABLE 16.3. Vitamin and mineral composition of distillers dried grains with solubles (corn DDGS).

	Amount
Vitamins	
A (IU/kg)	1,001.0
D (IU/kg)	605.0
E (IU/kg)	44.0
B_{12} (mg/kg)	0.0015
Biotin (mg/kg)	0.3-0.9
Choline (mg/kg)	2,803.0
Folic acid (mg/kg)	1.0
Inositol (mg/kg)	3,200.0
Niacin (mg/kg)	35.0
Pantothenic acid (mg/kg)	11-16
Pyridoxine (mg/kg)	6.5
Riboflavin (mg/kg)	77.0
Thiamin (mg/kg)	3,500.0
Minerals	
Calcium (%)	0.05-0.16
Chloride (%)	0.18
Cobalt (mg/kg)	0.18
Copper (mg/kg)	3-10
Iron (mg/kg)	78-240
Magnesium (%)	0.14-0.39
Manganese (mg/kg)	9-27
Phosphorus (%)	0.5-0.99
Potassium (%)	0.45-1.27
Selenium (mg/kg)	0.42
Sodium (%)	0.04-0.27
Sulfur (%)	0.34-0.89
Zinc (mg/kg)	38-105

Use of DDGS in Finfish Diets

Distiller's dried grains with solubles have been added to fish diets since the late 1940s; however, until recently, inclusion levels have been fairly low. Phillips (1949) formulated a diet for rainbow trout

TABLE 16.4. Percentage protein, lipid, fiber, ash and digestible energy (DE) values of distillers dried grains with solubles (milo DDGS).

Nutrient	%
Protein[a]	27.0
Lipid[a]	7.0
Fiber[a]	8.0
Ash[a]	5.0
DE (swine; Mcal/kg)	2.79
DE (poultry; Mcal/kg)	2.62

[a]Dry-matter basis.

TABLE 16.5. Amino acid composition of distillers dried grains with solubles (milo DDGS).

Amino acid	%
Arginine	2.40
Cystine	0.30
Histidine	0.69
Leucine	4.98
Lysine	0.95
Methionine	0.50
Isoleucine	0.92
Phenylalanine	1.47
Threonine	0.92
Tyrosine	1.20
Valine	1.07

(Oncorhynchus mykiss) that contained a 1:1 mixture of dry ingredients (which included wheat midds, CSM, FM, and DDGS) and liver or beef spleen. Sinnhuber (1964) reported that the Oregon Fish Commission used 3 percent DDGS in the moist pellets fed to trout and salmonids. Phillips et al. (1964) reported that a dry diet formulated with 24 percent FM, 5 percent CSM, 10 percent brewer's yeast, and 21 percent DDGS not only provided in brown trout *(Salmo trutta)* growth similar to that in fish fed meat-meal mixtures, but also were

TABLE 16.6. Vitamin and mineral composition of distillers dried grains with solubles (milo DDGS).

	Amount
Vitamins	
A (IU/kg)	1,725.0
E (IU/kg)	9.0
B_{12} (mg/kg)	35.2
Biotin (mg/kg)	1.7
Choline (%)	2,253.0
Folic acid (%)	0.20
Niacin (mg/kg)	130.0
Pantothenic acid (%)	12.2
Pyrodoxine (%)	2.2
Riboflavin (mg/kg)	11.7
Thiamin (%)	3,300.0
Minerals	
Calcium (%)	0.15
Cobalt (mg/kg)	5.00
Copper (mg/kg)	26.0
Iodine (mg/kg)	0.05
Iron (mg/kg)	380.0
Magnesium (%)	0.35
Manganese (mg/kg)	45.0
Phosphorus (%)	0.78
Potassium (%)	0.94
Sodium (%)	0.10
Sulfur (%)	0.03
Zinc (mg/kg)	63.0

the first dry diets to produce viable trout eggs in broodstock females with 86 percent total hatch and 88 percent survival to the eyed stage. Previously, dry diets had provided for hatchability percentages between 4 and 5 percent.

In warmwater fish, Deyoe and Tiemeier (1969) reported that feeding studies indicated that a diet containing 10 percent DDGS and 25 percent protein could be fed to channel catfish *(Ictalurus punctatus).*

Tidwell et al. (1990) reported that channel catfish fed a 32 percent-protein diet containing 30 percent DDGS, 8 percent menhaden FM, 30 percent SBM, and 18 percent corn had similar final individual weight, feed conversion ratio (FCR), and total length as fish fed a diet containing 32 percent protein, 8 percent menhaden FM, 50 percent SBM, and 38 percent corn.

Webster et al. (1991) fed small (10 g) channel catfish diets containing various percentages (0, 35, and 70 percent) of DDGS for twelve weeks in aquaria. As the diet containing 70 percent DDGS would be deficient in lysine, 0.40 percent L-lysine was added. After twelve weeks, channel catfish fed a diet containing 70 percent DDGS were significantly ($P < 0.05$) smaller (44 g) compared to fish fed all other diets, but fish fed the diet containing 70 percent DDGS and 0.4 percent L-lysine had similar final individual weights (74 g) compared to fish fed diets containing 0 percent and 35 percent DDGS (62 and 79 g, respectively). Furthermore, fish fed a diet containing 70 percent DDGS had significantly lower percentage weight gain, specific growth rate (SGR), and FCR of 232 percent, 1.47 percent/day, and 4.0, respectively.

Addition of L-lysine to a diet containing 70 percent DDGS increased growth of channel catfish so that there was no significant difference ($P > 0.05$) between fish fed the lysine-supplemented diet and fish fed the control diet. Whole-body analysis of fish fed diets containing DDGS showed significantly ($P < 0.05$) higher lipid levels (24.7 percent) compared to fish fed the control diet without DDGS (14.6 percent lipid). This could be due to an increase in digestible energy supplied by DDGS.

Webster, Tidwell, Goodgame, Yancey, et al. (1992) stated that a diet containing a combination of plant-protein sources, instead of one protein source, such as the usual practice of replacing marine FM in fish diets, may allow for complete FM replacement. In a feeding study, channel catfish were fed a diet containing 0 percent FM, 49.9 percent SBM and 35 percent DDGS, while other fish were fed a diet containing 12 percent menhaden FM and 48 percent SBM. After twelve weeks, fish fed the diet without FM had growth and body composition similar to fish fed the diet containing FM. Lysine and methionine percentages were similar among all diets and met requirements for channel catfish.

As channel catfish grow less during cold, winter months compared to warm/hot summer months, diets can be formulated with less protein than that used during the growing season. In Kentucky, two diets containing 26 percent protein were formulated to contain either 90 percent DDGS and 4 percent SBM, or 12 percent menhaden FM, 30 percent SBM, and 53 percent corn. Fish were stocked in October and fed until April. At the conclusion of the feeding trial, no differences ($P > 0.05$) in pond yield, final individual weight, survival, percentage weight gain, dress-out percentages, and body composition were found between the two diets (Webster, Tidwell, Goodgame, Clark et al., 1992).

With the positive results from the aquarium studies, it was decided that a production trial should be conducted where channel catfish were fed practical, commercial diets in which DDGS was included at various percentages. Channel catfish were stocked into floating cages at a density of 320 fish/m^3 and fed to satiation twice daily for 110 days. The four floating (extruded) diets contained 0, 10, 20, and 30 percent DDGS. The diet with 0 percent DDGS was formulated similar to the commercial catfish diet and served as the control. At harvest, no significant differences ($P > 0.05$) were found in final individual weight, SGR, FCR, and dressing percentage among treatments and averaged 219 g, 1.53 percent/day 1.59, and 53.5 percent, respectively (Webster et al., 1993). Organoleptic evaluation of the fillets indicated that use of DDGS in the diet imparted no adverse taste to channel catfish up to 30 percent inclusion. Furthermore, use of all percentages of DDGS evaluated in the study resulted in no difficulty in producing a floating diet (Webster et al., 1993).

The inclusion of DDGS to warmwater fish diets extends beyond its use in channel catfish diets. Webster et al. (1999) reported that sunshine bass *(Morone chrysops* $\times$ *M. saxatilis)* fed a diet containing 0 percent FM, 29 percent SBM, 29 percent MBM, and 10 percent DDGS had similar final weight, percentage weight gain, survival, SGR, and FCR compared to fish fed the diet containing 30 percent FM. However, sunshine bass fed a diet with 30 percent SBM and 30.5 percent MBM had significantly lower ($P < 0.05$) growth performance and higher FCR compared to fish fed the diet containing 30 percent FM. This feeding trial confirmed an earlier report that use of two plant-protein sources may have greater success in providing for suitable fish growth when FM is totally replaced, than when a diet with one plant-protein source is used (Webster, Tidwell, Goodgame, Yancey et al., 1992). Use of DDGS in sunshine bass diets may have improved the palatability of the diet containing SBM and MBM.

Addition of DDGS to diets for tilapia has also been investigated. In a feeding trial involving hybrid tilapia *(Oreochromis niloticus × O. aureus),* four diets were formulated. Diet 1 contained 12 percent FM and 41 percent SBM; Diet 2 contained 30 percent DDGS, 8 percent FM, and 34 percent SBM; Diet 3 contained 30 percent DDGS, 0 percent FM, 26 percent MBM, and 16 percent SBM; and Diet 4 contained 30 percent DDGS and 46 percent SBM. After ten weeks, hybrid tilapia fed Diets 1-3 had similar ($P > 0.05$) average weight gain, SGR, and FCR; however, fish fed Diet 4 had significantly lower ($P < 0.05$) weight gain and SGR, and higher FCR than fish fed Diets 1-3 (Coyle et al., 2004).

While use of DDGS in many warmwater freshwater fish diets appears feasible, its lower protein level compared to FM, SBM, and animal by-product meals appears to preclude it from inclusion in cold-water carnivorous fish diets. However, Cheng and Hardy (2004a) fed rainbow trout diets containing various (7.5, 15, and 22.5 percent) percentages of DDGS replacing 25, 50, and 75 percent of FM on an isonitrogenous and isocaloric basis. A FM-based diet containing 30 percent herring meal served as the control. Since the diets with DDGS may not meet lysine and methionine requirements of rainbow trout, three additional diets were formulated to contain supplemental lysine and methionine. Fish (average weight of 50 g) were fed for six weeks.

After six weeks, the authors reported that a diet containing 22.5 percent DDGS had significantly ($P < 0.05$) lower final weight and weight gain, and significantly higher FCR, than fish fed the control diet and diets containing 7.5 percent and 15 percent DDGS (Cheng and Hardy, 2004a). However, when diets were supplemented with lysine and methionine, there was no significant difference in growth of rainbow trout. Indeed, fish fed the diet containing 15 percent DDGS with supplemental lysine and methionine had the highest apparent protein retention. Results of this feeding trial show that DDGS is a good protein source for rainbow trout and can be used at 15 percent inclusion to replace 50 percent of the FM on an isonitrogenous basis. When lysine and methionine are added, DDGS can be added at 22.5 percent of the diet (Cheng and Hardy, 2004a). Supplementation of crystalline amino acids may allow for use of DDGS in rainbow trout diets, along with other protein sources, that may completely replace FM.

If DDGS is to be added to fish diets to partially or totally replace FM, phosphorus content and availability will need to be investigated, especially if other plant-protein sources are used in the diet. This is because approximately two-thirds of the total phosphorus in plants is present as phytate, which is not readily available to fish. Since excessive dietary phosphorus will not be retained by fish, it will be excreted to the water and result in eutrophication of the culture water, which could be a problem. Addition of phytase, an enzyme specifically to hydrolyze phytate to available forms of phosphorus, to diets may allow for increased digestibility of phytate-phosphorus by fish. This may allow for diets to be formulated with reduced added phosphorus, thus reducing the total phosphorus excreted into the culture water, which reduces potential pollution.

Cheng and Hardy (2004b) added various levels (0, 300, 600, 900, and 1,200 microbial phytase units—FTU/kg of diet) of phytase to diets for rainbow trout (average weight of 129 g). These diets were fed to fish to determine apparent digestibility coefficients (ADCs) of various dietary components (protein, energy, lipid, dry matter, minerals, and amino acids). It was reported that the addition of phytase to the diet increased digestibility of nutrients; however, supplementation in excess of 900 FTU/kg of diet did not result in further improvement in terms of ADCs for nutrients, except that ADCs of phytate-phosphorus and manganese were increased at 1,200 FTU/kg of diet. These authors were the first to report ADCs of DDGS for rainbow trout, which will assist in the formulation of diets with DDGS and other plant-protein sources as partial or total replacement of FM.

Extrusion processing uses a combination of moisture, pressure, friction, and temperature to cook the diet, which increases the digestibility of plant-protein sources and starch. As a result, the extruded diet will have a higher nutrient digestibility and water stability. Extrusion cooking may also improve dietary fiber digestibility. As DDGS contains a moderate amount (7 to 11 percent) of fiber, if extrusion processing can enhance fiber digestibility of DDGS, potential of the ingredient for use in more carnivorous fish diets could be enhanced. In a feeding study using rainbow trout, diets were produced by either cold-pelleting or cooking extrusion. Diets contained a combination of corn gluten meal (CGM) and corn DDGS to replace 0, 25, 50, and 75 percent of the FM (Stone et al., 2005).

Results indicated that rainbow trout fed the control diet had significantly better growth performance, FCR, and protein and energy retention efficiencies than fish fed any of the cold-pelleted diets or extruded diets containing CGM and DDGS (Stone et al., 2005), but that inclusion of corn products up to 18 percent, which equates to 25 percent FM replacement, was practical. These data are in agreement with data reported in Cheng et al. (2003) who stated that a diet containing 18.5 percent DDGS could be added to a rainbow trout diet if 0.55 percent methionine hydroxyl analog (MHA) was also added. It was suggested that new DDGS processing techniques, possibly to alter the nonstarch polysaccharides (NSP) component so that available energy is increased, might be desirable so as to allow for higher rates of its inclusion in rainbow trout diets.

Use of DDGS in Crustacean Diets

With its moderate protein level DDGS appears to be a desirable ingredient for many formulated freshwater crustacean diets. Most freshwater crustaceans are herbivorous/omnivorous in nature and can utilize plant materials efficiently. One species of commercial interest globally is the freshwater Malaysian prawn *(Macrobrachium rosenbergii)*. There have been several reports on the use of DDGS in prepared diets for freshwater prawn. In one feeding trial, juvenile prawn were stocked into ponds and fed one of three diets containing 32 percent protein with either 0, 20, or 40 percent DDGS as partial replacement of SBM and corn. All three diets contained 15 percent FM. After 101 days of feeding, mean individual weight, survival, yield, and FCR were not significantly different ($P > 0.05$) among prawn fed any of the diets (Tidwell, Webster, Clark et al., 1993). Furthermore, organoleptic evaluation of prawn indicated no adverse effects on sensory characteristics of tail muscle at any level of DDGS inclusion. These results suggested that DDGS was a suitable ingredient for use in freshwater prawn diets.

While the previously mentioned study concerned partial replacement of SBM and corn in a prawn diet by DDGS, more economic impact would be possible if FM could be partially or totally replaced by DDGS. Therefore, in a second feeding study, juvenile prawn (0.5 g) were stocked into earthen ponds and fed one of three diets containing 32 percent protein and various percentages of marine FM (0, 7.5 and 15 percent) with 40 percent DDGS added to partially and totally replaced FM (Tidwell, Webster, Yancey, et al., 1993). Prawn were fed for four months. At harvest, individual final weight, survival, yield, and FCR were not different ($P > 0.05$) among all treatments, indicating that a diet containing 0 percent FM, 26 percent SBM, and 40 percent DDGS can be fed to prawn without adverse effects on growth and feed conversion. Totally replacing FM could allow for less expensive, more sustainable diets to be used to feed freshwater pawn.

Fatty acid and amino acid composition of eggs, tail muscle, and midgut gland of freshwater prawn fed a diet with 7.5 percent FM and 40 percent DDGS are different from the composition of small juvenile freshwater prawn, in that the amounts of lysine, methionine, arginine, histidine, and n-3 highly unsaturated fatty acids (HUFA) appear to be lower than possibly required. However, diets used in the feeding trial appear to be suitable for pond culture of prawn based upon growth data (Tidwell et al., 1998).

Tidwell et al. (1995) reported in a third feeding trial, prawn were fed a complete diet with 15 percent marine FM and 32 percent protein, or a diet containing 32 percent protein, 0 percent FM, and 40 percent DDGS (SBM was added to both diets) without vitamin or mineral supplements. Prawn were stocked into ponds at a density of 39,520/ha and grown in earthen ponds for four months. At harvest, prawn fed the complete diet had higher individual weight and yield than prawn fed the diet without vitamin and mineral supplements; however, growth of prawn fed the incomplete diet was 90 percent of those fed the complete diet. The data could be of use to producers who wish to feed a less costly diet and have markets for a slightly smaller prawn.

Although much of the literature regarding the use of DDGS in aquaculture diets has been on corn DDGS, milo-, wheat-, and barley-based DDGS can be included in aquaculture diets as well. Wu (1986) reported that barley-based distiller's grains (BDG) had approximately 33 percent crude protein, 6 percent lipid, 4.5 percent ash, and 16.6 percent crude fiber; similar composition to corn and milo DDGS. In a feeding trial with juvenile white shrimp *(Litopenaeus vannamei)*,

Molina-Poveda and Morales (2004) fed shrimp diets containing various (0, 3, 24, 44, and 65 percent) percentages of a wheat gluten (WG)-BDG blend replacing 33, 66, or 100 percent of the marine animal protein sources (shrimp head meal, FM, and squid meal). The blend had a WG:BDG ratio of 1:1 except for the control diet, which had 0 percent BDG included but 2.95 percent WG added.

After forty-five days, percentage weight gain of shrimp fed the control diet was not significantly different ($P > 0.05$) from shrimp fed the diet which replaced 33 percent of the marine animal proteins with WG-BDG; however, it was significantly higher ($P < 0.05$) than percentage weight gain of shrimp fed the diets replacing 66 percent and 100 percent of the marine animal proteins with WG and BDG (Molina-Poveda and Morales, 2004). Growth reduction in shrimp fed the latter two diets may have been due to a deficiency in EAAs and/or reduced palatability.

BREWER'S GRAINS (WITH YEAST)

Brewer's grains, brewer's grains with yeast (BGY), malt protein flour (MPF), and brewer's yeast are all by-products of the brewery industry. Brewer's grains and BGY both possess the grain from fermentation of brewery products (beer and malt liquor) without, or with, the yeast added. Nutrient composition of the two ingredients is dependent on the addition and amount of yeast (if any) added, as yeast has a higher protein level than grains. Typical nutrient composition of BGY is given in Tables 16.7 and 16.8. Malt protein flour is made from brewer's spent grain (BSG), which is low in protein and high in fiber. Mechanical removal of the husk of BSG with a rolling mill and sieve produces MPF, which has higher protein content (50 percent).

Brewer's yeast are single-cell protein (SCP) that are highly nutritious with high levels of protein, B-vitamins, pigments, and complex carbohydrates. Some yeast, like brewer's yeast *(Saccharomyces cerevisiae)* are also believed to possess some immunostimulating properties due to the complex carbohydrate components and nucleic acid content. Three components in brewer's yeast that may have immunomodulating effects include: (1) beta-glucans, which have been shown to enhance immune responses and disease resistance in several fish species (Gatlin, 2002); (2) nucleotides, which have been reported to

TABLE 16.7. Percentage moisture, protein, lipid, fiber, and amino acid composition of brewer's grains with yeast (BGY).

Nutrient	%
Moisture	10.0
Protein[a]	35.0
Lipid[a]	5.0
Fiber[a]	9.5
Arginine	1.83
Cystine	0.58
Histidine	0.85
Isoleucine	1.45
Leucine	3.46
Lysine	1.63
Methionine	0.62
Phenylalanine	2.03
Threonine	1.37
Tryptophan	0.38
Valine	2.05

[a]Dry-matter basis.

improve disease resistance of fish; and (3) chitin, which has been reported to have immunostimulating effects. While the addition of yeast to brewer's grains improves nutrient composition, use of pure brewer's yeast does have some disadvantages when included in modest to high levels in fish diets, other than its high cost. One problem is that yeast, as most SCP, is deficient in one or more EAA and/or has amino acid imbalances. This situation can be rectified by adding with other ingredients with elevated levels of the amino acid(s) to meet the EAA requirements of the species, or the addition of individual crystalline amino acids. However, use of crystalline amino acids may not be practical as some species, particularly crustaceans, utilize intact protein sources more efficiently than individual crystalline amino acids.

A second potential problem could be the consumption of nucleic acids. In humans, and many monogastric animals, consumption of an excess of dietary nucleic acid is toxic. This is due to the limited abil-

TABLE 16.8. Vitamin and mineral composition of brewer's grains with yeast (BGY).

	Amount
Vitamins	
E (IU/kg)	36.80
Biotin (mg/kg)	2.44
Folic acid (mg/kg)	7.8
Thiamine (mg/kg)	46.2
Choline (mg/kg)	3,401.0
Niacin (mg/kg)	245.5
Pantothenic acid (mg/kg)	59.2
Pyrodoxine (mg/kg)	22.0
Riboflavin (mg/kg)	18.3
Minerals	
Calcium (%)	0.22
Cobalt (mg/kg)	0.16
Copper (mg/kg)	29.6
Iodine (mg/kg)	0.24
Iron (mg/kg)	185.0
Magnesium (%)	0.22
Manganese (mg/kg)	21.3
Phosphorus (%)	1.07
Potassium (%)	1.03
Selenium (mg/kg)	1.1
Sodium (%)	0.15
Zinc (mg/kg)	36.1

ity to excrete uric acid, which is formed from the utilization of nucleic acids, leading to deposits of uric acid in the body and to other metabolic disorders. However, in fish, it has been found that this should not be of concern since fish have high levels of liver uricase, which degrades uric acid and thus, prevents buildup in the body.

Use of brewery by-products could allow for more flexible diet formulations in finfish. In a feeding study with rainbow trout, juvenile fish (10 g) were fed diets containing increasing percentages (17, 34, 51, 68, 79, and 85 percent) of MPF replacing FM. After six weeks, fish fed diets containing 17 to 51 percent MPF·had growth and diet intake similar to fish fed a control diet containing 60 percent FM; however, fish fed higher levels of MPF (68 to 85 percent) had reduced ($P < 0.05$) growth (Yamamoto et al., 1994). These data indicate that MPF appears to have potential for inclusion in rainbow trout diets that have partially replaced FM.

Use of brewer's dried grains (BDG) and brewer's yeast in rainbow trout diets could also be a means of reducing diet costs, decreasing the reliance on FM in diets, and improving environmental impact of aquaculture diets (reducing water pollution associated with excretion of phosphorus and nitrogen into aquatic environments). In a study to evaluate the digestibility values of BDG, brewer's yeast, and brewer's dried grain high-protein fraction (BDGHP), Cheng et al. (2004) fed rainbow trout diets containing these ingredients. The BDGHP was produced by milling BDG and then fractionating the product on a 120-mesh U.S. screen. This has the effect of increasing the crude protein content by sieving the course materials, while concomitantly reducing the fiber content. Brewer's dried grain high-protein fraction also was shown to have higher amino acid concentrations due to the increase in protein level.

When ADCs were measured for the brewer's products, it was reported that BDGHP had significantly higher ($P < 0.05$) ADCs for crude protein and all essential and nonessential amino acids compared to brewer's yeast, while having higher ($P < 0.05$) ADCs for protein, arginine, histidine, lysine, alanine, aspartic acid, glutamic acid, proline, and tyrosine compared to BDG (Cheng et al., 2004). These data indicate that nutrient composition of brewery by-products might be greatly improved with some simple processing techniques (milling and sieving). This may allow for higher inclusion rates in aquaculture diets and allow for nutritious, economical, and environmentally sustainable diets for many aquaculture species.

Brewer's grains with yeast have been shown to be a suitable replacement for marine FM in diets for crustaceans (Muzinic et al., 2004). Juvenile (<1 g) Australian red claw crayfish *(Cherax quadricarinatus)* were fed diets containing various (0, 10, 20, and 30 percent) percentages of BGY as partial or total replacement of FM. Soybean meal was the other protein source in the diets with BGY. After eight weeks, mean final weight, percentage weight gain, and SGR of red claw were not significantly $(P > 0.05)$ different among all treatments and averaged 11.5 g, 977 percent, and 3.1 percent/day, respectively. Survival and body composition of red claw were also not different among diets. The results indicate that practical diets containing BGY and SBM, completely replacing FM, are adequate for good growth and survival of red claw crayfish. This may allow for less expensive diet formulations for red claw and may allow for increase in profits for producers.

Use of Brewer's Yeast in Fish Diets

Use of brewer's yeast in fish diets has been evaluated in several fish species. Yeast has generally been added in low (<5 percent) percentages. Nutritionists have added yeast to improve palatability, amino acid composition, and/or to improve immune function of fish. However, addition of yeast to fish diets does not ensure positive results. Hoffman et al. (1997) reported that African sharptooth catfish *(Clarias gariepinus)* fed diets containing FM and yeast, either at 15 percent or 52 percent of the diet, had significantly lower $(P < 0.05)$ final weight and SGR compared to fish fed diets without yeast.

Lee (2002) reported that rockfish *(Sebastes schlegeli)* did not digest brewer's yeast as well as other protein ingredients. Apparent digestibility coefficients for dry matter, protein, and energy were calculated to be 59 percent, 78 percent, and 66 percent for grower-size rockfish, compared to 82, 92, and 93 percent ADC for anchovy FM. Likewise, apparent amino acid availabilities for brewer's yeast were dramatically lower $(P < 0.05)$ compared to other protein sources evaluated. The low digestibility of brewer's yeast could have been due to yeast cell walls being composed of complex heteropolysaccharide, which may not have been digested well by rockfish.

While Hoffman et al. (1997) reported reduced growth of African catfish, other fish appear to benefit from addition of yeast, or suffer from no effect on performance. Juvenile sea bass *(Dicentrarchus labrax)* with average weight of 12 g, were fed diets with brewer's yeast replacing FM. Fish were fed in tanks for twelve weeks (Oliva-Teles and Goncalves, 2001). At the conclusion of the study, there were no differences $(P > 0.05)$ in growth rate or diet intake among fish fed all diets. However, apparent protein and energy digestibility coefficients were significantly lower ($P < 0.05$) as the level of yeast increased.

While brewer's yeast can be used as an alternative protein source in diets, its high cost precludes most practical diets from containing high percentages. However, since there appears to be a strong possibility that brewer's yeast can be considered an immunostimulant for fish (Siwicki et al., 1994; Ortuno et al., 2002; Rodriguez et al., 2003), several studies have examined the use of brewer's yeast on growth, immune function, and disease resistance in fish. Addition of brewer's yeast and a probiotic to diets for hybrid striped bass (HSTB) has been shown to positively influence growth performance, enhanced resistance to *Streptococcus iniae* infection, and increased macrophage respiratory burst activity (Li and Gatlin, 2003).

In contrast, Li and Gatlin (2005) fed juvenile sunshine bass diets containing 0, 1, and 2 percent brewer's yeast for sixteen weeks. After sixteen weeks, head kidney samples from fish were pooled for macrophage isolation and assay of respiratory burst of leukocytes. Extracellular and intracellular superoxide anion, and whole blood neutrophil oxidative radial production were also measured. At the conclusion of the study, percentage weight gain of sunshine bass were not significantly different ($P > 0.05$); although, after twelve weeks, fish fed diets containing 1 and 2 percent brewer's yeast had significantly higher percentage weight gain than fish fed the basal diet. Variation within treatments may explain why there were not significant differences among final percentage weight gain, despite sunshine bass fed a diet containing 2 percent brewer's yeast had significantly higher weight gain when fish fed the basal diet until the last sampling date. However, there were not significant differences in blood neutrophil oxidative radial production, serum lysozyme, and intracellular superoxide anion production of head kidney macrophages in sunshine bass fed any of the diets, and there were no significant differences ($P > 0.05$) in cumulative survival percentages when hybrid striped bass were challenged with *Mycobacterium marinum* (Li and Gatlin, 2005). Thus, it appears that there are inconsistencies in growth and immune effects when brewer's yeast is added to a fish diet. Further research will need to be conducted to elucidate possible reasons as to these discrepancies.

REFERENCES

Atkinson, R.L., T.M. Ferguson, J.H. Quisenberry, and J.R. Couch (1955) Dehydrated alfalfa meal, condensed fish solubles, distillers dried solubles, and dried whey as supplements to an all-vegetable protein turkey laying diet. *Poultry Science* 34:730-735.

Cheng, Z.J. and R.W. Hardy (2004a) Nutritional value of diets containing distiller's dried grain with solubles for rainbow trout, *Oncorhynchus mykiss. Journal of Applied Aquaculture* 15(3/4):101-113.

Cheng, Z.J. and R.W. Hardy (2004b) Effects of microbial phytase supplementation in corn distiller's dried grain with solubles on nutrient digestibility and growth performance of rainbow trout, *Oncorhynchus mykiss. Journal of Applied Aquaculture* 15(3/4):83-100.

Cheng, Z.J., R.W. Hardy, and M. Blair (2003) Effects of supplementing methionine hydroxyl analogue in soybean meal and distiller's dried grain-based diets on the performance and nutrient retention of rainbow trout (*Oncorhynchus mykiss* Walbaum). *Aquaculture Research* 34:1303-1310.

Cheng, Z.J., R.W. Hardy, and N.J. Huige (2004) Apparent digestibility coefficients of nutrients in brewer's and rendered animal by-products for rainbow trout (*Oncorhynchus mykiss). Aquaculture Research* 35:1-9.

Coyle, S.D., G.J. Mengel, J.H. Tidwell, and C.D. Webster (2004) Evaluation of growth, feed utilization, and economics of hybrid tilapia, *Oreochromis niloticus* × *Oreochromis aureus,* fed diets containing different protein sources in combination with distillers dried grains with solubles. *Aquaculture Research* 35:1-6.

Cromwell, G.L., K.L. Herkelman, and T.S. Stahly (1993) Physical, chemical, and nutritional characteristics of distillers dried grains with solubles for chicks and pigs. *Journal of Animal Science* 71:679-686.

Deyoe, C.W. and O.W. Tiemeier (1969) Feeding channel catfish. In: *Proceedings of the 24th Distillers Feed Conference.* Distiller's Feed Research Council, Des Moines, IA.

Gatlin, D.M., III (2002) Nutrition and fish health. In: *Fish Nutrition,* J.E. Halver and R.W. Hardy (eds.), pp. 671-702. Academic Press, San Diego, CA.

Harms, R.H., R.S. Moreno, and B.L. Damron (1970) Evaluation of distillers dried grains with solubles in diets of laying hens. *Poultry Science* 48:1652-1655.

Hawkins, G.E. and J.A. Little (1967) Combined effects of type of forage fed, of concentrate ingredients and of pelleting concentrations on rumen fermentation, milk yield, and milk composition. *Journal of Dairy Science* 50:62-67.

Hoffman, L.C., J.F. Prinsloo, and G. Rukan (1997) Partial replacement of fish meal with either soybean meal, brewers yeast or tomato meal in the diets of African sharptooth catfish *Clarias gariepinus. Water SA* 23(2):181-186.

Horn, G.W. and W.M. Beeson (1968) Effects of corn distillers dried grains with solubles and dehydrated alfalfa meal on the utilization of urea nitrogen in beef cattle. *Journal of Animal Science* 28:412-417.

Lee, S.M. (2002) Apparent digestibility coefficients of various feed ingredients for juvenile and grower rockfish *(Sebastes schlegeli). Aquaculture* 207:79-95.

Li, P. and D.M. Gatlin, III (2003) Evaluation of brewers yeast *(Saccharomyces cerevisiae)* as a feed supplement for hybrid striped bass *(Morone chrysops* × *M. saxatilis). Aquaculture* 219:681-692.

Li, P. and D.M. Gatlin, III (2005) Evaluation of the prebiotic GroBiotic-A and brewers yeast as dietary supplements for hybrid striped bass *(Morone chrysops* × *M. saxatilis)* challenged in situ with *Mycobacterium marinum. Aquaculture* 248: 197-205.

Manley, J.M., R.A. Voitle, and R.H. Harms (1978) The influence of distillers dried grains with solubles (DDGS) in the diet of turkey breeder hens. *Poultry Science* 57:726-728.

Marvel, J.A., C.W. Carrick, R.E. Roberts, and S.M. Hauge (1946) Distillers dried solubles in chick rations containing corn and vegetable protein supplements. *Poultry Science* 24:252-258.

Molina-Poveda, C. and M.E. Morales (2004) Use of a mixture of barley-based fermented grains and wheat gluten as an alternative protein source in practical diets for *Litopenaeus vannamei. Aquaculture Research* 35:1158-1165.

Muzinic, L.A., K.R. Thompson, A. Morris, C.D. Webster, D.B. Rouse, and L. Manomaitis (2004) Partial and total replacement of fish meal with soybean meal and brewer's grains with yeast in practical diets for Australian red claw crayfish *Cherax quadricarinatus. Aquaculture* 230:359-376.

NRC (National Research Council) (1993) *Nutrient Requirements of Fish.* National Academy Press, Washington, DC.

Oliva-Teles, A. and P. Goncalves (2001) Partial replacement of fishmeal by brewers yeast *(Saccharomyces cerevisiae)* in diets for sea bass *(Dicentrarchus labrax)* juveniles. *Aquaculture* 202:269-278.

Ortuno, J., A. Cuesta, A. Rodriguez, M.A. Esteban, and J. Meseguer (2002) Oral administration of yeast, *Saccharomyces cerevisiae,* enhances the cellular innate immune response of gilthead seabream *(Sparus aurata). Veterinary Immunology and Immunopathology* 85:41-50.

Phillips, A.M. (1949) *Fisheries Research Bulletin No. 13.* Cortland Hatchery Report No. 18. Cortland, NY.

Phillips, A.M., G.L. Hammer, J.P. Edwards, and H.F. Hosking (1964) Dry concentrates as complete trout foods for growth and egg production. *Progressive Fish-Culturist* 26:155-159.

Rodriguez, A., A. Cuesta, J. Ortuno, M.A. Esteban, and J. Meseguer (2003) Immunostimulant properties of a cell wall-modified whole *Saccharomyces cerevisiae* strain administered by diet to seabream *(Sparus aurata). Veterinary Immunology and Immunopathology* 96:183-192.

Sinnhuber, R.O. (1964) Pelleted fish food. *Feedstuffs* 36(28):16.

Siwicki, A.K., D.P. Anderson, and G.L. Rumsey (1994) Dietary intake of immunostimulants by rainbow trout affects non-specific immunity and protection against furunculosis. *Veterinary Immunology and Immunopathology* 41:125-139.

Stone, D.A.J., R.W. Hardy, F.T. Barrows, and Z.J. Cheng (2005) Effects of extrusion on nutritional value of diets containing corn gluten meal and corn distiller's dried grain for rainbow trout, *Oncorhynchus mykiss. Journal of Applied Aquaculture* 17(3):1-20.

Tidwell, J.H., C.D. Webster, J.A. Clark, and L.R. D'Abramo (1993) Evaluation of distillers dried grains with solubles as an ingredient in diets for pond culture of the freshwater prawn *Macrobrachium rosenbergii. Journal of the World Aquaculture Society* 24:66-70.

Tidwell, J.H., C.D. Webster, S.D. Coyle, W.H. Daniels, and L.R. D'Abramo (1998) Fatty acid and amino acid composition of eggs, muscle, and midgut glands of freshwater prawns, *Macrobrachium rosenbergii* (de Man), raised in fertilized ponds, unfertilized ponds or fed prepared diets. *Aquaculture Research* 29:37-45.

Tidwell, J.H., C.D. Webster, J.D. Sedlacek, P.A. Weston, W.L. Knight, S.J. Hill, L.R. D'Abramo et al. (1995) Effects of complete and supplemental diets and organic pond fertilization on production of *Macrobrachium rosenbergii* and associated benthic macroinvertebrate populations. *Aquaculture* 138:169-180.

Tidwell, J.H., C.D. Webster, and D.H. Yancey (1990) Evaluation of distillers grains with solubles in prepared channel catfish diets. *Transactions of the Kentucky Academy of Science* 51:135-138.

Tidwell, J.H., C.D. Webster, D.H. Yancey, and L.R. D'Abramo (1993) Partial and total replacement of fish meal with soybean meal and distillers' by-products in diets for pond culture of the freshwater prawn *(Macrobrachium rosenbergii). Aquaculture* 118:119-130.

Webster, C.D., J.H. Tidwell, L.S. Goodgame, J.A. Clark, and D.H. Yancey (1992) Winter feeding and growth of channel catfish fed diets containing varying percentages of distillers' grains with solubles as a total replacement of fish meal. *Journal of Applied Aquaculture* 1(4):1-14.

Webster, C.D., J.H. Tidwell, L.S. Goodgame, and P.B. Johnson (1993) Growth, body composition, and organoleptic evaluation of channel catfish fed diets containing different percentages of distillers' grains with solubles. *Progressive Fish-Culturist* 55:95-100.

Webster, C.D., J.H. Tidwell, L.S. Goodgame, D.H. Yancey, and L. Mackey (1992) Use of soybean meal and distillers grains with solubles as partial or total replacement of fish meal in diets for channel catfish, *Ictalurus punctatus. Aquaculture* 106:301-309.

Webster, C.D., J.H. Tidwell, and D.H. Yancey (1991) Evaluation of distillers' grains with solubles as a protein source in diets for channel catfish. *Aquaculture* 96:179-190.

Webster, C.D., L.G. Tiu, A.M. Morgan, and A. Gannam (1999) Effect of partial and total replacement of fish meal on growth and body composition of sunshine bass *Morone chrysops* × *M. saxatilis* fed practical diets. *Journal of the World Aquaculture Society* 30:443-453.

Wu, Y.U. (1986) Fractionation and characterization of protein-rich material from barley after alcohol distillation. *Cereal Chemistry* 63:142-145.

Yamamoto, T., P.A. Marcouli, T. Unuma, and T. Akiyama (1994) Utilization of malt protein flour in fingerling rainbow trout diets. *Fisheries Science* 60(4):455-460.

Chapter 17

Use of Diet Additives to Improve Nutritional Value of Alternative Protein Sources

Delbert M. Gatlin III
Peng Li

INTRODUCTION

Diet additives have been defined as nonnutritive ingredients or non-nutritive components of ingredients that are included in diet formulations to influence physical or chemical properties of the diet or affect fish performance or quality of resulting products (Barrows, 2000). Additives that affect diet quality include binders, preservatives (such as antimicrobial compounds and antioxidants), and feeding stimulants. Other additives that may have more direct effects on fish performance and product quality include exogenous enzymes, prebiotics, probiotics, hormones, chemotherapeutants, and pigments. The use of many such compounds in aquafeeds may be controlled by various regulatory authorities in different countries (Guillaume and Metailler, 2001). This chapter will primarily focus on exogenous enzymes, prebiotics, and probiotics, as well as the feeding stimulants or palatability enhancers because such additives tend to have the greatest potential application in aquafeeds when feedstuffs of marine origin are replaced with plant feedstuffs. In addition to the additives mentioned previously, this chapter also will discuss other purified nutrient supplements such as amino acids, lipids/fatty acids, and minerals that may be added

Alternative Protein Sources in Aquaculture Diets
© 2008 by The Haworth Press, Taylor & Francis Group. All rights reserved.
doi:10.1300/5892_17

to diets to improve their nutritional value. Although these nutrient supplements do not strictly meet the classification of nonnutritive additives, their inclusion in diet formulations containing alternative protein feedstuffs may be of particular importance.

This chapter will focus primarily on the application of nonnutritive and nutritive additives in the diets of carnivorous fish species because of their traditional reliance on relatively large amounts of marine feedstuffs and their general tendency to use plant feedstuffs to a more limited extent than omnivorous fish species. However, some of the additives to be considered have been evaluated rather thoroughly with some of the omnivorous species that are well established in commercial aquaculture.

NUTRITIONAL SUPPLEMENTS

Practical fish diets are largely composed of various feedstuffs and a few purified nutrients, such as vitamins and trace minerals in premix form, to meet the nutritional requirements of the target species. The replacement of marine feedstuffs, such as fish meal, with plant feedstuffs may necessitate supplementation of specific nutrients including amino acids, fatty acids, and minerals to maintain nutritional adequacy.

Amino Acids

Most plant feedstuffs contain lower concentrations of crude protein and relative amounts of indispensable amino acids compared to fish meals and other protein feedstuffs of marine origin. In particular, the sulfur amino acids methionine and cysteine, as well as lysine, may become increasingly limited as the quantity of marine feedstuffs in the diet decreases relative to feedstuffs of plant origin. The commercial availability of various lysine and methionine supplements has allowed cost-effective supplementation of these amino acids in diet formulations for swine and poultry. Some of the early attempts to supplement practical aquafeeds with crystalline amino acids to overcome deficiencies were not successful due to apparent limited utilization of the supplemented amino acids. Indeed, the utilization of crystalline amino acids by various fish species generally has been observed to be inferior to that of intact protein and has been attributed to various factors such as excessive leaching into the water, rapid absorption into the blood resulting in catabolism or excretion, and uncoordinated uptake with amino acids from intact protein for protein synthesis (Zarate and Lovell, 1997). However, several recent studies have shown that various fish species can utilize crystalline amino acids when supplemented in amino acid-deficient diets (Robinson and Li, 1994; Keembiyehetty and Gatlin, 1997; Zarate and Lovell, 1997; Wilson, 2002). Numerous studies over the past decade have shown that balancing dietary amino acid composition by supplementing crystalline amino acids when marine feedstuffs are replaced with other feedstuffs can be an effective strategy to maintain the nutritional value of the diet (Keembiyehetty and Gatlin, 1997; McGoogan and Gatlin, 1997; Gomez-Requeni et al., 2004). Therefore, if the cost and/or availability of marine protein feedstuffs dictate more limited incorporation in aquafeeds, then the use of crystalline amino acids to balance diet formulations may become more commonly prevalent.

In addition, research with various terrestrial livestock has demonstrated the utility of optimizing the amino acid composition of diets under various production conditions by applying the ideal protein concept (Lopez et al., 1994; Mack et al., 1999). This ideal protein concept is based on the fact that, although amino acid requirements may vary due to many different factors, the ratio of the various indispensable amino acids to each other should remain constant. Lysine has been the primary reference amino acid in terrestrial livestock for which requirements under various conditions have been determined. This is because absorbed lysine is used solely for protein accretion, its concentration in feed ingredients is often limited, and it is relatively easy to analyze. Once lysine requirements have been determined under various conditions, the concentrations of the other amino acids can be adjusted based on their constant ratio to lysine. Although application of the ideal protein concept to various aquatic species has been limited to date, further advancements in this area may increase the opportunity for supplementation of crystalline amino acids in practical diet formulations. Such manipulation of dietary amino acids may be of particular value in fish to maximize protein accretion and minimize nitrogen excretion.

Fatty Acids

Protein feedstuffs of marine origin not only provide a rich and balanced supply of indispensable amino acids, but also typically contain appreciable quantities of long-chain highly unsaturated fatty acids of the n-3 series (n-3 HUFA). As such, the replacement of marine proteins with alternative feedstuffs may limit the quantity of n-3 HUFA in the diet. Such a dietary modification may reduce the tissue concentration of n-3 HUFA in the cultured species and potentially limit its consumer appeal, or even lead to an essential fatty acid (EFA) deficiency unless adequate quantities of supplemental lipid or fatty acids are provided. Dietary supplementation of marine oils has been the traditional means of satisfying the EFA requirements of various marine fishes. However, the demand for and cost of oils rich in n-3 HUFA from various marine fish and invertebrates is projected to steadily increase as the aquaculture industry continues to expand, prompting an interest in partially substituting marine oils with various animal and vegetable oils in aquafeeds.

The urgency of evaluating such alternatives has increased with some recent estimates that within a decade or so there will not be enough fish oil to support global aquacultural production (New and Wijkstroem, 2002). Strategies of replacing a considerable amount of dietary fish oil with plant oils recently have been shown to work satisfactorily in terms of maintaining adequate growth and health of various marine fish species as long as the EFA requirements of those species are met (Bell et al., 2003; Glencross, Hawkins et al., 2003; Regost et al., 2003). However, the fatty acid composition of the target species may be substantially altered by such dietary manipulation, and may require some period of finishing with a diet high in n-3 HUFA to restore tissue levels of these desirable fatty acids to maintain consumer appeal. Such strategies are becoming more refined, and the time of refeeding to restore n-3 HUFA levels more precisely defined (Robin et al., 2003; Jobling, 2004).

In addition to more judicious use of marine oils in aquafeeds, alternative supplies of n-3 HUFA also should be readily sought. Various species of microalgae have been evaluated for their ability to produce n-3 HUFA. Another group of nonphotosynthetic, heterotrophic microorganisms known as Thraustochytrids also has been shown to produce fatty acids rich in docosahexaenoic acid (DHA; Leano and Liao, 2004). These organisms have the potential for mass production to supply a rich source of n-3 HUFA for aquafeeds. A variety of other molecular techniques are currently being explored to produce n-3 HUFA to limit the dependence on marine oils. However, to date these various products have not been thoroughly evaluated with fish and therefore must undergo extensive testing before their incorporation into aquafeeds will be readily accepted.

Minerals

The mineral fraction is another component that is typically abundant in marine protein feedstuffs such as fish meal compared to most plant feedstuffs. As such, the replacement or reduction of fish meal in aquafeeds generally results in reduced levels of various micro and macrominerals. A reduction in the quantities of trace or microminerals typically is not of concern as nutritionally complete diets contain a trace mineral premix that generally exceeds the established requirements of most fish species. In regard to macrominerals, calcium and phosphorus are most abundant in fish meals, and thus dietary concentrations of these minerals can be greatly affected by altering the composition of marine feedstuffs. Reduction of dietary calcium due to fish meal replacement should not pose any particular concern and in fact could improve the availability of various trace minerals due to its inhibitor effect at high levels (Gatlin, 2000). Dietary phosphorus has received considerable attention in recent years due to it potential contribution to eutrophication of aquatic systems (Dosdat et al., 1995; Gatlin and Hardy, 2002). Reducing phosphorus excretion of fish has been achieved largely by decreasing the level of phosphorus in aquafeeds or increasing the availability of dietary phosphorus to the fish because dietary phosphorus that is not absorbed and deposited in body tissues is released into the water (Ketola, 1985).

A primary form of phosphorus in plant ingredients is phytic acid (phytate), which is unavailable to monogastric animals due to the absence of the digestive enzyme phytase. Thus, replacement of marine protein feedstuffs with plant feedstuffs will typically increase the quantity of indigestible phosphorus in the diet. Use of the microbial enzyme phytase to cleave phosphate molecules from phytate has been demonstrated to improve phosphorus availability in numerous monogastric animals and will be discussed in the following section.

Exogenous Digestive Enzymes

Numerous feedstuffs of plant origin contain various nutrients that are not readily digested by monogastric animals due to the lack of specific digestive enzymes. Supplementation of exogenous enzymes to diets containing such feedstuffs has been readily employed in recent years, especially with terrestrial livestock such as swine and poultry, to increase the utilization of nutrients in these plant feedstuffs (Bedford, 2000).

As previously noted, phytate phosphorus in plant feedstuffs has been of primary concern in fish nutrition because of its limited utilization by fish, resulting in excretion and potential enrichment of receiving waters. Up to 70 percent of the phosphorus in plant feedstuffs may be in the form of phytate, the primary storage form of phosphorus in plants. Increasing the utilization of dietary phytate through supplementation of the microbial enzyme phytase has been a primary focus in fish nutrition due to concerns over limiting phosphorus waste in aquaculture.

Several studies have shown that dietary supplementation of phytase from *Aspergillus niger* can significantly improve phosphorus availability in numerous fish species. Specific responses of various species to dietary phytase supplementation are summarized in Table 17.1. Omnivorous fish species such as common carp *(Cyprinus carpio),* African catfish *(Clarius gariepinus),* and channel catfish *(Ictalurus punctatus)* have been shown to exhibit increased phosphorus utilization from soybean-meal-based diets with the supplementation of phytase (Schafer et al., 1995; Jackson et al., 1996; Eya and Lovell, 1997; Van Weerd et al., 1999; Yan and Reigh, 2002). The addition of 500-1,000 units of phytase per kilogram diet generally has been sufficient to increase phytate utilization as measured by enhanced bone ash, bone phosphorus, phosphorus retention, and decreased fecal phosphorus. Phytase supplementation has typically enhanced weight gain of fish fed diets limiting in phosphorus. However, increased protein digestibility has not been consistently observed in various studies. Yan and Reigh (2002) reported that weight gain and dietary protein utilization

TABLE 17.1. Summary of fish responses to dietary phytase supplementation.

Species	Dietary phytase concentration (Unit/kg)	Weight gain	Feed efficiency	Nutrient digestibility			Bone mineral deposition or retention		Reference
				Protein	Phosphorus	Ash	Phosphorus	Ash	
African catfish	1,000	0	0	+	+		+		Van Weerd et al. (1999)
Atlantic salmon	Pretreatment		+	+	+	+	+	+	Storebakken et al. (1998)
Common carp	500 or 1,000	+	+	0	+	+	+		Shafer et al. (1995)
Channel catfish	500, 1,000, 2,000, 4,000	+	+	0		+	+	+	Jackson et al. (1996)
Channel catfish	500, 1,000, 2,000, 4,000, 8,000	+	0	0		+	+	+	Yan and Reigh (2002)
Rainbow trout	1,000	+	0	+		+			Rodehutscord and Pfeffer (1995)
	Pretreatment	+	+			+			Vielma et al. (2002)
	500			0	+				Cheng and Hardy (2002)
	500, 1,000, 2,000, 4,000			+	+				Chang and Liu (2002)
European seabass	1,000 or 2,000	0	0	0	+				Oliva-Teles et al. (1998)

Note: Designations of (0) and (+) respectively, indicate neutral and positive responses to phytase supplementation.

of channel catfish were not improved by phytase addition; however, elevated phytase inclusion at 8,000 units per kilogram improved utilization of endogenous trace minerals such as zinc.

Similar improvements in utilization of dietary phosphorus from soybean-based feedstuffs have been observed with various carnivorous fish species such as rainbow trout *(Oncorhynchus mykiss)* (Cain and Garling, 1995; Rodehutscord and Pfeffer, 1995; Lanari et al., 1998; Forster et al., 1999; Sugiura et al., 2001; Cheng and Hardy, 2002; Vielma et al., 2002; Cheng et al., 2004), Atlantic salmon *(Salmo salar)* (Storebakken et al., 1998), striped bass *(Morone saxatilis)* (Hughes and Soares, 1998; Papatryphon et al., 1999; Papatryphon and Soares, 2001), and European seabass *(Dicentrarchus labrax)* (Oliva-Teles et al., 1998). Apparent digestibility of phosphorus, calcium, magnesium, iron, zinc, and energy in barley, canola meal, wheat, and wheat middlings by rainbow trout also was increased with the addition of phytase.

Improved protein digestibility and/or nitrogen utilization with dietary phytase supplementation has not been consistently observed in these species, although dietary phytic acid has the potential to interfere with protein digestion. In general, pretreatment of soy proteins with phytase has typically resulted in better weight gain and diet utilization by fish than addition of phytase prior to or after pelleting, and has been attributed to hydrolyze phytate more efficiently (Vielma et al., 2002). In a recent study with Atlantic salmon (Sajjadi and Carter, 2004), the addition of purified phytic acid to semipurified diets was shown to reduce protein digestibility but it did not affect feed intake, weight gain, or trypsin activity in pyloric caeca. Inclusion of phytase in the diet did not affect feed intake, but significantly improved weight gain in the presence or absence of purified phytic acid. Thus, dietary supplementation of phytase may not only enhance the utilization of phosphorus but also other nutrients including protein and amino acids. One of the most recent studies (Cheng et al., 2004) reported the addition of phytase at 500 units/kg increased the apparent digestibility coefficients of soybean-based diets for dry matter, crude protein, and most amino acids (excluding tryptophan and tyrosine). In addition, the availability of calcium, magnesium, total phosphorus, phytate phosphorus, manganese, and zinc was increased with phytase supplementation.

Dietary protein is the primary source of nitrogen, which is the other enriching nutrient of major concern in aquaculture (Handy and Poxton, 1993). Improving the utilization of dietary protein is desirable; however, attempts to do so by exogenous enzyme supplementation have been rather limited to date. Dabrowski and Glogowski (1977) reported one of the earliest attempts to provide exogenous proteolytic enzymes to common carp fry by supplementing the diet with bovine trypsin. This diet manipulation did increase proteolytic activity but otherwise had limited effects on fish performance. Kolkovski (2001) reviewed the potential use of enzymes in diet formulations for larval and juvenile fish and concluded that addition of proteolytic enzymes has not consistently improved diet digestibility or fish performance. The addition of four commercial proteases to the diet of Pacific white shrimp, *Litopenaeus vannamei,* did not enhance growth performance although in vitro proteolytic activity in the diet was maintained (Divakaran and Velasco, 1999). More research in this area is warranted as the contribution of exogenous proteinase and carbohydrase enzymes from various prey organisms can make substantial contributions to the digestive processes in fish (Kuz'mina and Golovanova, 2004).

The evaluation of other exogenous enzymes besides phytase in the diet of fish has been extremely limited compared with that of terrestrial livestock. Various exogenous carbohydrase enzymes, such as α-amylase, xylanase, and β-glucanase, have been primarily evaluated with swine and poultry (Bedford, 2000). Supplementation of some enzymes, such as α-amylase, to the diet of terrestrial livestock has been used to increase the digestibility of starch in nonviscous grains such as corn and sorghum, whereas the addition of other enzymes are intended to reduce the adverse effects of nonstarch polysaccharides (NSP) in viscous grains such as barley, oats, rye, and wheat. These enzymes include β-mannanase that hydrolyzes β-mannans in various plant feedstuffs; xylanase that hydrolyzes xylans found principally in wheat; and β-glucanase that hydrolyzes β-glucans that are prominent in barley (Lobo, 1998). The NSP in these viscous grains typically increase intestinal viscosity, reduce rates of digestion, and result in the production of wet, viscous manure. Therefore, addition of appropriate carbohydrases to diets containing viscous grains can facilitate more rapid and complete digestion of NSP. Another emerging advantage of using exogenous carbohydrases in the diets of monogastric livestock is the tendency for reduced variation in nutrient quality of various cereal grains (Bedford, 2000). Other enzymes, such as α-galactosidase, can hydrolyze oligosaccharides in soybean meal and thus increase its digestibility (Lobo, 1998).

Application of these enzymes to fish diets may enhance digestion of protein and organic matter, thus reducing the biochemical oxygen demand in aquaculture systems as well as settleable solids in effluent waters, which are also of concern (Johnsen et al., 1993; Kelly and Karpinski, 1994). However, evaluation of these enzymes with fish has been extremely limited to date. In one of the few studies with fish, Glencross, Boujard, et al. (2003) evaluated the addition of α-galactosidase to diets containing lupin meal because of its relatively high level of various α-galactosyl homologues of sucrose such as raffinose, stachyose, and verbascose. Supplementation of α-galactosidase significantly improved protein digestion and tended to increase the digestibility of energy and nitrogen-free extract by rainbow trout. Addition of an enzyme mixture to a diet containing canola meal increased diet digestibility and growth of black tiger prawn, *Penaeus monodon,* apparently due to removing the negative effects of glucosides and other antinutritional factors in canola meal (Buchanan et al., 1997). The addition of exogenous amylase to the diet of kuruma prawn, *Marsupenaeus japonicus,* also enhanced its growth and feed utilization (Maugle et al., 1983). Such application of carbohydrases to other plant feedstuffs containing undesirable NSP may increase their nutritive value to fish and shellfish. Therefore, more research in this area should be undertaken in the future to possibly improve the nutritive value of various plant feedstuffs for various aquacultured organisms.

One current constraint to dietary inclusion of exogenous enzymes is that most of these enzymes are not thermally stable at temperatures above approximately 95°C; therefore, they cannot withstand the harsh conditions of preconditioning prior to pelleting or extrusion processing. Thus, postextrusion application of liquid enzymes is currently required until more thermally tolerant forms become available. Vacuum infusion is one technique that may allow efficient application of enzymes after pellet formation. Encapsulation is another possible technique currently being explored to improve the heat resistance of supplemental enzymes.

Palatability Enhancers

Feedstuffs of marine origin, such as fish meal, krill meal, shrimp meal, fish solubles, fish oil, and various protein hydrolysates are noted for being highly palatable to various aquatic organisms (Barrows, 2000). The substitution of plant feedstuffs for these highly palatable marine feedstuffs has been shown to reduce diet acceptance and consequently performance of several cultured species, especially those with more carnivorous feeding habits (Hardy and Barrows, 2002). Therefore, palatability enhancement of diets based on plant feedstuffs is an area that should receive increasing attention in the coming years.

A variety of natural and synthetic ingredients have been identified to limit the adverse effects of reduced diet palatability caused by substitution of marine feedstuffs. Such ingredients may be referred to as palatability enhancers, gustatory stimulants, or attractants, although in the strictest sense attractants are defined as substances that orient animals at a distance toward a potential prey item (Guillaume and Metailler, 2001). A type of attractant that leads to ingestion is also commonly referred to as a stimulant. Many compounds that may stimulate diet intake by fish have been characterized to be of low molecular weight, nonvolatile, water soluble, and nitrogenous (Carr, 1982). Guillaume and Metailler (2001) further classified fish attractants into three primary groups consisting of (1) L-amino acids, (2) betaine or other molecules with a pentavalent nitrogen atom, and (3) nucleosides and nucleotides. The feedstuffs of marine origin that were previously mentioned as palatability enhancers readily fit into these categories as various fish meals and fish solubles are rich in nucleotides, fish protein hydrolysates contain free amino acids, and invertebrate meals are rich in nitrogenous bases.

The amino acid glycine has been shown to be stimulatory in several carnivorous fish species, including sea bream *(Pagrus major)* (Fuke et al., 1981; Shimizu et al., 1990), chinook salmon *(Oncorhynchus tshawytscha)* (Hughes, 1985), and red drum *(Sciaenops ocellatus)* (McGoogan and Gatlin, 1997). The addition of glycine at 2 percent of diet to soybean-meal-based diets consistently provided greater weight gain of red drum compared to the addition of menhaden fish solubles at 4 percent of diet (dry matter basis) or shrimp head meal at 5 percent of diet (McGoogan and Gatlin, 1997).

Betaine and various combinations of betaine, glycine, and other amino acids have been reported to enhance palatability of diets containing various feedstuffs for different fish species including gibel carp *(Carassius auratus gibelio)* (Xue and Cui, 2001), striped bass (Papatryphon and Soares, 2000), sole *(Solea solea)* (Reig et al., 2003), European sea bass (Dias et al., 1997), and yellow perch *(Perca flavescens)* (Gould et al., 2003). Betaine and amino acids also have been reported to improve osmoregulation and performance of rainbow trout (Virtanen et al., 1994) and coho salmon *(Oncorhynchus kisuch)* in seawater (Castro et al., 1998).

Xue and Cui (2001) reported that several different compounds promoted feeding enhancement of gibel carp fed diets in which some fish meal was replaced with meat and bonemeal. The most effective concentrations of these various compounds were (as percent of diet): 0.5 percent betaine, 0.1 percent glycine, 0.25 percent L-lysine, 0.1 percent L-methionine, 0.25 percent L-phenylalanine, and 0.1 percent squid extract, which provided the greatest efficacy of all compounds evaluated. Squid meal and krill meal included at 5 percent in soybean-meal-based diets were more effective than betaine at 0.5 percent of diet for enhancing diet consumption and weight gain of yellow perch (Gould et al., 2003). A mixture of L-alanine, L-serine, inosine 5'-monophosphate, and betaine included at 2 and 4 percent of diet was effective in improving weight gain and feed intake of striped bass fed a soybean-based diet (Papatryphon and Soares, 2000). Similar results were obtained when European sea bass were fed diets containing soy protein concentrate and supplemented with an amino acid attractant mixture (Dias et al., 1997).

Dietary nucleotides have most recently received considerable attention as immunomodulating compounds for various fish species (Burrells, William, and Forno, 2001; Burrells, William, Southage et al., 2001; Sakai et al., 2001; Li et al., 2004). However, these compounds also may influence diet intake of fish. Inosine and inosine 5'-monophosphate were shown to be effective in improving diet intake of turbot *(Scophthalmus maximus)* (Mackie and Adron, 1978). In another study with largemouth bass *(Micropterus salmoides),* inosine 5'-monophosphate significantly enhanced intake of a soybean-meal-based diet, whereas inosine, glycine, other amino acids, and betaine were not effective (Kubitza et al., 1997).

.

Probiotics and Prebiotics

This last group of additives to be discussed may have application in various aquafeeds based on animal or plant feedstuffs. Probiotics, originally defined as live microbial diet supplements that improve health of human and terrestrial livestock, have recently attracted heightened attention in aquaculture (Gatesoupe, 1999; Gatlin, 2002; Irianto and Austin, 2002a). These dietary supplements alter the microbiota in the gastrointestinal (GI) tract, which is recognized as playing important roles in the growth, digestion, immunity, and disease resistance of the host organism. Although some probiotics are designed to treat the aquatic environment for competitive exclusion of potential pathogens, most probiotics are delivered to hosts by dietary supplementation.

In recent years many exciting studies have shown the potential use of probiotics in aquaculture through diet manipulation to enhance growth, immunity, and resistance of aquatic animals to various diseases (Table 17.2). Besides various strains of bacteria, other organisms such as bacteriphages, fungi, and microalgae also have been incorporated in diets as probiotics. It is anticipated that dietary supplementation of probiotics will become increasingly important to provide an effective strategy to improve the health of aquacultured organisms.

Two other related dietary supplements emerging in recent years include prebiotics and inactivated probiotics. Prebiotics are defined as nondigestible dietary ingredients that beneficially affect the host by selectively stimulating the growth of and/or activating the metabolism of health-promoting bacteria in the GI tract. Use of these supplements is a novel concept in aquaculture. In a recent study with hybrid striped bass *(Morone chrysops × M. saxatilis),* supplementation of a fish meal-based diet at 2 percent with the commercial prebiotic Grobiotic-A, a mixture of partially autolyzed brewers yeast, dairy ingredient components, and dried fermentation products, significantly improved growth, nonspecific immune responses, and resistance to *Streptococcus iniae* infection compared to fish fed the basal diet (Li and Gatlin, 2004). However, specific responses of microbes in the GI tract were not determined in this study and will require further delineation. Dietary fatty acids and carbohydrates have been previously shown to alter the bacterial flora of the GI tract of fish and thus may have potential application as prebiotics (Ringø et al., 1998; Ringø

TABLE 17.2. Summary of fish responses to dietary supplementation with various probiotics.

Probiotic	Species	Dose and timing	Measured responses	Reference
Live bacteriophage	Ayu	10^7 CFU/g diet once	Resistance to *Pseudomonas plecoglossicida* +	Park et al. (2000)
Bacillus subtilis and *B. licheniformis*	Rainbow trout	4×10^4 spore/g diet for 42 days	Resistance to *Yersinia ruckeri* +	Raida et al. (2003)
Bacillus sp.	Black tiger shrimp	100 days	Weight gain + Survival + Resistance to *Vibrio harveyi* +	Rengpipat et al. (1998)
Bacillus S11	Black tiger shrimp	10^{10} CFU/g diet for 30 days	Survival +	Meunpol et al. (2003)
Carnobacterium divergens	Atlantic cod	21 days	Resistance to *Vibrio anguillarum* +	Gildberg et al. (1997)
Carnobacterium sp.	Rainbow trout	5×10^7 cells/g diet for 14 days	Resistance to *A. salmonicida* +; *V. ordalii* +; and *Y. ruckeri* + Resistance to *V. anguillarum* 0	Robertson et al. (2000)
Carnobacterium inhibens	Rainbow trout	10^6-10^8 cell/g diet for 7-14 days	Resistance to *Aeromonas salmonicida* + Immune response +	Irianto and Austin (2002b)
Debaryomyces hansenii	European sea bass	7×10^5 CFU/g diet	Amylase secretion + Survival + Growth −	Tovar et al. (2002)
Enterococcus faecium	European eel	0.1% diet for 14 days	Resistance to *Edwardsiella tarda* +	Chang and Liu (2002)

Lactobacillus rhamnosus	Rainbow trout	10^9 cell/g diet for 51 days	Resistance to *Aeromonas salmonicida* +	Nikoskelainen et al. (2001)
Lactobacillus rhamnosus	Rainbow trout	9×10^4, 2.1×10^6, 2.8×10^8, 9.7×10^{10} CFU/g diet	Immune responses +	Nikoskelainen et al. (2003)
Saccharomyces cerevisiae	Nile tilapia	0.1% diet for 63 days	Weight gain and feed efficiency +	Lara-Flores et al. (2002)
Vibrio fluvialis	Rainbow trout	10^6-10^8 cell/g diet for 7-14 days	Resistance to *Aeromonas salmonicida* + Immune response +	Irianto and Austin (2002b)
Inactivated bacteria				
Unidentified gram-positive coccus A1-6, *V. fluvialis* A3-47S, *Aeromonas hydrophila* A3-51, and *Carnobacterium* BA211	Rainbow trout	10^7 cell/g diet for 14 days	Weight gain + Resistance to *Aeromonas salmonicida* + No detected antibody	Irianto and Austin (2003)
Aeromonas hydrophila	Goldfish	2×10^7 cell/g diet	Resistance to *Aeromonas salmonicida* + Immunocyte number +	Irianto et al. (2003)

Note: Designations of (0), (+), and (−) respectively, indicate neutral, positive, and negative responses associated with the various measurements.

and Olsen, 1999). Development of prebiotics and combined prebiotic/probiotic supplementation in aquafeeds may have various applications in aquaculture that need to be further evaluated.

Inactivated probiotics also have shown protection against infectious diseases in some fish, including rainbow trout (Irianto and Austin, 2003) and goldfish (Irianto et al., 2003). In these studies, formalin-inactivated cells applied as a dietary supplement have assisted in controlling bacterial infection. Based on these studies, the classification of inactivated cells as probiotics, immunostimulants, or oral vaccines is uncertain. Nonetheless, the fact that there is obvious benefit in using dead cells as a diet additive for aquaculture should receive further consideration in the future.

CONCLUSIONS

This chapter has described several nutritive and nonnutritive additives, and their potential applications in aquafeeds. The inclusion of these additives in diets of various aquacultured organisms is anticipated to increase as feedstuffs of marine origin are progressively replaced with those of plant origin. It is anticipated that further research will allow development and refinement of various additives to enhance growth performance, health, and product quality of numerous cultured species.

REFERENCES

Barrows, F.T. (2000) Feed additives. In *Encyclopedia of Aquaculture,* R.R. Stickney (ed.). Hoboken, NJ: John Wiley & Sons, pp. 335-340.

Bedford, M. (2000) Exogenous enzymes in monogastric nutrition—Their current value and future benefits. *Animal Feed Science and Technology* 86:1-13.

Bell, J.G., D.R. Tocher, R.J. Henderson, J.R. Dick, and V.O. Crampton (2003) Altered fatty acid compositions in Atlantic salmon *(Salmo salar)* fed diets containing linseed and rapeseed oils can be partially restored by a subsequent fish oil finishing diet. *Journal of Nutrition* 133:2793-2801.

Buchanan, J., H.Z. Sarac, D. Poppi, and R.T. Cowan (1997) Effects of enzyme addition to canola meal in prawn diets. *Aquaculture* 151:29-35.

Burrells, C., P.D. William, and P.F. Forno (2001) Dietary nucleotides: A novel supplement in fish feeds. 1. Effects on resistance to diseases in salmonids. *Aquaculture* 199:159-169.

Burrells, C., P.D. William, P.J. Southage, and S.L. Wadsworth (2001) Dietary nucleotides: A novel supplement in fish feeds. 2. Effects on vaccination, salt water transfer, growth rate and physiology of Atlantic salmon. *Aquaculture* 199:171-184.

Cain, K.D. and D.L. Garling (1995) Pretreatment of soybean meal with phytase
for salmonid diets to reduce phosphorus concentrations in hatchery effluents.
Progressive Fish-Culturist 57:114-119.

Carr, W.E.S. (1982) Chemical stimulation of feeding behavior. In *Chemoreception
in Fishes,* T.J. Hara (ed.). Amsterdam: Elsevier, pp. 259-273.

Castro, H., J. Battaglia, and E. Virtanen (1998) Effects of FinnStim on growth and
sea water adaptation of coho salmon. *Aquaculture* 168:423-429.

Chang, C. and W. Liu (2002) An evaluation of two probiotic bacterial strains,
Enterococcus faecium SF68 and *Bacillus toyoi,* for reducing Edwardsiellosis in
cultured European eel, *Anguilla anguilla* L. *Journal of Fish Diseases* 25:311-315.

Cheng, Z.J. and R.W. Hardy (2002) Effect of microbial phytase on apparent nutrient
digestibility of barley, canola meal, wheat and wheat middlings, measured *in vivo*
using rainbow trout *(Oncorhynchus mykiss). Aquaculture Nutrition* 8:271-277.

Cheng, Z.J., R.W. Hardy, V. Verlhac, and J. Gabaudan (2004) Effects of microbial
phytase supplementation and dosage on apparent digestibility coefficients of
nutrients and dry matter in soybean product-based diets for rainbow trout
(Oncorhynchus mykiss). Journal of the World Aquaculture Society 35:1-15.

Dabrowski, K. and J. Glogowski (1977) A study of the application of proteolytic
enzymes to fish food. *Aquaculture* 12:349-360.

Dias, J., E.F. Gomes, and S.J. Kaushik (1997) Improvement of feed intake through
supplementation with an attractant mix in European seabass fed plant protein
rich diets. *Aquatic Living Resources* 10:385-389.

Divakaran, S. and M. Velasco (1999) Effect of proteolytic enzyme additions to a
practical feed on growth of the Pacific white shrimp, *Litopenaeus vannamei*
(Boone). *Aquaculture Research* 30:335-339.

Dosdat, A., F. Gaumet, and H. Chartois (1995) Marine aquaculture effluent monitoring: Methodological approach to the evaluation of nitrogen and phosphorus
excretion by fish. *Aquaculture Engineering* 14:59-84.

Eya, J.C. and R.T. Lovell (1997) Net absorption of dietary phosphorus from various
inorganic sources and effect of fungal phytase on net absorption of plant phosphorus by channel catfish *Ictalurus punctatus. Journal of the World Aquaculture
Society* 28:386-391.

Forster, I., D.A. Higgs, B.S. Dosanjh, M. Rowshandeli, and J. Parr (1999) Potential
for dietary phytase to improve the nutritive value of canola protein concentrate
and decrease phosphorus output in rainbow trout *(Oncorhynchus mykiss)* held in
11°C freshwater. *Aquaculture* 179:109-125.

Fuke, S., S. Konosu, and I. Kazuo (1981) Identification of feeding stimulants for red
sea bream in the extract of marine worm *Perinereis brevicirrus. Bulletin of the
Japanese Society of Scientific Fisheries* 47:1631-1635.

Gatesoupe, F.J. (1999) The use of probiotics in aquaculture. *Aquaculture* 180:147-165.

Gatlin, D.M., III (2000) Minerals. In *Encyclopedia of Aquaculture,* R.R. Stickney
(ed.). Hoboken, NJ: John Wiley & Sons, pp. 532-540.

Gatlin, D.M., III (2002) Nutrition and fish health. In *Fish Nutrition,* 3rd edn, J.E. Halver and R.W. Hardy (eds.). San Diego, CA: Academic Press, pp. 671-702.

Gatlin, D.M., III and R.W. Hardy (2002) Manipulations of diets and feeding to reduce losses of nutrients in intensive aquaculture. In *Aquaculture and the Environment in the United States,* J.R. Tomasso (ed.). Baton Rouge, LA: U.S. Aquaculture Society, A Chapter of the World Aquaculture Society, pp. 155-165.

Gildberg, A., H. Mikkelsen, E. Sandaker, and E. Ringø (1997) Probiotic effect of lactic acid bacteria in the feed on growth and survival of fry of Atlantic cod *(Gadus morhua). Hydrobiologia* 352:279-285.

Glencross, B.D., T. Boujard, and S.J. Kaushik (2003) Influence of oligosaccharides on the digestibility of lupin meals when fed to rainbow trout, *Oncorhynchus mykiss. Aquaculture* 219:703-713.

Glencross, B.D., W.E. Hawkins, and J.G. Curnow (2003) Restoration of the fatty acid composition of red seabream *(Pagrus auratus)* using a fish oil finishing diet after grow-out on plant oil-based diets. *Aquaculture Nutrition* 9:409-418.

Gomez-Requeni, P., M. Mingarro, J.A. Calduch-Giner, F. Medale, S.A.M. Marin, D.F. Houlihan, S. Kaushik et al. (2004) Protein growth performance, amino acid utilization and somatotropic axis responsiveness to fish meal replacement by plant protein sources in gilthead sea bream *(Sparus aurata). Aquaculture* 232: 493-510.

Gould, N.L., M.M. Glover, L.D. Davidson, and P.B. Brown (2003) Dietary flavor additives influence consumption of feeds by yellow perch *Perca flavescens. Journal of the World Aquaculture Society* 34:412-417.

Guillaume, J. and R. Metailler (2001) Raw materials and additives used in fish foods. In *Nutrition and Feeding of Fish and Crustaceans,* J. Guillaume, S. Kaushik, P. Bergot, and R. Metailler (eds.). Chichester, UK: Praxix Publishing, pp. 281-295.

Handy, R.D. and M.G. Poxton (1993) Nitrogen pollution in mariculture: Toxicity and excretion of nitrogenous compounds by marine fish. *Reviews in Fish Biology and Fisheries* 3:205-241.

Hardy, R.W. and F.T. Barrows (2002) Diet formulation and manufacture. In *Fish Nutrition,* J.E. Halver and R.W. Hardy (eds.). San Diego, CA: Academic Press, pp. 505-600.

Hughes, K.P. and J.H. Soares (1998) Efficacy of phytase on phosphorus utilization in practical diets fed to striped bass *Morone saxatilis. Aquaculture Nutrition* 4:133-140.

Hughes, S.G. (1985) Single-feeding response of chinook salmon fry to potential feed intake modifiers. *Progressive Fish-Culturist* 55:40-42.

Irianto, A. and B. Austin (2002a) Probiotics in aquaculture. *Journal of Fish Diseases* 25:633-642.

Irianto, A. and B. Austin (2002b) Use of probiotics to control furunculosis in rainbow trout, *Oncorhynchus mykiss* (Walbaum). *Journal of Fish Diseases* 25:333-342.

Irianto, A. and B. Austin (2003) Use of dead probiotic cells to control furunculosis in rainbow trout, *Oncorhynchus mykiss* (Walbaum). *Journal of Fish Diseases* 26:59-62.

Irianto, A.P.A., W. Robertson, and B. Austin (2003) Oral administration of forma-lin-inactivated cells of *Aeromonas hydrophila* A3-51 controls infection by atypi-cal *A. salmonicida* in goldfish, *Carassius auratus* (L.). *Journal of Fish Diseases* 26:117-120.

Jackson, L.S., M.H. Li, and E.H. Robinson (1996) Use of microbial phytase in chan-nel catfish *Ictalurus punctatus* diets to improve utilization of phytate phospho-rus. *Journal of the World Aquacutlure Society* 27:309-313.

Jobling, M. (2004) Are modifications in tissue fatty acid profiles following a change in diet the result of dilution? Test of a simple dilution model. *Aquaculture* 232:551-562.

Johnsen, R.I., O. Grahl-Nielsen, and B.T. Lunestad (1993) Environmental distribu-tion of organic waste from a marine fish farm. *Aquaculture* 118:229-244.

Keembiyehetty, C.N. and D.M. Gatlin III (1997) Performance of sunshine bass fed soybean-meal-based diets supplemented with different methionine compounds. *Progressive Fish-Culturist* 59:25-30.

Kelly, L.A. and A.W. Karpinski (1994) Monitoring BOD outputs from land-base fish farms. *Journal of Applied Ichthyology* 10:368-372.

Ketola, H.G. (1985) Mineral nutrition: Effects of phosphorus in trout and salmon feeds in water pollution. In *Nutrition and Feeding in Fish,* C.B. Cowey, A.M. Mackey, and J.G. Bell, (eds.). New York: Academic Press, pp. 465-473.

Kolkovski, S. (2001) Digestive enzymes in fish larvae and juveniles—Implications and applications to formulated diets. *Aquaculture* 200:181-201.

Kubitza, F., L.L. Lovshin, and R.T. Lovell (1997) Identification of feed enhancers for juvenile largemouth bass *Micropterus salmoides. Aquaculture* 148:191-200.

Kuz'mina, V.V. and I.L. Golovanova (2004) Contribution of prey proteinases and carbohydrases in fish digestion. *Aquaculture* 234:347-360.

Lanari, D., E. D'Agaro, and C. Turri (1998) Use of nonlinear regression to evaluate the effects of phytase enzyme treatment of plant protein diets for rainbow trout *(Oncorhynchus mykiss). Aquaculture* 161:345-356.

Lara-Flores, M., M.A. Olvera-Novoa, B.E. Guzmán-Méndez, and W. López-Madrid (2002) Use of the bacteria *Streptococcus faecium* and *Lactobacillus acidophilus,* and the yeast *Saccharomyces cerevisiae* as growth promoters in Nile tilapia *(Oreochromis niloticus). Aquaculture* 216:193-201.

Leano, E.M. and I.C. Liao (2004) Thraustochytrids: Potential DHA source for marine fish nutrition. *Global Aquaculture Advocate* 7:87-88.

Li, P. and D.M. Gatlin III (2004) Dietary brewers yeast and the prebiotic Grobiotic™ AE influence growth performance, immune responses and resistance of hybrid striped bass *(Morone chrysops × M. saxatilis)* to *Streptococcus iniae* infection. *Aquaculture* 231:445-456.

Li, P., D.H. Lewis, and D.M. Gatlin III (2004) Dietary oligonucleotides from yeast RNA influence immune response and resistance of hybrid striped bass *(Morone chrysops × M. saxatilis)* to *Streptococcus iniae* infection. *Fish and Shellfish Immunology* 16:561-569.

Lobo, J.P. (1998) How can enzymes benefit me? *Feed Management* 49:12-16.

Lopez, J., R.D. Goodband, G.L. Allee, G.W. Jesse, J.L. Nelssen, M.D. Tokach, D. Spiers et al. (1994) The effects of diets formulated on an ideal protein basis on growth performance, carcass characteristics, and thermal balance of finishing gilts housed in a hot, diurnal environment. *Journal of Animal Science* 72:367-379.

Mack, S., D. Bercovici, G. DeGroote, B. Leclecq, M. Lippens, M. Pack, J.B. Schutte et al. (1999) Ideal amino acid profile and dietary lysine specification for broiler chickens of 20 to 40 days of age. *British Poultry Science* 40:257-265.

Mackie, A.M. and J.W. Adron (1978) Identification of inosine and inosine 5'-monophosphate as the gustatory feeding stimulants for the turbot *Scophthalmus maximus*. *Comparative Biochemistry and Physiology* 60A:79-83.

Maugle, P.D., O. Deshimaru, T. Katayama, and K.L. Simpson (1983) The use of amylase supplements in shrimp diets. *Journal of the World Mariculture Society* 14:25-37.

McGoogan, B.B. and D.M. Gatlin III (1997) Effects of replacing fish meal with soybean meal in diets for red drum *Sciaenops ocellatus* and potential for palatability enhancement. *Journal of the World Aquaculture Society* 28:374-385.

Meunpol, O., K. Lopinyosiri, and P. Menasveta (2003) The effects of ozone and probiotics on the survival of black tiger shrimp *(Penaeus monodon)*. *Aquaculture* 220:437-448.

New, M.B. and U.N. Wijkstroem (2002) Use of fishmeal and fish oil in aquafeeds. Further thoughts on the fishmeal trap. FAO Fish. Circ. No. 975, 61 pp.

Nikoskelainen, S., A.C. Ouwehand, G. Bylund, S. Salminen, and E. Lilius (2003) Immune enhancement in rainbow trout *(Oncorhynchus mykiss)* by potential probiotic bacteria *(Lactobacillus rhamnosus)*. *Fish and Shellfish Immunology* 15:443-452.

Nikoskelainen, S., A.C. Ouwehand, S. Salminen, and G. Bylund (2001) Protection of rainbow trout *(Oncorhynchus mykiss)* from furunculosis by *Lactobacillus rhamnosus*. *Aquaculture* 198:229-236.

Oliva-Teles, A., J.P. Pereira, A. Gouveia, and E. Gomes (1998) Utilisation of diets supplemented with microbial phytase by seabass *(Dicentrarchus labrax)* juveniles. *Aquatic Living Resources* 11:255-259.

Papatryphon, E., R.A. Howell, and J.H. Soares (1999) Growth and mineral absorption by striped bass *Morone saxatilis* fed a plant feedstuff-based diet supplemented with phytase. *Journal of the World Aquaculture Society* 30:161-173.

Papatryphon, E. and J.H. Soares, Jr. (2000) The effect of dietary feeding stimulants on growth performance of striped bass, *Morone saxatilis,* fed a plant feedstuff-based diet. *Aquaculture* 185:329-338.

Papatryphon, E. and J.H. Soares, Jr. (2001) The effect of phytase on apparent digestibility of four practical plant feedstuffs fed to striped bass, *Morone saxatilis*. *Aquaculture Nutrition* 7:161-167.

Park, S.C., I. Shimamura, M. Fukunaga, K. Mori, and T. Nakai (2000) Isolation of bacteriophages specific to a fish pathogen, *Pseudomonas plecoglossicida,* as a candidate for disease control. *Applied and Environmental Microbiology* 66:1416-1422.

Raida, M.K., J.L. Larsen, M.E. Nielsen, and K. Buchmann (2003) Enhanced resistance of rainbow trout, *Oncorhynchus mykiss* (Walbaum), against *Yersinia ruckeri* challenge following oral administration of *Bacillus subtilis* and *B. licheniformis* (BioPlus2B). *Journal of Fish Diseases* 26:495-498.

Regost, C., J. Arzel, M. Cardinal, G. Rosenlund, and S.J. Kaushik (2003) Total replacement of fish oil by soybean or linseed oil with a return to fish oil in turbot *(Psetta maxima)* 2. Flesh quality properties. *Aquaculture* 220:737-747.

Reig, L., M. Ginovart, and R. Flos (2003) Modification of the feeding behavior of sole *(Solea solea)* through the addition of a commercial flavour as an alternative to betaine. *Aquatic Living Resources* 16:370-379.

Rengpipat, S., W. Phianphak, S. Piyatiratitivorakul, and P. Menasveta (1998) Effects of a probiotic bacterium in black tiger shrimp *Penaeus monodon* survival and growth. *Aquaculture* 167:301-313.

Ringø, E., H.R. Bendiksen, S.J. Gausen, A. Sundsfjord, and R.E. Olsen (1998) The effect of dietary fatty acids on lactic acid bacteria associated with the epithelial mucosa and from faecalia of Arctic charr, *Salvelinus alpinus* L. *Journal of Applied Microbiology* 85:855-864.

Ringø, E. and R.E. Olsen (1999) The effect of diet on aerobic bacterial flora associated with intestine of Arctic charr, *Salvelinus alpinus* L. *Journal of Applied Microbiology* 86:22-28.

Robertson, P.A.W., C. O'Dowd, C. Burrells, P. Williams, and B. Austin (2000) Use of *Carnobacterium* sp. as a probiotic for Atlantic salmon *(Salmo salar* L.) and rainbow trout *(Oncorhynchus mykiss,* Walbaum). *Aquaculture* 185:235-243.

Robin, J.H., C. Regost, J. Arzel, and S.J. Kaushik (2003) Fatty acid profile of fish following a change in dietary fatty acid source: Model of fatty acid composition with a dilution hypothesis. *Aquaculture* 225:283-293.

Robinson, E.H. and M. Li (1994) Use of plant protein in catfish feeds: Replacement of soybean meal with cottonseed meal and replacement of fish meal with soybean meal and cottonseed meal. *Journal of the World Aquaculture Society* 25:271-276.

Rodehutscord, M. and E. Pfeffer (1995) Effects of supplemental microbial phytase on phosphorus digestibility and utilization in rainbow trout *(Oncorhynchus mykiss). Water Science Technology* 31:143-147.

Sajjadi, M. and C.G. Carter (2004) Effect of phytic acid and phytase on feed intake, growth, digestibility and trypsin activity in Atlantic salmon *(Salmo salar,* L.). *Aquaculture Nutrition* 10:135-142.

Sakai, M., K. Taniguchi, K. Mamoto, H. Ogawa, and M. Tabata (2001) Immuno-stimulant effects of nucleotide isolated from yeast RNA on carp, *Cyprinus carpio* L. *Journal of Fish Disease* 24:433-438.

Schafer, A., W.M. Koppe, K.-H. Meyer-Burgdorff, and K.D. Gunther (1995) Effects of a microbial phytase on the utilization of native phosphorus by carp in a diet based on soybean meal. *Water Science Technology* 31:149-155.

Shimizu, C.A., A. Ibrahim, T. Tokoro, and Y. Shirakawa (1990) Feeding stimulation in sea bream, *Pagrus major,* fed diets supplemented with Antarctic krill meals. *Aquaculture* 89:43-53.

Storebakken, T., K.D. Shearer, and A.J. Roem (1998) Availability of protein, phosphorus, and other elements in fish meal, soy protein concentrate and phytase-treated soy-protein-concentrate-based diets to Atlantic salmon, *Salmo salar. Aquaculture* 161:365-379.

Sugiura, S.H., J. Gabaudan, F.M. Dong, and R.W. Hardy (2001) Dietary microbial phytase supplementation and the utilization of phosphorus, trace minerals and protein by rainbow trout [Oncorhynchus mykiss (Walbaum)] fed soybean meal-based diets. *Aquaculture Research* 32:583-592.

Tovar, D., J. Zambonino, C. Cahu, F.J. Gatesoupe, R. Vazquez-Juarez, and R. Lesel (2002) Effect of yeast incorporation in compound diet on digestive enzyme activity in sea bass *(Dicentrarchus labrax)* larvae. *Aquaculture* 204:113-123.

Van Weerd, J.H., K.H.A. Khalaf, F.J. Aartsen, and P.A.T. Tijssen (1999) Balance trials with African catfish *Clarias gariepinus* fed phytase-treated soybean meal-based diets. *Aquaculture Nutrition* 5:135-142.

Vielma, J., K. Ruohonen, and M. Peisker (2002) Dephytinization of two soy proteins increases phosphorus and protein utilization by rainbow trout, *Oncorhynchus mykiss*. *Aquaculture* 204:145-156.

Virtanen, E., R. Hole, J.W. Resink, K.E. Slinning, and M. Junnila (1994) Betaine/amino acid additive enhances the seawater performance of rainbow trout *(Oncorhynchus mykiss)* fed standard fish meal-based diets. *Aquaculture* 124:220.

Wilson, R.P. (2002) Protein and amino acid requirements. In *Fish Nutrition,* 3rd edn, J.E. Halver and R.W. Hardy (eds.). San Diego, CA: Academic Press, pp. 671-702.

Xue, M. and Y. Cui (2001) Effect of several feeding stimulants on diet preference by juvenile gibel carp *(Carassius auratus gibelio),* fed diets with or without partial replacement of fish meal by meat and bone meal. *Aquaculture* 198:281-292.

Yan, W. and R.C. Reigh (2002) Effects of fungal phytase on utilization of dietary protein and minerals, and dephosphorylation of phytic acid in the alimentary tract of channel catfish *Ictalurus punctatus* fed an all-plant-protein diet. *Journal of the World Aquaculture Society* 33:10-22.

Zarate, D.D. and R.T. Lovell (1997) Free lysine (L-lysine HCl) is utilized for growth less efficiently than protein-bound lysine(soybean meal) in practical diets by young channel catfish *(Ictalurus punctatus). Aquaculture* 159:87-100.

Chapter 18

Extrusion of Aquafeeds Formulated with Alternative Protein Sources

Mian N. Riaz

WHAT IS EXTRUSION?

Extrusion is simply the operation of shaping a plastic or doughlike material by forcing it through a restriction or die. Examples of hand operations for extruding foods include the rolling of noodles and pie crust dough, finger-stuffing of chopped meats through animal horns into natural casings, pressing of soft foods through hand ricers to produce stringlike particles, and cranking of hand-powered meat grinders. Mechanically powered extrusion devices include wire-cut cookie dough depositors, pasta presses, continuous mixing and scaling systems used in automated bakeries, pneumatic (batch) and continuous (pump) sausage stuffers, hamburger patty formers, and pellet mills used to prepare animal feeds. Rossen and Miller (1973) have offered the practical definition: "Food extrusion is a process in which a food material is forced to flow, under one or more of a variety of conditions of mixing, heating, and shear, through a die which is designed to form and/or puff-dry the ingredients."

A food-extruder is a device that expedites the shaping and restructuring process of food or feed ingredients. Extrusion is a highly versatile unit operation that can be applied to a variety of food and feed processes. Extruders can be used to cook, form, mix, texturize, and shape food and feed products under conditions that favor quality retention, high productivity, and low cost. The use of cooker-extruders

Alternative Protein Sources in Aquaculture Diets
© 2008 by The Haworth Press, Taylor & Francis Group. All rights reserved.
doi:10.1300/5892_18

has been expanding rapidly in the food and feed industries over the past few years.

Extrusion cooking is the technology that most aquaculture feed producers are most familiar with. This process uses high levels of thermal and mechanical forces to cook or work the ingredients mixture, and then extrudes pellets through the terminal die system. The moisture content before exiting the die is typically 25 percent. This relatively low moisture content means that a relatively low proportion of wet ingredients can be used. For the pet food industry, it may not be a problem, but for aquafeeds it may cause a problem when trying to make floating pellets.

ADVANTAGES OF EXTRUSION

The principal advantages of extrusion technology, as compared to the other traditional food and feed processing methods based on Smith (1969), Smith (1971), Riaz (2000), Riaz (2001), and Riaz (2003a), with modifications include:

Adaptability: An ample variety of products is feasible by changing the formulation/ingredients and the operation conditions of the extruder. The extrusion process is remarkably adaptable in accommodating the demand by consumers for new products.

Product characteristics: A variety of shapes, texture, color, and appearances can be produced, which is not easily done using other production methods.

Energy efficient: Extruders operate with relatively low moisture while cooking food products, therefore, less re-drying is required.

Low cost: Extrusion has a lower processing cost than other traditional cooking and forming processes, such as pelleting and can save 19 percent on raw materials, 14 percent on labor, and 44 percent on capital investment (Darrington, 1987). Extrusion processing also needs less space per unit of operation than traditional cooking systems.

New products: Extrusion can modify animal and vegetable proteins, starches, and other food material to produce a variety of new and unique snack food products.

High productivity and automated control: An extruder provides continuous high-throughput processing and can be fully automated.

High product quality: Since extrusion is a high-temperature short-time (HTST) heating process, it minimizes degradation of food nutrients while improving the digestibility of proteins (by denaturing) and starches (by gelatinizing). Extrusion cooking at a high temperature also destroys antinutritional compounds, such as trypsin inhibitors, and undesirable enzymes, such as lipases and lipoxidases, and micro-organisms.

No effluent: This is a very important advantage for food and feed industries, since new environmental regulations are stringent and costly. Extrusion produces little or no waste streams.

Process scale up: Data obtained from the pilot plant can be used to scale up the extrusion system for production.

Use as a continuous reactor: Extruders are being used as continuous reactors in several countries for deactivation of some aflatoxin in peanut meals, and destruction of antinutritional factors such as allergens and toxic compounds in castor seed meal and other oilseed crops (Rhee, 1987).

AQUACULTURE FEED MANUFACTURING

The environmental pollution from fish farming is a major concern in many countries. Fish farmers throughout the world are facing serious problems with pollution. Pollution has a bad effect on water quality, which results in inferior growth performance, diseases, mortality, and in some cases, complete loss of species (Botting, 1991). Effluent treatment is often costly. Feed millers and fish farmers should be looking at prevention rather than a temporary remedy. By using good feeding management and processing practices some of these problems can be reduced.

Characteristics of Aquaculture Feeds

Several varieties and forms of fish feeds are available in the market. A feed must fulfill the nutritional need of each fish species. Young fish may not have a completely developed digestive system, so their feed must be of appropriate shape and size so that it allows for easy consumption and digestion. Juvenile fish are slow eaters and thus require a more water-stable feed than in the case of mature fish. Thus, the feed provided to fish must be produced so that its floats and maintains its integrity for a longer duration in the water. Fish feeds need to become more environmentally friendly. A better understanding of protein:energy ratio and nutrient requirements is also needed. To fulfill these needs, fish feed should be produced from materials which have good digestibility and nutrient absorption.

Methods for Preparing Aquaculture Feeds

The traditional and popular method of preparing feed for the aquaculture industry is steam pelleting. A second method of processing is extrusion, which is relatively new to the industry. This chapter compares both methods for fish feed production.

Steam Pelleting

Pelleting is still considered an art rather than a science. It can be defined as a process in which finely ground feed ingredients are formed into larger particles (sinking pellets) by using heat, moisture, and pressure. There are several factors that influence the pellet quality, including composition of the formulae, moisture, temperature, ingredient composition, ingredient grinding quality, die condition, and the pellet mill operator.

Extrusion

Extrusion is a process in which material is forced through a die under a variety of controlled conditions. It has become one of the most important processes in the feed industry. It is a unique system where high levels of moisture, heat, and pressure can apply to institute physical and chemical changes in feed ingredients at an accelerated rate. Mixing, shearing, kneading, and cooking of feed ingredients can be completed in one step during the extrusion process.

The following is the comparison between pelleting and extrusion technology adapted from Riaz (1998) with some modifications:

1. Both processes reduce separation of microingredients and assure consistent dispersion throughout the mixture during feed production.
2. In the extrusion process, feed ingredients are generally cooked at a higher temperature than during pelleting, which eliminates

the chances of bacteria being able to survive in the finished feeds. Pelleting may destroy certain microorganisms during the process.

3. Steam-pelleted feeds are easy to handle in bins and automatic feeding systems. Extruded feeds are comparatively more resistant to breakage during transportation and handling.

4. Extruded feeds have an internal matrix system that tends to increase resistance to the mechanical handling of the feed, while in pelleted feeds ingredients are compressed and squeezed together.

5. In extrusion, the moisture content can be as high as 55 percent during the process, while in the pelleting process, moisture content cannot exceed 16 to 17 percent.

6. Both processes increase bulk density, which help transportation, storage, and feeding; there is higher density for sinking feeds relative to floating pellets.

7. With extrusion, different shapes and sizes of feed can be produced. Feed mills can use extruders to process grains and full-fat soybeans. Raw aquaculture waste can be coextruded with cereals as an ingredient. In a pellet mill, raw ingredients are very difficult to process.

8. Both processes enhance the palatability of feeds though allowing less-palatable ingredients to be used.

9. In the extrusion process, 90 percent or higher of starch in the feed is gelatinized, while in the pelleting process, approximately 50 percent is gelatinized, with the combination of multiple conditioners.

10. With extrusion, one can control the density of the product, which allows the feed mill to produce floating, sinking, or slow-sinking fish feeds. In the pelleting process, it may be impossible to produce floating or slow-sinking feeds.

11. In the extrusion process, capital expenditure and utility use are higher than in the pelleting process.

12. In the extrusion process, fat levels of 18 to 20 percent in the formulation can be used, while the total fat level in the pelleting process is less than 10 percent.

13. Heat treatment of ingredients can cause increase or decrease in nutrient bioavailability.

14. Feeds processed by an extruder absorb water well, retain their shape for a longer period of time, and precipitate slowly (Hilton et al., 1981). Extruded feeds produce less adverse effect on water quality of the pond as compared to pelleted feeds, which sink rapidly and do not retain their shape for any great length of time, thus causing pollution and water quality problems. The average acceptable stability time is approximately four hours (except for some species); depending on the formulae, some extruded feeds have shown stabilities of up to twenty-four hours (Kearns et al., 1988). Water stability of compressed pellets can be improved by using some extra additives and binders.

15. The pelleting process requires finer grinding (approximately sixty mesh) than the extrusion process (approximately twenty mesh).

16. Extrusion-cooked feeds produce approximately 1 to 2 percent fines, while pelleted feeds normally generate 5 to 8 percent fines in the handling of the feeds in bulk or in bags. Thus, extrusion reduces by 75 percent the amount of fine that enters the water that could decay on the bottom of the pond. Decreasing the fines will increase feed efficiency as well as water quality (Kearns, 1989).

17. Dry-blended feeds will lose water-soluble nutrients by leaching without special protection. Occasionally, pellets disintegrate in water before they reach the fish. These problems can be solved by using special feed formulations and feed conditioning (Wood, 1980). The fat coating also helps prevent leaching losses from the feed. Much greater amounts of fat can be added to extruded feeds than steam-pelleted feeds.

18. Feeds produced by both methods are fairly stable in dry states and can be stored for longer periods without nutrient deterioration, if stored properly such as in a cool dry place.

19. Conversion rates of extrusion-cooked feeds are higher due to the cooking process, which causes an increase in digestibility of the raw material, specifically the starch (Wood, 1980).

20. Regarding the design of the feed mill, both processes (extrusion and pelleting) are similar in layout.

MAJOR PARAMETERS AFFECTING
EXTRUSION PROCESSES

Aquaculture production has been expanding at a rate of 15 percent per year and is predicted to grow at this rate for at least the next decade (Tacon, 1997). According to Tacon (1997), the predicted increase in aquaculture production will require twenty-six million metric tons (mmt) of additional feed production, which in turn will require an additional 7 to 9 mmt of protein for this feed. Grain and oilseed by-products, rendered products, and seafood processing waste, make up the short list of product options to replace fish meal in feeds. However, when fish meal is replaced with other vegetable or animal protein in the formulation, the processing parameters of extrusion changed dramatically. Therefore, to obtain better quality products, it is necessary to understand the following parameters: raw material, hardware, processing conditions, and final product specification.

Raw Material

Ingredient selection will have a tremendous impact on final aquafeed quality, texture, uniformity, nutritional value, and economics. The main components of any aquafeed formulation are protein, fat, carbohydrates, and some other microingredients. Each of these ingredients will have a different effect on extrusion if changed in the formulation.

Protein: Proteineceous ingredients are the most important part of an aquatic feed. The protein can range from 25 to 60 percent in fish feed formulations depending on the species and fish size. Protein is not only important nutritionally, but it also has some functional effect on the pellet, such as water absorption, elasticity, and binding characteristics of pellets. Extrusion has several effects on protein (Rokey, 2004), including texturization, increased digestibility, protein lipid complexes, denaturation of protein, inactivation of antinutritional factors, and rearrangement of polymer bonds.

Protein sources can be classified into two groups: plant and animal sources. Plant-protein sources include oilseed products, (for example, soybeans, cottonseed, canola or rapeseed, and sunflower meal), legumes, wheat and corn gluten meal, and other minor oilseed crops. In general, plant proteins are cheaper and have good functional properties (water absorption and binding characteristics) during extruding.

Animal sources of protein are fish meal, meat meal, meat and bone-meal, poultry meal, poultry by-product meal, and blood meal. They have poor functional properties unless used fresh or spray dried. Animal proteins are usually more expensive than plant proteins, but have good amino acid profiles for most fish species.

Starch: Starches are primarily used in aquafeed formulations as an energy source. They also assist in expansion during extrusion, and improve binding and pellet durability. Their ranges of inclusion vary from 10 to 60 percent in aquafeed formulations. The recommended level for floating feed is approximately 20 percent. Extrusion has several effects on starch (Rokey, 2004), including gelatinization, increased digestibility, starch/lipid complexes, increased water solubility, and absorption.

Fats or lipids: Lipids are an essential component in aquatic diets, and are an excellent source of energy. Lipid levels can exceed 30 percent in some feeds, but usually less than 8 to 10 percent lipid is added during extrusion, and the rest of the lipid is sprayed after extrusion. The addition of lipid is critical to extrusion. Lipid should be added close to discharge, and should be heated to 60°C in order to facilitate the extrusion. The source of the lipid will also have an effect on extrusion, that is, indigenous lipids versus added lipids. The example of the indigenous lipid is full-fat soy meal. This type of lipid will have less effect on extrusion as compared to added lipid in the formulation. Floating feeds can absorb 30 percent fat, whereas sinking feeds can absorb a maximum of .15 percent lipid during coating. Lipid provides lubricity and plasticity, and above the 8 percent level begins to weaken the dough strength of the extrudates. Extrusion has several effects on fat/lipids when used in feed formulations (Rokey, 2004): it catalyzes oxidative rancidity, complexes with starch and protein, controls enzymes causing enzymatic rancidity, and provides lubricity to the extruder.

Minor ingredients: Extrusion appears to have minor effects on mineral availability. Some increases in metal content may be found due to metal transition during extrusion. Lipid-soluble vitamins (A, D, E, and K) are generally considered to be stable during extrusion, but losses of 15 to 25 percent can be expected with vitamin A. Water-soluble vitamins (riboflavin and ascorbic acid) are not as heat stable (Rokey, 2004). Pigments, such as carophyl pink, should be treated as vitamins during extrusion. This pigment rarely showed more than 10 to 15 percent loss during extrusion (Rokey, 1994).

As long as one can understand the effect of extrusion on these ingredients he or she can manipulate feed formulations using animal or oilseed proteins rather than fish meal. By changing raw materials, one will know how much more lipids or proteins are being added in the formulation, and adjustments to the hardware and processing conditions can be made accordingly.

HARDWARE

The selection of the appropriate hardware system is important in feed making. Some examples of hardware include single- versus twin-screw extruder, delivery systems, preconditioners, barrels, screws, (single or double flight, cut flights, screws with different pitch), shear locks, dies, and cutters. One can make a hardware selection based on the final specification of the aquafeed and the raw materials.

Single versus Twin Screw

There are several reasons why single- or twin-screw extruders are needed for aquafeeds. According to Riaz (2003b), a twin-screw extruder is the equipment of choice when high fat-feeds (>17 percent) are being produced, high levels of fresh meats or slurry (>30 percent) are extruded, uniform sizes and shapes of the feed are required, small-size pellets (less than 1.5 mm diameter) need to be produced, when a feed with a commercial value justifies the costs of twin-screw processing, when aquaculture feeds are made in plants where each production run is small and there are significant variations in formulations, and frequent product changes over a short-term production schedule. Typical single- and twin-screw extruders are shown in Figures 18.1 and 18.2.

Preconditioner

Material from the delivery system is fed into the section of the extruder called the preconditioner. It is not important that every kind of extruder has a preconditioner. Most of the dry extruders for full-fat

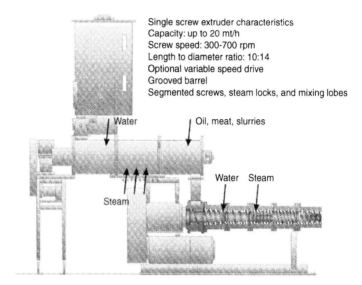

FIGURE 18.1. A typical single-screw extruder. (Courtesy of Wenger Manufacturing, Sabetha, KS.)

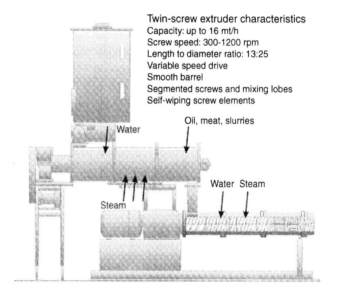

FIGURE 18.2. A typical twin-screw extruder. (Courtesy of Wenger Manufacturing, Sabetha, KS.)

soybean production do not have a preconditioner, but the majority of the aquaculture feed extruders have a preconditioner. There are several advantages of using a preconditioner in aqua feed extrusion. Recently, some of the dry extruder manufacturers realized the benefits of a preconditioner and started offering a preconditioning unit with their dry extrusion system. In general, preconditioning will enhance any extrusion process that would benefit from higher moisture and longer retention time.

The main function of the preconditioner is to mix the ingredients (liquid and dry), hydrate the ingredients, and precook the ingredients to a certain temperature. By doing this, one can increase production output by precooking the formulation, increase the life of the extruder components by premoistening the material, and properly cook the raw materials. The preconditioner provides the feed mill a way to increase the thermal energy in the extrusion system, which is cheaper than mechanical energy (Riaz, 2003b).

There are several different types of preconditioners available from the extruder manufacturers. The two main types of preconditioning systems include (1) the pressurized preconditioner and (2) the atmospheric preconditioner. From a practical point of view, pressurized preconditioners are difficult to operate and need a lot of extra devices to maintain the pressure inside the preconditioner, whereas atmospheric preconditioners are easy to operate and maintain. There are three different types of atmospheric preconditioners available in the market and they are (1) single- shaft preconditioner, which has only one shaft for mixing, and has no control over holding time, (2) double-shaft preconditioner, which has two shafts and facilitates better mixing and retention time, and (3) differential-diameter preconditioner, which has two shafts, with one larger than the other in diameter. The larger one is used for retention time, whereas the smaller one is used for mixing the raw material. The smaller shafts run at twice the speed of the larger one.

Extruder Barrel

This is the heart of the extrusion system. Most of the work is done by this section of the extrusion system. This component consists of screws, sleeves, barrel head, and dies. This section of extruder consists of either single-screw or twin-screw extruder. In a twin-screw extruder the two shafts are parallel to one another and in a single-screw extruder there is only one shaft. The main purpose of this section is to mix and cook the ingredients, and at the same time, shape the ingredients into predetermined sizes, which can be controlled by die(s) at the end of the barrel. Water and steam can be injected into the barrel for cooking ingredients. Heat can also be added to the barrel by injecting steam or heating oil in the jacked heads to heat the ingredients. Different combinations of screws (single, double, cut flight screws with different pitches as shear locks) can be used to make different aquaculture feeds. The pressure, temperature, moisture, and resulting viscosity of the extruded feed is affected by both the system configuration and processing conditions. Different screws are required to produce floating feeds, sinking feeds, and high-fat aquaculture feeds (Riaz, 2003b, 2004). Liquid ingredients can also be added in this section, and a vacuum system can also be used to produce dense feeds, such as sinking pellets. Segmented barrels give the option to increase or decrease the length of the barrel for different products.

Knife Assembly

This is the last part of the extrusion system. The main function of this component is to cut the product to the desired length and shape. Knife assemblies can be of different designs depending upon the manufacturer of the extrusion system. There is the option to use two, four, or six knives in the assembly.

PROCESSING CONDITIONS (SOFTWARE)

Proper control of the extrusion processing variables is important to the success of producing aquaculture feeds. It is critical to understand what variables can be controlled directly by the extruder operator, and which processing variable is simply a result of what is controlled by the operator. To understand the interaction of the extrusion processing variables, these variables can be divided into two main categories (Huber, 2000): (1) independent variables and (2) dependent variables of the aquaculture feed extrusion process.

Independent Variables

These variables are the parameters that can directly be controlled by the aquafeed extruder operator. The exact nature of these variables will vary with the sophistication of the control system utilized with the aquafeed extruder (Riaz, 2003a). These variables include:

1. *Formula:* An operator can suggest the change of feed formulation such as adjusting the level of starch, fat, and protein.
2. *Feeding rate:* An operator can adjust the feeding-screw speed to increase or decrease the rate at which the dry ingredients are added to the extruder.
3. *Water injected into the preconditioner:* The operator can alter the amount of water injected into the systems.
4. *Steam injected into the preconditioner:* The operator has direct control on the steam injection into the preconditioner.
5. *Preconditioner shaft speed:* The operator has direct control on the preconditioner's shaft speed. By increasing or decreasing the speed of the preconditioner shaft, volume and retention time in the preconditioner can be changed.
6. *Preconditioner configuration:* The operator can select the configuration of the preconditioner's shaft which can increase or decease the retention time in the preconditioner.
7. *Water injected into the extruder barrel:* The operator has direct control on the water injection systems for the extruder barrel to increase or decrease moisture content of mix ingredients.
8. *Steam injected into the extruder barrel:* The operator has direct control on the steam injection systems for the extruder barrel to increase or decrease temperature and moisture.
9. *Extruder configuration:* The operator has direct control on choosing the extruder configuration for different formulations. He or she can make different configurations by choosing different screws (single, double or cut flight), different shears, or steam locks (different thickness and sizes, different angles, etc.).
10. *Extruder shaft speed:* The operator has direct control on the extruder shaft speed. He or she can run the shaft at any revolutions per minute (RPM) required.

11. *Extruder barrel heating element or thermal fluid temperature:* The operator has direct control on the heating elements or thermal fluid temperature control system. The heating element of thermal fluid will heat all of these zones at set temperature points. By doing so cooking temperature of the barrel can be increased or decreased.

12. *Die configuration:* The operator has a direct control for choosing the die. He or she can pick any die (single, double, or multiple holes; die with one or two spacers). Dies have direct effect on expansion as well as floating or sinking characteristics. Also, production rate can be influenced by the die opening.

Dependent Variables

These variables are parameters that change as a result of changing one or more of the independent variables. Dependent variables include:

1. *Retention time in the preconditioner:* This is a result of the paddle configuration and shaft RPM adjusted by an operator. Parameter adjustments for the preconditioner (shaft speed and configuration) influence retention and mixing time. Increasing the retention time in the preconditioner will affect the final cook of the diet and also the floating characteristics.

2. *Temperature in the preconditioner:* This is a result of the steam injection in the preconditioner.

3. *Moisture in the preconditioner:* This is a result of the water injection in the preconditioner.

4. *Retention time in the extruder:* This is a result of the extruder shaft RPM and extruder barrel configuration.

5. *Temperature in the extruder:* This is a result of the steam injection in the extruder barrel.

6. *Moisture in the extruder:* This is a result of the moisture injection in the extruder barrel.

7. *Pressure in the extruder:* This is a result of the dies selection and extruder barrel configuration.

8. *Mechanical energy input to the extruder:* This is a result of the extruder barrel configuration.

There are four critical parameters in the extrusion processing that will determine the final aquafeed product characteristics. These parameters are actual moisture, thermal energy and mechanical energy in the extrusion process, and retention time in the extruder barrel.

MAKING FEEDS WITH ALTERNATIVE
SOURCES OF PROTEIN

Use of Plant Proteins as Sources of Alternatives to Fish Meal

Once the effect of raw material, hardware, processing conditions, and final product specifications on extrusion is understood, one can use plant proteins to replace fish meal during extrusion. Table 18.1 shows the maximum level of vegetable protein that can be added in salmon, trout, and shrimp feed without affecting the physical quality of feeds. In making feeds with plant-protein sources, one must know the level of protein, starch, and fat contents in the protein sources, and understand the effect of these ingredients on extrusion. A 25 to 40 percent change in the formulation will affect the extrusion processing so it will be necessary to reorganize extrusion processing parameters based upon the new formulations.

There are some advantages of using plant-protein ingredients in aquafeed extrusion. The first is that there is more pellet expansion,

TABLE 18.1. Maximum level of vegetable protein that can be added in salmon, trout, and shrimp diets.

Ingredient	Maximum substitution for fish meal (%)	Disadvantage
Maize gluten meal	40	Yellow pigmentation of flesh
Wheat gluten	25	High cost
Soybean meal	50	Palatability and growth inhibitors
Soy concentrate	75	High cost
Canola meal	67	Low protein content

Source: Courtesy of Wenger Manufacturing, Sabetha, KS.

which allows for a better-floating pellet. Second, binding properties of the feed are improved, which means a more water-stable feed. Third, use of plant proteins reduce the incidence of white mineral deposits in the screw and die areas of the extruder, which can plug the die. Fourth, higher levels of lipids can be added to the feeds because of the greater absorption properties of most plant-protein ingredients. Last, there is a reduced reliance on fish meal and a lower feed cost due to the lower cost of plant-protein ingredients. Figure 18.3 shows the effect of the vegetable protein level on extrusion moisture content.

Uses of Animal and Fish By-Products as Alternatives to Fish Meal

Until recently, the common practice of animal by-product mortality disposal has been through landfill dumps and incineration. Both practices represent serious problems economically and environmentally, and stringent regulations may eliminate these disposal options in the near future. One option for utilization of this waste is to convert, through extrusion, these materials into useful feed ingredients that could be utilized as protein sources. A second option is to convert, through extrusion, these waste materials into a complete feed. This can be accomplished by blending by-product materials with an otherwise nutritionally complete feed formulation and extrude the final product. Either option employs extrusion cooking as the means of imparting

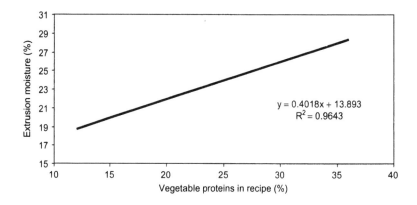

FIGURE 18.3. Effect of vegetable protein levels on extrusion moisture.

energy, both mechanical and thermal, into the finished product. The final product yields a value-added feed or feed ingredient that is shelf-stable and free of pathogenic organisms. There is almost no limitation to the types of processing wastes and by-products that can be utilized in this manner.

Raw Material Specification

Raw material utilization and cost-effective formulations are key operational characteristics. As with production of all feeds, raw materials are vital to achieve correct end products. To utilize raw, wet-processing waste and by-products, the following processing steps must be accomplished.

By-Product as a Component of Finished Feed

Prior to extrusion, the raw, wet by-products are initially processed through a meat grinder containing 2.5 mm die openings to reduce the particle size of bones and other hard components and obtain a more uniform product.

The dry ingredient portion of the feed, prior to extrusion, also should be reduced in particle size. This will promote a uniform moisture uptake by all particles. A recommended dry feed particle size analysis is given in Table 18.2.

In the feed industry there are presently two different approaches for using animal by-products: addition of by-product to the feed; and

TABLE 18.2. A recommended dry diet particle size analysis.

U.S. standard seive number	Openings (μm)	Percent on sieve
20	850	1.00
30	600	3.00
40	425	28.00
50	300	29.00
60	250	21.00
80	180	5.00
Pan	0	14.00

Source: Courtesy of Wenger Manufacturing, Sabetha, KS.

processing these by-products with any dry meal to make high-protein ingredients.

Both of these approaches have their advantages and disadvantages as well as limitations.

Addition of Slurry Directly to the Feed

Some of the aquafeed and pet food industries are using this approach, based upon single- and twin-screw extrusion operation. Different levels of slurry can be added in the aquaculture feed during extrusion. Tables 18.3 and 18.4 show the maximum level of slurry that can be pumped into a single- and twin-screw extruder. Figure 18.4 shows a typical set up for pumping the meat slurry into the extruder. There are some important points that need to be considered when using animal and fisheries by-products. One is that the maximum particle size in the slurry should not exceed 1.5 mm (microcutter). A second is that the fish by-product slurries can be pumped into a preconditioner in a chilled, emulsified form (5-10°C) or can be heated to 40-60°C, but the latter requires steam-jacketed kettles. A third point is that fat/oil slurries can be heated to 60°C. Last, high moisture content is the limiting factor for most slurry additions.

TABLE 18.3. Maximum level of meat slurry that can be added to single-screw extrusion system.

% Moisture in meat	Maximum meat slurry addition (% of total)	Maximum meat slurry addition (% of dry)	% Meat in final dry product
66.7	25.0	33.3	10.9
40.0	41.8	71.8	32.4
50.0	33.4	50.0	21.8
60.0	27.8	38.5	14.6
70.0	23.9	31.4	9.5
80.0	20.9	26.4	5.6

Source: Courtesy of Wenger Manufacturing, Sabetha, KS.

Note: Values are based on maximum additional moisture (16.7%) that can be added to single-screw extrusion systems.

TABLE 18.4. Maximum meat slurry that can be added to twin-screw extrusion system.

% Moisture in meat	Maximum meat slurry addition (% of total)	Maximum meat slurry addition (% of dry)	% Meat in final dry product
66.7	30.0	43.0	13.7
40.0	50.0	100.0	40.0
50.0	40.0	66.7	26.3
60.0	33.3	50.0	18.1
70.0	28.6	40.1	11.8
80.0	25.0	33.3	6.9

Source: Courtesy of Wenger Manufacturing, Sabetha, KS.

Note: Values are based on maximum additional moisture (20.0%) that can be added to twin screw extrusion systems.

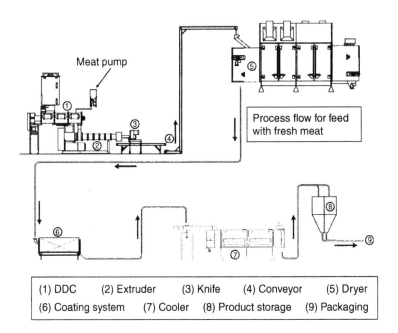

FIGURE 18.4. A typical set up for pumping the meat slurry into extruder. (Courtesy of Wenger Manufacturing, Sabetha, KS.)

This process has some advantages and disadvantages, which should be taken into consideration when using this approach. Advantages of this process are that the protein is more functional, no heat treatments are required, and that it is a wiser nutritional choice than dried protein. Disadvantages include the additional equipment cost, the requirement of cold and more storage spaces, high labor cost for handling, fat and moisture variations, demand for more attention to the extrusion processing (because the moisture and fat variations can change the final pellet characteristics), and microbial contamination.

Processing of High-Protein Ingredients from By-Products

Several of the small-scale aquafeed manufacturers are using this approach. A typical flow diagram of all the steps involved in this approach is shown in Figure 18.5. In this process, the wet materials (fisheries by-products, dead fish, dead birds, food processing waste, restaurant waste, or any other waste which contains protein) are ground. The main purpose of grinding is to decrease the particle size and to make a homogeneous dispersion. At the same time, bone and other hard particles need to be reduced in size to avoid any problem with die plugging. The raw, wet by-products are processed through a meat grinder containing 2.5 mm die openings. Dry diet ingredients do not require more grinding than the complete food formulation. A recommended dry diet particle size analysis for production of a diet ingredient is given in Table 18.5.

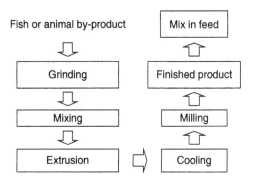

FIGURE 18.5. A typical flow diagram for processing animal and fish by-products by extrusion.

TABLE 18.5. A recommended dry diet particle size analysis for production of an ingredient.

U.S. standard sieve number	Openings (μm)	Percent retained on sieve
8	2,500	0.00
12	1,700	6.24
20	850	35.71
30	600	11.54
40	452	21.67
50	300	13.83
60	250	4.41

Source: Courtesy of Wenger Manufacturing, Sabetha, KS.

The wet material can be mixed with any available dry ingredients. Some of the dry materials that can be used are soybean meal, corn meal, wheat middling, cassava flour, wheat flour, barley flour, or any other local material.

The two limiting factors on how much by-product or processing waste can be added to the extruder are the moisture and fat contributed by the addition of these by-products. Usually the moisture content of these products ranges from 60 to 87 percent and, is the primary limiting factor. A general rule of thumb is to add as much processing waste or by-product to the extruder which will not exceed a total contributed moisture of 30 percent.

When a diet ingredient is extruded, the limiting factor remains moisture content. However, since product durability is not required, the percent of by-product that can be incorporated can be substantially increased to facilitate uniform temperature rise throughout the extrudate while retaining complete pasteurization. This level of incorporation is typically 50 percent by-product and 50 percent carrier material (i.e., defatted soy, mill by-products, or grains).

The next step is the mixing of the wet and dry ingredient together to make a flowable raw material which can pass through the extruder. This mixing ratio may vary if the raw material contains a lot of fat. The goal is to obtain raw material mixture with moisture content of 30 percent. The next step is extrusion of this mixture. In this process, a dry extruder is mostly used, although wet extruders are also being used sometimes. The processing time ranges from 30 to 35 seconds at 140-160°C. Table 18.6 shows the effect of extrusion on the microorganism.

After extrusion, the resulting product is cooled and dried. Most by-products coextruded with dry material are being extruded at moisture content between 26 and 38 percent. A portion of this moisture is lost due to flash evaporation as the superheated extrudate exits the die and expands. Furthermore, moisture is lost through evaporative cooling. Drying this material under ambient conditions is possible, but is very slow requiring large spaces and the possibility of recontamination with microorganisms is increased. Most extruded products are subject to heat drying to reduce the moisture content to no more than 10 to 12 percent, otherwise, spoilage will occur. Reducing the moisture level after extrusion increases the shelf-life stability. High-temperature drying eliminates microbial contamination because a pathogen-free environment is retained at air temperature of 94-150°C. Following drying, the product temperature should be reduced to 30-40°C prior to storage or packaging.

The goal of extrusion of by-products is to economically transform waste material into a value-added product. Extrusion processing of ingredients for further processing offers its own advantages. Adding the by-product directly to an otherwise complete feed eliminates the need for reprocessing the material at a later time. Extruding a feed

TABLE 18.6. Effect of extrusion on microbial populations.

Microbe	Before extrusion	After extrusion
TPC[a] (CFU/g)[b]	240,000	9,300
Coliform	22,600	<10
Mold count	54,540	<10
Clostridium	16,000	<10
Listeria	Positive	Negative
Salmonella	Negative	Negative

Source: Rokey, 2000.

[a]Total plate count.

[b]Colony forming unit.

ingredient requires a lower capital investment while allowing elevated incorporation of by-product materials. There is almost no limitation to the methods of processing wastes and by-products that can be utilized in this manner. However, one must be concerned about those by-products, which obviously have little or no value. There are some advantages and disadvantages associated with this process. Some of the advantages are: there is no microbial contamination, no special storage is needed, ingredients can be used as needed, and recycled materials can be used. Some of the disadvantages are: the need for two extruders for continuous extrusion, fat and moisture as limiting factors of the addition of wet material, need for a separate facility and equipment to make these high-protein ingredients, and the need to grind extruded ingredients for use back in the feed.

CONCLUSION

Extrusion cooking is a very flexible processing tool that can produce a wide range of floating, sinking, and high-fat feed when proper raw materials, hardware, and processing techniques are used. Different sources of protein (vegetable or animal) can be successfully used in the formulation without affecting the quality of the aquaculture feed.

REFERENCES

Botting, C.C. (1991). Extrusion technology in aquaculture feed processing. In *Proceedings of the Aquaculture Feed Processing and Nutrition Workshop,* D.M. Akiyama and R.K.H. Tan (eds.). Singapore: American Soybean Association.

Darrington, H. (1987). A long-running cereal. *Food Manufacturing* 3:47-48.

Hilton, J.W., C.Y. Cho, and S.J. Slinger (1981). Effect of extrusion processing and steam pelleting feeds on pellet durability, pellet water absorption, and the physiological response of rainbow trout *(Salmo gairdneri r.). Aquaculture* 25:185-194

Huber, G. (2000) Twin screw extruders. In *Extruders in Food Application.* M. Riaz (ed.). Boca Raton, FL: LRC Press, pp. 81-113.

Kearns, J.P. (1989). Advantages of extrusion cooking and comparisons with the pelleting process for aquatic feeds. Aquaculture Workshop for the People's Republic of China, Book of Abstracts, St. Louis, MO: American Soybean Association.

Kearns, J.P., G.R. Huber, W.G. Dominy, and D.W. Freeman (1988). Properties of extrusion-cooked shrimp feeds containing various commercial binders. Paper presented at the World Aquaculture Society Meeting. Hawaii, January 4-8.

Rhee. K.C. (1987). Development of a caster meal detoxification and deallergenation technology. Annual Progress Report to the Natural Fiber and Food Protein Commission. pp. 247-257.

Riaz, M.N. (1998). Pro's and con's of extruding and pelleting aquatic feeds. *Feed Technology* 3(2):39-41.

Riaz, M.N. (2000). Introduction to extruders and their principles. In *Extruders in Food Application.* M.N. Riaz (ed.). Boca Raton, FL: CRC Press, pp. 1-23.

Riaz, M.N. (2001). Selecting the right extruder. In *Extrusion Cooking, Technologies and Application.* J. R. Guy (ed.). Boca Raton, FL: CRC Press, pp. 29-50

Riaz, M.N. (2003a). Extrusion basic. In *Petfood Technology.* Tim Phillips (ed.). Mt. Morris. IL: Watt Publishing, pp. 347-360.

Riaz, M.N. (2003b). Extrusion of aquafeed using alternative protein sources. Paper presented at the Alternative Protein Sources for Aquafeeds Workshop. Honolulu, HI. November 3-6.

Riaz, M. N. (2004). Application of extruders in food and feed industries. Paper presented at the seminar on Application of Extruders in Feed Industries at Karachi, Pakistan. Dec. 17.

Rokey, G. (1994). Petfood and fishfood extrusion. In *The Technology of Extrusion Cooking.* M. D. Frame (ed.). CITY, UK: Blackie Academic and Professional pp. 144-189.

Rokey, G. (2004). Raw material for extrusion processing. In *Feeds and Petfood Extrusion Manual.* M.N. Riaz and M. Barron (eds). College Station, TX: Texas A&M University, pp. 5

Rossen, J. L. and R.C. Miller (1973). Food extrusion. *Food Technology* 27(8):46-53.

Smith, O.B. (1969). History and status of specific protein-rich foods: Extrusion-processed cereal foods. In *Protein-Enriched Cereal Foods for World Needs,* M. Milner (ed.). St. Paul, MN: American Association of Cereal Chemists, pp. 140-153.

Smith, O.B. (1971). Why use extrusion. Symposium on "Extrusion: Process and Product Development". American Association of Cereal Chemists. St. Paul, Minnesota.

Tacon, A.G. (1997). Global trends in aquaculture and aquafeed production 1984-1995. *International Aquafeed Directory* 1997/1998.

Wood, J.F. (1980). The preparation of water-stable fish feeds 1. Methodology of feed preparation—A brief resume. *Tropical Science* 22(4):351-355.

Index

Page numbers followed by an *f* or *t* indicate figures or tables.

AA. *See* Arachidonic acid
Abalone, peanut meal and, 445
Acacia meal, amino acids and, 170*t*
Adaptability, extrusion and, 524
ADCs. *See* Apparent digestibility
    coefficients
Additives
    amino acids and, 502-504
    conclusions, 516
    fatty acids and, 504-505
    introduction to, 501-502
    minerals and, 505-506
    probiotics/prebiotics and, 513-516
    taste and, 511-512
Aeromonas hydrophila, additives and,
    514-515*t*
AFCO. *See* American Feed Control
    Officials
Aflatoxin
    distillers' grains and, 481
    palm kernel meal and, 464-465
African catfish. *See* Catfish
AFW. *See* Apparent feed waste
Age considerations, apparent digestibility
    of poultry by-product meal and,
    56
Alanine, distillers' grains and, 481*t*
Alaska walleye pollock. *See* Pollock
Alcohol by-products. *See* Distillers'
    grains
Alfalfa, crustacean diets and, 166
Algae meals, crustacean diets and,
    192-193
Alkaloids, lupins and, 414*t*
American Feed Control Officials, 52

Amino acids
    as additive, 502-504
    additives and, 512
    apparent digestibility of, 56-59,
        57-58*t*, 59*t*
    bioavailability and, 138-139
    canola protein and, 343, 354,
        355-357*t*, 358*t*, 390-391, 392*t*
    cottonseed meal and, 314, 329*t*,
        332*t*, 333*t*
    crustacean diets and, 170*t*, 171*t*, 172*t*
    DEAA profile vs. EAA
        requirements, 60*t*, 75*t*
    demand for alternate protein
        ingredients and, 9-10, 10*t*
    digestibility and, 239-243*t*
    dissolved nitrogen waste and, 22-25
    distillers' grains and, 481*t*, 483*t*, 492*t*
    feather meal and, 73-74
    importance of, 83-84
    lupins and, 412, 413*t*
    meatpacking by-products and, 96,
        101-102, 101*t*
    palm kernel meal and, 463-464
    peanut meal and, 442-443, 445
    plant proteins and, 165, 478
    poultry by-product meal and, 52, 53*t*
    protein sources and, 7
    quality and, 105
    requirements for, 476
    retention of, 32-33
    soybeans and, 208-209, 209*t*,
        262-265, 263*t*, 264*t*, 265*f*,
        284, 285*t*, 291-294, 292*t*,
        293*t*, 303
    sunflower seed meal and, 446-447